Accretion Flows in Astrophysics

Astrophysics and Space Science Library

More information about this series at http://www.springer.com/series/5664

Nikolay Shakura
Editor

Accretion Flows in Astrophysics

Springer

Editor
Nikolay Shakura
Relativistic Astrophysics Department
Sternberg Astronomical Institute
Moscow, Russia

ISSN 0067-0057 ISSN 2214-7985 (electronic)
Astrophysics and Space Science Library
ISBN 978-3-319-93008-4 ISBN 978-3-319-93009-1 (eBook)
https://doi.org/10.1007/978-3-319-93009-1

Library of Congress Control Number: 2018953182

Cover illustration: Visualization of a twisted disc according to a solution by Scheuer and Feiler (1996). The action of viscosity on a differentially precessing disc ensures that the inner portions of the disc are aligned with the angular momentum of the black hole; further out, the disc is warped where the transition from aligned to misaligned disc occurs (Bardeen and Petterson 1975), Copyright Galina Lipunova. Background: Astronomy Blue Bright Clouds, Photographer: Albin Berlin, https://www.pexels.com/photo/astronomy-blue-bright-clouds-340901/, Creative Commons Zero (CC0) license

This Springer imprint is published by the registered company Springer Nature Switzerland AG
The registered company address is: Gewerbestrasse 11, 6330 Cham, Switzerland

Preface

The word "accretion" has a Latin origin (*accretio*) and means an augmentation or increment of an initial amount by the addition of new portions. Astronomers use this term to describe the falling of diffused matter onto a center of gravity. Accretion onto compact stellar objects, for example, neutron stars and black holes, is accompanied by an enormous output of energy. In the 1970s, the study of such processes became of special importance. It was the time when the American UHURU satellite discovered X-ray emission from accreting black holes and neutron stars in binary stellar systems. Some time earlier, in the end of the 1960s, when I was a graduate student at the physics faculty of the Moscow State University (MSU), my scientific advisor academician Ya. B. Zeldovich suggested to me to calculate the structure and radiation spectra of the shock wave arising when gas accretes onto a neutron star. The choice of this particular scientific problem was triggered by the following circumstances.

In 1962, a group of American scientists led by Prof. Riccardo Giacconi discovered the first X-ray sources. Before that, astronomers had known only one X-ray source of extraterrestrial origin, namely, the solar corona. The coronal gas, heated to a million degrees by some then unknown mechanism, was known to produce X-ray emission. The luminosity of the solar corona in X-rays is approximately one millionth of the optical luminosity of the Sun (4×10^{33} erg/s). It was thus natural to assume that other stars are also surrounded by hot coronae. However, simple calculations showed that detectors available at that time could not detect coronae even around the nearest stars located at distances of a few parsecs.

Nevertheless, astronomers tried to detect X-ray radiation from the Moon! The Moon has no atmosphere, but perhaps some radiation could be produced by fluorescence as the Moon's surface is illuminated by X-rays from the Sun. To investigate this, precisely at midnight of June 18, 1962, when a full moon was shining, the Aerobee rocket was launched. It reached a height of 225 km. Its flight continued for 350 s and was quite successful: two of the three Geiger counters, with large surface and good sensitivity in the range 1.5–6 keV, were operating during the flight. In this energy range, the Earth's atmosphere is totally opaque. Suddenly, instead of X-ray radiation from the Moon, a bright and before unknown source was discovered, which was far beyond the solar system in the direction of the constellation Scorpius. It was named Sco X-1.

Ya. B. Zeldovich. Credit: Photo-Archive of the Sternberg Astronomical Institute

In the following years, new rocket flights brought more and more discoveries of new X-ray sources. Gradually, a sky map covered with X-ray sources of different nature was created. The first sources got their names according to their location in the night sky (Cyg X-1, Cyg X-2, Her X-1, Cen X-3, and so on). Later it was revealed that their X-ray luminosities were thousands or even tens of thousand times stronger than the Sun's luminosity in visual light. The epoch of X-ray astronomy, an epoch of stunning discoveries in the universe, began.

According to simple estimates made by Ya. B. Zeldovich himself, the shock wave arising when the gas surrounding a neutron star falls onto its surface should produce radiation primarily in the X-ray range. My goal was to carry out a full calculation and investigate the process in detail. The main difficulty was connected to the following property: the mean free path of a falling particle near the surface of the neutron star is much greater (tens of times) than the characteristic scale of interaction between matter and radiation. In many such problems, it is not necessary to calculate the structure of the shock wave: it is sufficient to specify the changes in density, pressure, temperature, and other physical parameters depending on the velocity and the adiabatic index of the falling gas. In my problem, the density, temperature, and other parameters depended on the energy release in the braking zone. Moreover, plasma oscillations may arise in this zone. To describe these, the use of kinetic plasma equations is required rather than ordinary hydrodynamics. In the end, however, I managed to show that shock wave emission spectra from accreting neutron stars could explain the observational data obtained with the recently launched instruments.

The first identifications of cosmic X-ray sources with their optical counterparts appeared in the 1960s, allowing a determination of their luminosities and distances to them. It became clear that the large luminosities of accreting neutron stars could be provided only in close binary systems with mass flowing from the stellar component to the neutron star.

When I was a student of astronomy, I attended a course in astrophysics given by the director of the Sternberg Astronomical Institute, Prof. D. Ya. Martynov. In his lectures, he paid special attention to the processes of mass exchange in binary stellar systems through the inner Lagrangian point, and explained how due to the relative orbital motion of the components, a stream of gas forms a disc-like envelope around one of the stars. So, I decided to place a neutron star or even a black hole as an accreting component in a binary system and found that in this case, a new type of accretion (namely, disc accretion!) is possible since the matter, which falls onto such a powerful center of gravity as a neutron star or a black hole, possesses a large angular momentum that prevents it from falling radially inwards. In a first approximation, matter in the disc moves along nearly Keplerian orbits. Slow radial movement of the disc matter toward the center of gravity accompanied by a large energy output (disc accretion) can only be triggered by exchange of angular momentum between adjacent layers of the differentially rotating disc. The reason for such exchange could be turbulence and/or magnetic fields.

In 1969, the article with the calculation and description of the shock wave structure was published in "Astronomicheskii Zhurnal" and became my diploma. This year, I became a postgraduate student at the physics faculty of the MSU. Academician Ya. B. Zeldovich became my scientific advisor.

D. Ya. Martynov. Credit: Photo-Archive of the Sternberg Astronomical Institute

As a postgraduate student, I continued to study the structure and spectra of accretion discs that form around accreting neutron stars and black holes in close binaries due to mass flow from the surface of an optical star.

The foundations of the theory of disc accretion were published, also in "Astronomicheskii Zhurnal", in 1972. The main part of the work was done in collaboration with R. A. Sunyaev. Together we developed the so-called standard model of disc accretion. The work was presented at the 55th symposium of the International Astronomical Union in Madrid in 1972 (Shakura and Sunyaev 1973b). It was there that the first observational results from the UHURU satellite were presented and the first theoretical models of compact X-ray sources in stellar binaries discovered by UHURU were reported. Our report in Madrid was an introduction to a highly influential article published in "Astronomy and Astrophysics" in 1973 (Shakura and Sunyaev 1973a). On the basis of this article, I. D. Novikov and K. S. Thorne calculated the relativistic corrections required by general relativity (Novikov and Thorne 1973).

The pioneering work made together with R. A. Sunyaev is still topical today. According to the NASA ADS data system, the number of references to this article exceeds 8400 (as of April of 2018). It is our great pleasure to present to you this book covering some of the most principal and important areas of modern theory of disc and quasi-spherical accretion onto black holes and magnetized neutron stars.

In Chap. 1, the authors (G. V. Lipunova, K. L. Malanchev, and N. Shakura) present the equations of disc accretion in the framework of the standard model, the basics of the phenomenological theory of turbulent viscosity, and the properties of thin accretion discs and their structure along the radial and vertical directions. The

N. Shakura and R. A. Sunyaev in 1973 and 2017

authors describe analytical solutions to the basic equation of evolution of a non-stationary viscous accretion disc, in the case of infinitely large discs and for discs in binary systems enclosed within their Roche lobes. It is shown how the characteristic time scale of variability in non-stationary disc accretion allows us to determine the level of developed turbulence in accretion discs. A method for a joint numerical solution of the evolution equation and the equations of vertical disc structure is presented.

Chapter 2 (by N. Shakura) is devoted to motion of particles along spherical geodesics around rotating black holes. Such motion is possible if the plane of the outer parts of the accretion disc is tilted toward the equatorial plane of the rotating black hole. A study of this motion is necessary for understanding the structure of a warped disc. This chapter uses a special approach to determine how the quantities, which are measured in the local frame of a fiducial observer in the axially symmetric gravitational field, are related to each other. This approach allows us to better understand the basic principles for measuring physical quantities in GR. These basic principles, which are systematically presented in the next chapter, are required for a more comprehensive understanding of the structure of relativistic accretion discs.

Chapter 3 (by V. V. Zhuravlev) presents a self-consistent model of a standard relativistic accreion disc. The disc is aligned with the equatorial plane of a rotating black hole, and calculation is performed taking full account of relativistic effects. In the first half of the chapter, the author describes in detail how relativistic corrections to the disc structure are deduced using a tetrad basis that is carried by an observer comoving with the rotating matter. Further, using the basic simplifying assumptions of the standard accretion disc model, the relativistic hydrodynamic equations are projected onto the tetrad basis. After that, the author presents an explicit relativistic generalization of radial profiles of the viscous stress and the energy flux from the disc surface.

In Chap. 4, V. V. Zhuravlev outlines the theory of twisted relativistic accretion discs. A warped disc forms around a rotating black hole if the outer parts of the disc are not aligned with the black hole's equatorial plane. The author derives the equations describing the evolution of the shape of a twisted disc and the perturbations of density and velocity necessarily arising in such a disc. This is done under some simplifying assumptions (namely, a small aspect ratio of the disc, slow rotation of the black hole, and a small tilt angle of the disc rings with respect to the black hole equatorial plane), nevertheless including all general relativity effects. The author further presents an analysis of particular regimes of nonstationary twist dynamics (the wave and diffusion regimes), both in the framework of Newtonian dynamics and taking into account Einstein's relativistic precession. At the end of the chapter, a calculation of the shape of a stationary relativistic twisted accretion disc for different values of free parameters of the model is presented.

In Chap. 5, the authors (P. K. Abolmasov, N. Shakura, and A. A. Chashkina) examine the structure of accretion discs in distant quasars from the point of view of the spatial information obtained with the help of quasar microlensing. This exotic effect appears when strong lensing by a foreground galaxy is accompanied by microlensing on individual stars in it. The authors of this chapter aim to give a general introduction to QSO microlensing and to show the opportunities of the method, providing a review of the recent results in this area. It is also shown that the typical variability of the radiation (observed in different spectral ranges) caused by microlensing allows us to study the structure of both subcritical and supercritical (super-Eddington) accretion discs. The latter are characterized by outflow of matter from the inner parts of the disc due to strong radiation pressure. As a consequence, a quasi-spherical envelope forms with a radius determined by processes of scattering by free electrons. This radius has different, presumably weaker, dependence on wavelength, whereas the effective radius of the standard subcritical disc is proportional to the wavelength as $r \sim \lambda^{4/3}$.

Chapter 6 (by V. V. Zhuravlev and D. N. Razdoburdin) is focused on the study of transient growth of small perturbations in spectrally stable rotating shear flows, in particular, those with a Keplerian profile of angular velocity. The mechanism of perturbation growth is discussed in the simplest model of local two-dimensional adiabatic perturbations in a spatially homogeneous flow. Furthermore, special emphasis is placed on mathematical methods that make it possible to perform a rigorous analysis of transient dynamics in disc models of various sophistication. The transient growth of perturbations seems to be capable of transferring energy from a background flow to perturbations in a homogeneous Keplerian flow (in the absence of a magnetic field). Without this energy transport, the emergence of turbulence and/or enhanced angular momentum flux towards the disc outskirts would not be possible.

In Chap. 7, the authors (N. Shakura, K. A. Postnov, A. Yu. Kochetkova, and L. Hjalmarsdotter) examine the theoretical model of quasi-spherical subsonic accretion onto slowly rotating magnetized neutron stars. In this case, the accreting matter slowly, with subsonic velocity, settles onto the rotating magnetosphere of the neutron star, forming an elongated quasi-spherical envelope. The angular momentum transfer in the envelope is effected through large-scale convective motions, implying that the differential rotation law in envelopes above magnetospheres of actual X-ray pulsars corresponds to an approximately isomomentum distribution. The accretion rate in the envelope depends on the ability of the plasma to penetrate into the magnetosphere due to the Rayleigh–Taylor instability, if cooling processes are taken into account. Subsonic infall of matter may occur at moderate X-ray luminosities corresponding to accretion rates of $\dot{M} \lesssim 4 \times 10^{16}$ g/s. In the case of higher accretion rates, a region of free-falling matter arises in the flow above the magnetosphere due to fast Compton cooling, making accretion highly nonstationary. One can determine the basic parameters of the model and estimate the magnetic field of the neutron star when observing acceleration and slowdown in the rotation periods of equilibrium X-ray pulsars with known orbital periods, such as GX 301-2 and Vela X-1, in which quasi-spherical accretion from the stellar wind occurs. It is possible to estimate the velocity of the stellar wind emitted by the optical counterpart of an equilibrium pulsar in a binary, without conducting complex spectroscopic measurements, if an independently measured magnetic field is known for the neutron star. There is a maximal possible value for the slowdown rate of the neutron star for accretion onto a nonequilibrium pulsar. Examples of such pulsars are GX 1+4, SXP 1062 and 4U 2206+54. Knowing the slowdown rate of the rotation of such a pulsar and its X-ray luminosity, we may estimate a lower limit on the magnetic field of the neutron star, which always turns out to be close to the standard value and corresponds to the observed cyclotron peculiarities in measured spectra. The model explains why rotation in nonequilibrium pulsars accelerates and slows down on long timescales and why the pulsar frequency varies on short timescales. In different binaries, these variations may display either a correlation or an anticorrelation with the observed fluctuations in X-ray flux.

The authors of Chap. 8 (N. Shakura and K. A. Postnov) examine the conditions under which the Velikhov–Chandrasekhar magneto-rotational instability (MRI) in ideal and nonideal plasma may arise. In the presence of magnetic fields, this instability arises in an axially symmetric hydrodynamic flow if the angular velocity of the flow decreases outwards, whereas the angular momentum increases. The growth rate of MRI decreases if the magnetic field becomes stronger; there is a critical value of the magnetic field, above which the exponential growth gives way to oscillations. The influence of viscosity and electrical conductivity of the plasma on the development of MRI is studied. The limiting values (lower limits) of ion mean free paths, for which MRI is still possible in thin discs obeying Kepler's law of rotation, are obtained.

On the other hand, the authors show that the MHD mode becomes stabilized in a hydrodynamic flow, which is unstable according to the Rayleigh criterion (the angular momentum decreases outwards), for small perturbation wavelengths.

Many excellent books have been published about accretion in astrophysics. We would like to mention, in particular, *Black-Hole Accretion Disks* by S. Kato, J. Fukue, and S. Mineshige; *Oscillations of Disks* by S. Kato; and *Accretion Power in Astrophysics* by J. Frank, A. King, and D. Raine. We hope that the reader will find the present volume useful as well.

Moscow, Russia
April 2018

Nikolay Shakura

References

Novikov ID, Thorne KS (1973) Astrophysics of black holes. In: DeWitt C, DeWitt BS (eds) Black holes (Les Astres Occlus). Gordon and Breach, New York, pp 343–450

Shakura NI (1973) Disk model of gas accretion on a relativistic star in a close binary system. Sov Astron 16:756 (in Russian: 1972, Astronomicheskii Zhurnal 49:921)

Shakura NI, Sunyaev RA (1973a) Black holes in binary systems. Observational appearance. Astron Astrophys 24:337–355

Shakura NI, Sunyaev RA (1973b) Black holes in binary systems: observational appearances. In: Bradt H, Giacconi R (eds) X- and gamma-ray astronomy, IAU symposium, vol 55, p 155

Zel'dovich YB, Shakura NI (1969) X-ray emission accompanying the accretion of gas by a neutron star. Sov Astron 13:175

Acknowledgements

The work of the authors was partially supported by the grant No. 14-12-00146 of the Russian Science Foundation. We thank Linnea, our translator, for her hard work. We also thank Natalia Lipunova and Natalia Shakura for their help in preparing the current edition. We are also grateful to Sergei Popov for his friendly concern and valuable suggestions.

Contents

Chapter 1
The Standard Model of Disc Accretion

Galina Lipunova, Konstantin Malanchev, and Nikolay Shakura

Abstract Accretion discs are powerful energy factories in our Universe. They effectively transform the potential energy of gravitational interaction to emission, thereby unraveling the physics of distant objects. This is possible due to the presence of viscosity, driven by turbulent motions in accretion discs. In this chapter, we describe the equations for disc accretion in the framework of the standard model. We outline basic elements of the theory of turbulent viscosity and the emergence of the α-parameter. We further describe the radial and vertical structure of thin stationary accretion discs, and present analytical solutions to the basic equation of the evolution of a viscous accretion disc for both an infinite disc and for a disc in a binary system. Finally, we present a numerical method to solve the equations of disc evolution and vertical structure simultaneously.

1.1 Introduction

The theory of disc accretion has tremendously broad applications in astrophysics—it is used to study for example bright objects at a wide spectral range in our own Galaxy, the luminous centres of other active galaxies, relativistic jets from compact objects, protostars and the formation of planetary systems, and to explain the most luminous sources of the universe, the gamma-ray bursts.

G. Lipunova (✉)
Sternberg Astronomical Institute, Lomonosov Moscow State University, Moscow, Russia
e-mail: galja@sai.msu.ru

K. Malanchev
Sternberg Astronomical Institute, Lomonosov Moscow State University, Moscow, Russia
e-mail: malanchev@sai.msu.ru

National Research University Higher School of Economics, Moscow, Russia

N. Shakura
Sternberg Astronomical Institute, Lomonosov Moscow State University, Moscow, Russia

Kazan Federal University, Kazan, Russia

N. Shakura (ed.), *Accretion Flows in Astrophysics*, Astrophysics and Space Science Library 454, https://doi.org/10.1007/978-3-319-93009-1_1

The bases for the theory of standard disc accretion are found in the papers by Shakura (1973) and Shakura and Sunyaev (1973). Other important early works include the papers by Gorbatskii (1965), Lynden-Bell (1969), and Pringle and Rees (1972). The development of the theories for the various processes connected to disc accretion can be found in the textbooks by Kato et al. (1998), Frank et al. (2002) and in the overview by Abramowicz and Fragile (2013). Galactic discs, discs in close binaries and in protoplanetary systems are discussed in the textbook by Morozov and Khoperskov (2005). Bisikalo et al. (2013) studied the gas dynamics of mass-transfer in close binary systems. A short and comprehensive overview of standard disc accretion, including aspects of discs in dwarf and X-ray novae, can be found in Lasota (2015).

In this chapter we consider the basic properties of stationary and non-stationary discs in the framework of the standard model of disc accretion, touching only lightly upon relativistic effects. We derive the basic equations describing non-radial infall of matter in astrophysical situations, where the effects of viscous stresses lead to heating of the matter and subsequent emission of thermal energy that can be observed by astronomical instruments from enormous distances.

Discs are formed around stars as a result of matter with non-zero angular momentum being captured by the star's gravitational field. The matter may originate from the interstellar medium or be transferred from a close companion star. If the matter is rotating in approximately a single plane, the structure is called an accretion disc. As a result of transfer of angular momentum the matter moves towards the central object and thereby releases its gravitational energy. This energy is transferred to kinetic energy, increasing towards the centre, and to thermal energy of the plasma. If the thermal energy can be emitted effectively, the disc is relatively thin.

We here consider geometrically thin 'flat' accretion discs. In a geometrically thin disc, the half-thickness in the direction perpendicular to the disc plane is much smaller than the distance to the centre at a given point in the disc. If a geometrically thin disc has an optical depth much exceeding unity ($\tau \gg 1$) in the direction perpendicular to the disc plane, the equations of energy balance can be written in a rather simple form. In this case the photons are absorbed and dissipated or scattered many times before they leave the disc and we can assume local thermodynamic equilibrium. In a geometrically thin disc we may also neglect radial advection (the transfer of heat with matter moving radially). The condition of local thermodynamic equilibrium suggests equal temperature of electrons and ions in the plasma. Moreover, the standard models do not take into account mass loss from the disc surface: the matter leaves the disc only through its inner boundary. In reality, or rather in the current largely consistent picture of accretion discs, these conditions are satisfied at distances far from the disc centre.

It is important to note that the accretion process is driven by viscosity caused by turbulent motions of the matter in the disc. The characteristic time scale for changes in the radial structure of the disc is called the viscous time scale. The viscous time scale is related to the radial motion of matter in the disc. In the framework of the standard equations for accretion discs, discussed in this section, the characteristic viscous time scale $\tau_{\rm vis}$ is much longer than the dynamic time scale $\tau_{\rm dyn}$, set by the orbital velocity of matter in the disc. The viscous time scale is also much longer than the 'hydrostatic' time scale τ_{hyd}, on which the thickness of the disc changes with pressure, and much longer than the 'thermal' time scale τ_{th}, that is, the time for a given patch of the disc to radiate the stored thermal energy and to change the temperature:

$$\tau_{\mathrm{vis}} \gg \tau_{\mathrm{dyn}}\,; \qquad \tau_{\mathrm{vis}} \gg \tau_{\mathrm{hyd}}\,; \qquad \tau_{\mathrm{vis}} \gg \tau_{\mathrm{th}}\,.$$

As in stars, the disc equilibrium structure depends on its luminosity. For a wide range of accretion rates, the disc luminosity is proportional to the rate with which matter flows into the disc. There is, however, a critical luminosity close to which radiation pressure starts to play a decisive role for the formation of the disc structure. This is the Eddington luminosity limit. Like in stars, the Eddington luminosity is determined from the balance between the forces of radiation pressure and gravitational forces acting on the proton. In the case of spherical symmetry:

$$L_{\mathrm{Edd}} = \frac{4\,\pi\,c\,G\,M\,m_{\mathrm{p}}}{\sigma_{\mathrm{T}}} \approx 1.25 \times 10^{38}\,\frac{M}{M_{\odot}}\;\mathrm{erg/s}\,. \tag{1.1}$$

We use the following notations: the universal gravitational constant G, the mass of the central body M, the mass of the Sun $M_{\odot} \approx 2 \times 10^{33}$ g, the proton mass m_{p}, the Thomson cross section for electron scattering σ_{T}. Using the expression for the effective luminosity in the accretion process $L = \eta_{\mathrm{accr}}\,\dot{M}\,c^2$, we obtain the critical accretion rate in the disc:

$$\dot{M}_{\mathrm{Edd}} = \frac{4\,\pi\,G\,M}{c\,\eta_{\mathrm{accr}}\,\kappa_{\mathrm{T}}} \approx 1.4 \times 10^{18}\,\frac{M}{M_{\odot}}\;\mathrm{g/s}\,,$$

where we have set the energy conversion efficiency of accretion $\eta_{\mathrm{accr}} = 0.1$ ($\eta_{\mathrm{accr}} = 1/12$ in the Newtonian metric for a disc with the inner boundary at radius $6\,G\,M/c^2$), and the Thomson cross section per gram $\kappa_{\mathrm{T}} \approx 0.4\,\mathrm{cm^2/g}$. In disc models the accretion rate is often normalised to this value. However, it is only an approximate evaluation of the accretion rate, above which the disc becomes thick. The thin disc approximation is no longer valid in a region, the radius of which is proportional to the accretion rate, and this region may experience outflow of material from the disc surface.

Various disc instabilities may arise at accretion rates lower than the critical one. For example, at temperatures and densities corresponding to the conditions for recombination of ions in the plasma, thermal instability arises which results in a change in the vertical structure of the disc on thermal time scales (Meyer and Meyer-Hofmeister 1981). In particular, this instability leads to outbursts in dwarf novae. Close to the disc centre, if the radiation pressure exceeds the gas pressure, viscous and thermal instabilities arise (Lightman and Eardley 1974; Shibazaki and Hōshi 1975; Shakura and Sunyaev 1976). Nevertheless, there exists a wide range of accretion rates at which the structure of the accretion disc can be treated as quasi stationary.

1.2 Disc Equations

When examining geometrically thin discs, it is convenient to work in cylindrical coordinates $(r,\ \varphi,\ z)$. We assume that accretion discs are axially symmetric. This leads to the disappearance of all derivatives with respect to φ. For thick discs or for a study of the structure of outflowing matter, spherical coordinates should be used.

1.2.1 Important Note

In the standard theory of disc accretion, as we outline it here following Shakura and Sunyaev (1973), the viscous stress tensor is frequently written as a physically positive value. In Chap. 1 this value appears under notation $w^{\mathrm{t}}_{r\varphi} = -w_{r\varphi}$.

1.2.2 Continuity Equation

The continuity equation in cylindrical coordinates in the axial symmetric case takes the form:

$$\frac{\partial \rho}{\partial t} + \frac{1}{r}\frac{\partial}{\partial r}(\rho\, v_r\, r) + \frac{\partial}{\partial z}(\rho\, v_z) = 0\,. \tag{1.2}$$

1.2.3 Equations of Motion

The equations of motion in cylindrical coordinates in the axial symmetric case are written as:

$$\frac{\partial v_r}{\partial t} + v_r\frac{\partial v_r}{\partial r} + v_z\frac{\partial v_r}{\partial z} - \frac{v_\varphi^2}{r} = -\frac{\partial \Phi}{\partial r} - \frac{1}{\rho}\frac{\partial P}{\partial r} + N_r\,, \tag{1.3}$$

$$\frac{\partial v_\varphi}{\partial t} + v_r\frac{\partial v_\varphi}{\partial r} + v_z\frac{\partial v_\varphi}{\partial z} + \frac{v_r v_\varphi}{r} = N_\varphi\,, \tag{1.4}$$

$$\frac{\partial v_z}{\partial t} + v_r\frac{\partial v_z}{\partial r} + v_z\frac{\partial v_z}{\partial z} = -\frac{\partial \Phi}{\partial z} - \frac{1}{\rho}\frac{\partial P}{\partial z} + N_z\,, \tag{1.5}$$

where Φ is the gravitational potential, P is the pressure, and N_r, N_φ, and N_z are the components of the viscous force $\boldsymbol{N}$ per unit mass. We write the components of $\boldsymbol{N}$ in the case of axial symmetry as:

$$\rho N_r = \frac{1}{r}\frac{\partial}{\partial r}(r w_{rr}) - \frac{w_{\varphi\varphi}}{r} + \frac{\partial w_{rz}}{\partial z}, \tag{1.6}$$

$$\rho N_\varphi = \frac{1}{r^2}\frac{\partial}{\partial r}(r^2 w_{\varphi r}) + \frac{\partial w_{\varphi z}}{\partial z}, \tag{1.7}$$

$$\rho N_z = \frac{1}{r}\frac{\partial}{\partial r}(r w_{zr}) + \frac{\partial w_{zz}}{\partial z}, \tag{1.8}$$

where w_{ik} are the components of the viscous stress tensor. We write these components:

$$w_{rr} = 2\eta\frac{\partial v_r}{\partial r} + \left(\zeta - \frac{2}{3}\eta\right)\operatorname{div}\boldsymbol{v}, \tag{1.9}$$

$$w_{r\varphi} = w_{\varphi r} = \eta\left[r\frac{\partial}{\partial r}\left(\frac{v_\varphi}{r}\right)\right], \tag{1.10}$$

$$w_{rz} = w_{zr} = \eta\left(\frac{\partial v_z}{\partial r} + \frac{\partial v_r}{\partial z}\right), \tag{1.11}$$

$$w_{\varphi\varphi} = 2\eta\frac{v_r}{r} + \left(\zeta - \frac{2}{3}\eta\right)\operatorname{div}\boldsymbol{v}, \tag{1.12}$$

$$w_{\varphi z} = w_{z\varphi} = \eta\frac{\partial v_\varphi}{\partial z}, \tag{1.13}$$

$$w_{zz} = 2\eta\frac{\partial v_z}{\partial z} + \left(\zeta - \frac{2}{3}\eta\right)\operatorname{div}\boldsymbol{v}, \tag{1.14}$$

where

$$\operatorname{div}\boldsymbol{v} = \frac{1}{r}\frac{\partial}{\partial r}(r v_r) + \frac{\partial v_z}{\partial z}.$$

Here η is the dynamic coefficient of the shear viscosity due to the relative motion of different layers of the flow, and ζ is the second viscosity (Landau and Lifshitz 1959). In the following, we will omit the effects of second viscosity.

For thin accretion discs, the only significant component of the viscous stress tensor is $w_{r\varphi}$. As a result we have that

$$\rho N_\varphi = \frac{1}{r^2}\frac{\partial}{\partial r}(r^2 w_{r\varphi}), \tag{1.15}$$

$$w_{r\varphi} = \eta r\frac{\partial}{\partial r}\frac{v_\varphi}{r} = \eta r\frac{\partial \omega}{\partial r}, \tag{1.16}$$

where $\omega = v_\varphi / r$ is the angular velocity of matter in the disc.

We will consider thin stationary discs for which the partial derivatives with respect to time in the equation of motion (1.3)–(1.5) become zero. For such discs the most important terms in Eq. (1.3) are v_φ^2/r and the gravitational potential gradient. For a Newtonian gravitational potential $\Phi = -GM/r$, neglecting the self-gravitation of the disc, we obtain Kepler's law:

$$\omega_K = \sqrt{\frac{GM}{r^3}}. \tag{1.17}$$

In this case the radial component of the friction force and the pressure gradient are negligible comparing to the gravitational force from the central body.

In the direction perpendicular to the disc plane, hydrostatic equilibrium is established, in which the vertical gravity component is balanced by the vertical pressure gradient. From (1.5) we have:

$$-\frac{1}{\rho}\frac{\partial P}{\partial z} = \frac{GM}{r^3}\, z\,. \tag{1.18}$$

1.2.4 Energy Conservation Equation

The energy conservation equation for the general case is written in the following way (Landau and Lifshitz 1959; Kato et al. 2008):

$$\frac{\partial E}{\partial t} + \mathrm{div}\Big[(E + P)\,\boldsymbol{v} - (\boldsymbol{v}\, w_{ik}) + \boldsymbol{F}_{\mathrm{th}}\Big] = \rho \epsilon_{\mathrm{mass}}\,, \tag{1.19}$$

where $E = \rho\,(e + v^2/2 + \Phi)$ is the sum of the thermal, kinetic and potential energy per unit volume. Its change over time is a result of the energy flux arising due to motion of the medium, the work of pressure and viscosity forces, and other possible energy flows. In a thin plane disc, the energy flux connected to viscous forces is radial and equals $(-v_\varphi\, w_{r\varphi})$. The vector $\boldsymbol{F}_{\mathrm{th}}$ contains other types of thermal energy flows: radiative, conductive and convective. In a thin disc the main contribution is given by F_z, which includes the radiative flux transferring energy to the radiating disc surface. In general, there could be other heating or cooling mechanisms operating within a unit volume (for example, nuclear reactions, Joule dissipation or radiative cooling of the optically thin medium). The rate of such heating or cooling per unit mass is ϵ_{mass}.

1.2.5 Energy Dissipation

Let us consider the change of kinetic energy in the flow expressing the total velocity derivative with the help of the Navier-Stokes equation (Eqs. (1.3)–(1.5)):

$$\frac{\mathrm{d}}{\mathrm{d}t}\left(\frac{v^2}{2}\right) = \boldsymbol{v}\left(-\nabla\Phi - \frac{1}{\rho}\nabla P + \boldsymbol{N}\right). \tag{1.20}$$

Subtracting this equation from the equation for the total energy conversion (1.19) and using the first law of thermodynamics

$$T\mathrm{d}s = \mathrm{d}e + P\,\mathrm{d}\left(\frac{1}{\rho}\right),$$

where s is the specific entropy, we arrive at the following equation for the thermal balance:

$$\rho\, T\,\frac{\mathrm{d}s}{\mathrm{d}t} = \varepsilon + \rho\,\epsilon_{\text{mass}} - \mathrm{div}\boldsymbol{F}_{\text{th}}, \tag{1.21}$$

for a gravitational potential Φ constant in time, where

$$\varepsilon = w_{ik}\frac{\partial v_i}{\partial x_k} \tag{1.22}$$

is the dissipated energy per unit volume per unit time due to viscosity (summation over indices).

In cylindrical coordinates for an axisymmetric flow:

$$\varepsilon = \eta\left[4\left(\frac{\partial v_r}{\partial r}\right)^2 + 2\left(\frac{\partial v_z}{\partial z}\right)^2 + \left(\frac{\partial v_\varphi}{\partial r} - \frac{v_\varphi}{r}\right)^2 + \left(\frac{\partial v_\varphi}{\partial z}\right)^2 + \right.$$
$$\left. + \left(\frac{\partial v_z}{\partial r} + \frac{\partial v_r}{\partial z}\right)^2\right] - \frac{2}{3}\eta\left(\frac{\partial v_r r}{r\partial r} + \frac{\partial v_z}{\partial z}\right)^2. \tag{1.23}$$

In thin accretion discs, v_φ dominates significantly over other velocity terms. We note that v_φ does not change with z in a thin disc. Thus, the dominant component in the energy dissipation has the form:

$$\varepsilon = w_{r\varphi}\left(\frac{\partial v_\varphi}{\partial r} - \frac{v_\varphi}{r}\right) = \eta\left(\frac{\partial v_\varphi}{\partial r} - \frac{v_\varphi}{r}\right)^2 = \eta\, r^2\left(\frac{\partial \omega}{\partial r}\right)^2. \tag{1.24}$$

1.2.6 Energy Source in the Disc

The main source of energy which dissipates in the disc due to friction, and which in principle can be radiated, is the released potential energy as the matter moves progressively closer to the gravitating body. Let us illustrate this for a thin Keplerian disc.

With the help of expressions (1.7), (1.10) and (1.24), it can be shown that the identity

$$\rho\, v_\varphi\, N_\varphi + \varepsilon = \frac{1}{r}\frac{\partial}{\partial r}(r\, v_\varphi\, w_{r\varphi})\,. \tag{1.25}$$

holds in a thin disc.

To write down $\rho\, v_\varphi\, N_\varphi$, we use (1.20). Let the mass of the central object and its gravitational potential be constant in time. We have:

$$\rho\, v_r\, \frac{\partial}{\partial r}\left(\frac{v^2}{2} + \Phi\right) = \frac{1}{r}\frac{\partial(r\, v_\varphi\, w_{r\varphi})}{\partial r} - \varepsilon\,. \tag{1.26}$$

Here, we omit the term $v_r\, \partial P/\partial r$, which is small compared to the other terms, i.e. we neglect the work performed by pressure forces in a Keplerian disc.

And thus, the energy from the gravitational interaction extracted as the matter moves in the disc progressively towards the centre is transformed to kinetic energy of orbital motion, then redistributed in the disc due to viscous forces transferring angular momentum, and finally spent on heating of the disc (Lynden-Bell and Pringle 1974; Shakura and Sunyaev 1976).

1.3 Viscosity in Accretion Discs

The key hypothesis in models for accretion discs is the turbulent nature of their viscosity (Shakura 1973; Shakura and Sunyaev 1973; Zeldovich 1981; Fridman 1989; Dubrulle 1993; Balbus and Hawley 1998; Richard and Zahn 1999; Bisnovatyi-Kogan and Lovelace 2001; Marov and Kolesnichenko 2011). The values of the coefficients of molecular viscosity obtained from studies of the properties of liquids and gases cannot explain the observed properties of astrophysical discs. The required rate of transfer of mass towards the disc centre and the accompanying outward transfer of momentum can be explained only for very high values of viscosity in the disc matter, exceeding the molecular viscosity by several orders of magnitude. Turbulent motions of the matter may lead to significant stresses $w_{r\varphi}$ in the disc. In addition, the Reynolds number for accretion discs is very large and this in itself may serve as a basis for development of turbulence regardless of the specific mechanisms for it occurrence.

Differential rotation in Keplerian gaseous discs is considered the basic source of their turbulence. The angular momentum transfer by small-scale magnetic fields in accretion discs was suggested in Shakura and Sunyaev (1973). In the late 1950s and early 1960s, Velikhov (1959) and Chandrasekhar (1960) discovered the MHD-instability in shear flows with angular velocity falling outwards in the existence of a seed poloidal magnetic field. The importance of this instability for accretion discs was shown in the calculations by Balbus and Hawley (see the reviews by Balbus and Hawley (1991, 1998)). Disc accretion with the presence of magnetic fields was studied by many authors (see, for example, Eardley and Lightman 1975; Galeev et al. 1979; Coroniti 1981; Tout and Pringle 1992; Brandenburg et al. 1996).

At the present stage of development of theories of accretion discs, there is no full consensus regarding how to express viscous stresses in a viscous flow. Most authors describe the action of a small scale viscosity by a phenomenological α-prescription (Shakura 1973; Shakura and Sunyaev 1973).

In Sect. 1.2 it was assumed that the derived equations describe the average large-scale motions in the gas. Turbulence arises as a result of transfer of part of the energy of the large-scale motions to random perturbations on smaller scales. In general, such chaotic perturbations in the flow have a very complicated structure and an individual description does not seem possible. Numerical solutions to the associated non-linear equations cannot be achieved at present due to the unreasonably large computational power needed for such a task, and an analytical solution to the general spatially-unbounded problem with smooth initial conditions have not been found either. A solution to the Navier-Stokes equations is one of the seven Millennium Goals announced in 2000 by the Clay Mathematical Institute. In applied problems, methods based on various approximations are mainly used, for example the Reynolds method of averaging or large-eddy simulations (Deardorff 1970).

1.3.1 The Reynolds Equations and the Reynolds Tensor

Reynolds suggested a decomposition of the hydrodynamic fields in the real medium into two components: an average field and a fluctuating (chaotic) field, followed by an averaging of the equations. For example, for the velocity components we assume $v_i = \overline{v_i} + v_i'$, for the pressure $p = \overline{p} + p'$, etc. The average fields are always smooth and slowly changing. The fluctuating fields are chaotic in both space and time. Note that elsewhere in this chapter, we will use ρ without a bar for the local averaged density of the turbulized matter.

Averaging the Navier-Stokes equations according to the rules suggested by Reynolds leads to equations of motion for the average quantities—the Reynolds equations. The method of averaging is not very important. It may be over time, in space, or it may be a theoretical average over a statistical ensemble of various hydrodynamic flows with common boundary conditions (Monin and Yaglom 1971). Average quantities over space and time converge to theoretic-probabilistic mean values if the random process is stationary and spatially homogeneous. The second condition is always a mathematical idealization. In practice, we can only talk about homogeneity in some limited space and time domain. The general condition for convergence of the values that are averaged over space and time to probabilistic mean values is the condition of ergodicity.

Let us write down the Navier-Stokes equation in tensor notation:

$$\rho\left(\frac{\partial v_i}{\partial t}+v_k\frac{\partial v_i}{\partial x_k}\right)=f_i-\frac{\partial\,(P\,\delta_{ik})}{\partial x_k}+\frac{\partial w_{ik}}{\partial x_k}\,, \tag{1.27}$$

where f_i are the components of an external force acting on a unit volume of matter.

For an incompressible fluid ($\rho = const$) we use the equalities

$$\frac{\partial v_k}{\partial x_k}=0 \qquad \text{and} \qquad \frac{\partial\,(v_i\,v_k)}{\partial x_k}=v_k\,\frac{\partial v_i}{\partial x_k}\,,$$

in particular, replacing the second term on the left in (1.27) with $\partial(v_i\,v_k)/\partial x_k$.

We perform averaging according to the Reynolds rules (Monin and Yaglom 1971, their chapter 2), part of which looks like the following:

$$\overline{f'}=0\,,\quad \overline{\frac{\partial f}{\partial x}}=\frac{\partial\overline{f}}{\partial x}\,,\quad \overline{v_i\cdot v_k}=\overline{v_i}\cdot\overline{v_k}+\overline{v_i'\cdot v_k'}\,.$$

The average mass transfer due to turbulent motions is zero: $\overline{\rho\,v_i'}=0$.

In the Reynolds-averaged Navier–Stokes equation, we find the appearance of an additional term dependent on the pulsating velocity (with a prime) arising due to the non-linearity of the original equation:

$$\rho\left(\frac{\partial\,\overline{v_i}}{\partial t}+\frac{\partial\,\overline{v_i}\,\overline{v_k}}{\partial x_k}+\frac{\partial\,(\overline{v_i'v_k'})}{\partial x_k}\right)=\overline{f_i}-\frac{\partial\,(\overline{P}\,\delta_{ik})}{\partial x_k}+\frac{\partial\overline{w_{ik}}}{\partial x_k}\,.$$

To find out the meaning of the last term, we consider the average flow of momentum:

$$\overline{\Pi_{ik}}=\overline{P}\,\delta_{ik}+\rho\,\overline{v_i}\,\overline{v_k}-(\overline{w_{ik}}-\rho\,\overline{v_i'v_k'})\,. \tag{1.28}$$

The two first terms on the right-hand side are responsible for the reversible (mechanical) transfer of momentum by the average motion.

We see that for turbulent motion, the viscous tensor, dependent on the properties of the medium, is accompanied by the term connected with chaotic flows. Thus, turbulent motions lead to exchange of momentum between different regions of the fluid. In other words, turbulent mixing acts like viscosity. The following way of writing the Reynolds equations emphasizes this interpretation:

$$\frac{\partial \overline{v_i}}{\partial t} + \overline{v_k}\, \frac{\partial \overline{v_i}}{\partial x_k} = \frac{\overline{f_i}}{\rho} - \frac{1}{\rho}\frac{\partial (\overline{P}\, \delta_{ik})}{\partial x_k} + \frac{\partial}{\partial x_k}\left(\eta \frac{\partial \overline{v_i}}{\partial x_k} - \overline{v_i' v_k'}\right). \tag{1.29}$$

Here, we use the expression for the viscous stress tensor in an incompressible fluid $w_{ik} = \eta\,(\partial v_i/\partial x_k + \partial v_k/\partial x_i)$ (see, for example, Chapter 2 in Landau and Lifshitz 1959) and apply the incompressibility condition of the fluid.

The quantity

$$R_{ik} = -\rho\, \overline{v_i'\, v_k'} \tag{1.30}$$

is called the Reynolds tensor for turbulent viscosity. The form of this term is unknown and we should make more or less empirically based assumptions to solve the Reynolds equations. The main problem of the phenomenological theory of turbulence is finding the unknown turbulent flows (flow of momentum for the equations above) expressed in the averaged parameters of the properties of the medium. This problem is referred to as a closure problem.

1.3.1.1 Compressible Fluids

In the case of a compressible fluid, instead of the Reynolds average, the weighted average as suggested by Favre (1969) is used. The weighted average velocity is equal to $\widetilde{v_i} = \overline{\rho\, v_i}/\overline{\rho}$, where bars over the values indicate the Reynolds average (time average). The velocity of the flow is then represented by the sum of the weighted average and the fluctuating velocities: $v_i = \widetilde{v_i} + v_i''$. Now $\overline{v_i''} \neq 0$ (average over the ensemble) for $\widetilde{v_i''} = 0$ (weighted average, average over the ensemble), but, as before, the turbulent motions do not lead to transfer of mass, $\overline{\rho v_i''} = 0$ (see, for example, Marov and Kolesnichenko 2011, their chapter 3).

After such a representation of the hydrodynamic functions and averaging over time of the Navier-Stokes equation for $\rho \neq const$, we arrive at an equation of motion, which also can be written in compact form, analogous to (1.29), but with an additional term, which corresponds to the turbulent viscosity, of a more complicated form:

$$R_{ik}^{*} = -\overline{\rho v_i'' v_k''} + \eta \left(\frac{\partial \overline{v_i''}}{\partial x_k} + \frac{\partial \overline{v_k''}}{\partial x_i} - \frac{2}{3}\delta_{ik}\frac{\partial \overline{v_k''}}{\partial x_k}\right). \tag{1.31}$$

We ignore here fluctuations in the coefficient of molecular viscosity η. Thus, if density fluctuations are present in the medium, the viscosity tensor cannot be divided into two constituents, one dependent on the properties of the environment only (the viscosity η and the average velocity $\overline{v_i}$ of the laminar flow) and the other dependent only on the turbulent dynamics of the flow (terms with fluctuating velocity). It is expected, however, that the second term in the last expression, the term that includes the molecular viscosity, is significantly smaller than the first term (Pletcher et al. 1997).

1.3.2 The Closure Problem

The form of the Reynolds tensor cannot be found from the hydrodynamic equations. The second-order moment tensors for the velocity field $\overline{v_i' \, v_k'}$ can be expressed from the third- or higher-order moments ($\overline{v_i' \, v_j' \, v_k'}$, etc.), but the number of unknowns is always greater than the number of equations. The impossibility of finding a closed system of equations for a finite number of moments is a consequence of the non-linearity of the equations of hydrodynamics. In the case of weighted averaging, the problem becomes even more complicated (Marov and Kolesnichenko 2011, their chapter 3).

The need to solve practical problems have led to the performance of a large number of experiments regarding turbulent flows. Based on these studies, semi-empirical theories of turbulence have been worked out, which systematize the obtained results.

Important steps in this direction were taken by Boussinesq (in the end of the nineteenth century) and by Taylor, Prandtl and Karman in the 1920s and 1930s. The semi-empirical models of turbulence are based on the analogy between turbulence and molecular viscosity. An application of the simplest models allows us to close the very first equations for hydrodynamic fields—the ones for lower moments (the Reynolds equations). As a result, the Reynolds equations can be solved if R_{ik} is expressed from certain large-scale characteristics of the flow. These characteristics describe the transfer of heat and momentum through the turbulent medium. Large-scale characteristics of turbulence are to a great deal dependent on the geometry of the boundaries of the flow and the nature of external influences, which are always different in different situations. Therefore, on the one hand, we talk about the ambiguity of semi-empirical closing relations. On the other hand, using more complex closing relations leads to neither more general nor more exact solutions. Thus, in most cases, preference is given to the simpler models, and the limits to their applicability are studied (see Marov and Kolesnichenko 2011, their Sect. 1.1.6).

1.3.3 Coefficient of Turbulent Viscosity

The Reynolds equations can be solved only with the addition of closing relations, which use the averaged characteristics of the turbulent flow (pressure, density, temperature, and average velocity). This is the way semi-empirical models for

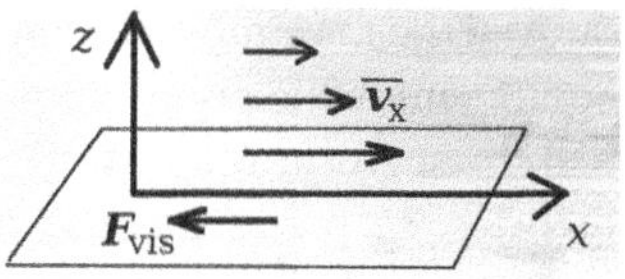

Fig. 1.1 Coordinates in a plane flow

turbulence are constructed. Most of these models are based on Boussinesq's gradient hypothesis (1897) which suggests that there is a linear connection between the turbulent viscous tensor and the shear tensor which in turn is a linear combination of the terms $\partial \bar{v}_j/\partial x_i$ together with certain local proportionality coefficients (coefficients of turbulent transfer). It is, however, necessary to make concrete assumptions regarding these coefficients.

Let us consider a small area inside a turbulent flow (Fig. 1.1). We consider this area to be flat and assume that the average motion is directed along the plane of the area (along the x-axis). Let the area be located in the plane $z = 0$. The frictional force acting on a unit area, directed along the x-axis is equal to:

$$\overline{w_{xz}} - \rho\, \overline{v'_x v'_z} = \rho\, \nu \,\frac{\partial \overline{v_x}}{\partial z} - \rho\, \overline{v'_x v'_z}\,.$$

According to what is called Bussinesq's gradient hypothesis, there is an analogy between the viscous and the turbulent flow of momentum and we may set:

$$-\rho\, \overline{v'_x v'_z} = \rho\, \nu_t \,\frac{\partial \overline{v_x}}{\partial z}\,, \tag{1.32}$$

introducing the proportionality constant ν_t. This approach allows us to solve the Reynolds equations using standard methods if we know the kinematic coefficient of turbulent viscosity ν_t that replaces the usual coefficient of molecular viscosity. The turbulent viscosity coefficient cannot be derived from microscopic considerations.

The gradient model works well for quasi-stationary flows. It is assumed that a local equilibrium is formed in the structure of developed turbulence, in which the characteristics of turbulence at every point of the flow are completely determined by the local characteristics of the field of the averaged flow in the vicinity of this point and by the local averaged parameters of the state of the medium itself.

In general, ν_t is significantly larger than ν. The turbulent viscosity coefficient, as opposed to the molecular viscosity coefficient, does not describe the physical properties of the fluid but the statistical properties of the fluctuations. Its value depends on the method of averaging over the ensemble of analogous flows. A semi-empirical model of turbulence can be constructed if ν_t is estimated in the course of experiments. For example, it is known that in the case of motion of a turbulent flow in a plane channel, ν_t cannot be constant since it has been established empirically that $\nu_t \to 0$ close to the walls. In an infinite turbulent flow, however, it is often quite reasonable to assume that $\nu_t = const$ (see Sect. 5.8 in Monin and Yaglom 1971).

And thus, we moved from the unknown Reynolds tensor to the turbulent viscosity coefficient, which is also unknown. Choosing this parameter is another separate task and, to solve it, other semi-empirical theories have been proposed in turn. These theories, in particular, use the concept of mixing length. This concept plays an important role in the theory of turbulent viscosity in accretion discs.

1.3.4 Mixing Length

The concept of mixing length introduced by Prandtl to the theory of turbulence (1925) allows us not only to express simply the coefficients of turbulent mixing (in particular, the turbulent viscosity coefficient), using the length of the mixing path, but also to obtain defining relations for turbulent flows in some particular cases. The mixing length is the distance which a unit volume of gas travels in a turbulent flow before this volume is mixed completely with the surrounding medium. This distance is in a sense analogue to the mean free path in kinetic gas theory.

Turbulent stresses are the result of transfer of momentum due to fluctuations of turbulent velocity. Prandtl's hypothesis is that vortices, shifting as 'trickles' along the z-axis for the path of the 'mixing length' ξ'_z, retain their momentum. This is similar to the picture of turbulent diffusion of impurities. At the height $z + \xi'_z$, a fluctuation v'_x may be represented as the difference between the proper velocity of a trickle $\overline{v}_x(z)$ (the average velocity at the initial level) and the velocity of the surrounding flow $\overline{v}_x(z + \xi'_z)$. Linearization of the profile of the average velocity $\overline{v}_x$ yields: $\overline{v_x}(z + \xi'_z) = \overline{v_x}(z) + \xi'_z\, \partial\overline{v_x}/\partial z$. We thus write the Prandtl relation for transfer of momentum as:

$$v'_x = -\xi'_z\, \partial\overline{v_x}/\partial z\,. \tag{1.33}$$

In the case of a plane shear flow, we get for the component of the Reynolds tensor (1.30):

$$R_{xz} \equiv -\rho\, \overline{v'_x\, v'_z} = \rho\, \overline{\xi'_z\, v'_z}\, \frac{\partial\overline{v_x}}{\partial z}\,. \tag{1.34}$$

If we define the kinematic coefficient of turbulent viscosity as

$$\nu_\mathrm{t} = \overline{\xi'_z\, v'_z}\,, \tag{1.35}$$

formula (1.34) is terminologically consistent with the gradient hypothesis (See expression (1.32)). Formula (1.35) is similar to the formula for the molecular viscosity coefficient: $\nu = l_\mathrm{m} v_\mathrm{m}$, where l_m is the mean free path of the molecules and v_m is the velocity of their thermal motion. The 'amount of exchange' in a turbulent flow ν_t is also a product of the distance and velocity at which turbulent exchange takes place—the mixing speed. The value of ξ'_z is essentially a random (fluctuating) quantity.

In order for formula (1.34) to be applied in practice, the mixing speed v'_z should also be estimated, which Prandtl does (1925). As a result of mixing, the mixing speed itself should decrease as the conditions (velocities) in the medium are leveled out. From this follows the assumption that the mixing speed should be proportional to the velocity gradient of the average motion $\partial \overline{v_x}/\partial z$. This simultaneously means that the fluctuations of velocity in different directions have similar absolute values, i.e. $v'_z \sim v'_x$ (Monin and Yaglom 1971).

We thus use (1.33), substituting in (1.34), and obtain:

$$R_{xz} = \rho\, \overline{(\xi'_z)^2} \left| \frac{\partial \overline{v_x}}{\partial z} \right| \frac{\partial \overline{v_x}}{\partial z},$$

where the modulus is inserted in order that the sign of the turbulent viscosity tensor be the same as for $\partial \overline{v_x}/\partial z$. This corresponds to the fact that momentum is transferred from layers moving faster to those moving more slowly. Then for the kinematic coefficient of the turbulent viscosity we obtain the Prandtl formula (1925):

$$\nu_t = \alpha^* \, L^2 \left| \frac{\partial \overline{v_x}}{\partial z} \right|, \tag{1.36}$$

where α^* is a dimensionless quantity of the order of 1. The local mixing path ξ'_z is a too uncertain quantity and cannot be measured. Here, the distance L, or the mixing length, is already not a random quantity. Its magnitude is of the order of $\sqrt{\overline{(\xi'_z)^2}}$ and characterizes the scale of turbulence. Now, what is left is to establish the dependence of L on the coordinates, for example, empirically.

The expression (1.36) may also be retrieved from dimensional considerations. For this we use the principle of local similarity of turbulent transfer (Sect. 3.3 in Marov and Kolesnichenko (2011)),—the coefficients of turbulent transfer in each point depend only on the properties of the medium in this point, the local size of the scale of turbulence L and on certain characteristics of the averaged flow. In other words, ν_t is a function of the quantities ν, L, and $\partial \overline{v_x}/\partial z$. The scale L characterizes the geometry of the turbulent flow or the characteristic size. Far away from the hard surface, ν can be excluded from the list of parameters, and the dimensional considerations yield the Prandtl formula (1.36).

The constant factor α^* is determined for each specific type of motion on the basis of experimental data.

1.3.5 *Turbulent Viscosity Parameter* α

Let us consider an accretion disc, with orbital motion in circular orbits and orbital velocities in the plane parallel to the disc symmetry plane. Placing an imaginary wall perpendicular to the radius in a given point at a distance $r_\star$ from the centre, we find the frictional force applied per unit area of the wall.

The averaged velocity of matter in the disc is tangential to the radius with great accuracy since the orbital velocity dominates over other components. Let the imaginary wall rotate around the centre with the averaged velocity of the flow. The frictional force is directed tangentially and is equal to the density of the flow of the φ-component of momentum in the radial direction. If we assume, as mentioned in Sect. 1.3.1.1 after formula (1.31), that the first term in the turbulent viscosity tensor, defined only by the dynamics of the flow, dominates over the others, which contain the molecular viscosity coefficient η, then the $r\varphi$-component for the frictional stress on the wall is equal to

$$(\overline{w_{r\varphi}} - \overline{\rho\, v_r''\, v_\varphi''})_{\,r=r_\star} = (\overline{\rho\,\nu}\, r\frac{\mathrm{d}\omega}{\mathrm{d}r} - \overline{\rho\, v_r''\, v_\varphi''})_{\,r=r_\star}\,,$$

where we used the expression for the component of the stress tensor in cylindrical coordinates (see Chapter 2 in the book by Landau and Lifshitz 1959).

According to common practice, we will define $w_{r\varphi}^{\mathrm{t}}$ as the quantity with opposite sign to the component of the viscous stress tensor in the disc $w_{r\varphi}$.[1] Using the gradient hypothesis (see Sect. 1.3.3 and (1.32)) we can write:

$$w_{r\varphi}^{\mathrm{t}} \equiv \overline{\rho\, v_r'' v_\varphi''} = -\overline{\rho}\;\nu_{\mathrm{t}}\, r\,\frac{\mathrm{d}\omega}{\mathrm{d}r}\,, \tag{1.37}$$

where ν_{t} is the kinematic coefficient of the turbulent viscosity [cm^2/s]. For a Keplerian disc, we get from (1.37):

$$w_{r\varphi}^{\mathrm{t}} = \frac{3}{2}\,\omega_{\mathrm{K}}\,\nu_{\mathrm{t}}\,\overline{\rho}\,. \tag{1.38}$$

As a consequence of the Prandtl hypothesis (1.35), $\nu_{\mathrm{t}} = \overline{v_t\, l_t}$, where v_t and l_t are the velocity and length of turbulent mixing, respectively, which both take random values in a turbulent flow.[2] Applying the Prandtl relation to describe the radial transport of turbulent velocity

$$v_t = l_t\, r\left|\frac{\mathrm{d}\omega}{\mathrm{d}r}\right|$$

(cf. (1.33)), we get, substituting in (1.37), that

$$w_{r\varphi}^{\mathrm{t}} = \overline{\rho}\;\overline{v_t^2} \equiv m_t^2\,\overline{\rho}\, v_{\mathrm{s}}^2\,,$$

[1] With this definition, $w_{r\varphi}^{\mathrm{t}}$ will be positive in accretion discs. In other literature on the subject, the definition $t_{r\varphi} = -w_{r\varphi}^{\mathrm{t}}$ is often used instead.

[2] They are analogous to the quantities v_x' and ξ_z' discussed in Sect. 1.3.4.

where the averaged turbulent velocity squared (fluctuating component of the velocity of matter in the disc) $\overline{v_t^2}$ is expressed using the sound speed v_s and the Mach number $m_t^2 = \overline{v_t^2}/v_s^2$.

The last formula can be rewritten as

$$w^{t}_{r\varphi} = \alpha P \,, \tag{1.39}$$

where the dimensionless quantity α is called the turbulent parameter and P is the total pressure (the sum of gas and radiation pressures).

Disc models, in which turbulence is taken to be the source of viscosity and where the connection (1.39) is used, are called α-discs. In the simplest models, this coefficient is considered fixed within the whole accretion disc. Its value may be found from a comparison with observations of transient phenomena, which are manifestations of viscous evolution of discs in the case of non-stationary accretion onto space objects.

Equating (1.38) to the quantity $\alpha\overline{\rho}\, v_s^2$, we obtain a relation between the dimensionless turbulence parameter and the kinematic viscosity coefficient in a Keplerian disc:

$$\nu_t = \frac{2}{3}\alpha\, v_s^2 \frac{1}{\omega_K} = \frac{2}{3}\alpha\, v_s\, z_{hyd} \,, \tag{1.40}$$

where we introduce the 'hydrostatic half-thickness' of the disc, which can be found from approximate integration of (1.18):

$$z_{hyd} \equiv \sqrt{\frac{p}{\rho}\frac{1}{\omega_K^2}} = \frac{v_s}{\omega_K} \,.$$

Using (1.35), which is a consequence of the Prandtl hypothesis, we may write:

$$\alpha = \frac{\overline{v_t\, l_t}}{\frac{2}{3}\, v_s\, z_{hyd}} \,.$$

From general considerations it is clear that the α-parameter is a quantity whose value does not exceed unity. Indeed, if the turbulent motions have velocities exceeding the sound speed, these motions are quickly quenched by shock-waves. The inequality $l_t > z_{hyd}$ would suggest that the turbulence has an anisotropic character since the transverse thickness of the disc is limited by the quantity $\sim z_{hyd}$.

The use of the α-parameter is justified in situations where it may be considered approximately constant. As proved during the last decades, such an approximation describes well a variety of observed phenomena in sources with disc accretion. Numerical modelling of outbursts in dwarf novae and X-ray transients demonstrates that the α-parameter can be considered constant for certain ranges of physical conditions in these astrophysical discs. Typical values from observations are 10^{-2} – 1 (Meyer and Meyer-Hofmeister 1984; Cannizzo 1998; Kotko and Lasota 2012).

1.4 Thin Discs

1.4.1 Equations of Radial Structure

Let us write down the equations for disc accretion for geometrically thin α-discs. We will neglect any dependence of the physical parameters in the disc on z, averaging (integrating) along the vertical. We will consider discs without radial advection (transfer of heat with matter moving radially) and without mass loss from the disc surface. In such discs, the angular velocity of the rotating matter at each radius r is approximately equal to the angular rotational velocity of a free particle. In other words, $v_r \ll v_\varphi$.

The parameters determining the structure of a geometrically thin disc are the mass of the gravitating centre M, the inner radius of the accretion disc $r_{\rm in}$ and the accretion rate $\dot{M}$.

1.4.1.1 Mass Conservation Equation

We introduce the surface density

$$\Sigma_0(t,r) = \int_{-z_0}^{+z_0} \rho(t,r,z)\,\mathrm{d}z\,, \tag{1.41}$$

where z_0 is the disc half-thickness at radius r. As agreed earlier, the velocities in thin discs are independent of z. Integrating (1.2) along the disc height, we obtain:

$$\frac{\partial \Sigma_0}{\partial t} = -\frac{1}{r}\frac{\partial}{\partial r}(\Sigma_0\, v_r\, r)\,. \tag{1.42}$$

The product within parentheses on the right-hand side of this equation, multiplied by 2π, is equal to the radial flow of the matter in the disc [g/s] through a cylindrical surface with radius r.

1.4.1.2 The r-Component of the Equation of Motion

For a thin stationary disc, the dominant terms in this equation are

$$\frac{v_\varphi^2}{r} = \frac{\partial \Phi}{\partial r}\,. \tag{1.43}$$

For a Newtonian potential, this equation corresponds to Kepler's law:

$$\omega_{\mathrm{K}} = \frac{\sqrt{G M}}{r^{3/2}} \, .$$

Other potentials that take into account the curvature of space around a Schwarzschild black hole are discussed in Sect. 1.4.4.

1.4.1.3 The φ-Component of the Equation of Motion

After multiplying by ρr^2, we integrate vertically (1.4) and obtain the law of conservation of the angular momentum

$$\Sigma_0 \, v_r \, r \, \frac{\partial \, (\omega \, r^2)}{\partial r} = -\frac{\partial}{\partial r} \, (W_{r\varphi} r^2) \, , \tag{1.44}$$

where

$$W_{r\varphi}(t, r) = \int_{-z_0}^{+z_0} w^{\mathrm{t}}_{r\varphi}(t, r, z) \, \mathrm{d}z \tag{1.45}$$

is the height-integrated component of the viscous stress tensor.

1.4.2 *Solution for a Constant Accretion Rate*

From the continuity equation (1.42) it follows that in the stationary regime

$$\Sigma_0 \, v_r \, r = const \, .$$

We determine the accretion rate as the mass of matter intersecting the surface of a cylinder with radius r per unit time:

$$\dot{M} \equiv -2\pi r \, v_r \, \Sigma_0 \, . \tag{1.46}$$

The minus sign is inserted in order to make a quantity $\dot{M}$ positive and to compensate for the fact that as matter moves towards the centre, $v_r < 0$.

For a constant accretion rate, the equation of motion (1.44) can be easily integrated:

$$\dot{M} \, \omega \, r^2 - 2\pi \, W_{r\varphi} \, r^2 = const \, . \tag{1.47}$$

This is the law of conservation of angular momentum for a stationary disc. The constant can be determined from the boundary conditions at the inner edge of the disc:

$$\dot{M}\,(h - h_{\rm in}) = F - F_{\rm in},$$

where F is the momentum of viscous forces between adjacent rings of the disc (the viscous torque, a positive quantity in our notation)

$$F = 2\pi\, W_{r\varphi}\, r^2\,, \tag{1.48}$$

and $h = \omega r^2$ is the specific angular momentum, where the subscript indicates quantities at the inner disc radius.

The equation of conservation of angular momentum can be written in the form:

$$W_{r\varphi} = \frac{\dot{M}\,\omega}{2\pi}\, f(r) \quad \text{or} \quad F = \dot{M}\, h\, f(r)\,, \tag{1.49}$$

where the function $f(r) = 1 - h_{\rm in}/h + F_{\rm in}/(\dot{M}\,h)$ contains information about inner boundary conditions for the viscous stress tensor (the form of $f(r)$ at $\dot{M}(r) \neq const$, see Sect. 1.5.3). For example, in the case of black holes, the viscous stress tensor is set to zero since the inner radius of the disc is determined by the radius of the last stable orbit, from which matter falls freely onto the black hole. Then, far away from the inner radius, $f(r) \approx 1$. For accretion onto a magnetized star, the stress tensor at the inner edge of the disc depends on the strength of the magnetic field and its radial distribution changes accordingly. For central objects with a sufficiently strong magnetic field, accretion may seize at the inner radius of the disc. Such discs are called disc reservoirs (Syunyaev and Shakura 1977). In a disc reservoir F is radially constant close to the inner boundary, and at large radii F is affected by the conditions at the outer boundary.

1.4.3 Radial Velocity of Matter in the Disc

Let us estimate the radial component of the velocity of matter in a disc in the stationary regime from the φ-component of the equation of motion. For this, we use (1.49) in the approximation $f(r) \sim 1$, which is valid in the quasi-stationary case, far away from the centre, and definition of accretion rate (1.46):

$$|v_r| = \frac{\dot{M}}{2\pi\, r\, \Sigma_0} = \frac{W_{r\varphi}}{\omega r\, \Sigma_0}\,.$$

Obviously, this velocity, with which matter approaches the gravitating object, depends on the value of the viscosity. We use the formula (1.38) and obtain:

$$|v_r| = \frac{3}{2}\frac{\nu_\mathrm{t}}{r}\,, \tag{1.50}$$

where $W_{r\varphi} \approx 2\,z_0\,w^\mathrm{t}_{r\varphi}$ and $\Sigma_0 \approx 2\,z_0\,\rho$ (cf. (1.41), and (1.45)).

The characteristic time scale for movement of the matter radially towards the centre is

$$\tau_\mathrm{vis} \sim \frac{r}{|v_r|} = \frac{2}{3}\frac{r^2}{\nu_\mathrm{t}}\,.$$

Making an assumption regarding the α-viscosity in the disc and using the relation (1.40) between ν_t and α, we re-write the obtained formulas in the form:

$$|v_r| = \alpha\,v_\mathrm{s}\,\frac{z_0}{r} = \alpha\,v_\varphi\left(\frac{z_0}{r}\right)^2\,, \tag{1.51}$$

$$\tau_\mathrm{vis} = \frac{1}{\alpha\,\omega_\mathrm{K}}\left(\frac{z_0}{r}\right)^{-2}\,. \tag{1.52}$$

In a geometrically thin disc, the viscous time scale is much larger than the characteristic dynamic time scale

$$\tau_\mathrm{dyn} \sim \frac{r}{v_\varphi} \sim \frac{1}{\omega_\mathrm{K}}\,. \tag{1.53}$$

1.4.4 Accretion Onto a Black Hole

In Chap. 3, devoted to relativistic standard discs, a theory will be presented, the basics of which were worked out by Novikov and Thorne (1973). For further acquaintance with the astrophysical aspects of this theory we also recommend the books by Shapiro and Teukolsky (1983), Thorne et al. (1986). Here, we outline only the basics of the behavior of an accretion disc around a black hole.

Close to the black hole the curvature of space-time plays a crucial role for the formation of an accretion disc. The thin-disc approximation, according to which matter rotates in approximately circular orbits, breaks down. The flow of matter onto the black hole speeds up, becomes highly supersonic in the radial direction and, starting from some certain radius, goes in the free-fall regime.

At free fall, the momentum of the in-falling matter is conserved. In this case, there is no outward flux flow of the viscous tensor, implying that it is equal to zero at the disc inner boundary: $F_\mathrm{in} = 2\pi\,W_{r\varphi}r^2_\mathrm{in} = 0$. This condition is confirmed by numerical one-dimensional calculations of the equations of hydrodynamics in a post-Newtonian potential (Shafee et al. 2008). It turns out that the conditions for the viscous stress tensor to be equal to zero are satisfied close to the innermost stable circular orbit.

For non-rotating black holes, the radius of the innermost stable circular orbit $r_{\rm ISCO} = 3\,R_{\rm g}$, where the Schwarzschild radius $R_{\rm g}$ is the event-horizon radius of a non-rotating black hole:

$$R_{\rm g} = 2\,G\,M/c^2\,.$$

The radius $r_{\rm ISCO}$ for a rotating black hole is determined in the Kerr space-time metric and given by formula (3.22) in Sect. 3.1.3.

At radii less than $3\,R_{\rm g}$, there is no energy release due to viscosity. We note that in this area radiation may be generated as a result of processes which involve plasma and magnetic fields.

Thus, for accretion onto a Schwarzschild black hole, the boundary condition at the inner radius is written as

$$W_{r\varphi}(r = 3\,R_{\rm g}) = 0\,.$$

We use Eq. (1.47) for $\dot{M} = const$ in the form

$$\dot{M}(\omega_{\rm in} - \omega) = 2\pi\;W_{r\varphi}\,,$$

or, for the viscous torque,

$$F = \dot{M}(h - h_{\rm in})\,, \tag{1.54}$$

where $h_{\rm in}$ is the specific angular momentum of the matter at the innermost orbit around the black hole.

If the viscous stress tensor is equal to zero at the inner boundary of a stationary infinite disc, the function in (1.49) is:

$$f(r) = 1 - h_{\rm in}/h(r) + F_{\rm in}/(\dot{M}\,h) = 1 - h_{\rm in}/h(r). \tag{1.55}$$

In the Newtonian approximation, $f(r) = 1 - \sqrt{r_{\rm in}/r}$ and

$$W_{r\varphi} = \frac{\dot{M}\,\omega}{2\pi}\,(1 - \sqrt{r_{\rm in}/r})\,.$$

To approximately take into account the effects of general relativity in the vicinity of non-rotating black holes, the Paczynski–Wiita potential can be used (Paczynsky and Wiita 1980):

$$\Phi_{\rm PW} = -\frac{G\,M}{r - R_{\rm g}}\,. \tag{1.56}$$

For free particles in circular orbits, the velocities can be found from (1.3):

$$\frac{v_\varphi^2}{r} = \frac{\mathrm{d}\Phi}{\mathrm{d}r} \,. \tag{1.57}$$

As a result, we obtain

$$\frac{v_\varphi^{\mathrm{PW}}}{c} = \frac{1}{\sqrt{2}} \frac{\sqrt{r\,R_\mathrm{g}}}{(r - R_\mathrm{g})} \,,$$

and the specific angular momentum of a test particle in the Paczynski-Wiita potential is:

$$h^{\mathrm{PW}} = v_\varphi^{\mathrm{PW}}\, r = \sqrt{\frac{G\,M\,r}{(1 - \frac{2\,G\,M}{c^2\,r})^2}} \,. \tag{1.58}$$

The modified potential (1.56) is often used (for example when substituting into (1.43)) since it fits quite well the curvature effects of the space-time metric around a Schwarzschild black hole. Other approximate potentials, in particular such applicable to the case of rotating black holes, can be found in the book by Kato et al. (1998).

Let us write down the Schwarzschild stationary metric as the square of the interval between two events separated in time and space

$$\mathrm{d}s^2 = -(1 - R_\mathrm{g}/r)\,\mathrm{d}t^2 + (1 - R_\mathrm{g}/r)^{-1}\,\mathrm{d}r^2 + r^2(\mathrm{d}\theta + \sin^2\theta\,\mathrm{d}\varphi) \,.$$

Here, $t,\ r,\ \theta,\ \varphi$ are the Schwarzschild coordinates. Due to the curvature of space-time near a black hole, the distance element $\mathrm{d}l$ along the radius, as measured by a local observer, is longer than the corresponding coordinate element $\mathrm{d}r$ (see Fig. 1.2):

$$\mathrm{d}l = \frac{\mathrm{d}r}{\sqrt{1 - R_\mathrm{g}/r}} \,.$$

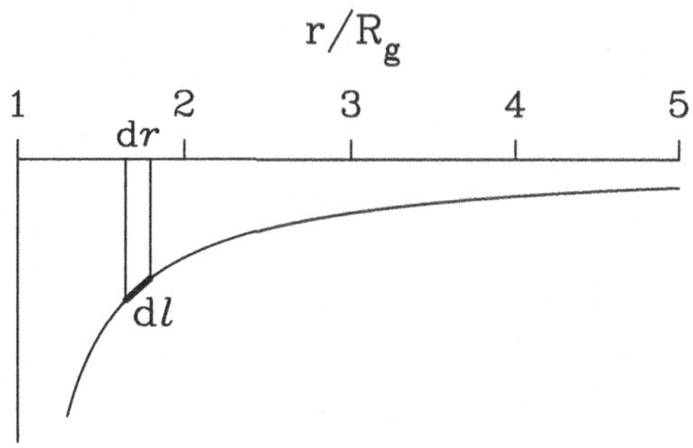

Fig. 1.2 Illustration of the 'shrinking' of a coordinate element $\mathrm{d}r$, corresponding to an element of distance $\mathrm{d}l$, measured by a stationary observer

To describe the relativistic motion in the vicinity of a Schwarzschild black hole, we may use the following 'logarithmic' potential (Landau and Lifshitz 1973; Thorne et al. 1986; Abramowicz 2016):

$$\Phi = \frac{c^2}{2} \ln\left(1 - \frac{R_g}{r}\right) = c^2 \ln\sqrt{1 - \frac{R_g}{r}}\,. \tag{1.59}$$

Here $\sqrt{1 - R_g/r}$ is the lapse function in the Schwarzschild metric. It determines the redshift of the signal emitted from the vicinity of the black hole and the difference between two time intervals, one of which, dt, is measured at infinity and the other, $d\tau_l$, by an observer in the local stationary reference frame:

$$d\tau_l/dt = \sqrt{1 - R_g/r}\,. \tag{1.60}$$

The time measured in the frame of moving particle is related to the time measured by the local stationary observer as

$$d\tau_p/d\tau_l = \sqrt{1 - v^2/c^2}\,. \tag{1.61}$$

The momentum $\boldsymbol{p}$ and the energy $E_{\rm local}$ of a relativistic particle with rest mass m_o relative to the local stationary observer are

$$\boldsymbol{p} = \frac{m_o\,\boldsymbol{v}}{\sqrt{1 - v^2/c^2}}\,, \quad \text{and} \quad E_{\rm local} = \frac{m_o\,c^2}{\sqrt{1 - v^2/c^2}}\,,$$

respectively, where $v^2 = v_r^2 + v_\varphi^2$ for particles moving in the equatorial plane. We may also introduce the notion of 'energy at infinity' E. This value remains unchanged along the particle trajectory. Let us determine it.

Consider a particle travelling past a stationary observer who is located at a distance r from the centre of a black hole. The equation of particle motion in the reference system of this observer looks as follows:

$$\frac{d\boldsymbol{p}}{d\tau_l} = -\frac{m_o}{\sqrt{1 - v^2/c^2}}\,\nabla\Phi\,. \tag{1.62}$$

Note that the potential Φ is spherically symmetric. On multiplying Eq. (1.62) by $\boldsymbol{v}$, we obtain

$$\boldsymbol{v}\,\frac{d}{d\tau_l}\left(\frac{m_o\boldsymbol{v}}{\sqrt{1 - v^2/c^2}}\right) = -\frac{m_o\,\boldsymbol{v}}{\sqrt{1 - v^2/c^2}}\,\nabla\Phi = -\frac{m_o\,\boldsymbol{v}\,\boldsymbol{e}_r}{\sqrt{1 - v^2/c^2}}\,\frac{d\Phi}{dl}\,, \tag{1.63}$$

where $\boldsymbol{e}_r$ is a unit radial vector in the Cartesian reference system of the local observer. Further, we differentiate the left-hand side of (1.63):

$$\frac{1}{2}\frac{m_o}{\sqrt{1-v^2/c^2}}\frac{\mathrm{d}v^2}{\mathrm{d}\tau_l}+\frac{1}{2}\frac{m_o\, v^2/c^2}{(1-v^2/c^2)^{3/2}}\frac{\mathrm{d}v^2}{\mathrm{d}\tau_l}=-\frac{m_o\,\boldsymbol{v}\,\boldsymbol{e}_r}{\sqrt{1-v^2/c^2}}\frac{\mathrm{d}\Phi}{\mathrm{d}l}\,.$$

When multiplying this by $(1-v^2/c^2)^{3/2}$, cancelling out the two equal terms with opposite signs in the left-hand part of the equation and using the equality $v_r = \mathrm{d}l/\mathrm{d}\tau_l$ for the radial velocity, we obtain

$$\frac{1}{2}\frac{\mathrm{d}}{\mathrm{d}\tau_l}(1-v^2/c^2)=(1-v^2/c^2)\frac{\mathrm{d}l}{\mathrm{d}\tau_l}\frac{\mathrm{d}}{\mathrm{d}l}\ln(1-R_\mathrm{g}/r)^{1/2}\,,$$

which is equivalent to the following equation

$$\frac{\mathrm{d}}{\mathrm{d}\tau_l}\ln(1-v^2/c^2)=\frac{\mathrm{d}}{\mathrm{d}\tau_l}\ln(1-R_\mathrm{g}/r)\,.$$

Finally, we obtain the following relationship:

$$(1-R_\mathrm{g}/r)\Big/(1-v^2/c^2)=const.$$

Hence, the value

$$E=\frac{m_o\,c^2}{\sqrt{1-v^2/c^2}}\sqrt{1-\frac{R_\mathrm{g}}{r}}=E_\mathrm{local}\sqrt{1-\frac{R_\mathrm{g}}{r}}=const\,, \qquad (1.64)$$

does not change for a freely moving particle, while the locally measured energy E_local varies in the gravitational field of the black hole. This value E is termed 'energy-at-infinity' (Thorne et al. 1986). For a photon, the rest mass of which is $m_o = 0$, Eq. (1.64) yields the relation between its frequency in the reference system of the local observer ν_o and its frequency detected at infinity $\nu_\infty = \nu_o\sqrt{1-R_\mathrm{g}/r}$, describing the redshift effect.

In the non-relativistic approximation, the expression for the energy $\mathscr{E}_\mathrm{N}$ of the particle has the well-known form

$$E-m_o\,c^2=\mathscr{E}_\mathrm{N}=m_o\,v^2/2-m_o\,G\,M/r\,.$$

Let us now determine the components of the particle velocity in the equatorial plane. A freely moving particle with mass m_o keeps its angular momentum unchanged

$$h_\mathrm{p}=\frac{m_o\,v_\varphi\,r}{\sqrt{1-v^2/c^2}}\,. \qquad (1.65)$$

When taking into consideration that $v^2 = v_r^2 + v_\varphi^2$, Eqs. (1.64) and (1.65) yield

$$\frac{v_r^2}{c^2} = 1 - \frac{m_o^2\, c^4}{E^2} \left(\frac{h_\mathrm{p}^2}{r^2\, m_o^2\, c^2} + 1 \right) \left(1 - \frac{R_\mathrm{g}}{r} \right) . \tag{1.66}$$

Multiplying this by a factor $E^2/(m_o^2\, c^4)$ and using (1.61) and (1.64) together with the relation

$$\frac{v_r^2}{c^2} = \frac{1}{c^2} \left(\frac{\mathrm{d}r}{\mathrm{d}\tau_\mathrm{p}} \right)^2 \frac{m_o^2\, c^4}{E^2} ,$$

we may rewrite (1.66). As a result, we obtain the law of motion for a particle with energy E, which is identical to the exact solution in GR, see Shapiro and Teukolsky (1983):

$$\frac{1}{c^2} \left(\frac{\mathrm{d}r}{\mathrm{d}\tau_\mathrm{p}} \right)^2 = \frac{E^2}{m_o^2\, c^4} - \left(\frac{h_\mathrm{p}^2}{r^2\, m_o^2\, c^2} + 1 \right) \left(1 - \frac{R_\mathrm{g}}{r} \right) .$$

Note that in the approximation of the Newtonian potential, this law of motion looks as follows:

$$v_r^2 = \frac{2}{m_o} \left(\mathscr{E}_\mathrm{N} + m_o \frac{G\, M}{r} \right) - \frac{h_\mathrm{N}^2}{r^2\, m_o^2} ,$$

where $h_\mathrm{N} = m_o\, v_\varphi\, r = const.$

Let us consider particles moving in circular orbits around a Schwarzschild black hole. For such motion, both v_r and $\mathrm{d}r/\mathrm{d}\tau_\mathrm{p}$ become zero. For the sake of convenience, we may introduce an effective potential (Shapiro and Teukolsky 1983):

$$V(r) = \left(\frac{h_\mathrm{p}^2}{r^2\, m_o^2\, c^2} + 1 \right) \left(1 - \frac{R_\mathrm{g}}{r} \right) .$$

For circular orbits, the first derivative of this potential becomes zero (the potential has an extremum). The system of equations

$$\frac{\mathrm{d}r}{\mathrm{d}\tau_\mathrm{p}} = 0 , \qquad \frac{\partial V(r)}{\partial r} = 0$$

yields the following angular momentum in a circular orbit:

$$h_\mathrm{p}^2 = \frac{m_o^2\, r\, R_\mathrm{g}\, c^2}{2 - 3 R_\mathrm{g}/r} . \tag{1.67}$$

After squaring (1.65), we derive the tangential velocity, as measured by the local observer, from (1.67):

$$\frac{v_\varphi}{c} = \frac{1}{\sqrt{2}}\sqrt{\frac{R_g}{r - R_g}}\,. \tag{1.68}$$

For the local observer, the angular velocity of a particle is

$$\omega_l = \frac{v_\varphi}{r} = \frac{c}{\sqrt{2}\,r}\sqrt{\frac{R_g}{r - R_g}}\,. \tag{1.69}$$

Using (1.60), we obtain for an observer at infinity:

$$\omega = \frac{c\sqrt{R_g}}{\sqrt{2}\,r^{3/2}} = \frac{\sqrt{G\,M}}{r^{3/2}}\,, \tag{1.70}$$

that is, the classical expression following from Kepler's law.

According to the Rayleigh criterion (Rayleigh 1917), stable orbits cannot exist where $\mathrm{d}h_p/\mathrm{d}r < 0$. This criterion implies that the last stable circular orbit has a radius $r_{\rm ISCO} = 3\,R_g$.

When substituting the velocity $v_\varphi = c/2$, which corresponds to $r_{\rm ISCO}$, into (1.64), we determine the energy of a particle rotating in the innermost possible stable orbit. The energy of this particle, $E = m_o\,c^2\,2\sqrt{2}/3$, is less than its rest energy at infinity, $m_0\,c^2$. This means that when a particle moves from infinity towards the Schwarzschild black hole, that is, in the process of accretion, the released energy is $(m_0\,c^2 - E) \approx 0.0572\,m_0\,c^2$. Thus, the energy conversion efficiency in the accretion process onto a non-rotating black hole is equal to $\sim 6\%$. A calculation using the Kerr metric shows that the binding energy of the particle is largest for an extremely-fast rotating black hole and equals to $1 - \sqrt{1/3} \approx 0.423$ times the rest energy (Kato et al. 2008).

Extracting the square root of (1.67), we find the specific angular momentum of a particle in circular orbit in the Schwarzschild metric:

$$h = \frac{h_p}{m_o} = \frac{\sqrt{G\,M\,r}}{\sqrt{1 - \frac{3\,G\,M}{c^2\,r}}}\,. \tag{1.71}$$

Figure 1.3 (upper panel) shows the dependence of the specific angular momentum of a test particle on the radius of the orbit in the gravitational field of the black hole. In addition, the respective dependencies are shown in the Newtonian potential (*dashed line*) and in the Paczynski–Wiita potential (*dotted line*). In the gravitational field of the Schwarzschild black hole, the specific angular momentum h becomes minimum at the radius of the innermost stable circular orbit $6\,G\,M/c^2$. In contrast to the case of the Newtonian potential, the first derivative of the specific angular momentum,

Fig. 1.3 Specific angular momentum h of a test particle in the gravitational field of a black hole (uppermost panel) and the viscous torque $F(h)$ in a stationary disc, normalised values (lower panels). The inner radius of the disc is $r_{\rm in} = 3\,R_{\rm g} = 6\,G\,M/c^2$. *Solid lines* show the dependence in the exact logarithmic potential (1.59), *dotted lines* show the same in the Paczynski–Wiita potential, dashed lines—in the Newtonian approximation. In the middle panel, a rectangular area is drawn, shown enlarged in the lower panel

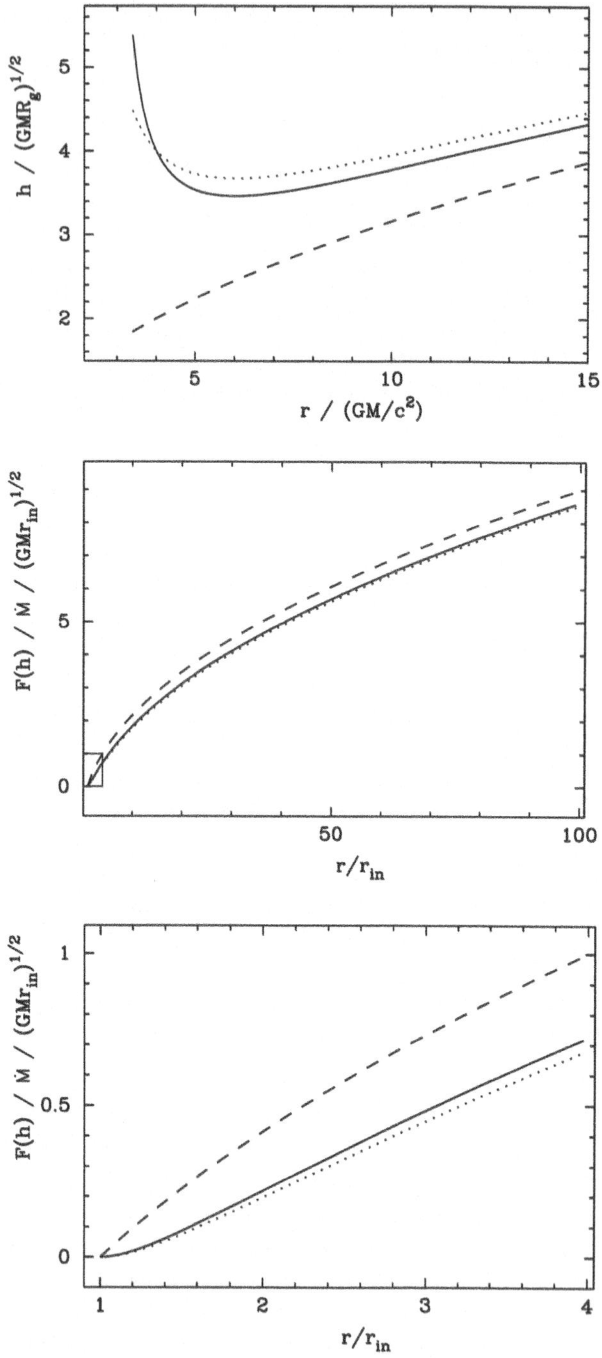

$\mathrm{d}h/\mathrm{d}r$, vanishes at this radius (see Fig. 1.3, upper panel). The innermost stable orbit is located at $3\,R_\mathrm{g}$ in both the approximate Paczynski–Wiita potential (1.56) and the exact potential (1.59). The binding energy in the Paczynski–Wiita potential, however, differs from the value in the Schwarzschild metric:

	$(m_0\,c^2 - E)/(m_0\,c^2)$
Newtonian potential:	$1/12$
Paczyński–Wiita potential:	$1/16$
Logarithmic potential and Schwarzschild metric:	$1 - 2\sqrt{2}/3$

Circular orbits exist only down to the radius where $v_\varphi = c$. In the logarithmic potential, the innermost circular orbit lies at $3\,R_\mathrm{g}/2$, which coincides with the exact value predicted by general relativity. In the Paczynski–Wiita potential, the innermost circular orbit is located at $2\,R_\mathrm{g}$.

Figure 1.3 also shows the viscous torques in the disc as functions of radius given by formula (1.54). Note that, for both the Paczynski–Wiita and the logarithmic potentials, the torque itself, as well as its first derivative, vanishes at the innermost stable orbit (see Fig. 1.3, lower panel).

1.4.5 Energy Release in Geometrically Thin Discs

Let us return to the study of discs in the Newtonian approximation. A detailed analysis of the energy balance equation is given, for example, in the appendix of the book by Kato et al. (1998). The energy dissipated in the disc per unit volume per unit time is equal to

$$\varepsilon = \rho\,\nu_\mathrm{t}\,r^2\left(\frac{\mathrm{d}\omega}{\mathrm{d}r}\right)^2 . \tag{1.72}$$

In the general case of optically thick discs, the energy release ε can be given in the form of a power-law function of temperature and density (Tayler 1980).

In the simplest approximation for a geometrically thin disc, all the energy released due to friction in a disc ring is radiated away from the top and bottom surfaces of this ring. The energy released per unit time per unit surface area of a geometrically thin disc in a calculation per one side of the disc is

$$Q_\mathrm{vis}(t,r) \equiv \int_0^{+z_0} \varepsilon(t,r,z)\,\mathrm{d}z = -\frac{1}{2}W_{r\varphi}\,r\,\frac{\mathrm{d}\omega}{\mathrm{d}r} . \tag{1.73}$$

Note that the last formula works also in the case of disc reservoirs (Syunyaev and Shakura 1977), in which the accretion rate is zero. In view of (1.49), we have for an accreting disc:

$$Q_{\rm vis} = -\frac{\dot{M}}{4\pi}\,\omega\, r\,\frac{{\rm d}\omega}{{\rm d}r}\,f(r)\,. \tag{1.74}$$

For a Keplerian disc, the above expressions can be re-written in the form (using (1.72)):

$$\varepsilon = \frac{3}{2}\,\omega_{\rm K}\, w^{\rm t}_{r\varphi} = \frac{9}{4}\,\rho\,\nu_{\rm t}\,\omega_{\rm K}^2\,, \qquad Q_{\rm vis} = \frac{3}{4}\,\omega_{\rm K}\, W_{r\varphi} = \frac{3}{8\pi}\,\dot{M}\,\frac{G\,M}{r^3}\,f(r)\,. \tag{1.75}$$

One can see that the viscous time scale (1.52) in a geometrically thin disc is much larger than the characteristic thermal time scale, on which the thermal energy in a unit volume changes:

$$\tau_{\rm th} \sim \frac{\rho\, v_{\rm s}^2}{\varepsilon} \sim \frac{1}{\alpha\,\omega_{\rm K}}\,, \tag{1.76}$$

where we have replaced $\nu_{\rm t}$ using (1.40).

For an accretion disc with a zero viscous torque at the inner boundary and with a Keplerian distribution of angular momentum, we have (see Eq. (1.55)):

$$Q_{\rm vis} = \frac{3}{8\pi}\,\dot{M}\,\frac{G\,M}{r^3}\left(1 - \sqrt{\frac{r_{\rm in}}{r}}\right)\,,$$

where $r_{\rm in}$ is the radius of the inner boundary of the disc.

The most general expression for the viscous heat in a Keplerian disc, including one with zero accretion rate, is:

$$Q_{\rm vis} = \frac{3}{8\pi}\,\frac{\omega_{\rm K}\,F}{r^2} \quad \text{or} \quad Q_{\rm vis} = \frac{3}{8\pi}\,F\,\frac{(G\,M)^4}{h_{\rm K}^7}\,, \tag{1.77}$$

where $h_{\rm K}$ is the specific angular momentum and F is the viscous torque (1.48).

The energy balance equation for geometrically thin discs reflects the fact that the thermal energy released due to viscosity at radius r is completely radiated away at the same radius:

$$Q_{\rm vis}(r) = Q_{\rm rad}(r)\,, \tag{1.78}$$

where $Q_{\rm rad}(r)$ is the radiated flux from one of the two surfaces of the accretion disc. The last equation requires a modification if the accretion rate is high, $\gtrsim \dot{M}_{\rm Edd}$. It turns out that the radial transport of heat should also be taken into account.

In the approximation of a disc radiating like a blackbody, it is possible to characterize its flux with an effective temperature:

$$Q_{\mathrm{rad}} = \sigma_{\mathrm{SB}}\, T_{\mathrm{eff}}^4\,, \qquad T_{\mathrm{eff}} \propto r^{-3/4}\,. \tag{1.79}$$

The effective temperature at the disc surface has its maximum $T_{\max}$ at radius

$$r_{\max} = \left(\frac{7}{6}\right)^2 r_{\mathrm{in}}\,,$$

and is equal to

$$T_{\max} = 2^{3/4}\left(\frac{3}{7}\right)^{7/4}\left(\frac{G\,M\,\dot{M}}{\pi\,\sigma_{\mathrm{SB}}\,r_{\mathrm{in}}^3}\right)^{1/4} = 2\left(\frac{3}{7}\right)^{7/4}\left(\frac{L_{\mathrm{d}}}{\pi\,\sigma_{\mathrm{SB}}\,r_{\mathrm{in}}^2}\right)^{1/4}.$$

We introduced in the last formula the total bolometric luminosity from both sides of the disc, equal to half the released gravitational energy of the matter falling from infinity to the gravitating centre:

$$L_{\mathrm{d}} = 4\,\pi\int_{r_{\mathrm{in}}}^{r_{\mathrm{out}}} Q_{\mathrm{rad}}\, r\,\mathrm{d}r = \frac{1}{2}\,\dot{M}\,\frac{G\,M}{r_{\mathrm{in}}}\,.$$

The specific potential energy of a particle moving from infinity to the inner edge of the disc decreases from zero to $-G\,M/r_{\mathrm{in}}$. Half of this energy heats the disc and is radiated and the other half goes into kinetic energy of rotation.

This 'virial theorem' does not apply to individual rings in the disc. We integrate the energy released from the disc at distances $r \gg r_{\mathrm{in}}$ from both sides of the disc:

$$2\int \frac{3}{8\pi}\,\dot{M}\,\frac{GM}{r^3}\,2\,\pi\,r\,\mathrm{d}r = \frac{3}{2}\,\dot{M}\,\frac{GM}{r}\,,$$

and find that it is three times as high as the amount of released gravitational energy. This happens since along with angular momentum, transferred outwards from the centre during the accretion process, a part of the energy is transferred as well.

Indeed, using the definitions of the integrated quantities (1.45), (1.46), and (1.73), let us multiply the energy balance equation (1.26) by $2\pi r$ and integrate it over disc thickness, keeping in mind that we agreed to use the positive value $w^{\mathrm{t}}_{r\varphi} = -w_{r\varphi}$ for accretion discs. We obtain

$$\dot{M}\,\frac{\partial}{\partial r}\left(\frac{v_\varphi^2}{2} + \Phi\right) = 2\,\pi\,r \times 2\,Q_{\mathrm{vis}} + \frac{\partial}{\partial r}(\omega_{\mathrm{K}}\,F)\,, \tag{1.80}$$

where $F = 2\,\pi\,r^2 W_{r\varphi}$ is the total viscous torque between neighboring rings in the disc, introduced in Sect. 1.4.2. And thus, the energy from gravitational interaction, released as matter moves towards the centre, is dissipated (radiated from both sides of the disc) and is redistributed over the disc as a result of the work of viscous forces transferring angular momentum.

Another important conclusion can be drawn from considering the last equation. The disc releases heat and radiates even if the accretion rate is zero. If the matter cannot pass through the inner boundary, the radial motion of matter towards the disc centre may be interrupted. This happens, for example, if the central object is a neutron star with a strong magnetic field. While $\dot{M} = 0$, the viscous forces do not stop working, the matter is heated up and the heat turns into radiation. The energy in such a disc, along with the angular momentum, comes from the neutron star through the inner boundary of the disc.

1.4.6 Disc Radiation

The radiative flux in a unit solid angle from a flat accretion disc at distance d from the disc is equal to

$$F_\nu = \frac{2\pi}{d^2} \cos i \int_{r_{\rm in}}^{r_{\rm out}} I_\nu \, r \, \mathrm{d}r \,, \tag{1.81}$$

where i is the inclination of the disc to the line of sight and $I_\nu(r)$ is the intensity of radiation from the disc surface.

In the disc photosphere, the following radiative processes are frequently considered (see, for example, Kato et al. 2008):

- Free-free and bound-free transitions,
- Scattering off free electrons,
- Compton scattering (scattering off cold electrons),
- Inverse Compton scattering (if the energy of the electrons and/or ions are very high),
- Line broadening caused by the rotation of the disc.

The Planck spectrum describes the spectral density of electromagnetic radiation emitted by an isothermal atmosphere if scattering is not taken into account. At every radius, the disc radiates like a blackbody of temperature $T_{\rm eff}$ with intensity:

$$B_\nu(T_{\rm eff}) = \frac{2\,h\,\nu^3}{c^2} \frac{1}{e^{h\nu/kT_{\rm eff}} - 1} \,. \tag{1.82}$$

The spectral flux integrated along the disc radius in shown in Fig. 1.4.

For a disc spectrum as shown in Fig. 1.4, the power-law distribution describes the middle interval. Let is determine the power-law index of this distribution. Almost the whole disc, with the exception of the central parts (which, however, give an overwhelming contribution to the total amount of the radiated energy),

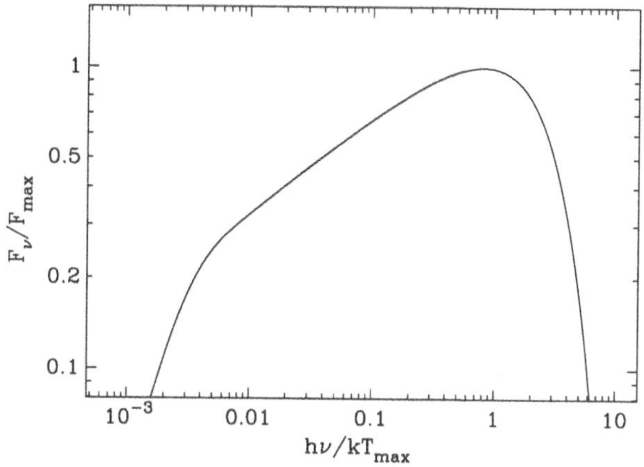

Fig. 1.4 Spectral distribution of radiative flux density from a standard optically thick, geometrically thin disc in the Newtonian metric. The horizontal axis shows the normalised radiation frequency. The vertical axis shows the spectral radiative flux density in units of [erg/Hz/cm^2/s] normalised to the maximum flux density at $h\,\nu/k\,T_{\rm max} \approx 0.8$. The maximum of distribution νF_ν is at $h\,\nu/k\,T_{\rm max} \approx 2.5$

may be characterized by the effective temperature in the form of a simple power-law function of radius (1.79). Substituting $r = r_0(T_0/T_{\rm eff})^{4/3}$ and (1.82) in the integral (1.81), we get

$$F_\nu = \frac{16\pi}{3\,d^2}\cos i\left(\frac{k\,T_0}{h}\right)^{8/3}\frac{h\,\nu^{1/3}}{c^2}\,r_0^2\int_{x_{\rm in}}^{x_{\rm out}}\frac{x^{5/3}}{e^x-1}\,{\rm d}x\,,$$

where we have made the substitution $x = h\nu/kT_{\rm eff} = (h\nu/kT_0)\,(r/r_0)^{3/4}$.

The radius r_0 can be chosen rather close to $r_{\rm in}$, implying that $T_0 \approx T_{\rm max}$ with fairly good accuracy. Then $x = (h\nu/kT_{\rm max})(r/r_{\rm in})^{3/4}$. At those frequencies where the conditions $x_{\rm in} \ll 1$ and $x_{\rm out} \gg 1$ are satisfied, the value of the integral in the last expression varies only little for different ν, and is approximately equal to the integral from zero to infinity when expressed with the help of the special gamma function and Riemann zeta function as $(10/9)\,\Gamma(2/3)\,\zeta(8/3) \approx 1.93$. Thus, for a wide frequency range $(r_{\rm in}/r_{\rm out})^{3/4} < h\nu/kT_{\rm max} < 1$, the spectral flux density of disc radiation depends on the frequency according to $F_\nu \propto \nu^{1/3}$.

For a homogeneous atmosphere where scattering is present, the spectrum will differ from that of a blackbody (Felten and Rees 1972):

$$I_\nu \simeq \sqrt{\frac{\varkappa_{\rm a}}{\varkappa_{\rm a}+\varkappa_{\rm sc}}}\,B_\nu(T_{\rm eff})\,,$$

where $\varkappa_{\rm a}$ is the absorption coefficient and $\varkappa_{\rm sc}$ is the coefficient for scattering off free electrons. If electron scattering dominates over absorption and if the disc spectrum is susceptible to Comptonization, the change in shape of the X-ray spectrum from a disc around a stellar mass compact object is approximately described by the spectral hardening factor f_c:

$$F_\nu = \frac{1}{f_c^4}\,\pi\,B_\nu(f_c\,T_{\rm eff})\,,$$

where F_ν is the flux from a unit surface into a half-space. The product $f_c\,T_{\rm eff}$ is called the colour temperature. The power of four is explained by the fact that the total radiated energy from the disc is independent of the spectral shape.

1.5 Stationary α-Discs

As we have seen in Sect. 1.4, the use of the continuity equation and the equation of motion, integrated (or averaged) along the vertical coordinate, enable us to find out the radial structure of thin stationary accretion discs. Is is possible to separately study the vertical and the radial structure of the disc because the characteristic time scales, namely, viscous and hydrostatic ones, are significantly different. The characteristic hydrostatic time scale corresponds to the time scale for changes in the thickness of the disc at a given radius as a result of a change of its central temperature. For dimensional reasons, this quantity is proportional to the disc half-thickness divided by the sound speed, $z_0/v_{\rm s} \sim 1/\omega_{\rm K} \sim \tau_{\rm dyn}$, and corresponds to the dynamical time which is much smaller than the viscous time in a thin disc (see Sect. 1.4.3).

The vertical structure of accretion discs in the general case (stationary as well as non-stationary) is described by a system of four ordinary differential equations, the exact solution to which, for given boundary conditions, can be found using numerical methods. In some sense, a calculation of the vertical structure of a disc is similar to the calculation of the internal structure of stars (Tayler 1980). The system of differential equations for the vertical structure of a disc was solved by a number of authors (see, for example, Meyer and Meyer-Hofmeister (1982), Shaviv and Wehrse (1986), Suleimanov (1992), Cannizzo (1992), Ketsaris and Shakura (1998), Hameury et al. (1998), Dubus et al. (1999)).

The disc can be divided into different zones (A, B, and C) according to the processes predominant in opacity formation and depending on comparative role of gaseous and radiative pressure (Shakura and Sunyaev 1973). A high temperature zone with main contribution from radiation pressure may arise in the central parts of the disc—the so-termed zone A. In this region, the opacity is determined by electron scattering. There are a number of studies devoted to the instabilities in this region (Lightman and Eardley 1974; Shibazaki and Hōshi 1975; Shakura and Sunyaev 1976). It was shown that zone A is thermally and viscously unstable. Its vertical structure can be described using the polytrope approximation. Convection plays an important role in the energy transfer to the disc surface (Bisnovatyi-Kogan and Blinnikov 1976; Shakura et al. 1978). In addition, the standard model should be modified since it is necessary to take into account non-Keplerian motion of gas in the disc due to a significant contribution of the pressure gradient in the equation of motion. It is also important to address the non-local character of the energy balance equation because the heat is effectively transported together with the radially moving matter (Paczynski and Bisnovatyi-Kogan 1981).

For quick estimates, one can use the following expressions. The boundary between zones A and B, where the gas pressure equals the radiative pressure, is located at

$$R_{\rm AB}/(3\,R_{\rm g}) \sim 80\,(m_x\,\alpha)^{2/21}\,(\dot{M}/\dot{M}_{\rm Edd})^{16/21}\,.$$

The boundary between zones B and C, where the cross-sections of absorption and scattering of photons are equal:

$$R_{\rm BC}/(3\,R_{\rm g}) \sim 330\,(\dot{M}/\dot{M}_{\rm Edd})^{2/3}\,.$$

The outer boundary of zone C, beyond which recombination of hydrogen starts:

$$R_{\rm C}/(3\,R_{\rm g}) \sim 10^5\,(\dot{M}/\dot{M}_{\rm Edd}/m_x)^{1/3}\,.$$

We have normalised here the accretion rate to its critical value $\dot{M}_{\rm Edd} = 1.4 \times 10^{18}\,m_x$ g/s (see Sect. 1.1), the radius, to the characteristic value of the inner radius of a disc around a compact object, $3\,R_{\rm g} \approx m_x \times 8.9 \times 10^5$ cm (see Sect. 1.4.4), and the mass of the central body, to the solar mass: $m_x = M/M_\odot$.

In this section, we consider only the stable zones of the disc where the standard model holds. In Sect. 1.5.1 we write down the standard disc equations (Shakura and Sunyaev 1973). In Sects. 1.5.2 and 1.5.3 we consider zones B and C, for which we present stationary solutions.

1.5.1 Equations of Vertical Structure

1.5.1.1 Equation of Hydrostatic Balance

The equation of hydrostatic equilibrium along the z-coordinate in the Newtonian metric in the case of a thin disc has the form:

$$\frac{1}{\rho}\frac{{\rm d}P}{{\rm d}z} = -\omega_{\rm K}^2\,z\,, \tag{1.83}$$

where $P(z)$ is the total pressure in the disc, equal to the sum of the radiation pressure $P_{\rm rad}(z) = aT^4/3$, where $a = 7.56\times 10^{-15}$ erg/cm^3/K^4 is the radiation constant, and the gas pressure $P_{\rm gas}(z)$, which is determined from the equation of an ideal gas:

$$P_{\rm gas} = \frac{\rho\,kT}{\mu\,m_{\rm p}}\,,$$

where μ is the mean molecular weight of matter in the disc, $T(z)$ is the temperature and $\rho(z)$ the density of the matter.

1.5.1.2 Energy Generation

The heat dissipated in the disc at a given radius between the plane of symmetry of the disc and a given level at height z is a function of the vertical coordinate z:

$$Q_{\rm vis}(z) = \int_0^z \varepsilon \, {\rm d}\tilde{z} \, .$$

The rate of energy generation ε [erg/cm^3/s] in a Keplerian disc is determined by the viscous stress tensor. From (1.75) we have:

$$\frac{{\rm d}Q_{\rm vis}}{{\rm d}z} = \frac{3}{2} \, \omega_{\rm K} \, w^{\rm t}_{r\varphi} \, . \tag{1.84}$$

The component of the turbulent viscosity tensor in the disc $w^{\rm t}_{r\varphi}(z)$ is locally expressed in terms of the total pressure in this location with the help of the α-parameter

$$w^{\rm t}_{r\varphi} = \alpha \, P \, .$$

These equations represent the simplest hypothesis regarding energy release in the disc. It is possible to model the disc vertical structure under more complicated assumptions. For example, Nakao and Kato (1995) study the case of a disc with turbulent diffusion determining the dependence of viscous heating, and the α-parameter itself, on z.

1.5.1.3 Radiative Transfer in the Disc

If the opacity in the disc does not exceed certain values, energy is transferred vertically towards the disc surfaces by electromagnetic radiation. Let us assume that the condition of local thermodynamic equilibrium (LTE) holds inside the disc, i.e. Kirchhoff's law applies, according to which (Sobolev 1969)

$$j_\nu = 4\pi \, \varkappa_{\rm a}(\nu) \, B_\nu(T) \, ,$$

where j_ν is the emission coefficient per gram [erg/Hz/s/g/sr], $\varkappa_{\rm a}(\nu)$ is the absorption coefficient per gram [cm^2/g], $B_\nu(T)$ is the Planck distribution [erg/Hz/cm^2/s/sr] and $T(z)$ is the temperature.

We write down the moments of the stationary equation for radiative transfer (Mihalas and Mihalas 1984), assuming that the medium is motionless in the direction of radiation propagation, along the z-axis. The zeroth moment of the transfer equation is given as a result of integrating the basic radiative transfer equation over all solid angles. After integrating over all frequencies we get

$$\frac{1}{\rho}\frac{\mathrm{d}Q_{\mathrm{rad}}(z)}{\mathrm{d}z} = 4\pi(\varkappa_{\mathrm{P}}\,B(T) - \varkappa_{\mathrm{a}}\,J(z))\,, \tag{1.85}$$

where $\varkappa_{\mathrm{a}}$ is the frequency-averaged absorption coefficient per gram, which is equal to the Planck mean opacity coefficient $\varkappa_{\mathrm{P}}$ (Mihalas and Mihalas 1984) at thermodynamic equilibrium, $Q_{\mathrm{rad}}(z)$ is the radiative energy flux along the z-axis, $B(T) = \sigma_{\mathrm{SB}}T^4/\pi$ the Planck function integrated over frequency and $J(z)$ the mean intensity of radiation entering the layer dz, integrated over frequency. The physical meaning of this equation is clear: the change in the flux of radiative energy is equal to the input of energy as a result of radiation of the matter (this term is written with the help of Kirchhoff's law) minus the energy absorbed by the matter.

The first moment of the equation of radiative transfer is obtained when we multiply it by the cosine of the angle to the unit area, divide by c, and integrate over all solid angles. This equation in principle expresses the conservation of the total momentum of radiation.

$$\frac{1}{\rho}\frac{\mathrm{d}P_{\mathrm{rad}}(\nu, z)}{\mathrm{d}z} = -(\varkappa_{\mathrm{a}}(\nu) + \varkappa_{\mathrm{s}}(\nu))\frac{Q_{\mathrm{rad}}(\nu, z)}{c}\,. \tag{1.86}$$

where $\varkappa_{\mathrm{s}}(\nu, z)$ is the scattering coefficient, which is generally frequency-dependent, $Q_{\mathrm{rad}}(\nu, z)$ is the radiative energy flux along the z-axis, and $P_{\mathrm{rad}}(\nu, z)$ is the radiation pressure at frequency ν. Thus, the radiation pressure force balances the change in momentum of the radiation caused by interaction with the matter.

If we consider the moments of the equation, we get rid of the angular coordinate. The mean intensity of the radiation J_ν is the zeroth moment of the intensity. The spectral flux of radiative energy Q_ν is the first moment, and the radiation pressure P_{rad} is the second moment. As is well known, every moment of the transfer equation contains a quantity a higher order. The solution to such systems of equations requires imposition of certain additional closing relations. The main closing method for an isotropic field is the Eddington approximation.

The mean intensity of radiation $J(z)$ is related by definition to the radiation energy density via the relation:

$$\varepsilon_{\mathrm{rad}} = \frac{4\pi J}{c}\,. \tag{1.87}$$

For an isotropic radiation field, there exists a simple relation between the radiation energy density and the radiation pressure:

$$P_{\mathrm{rad}} = \frac{\varepsilon_{\mathrm{rad}}}{3}\,. \tag{1.88}$$

This approximation works well in the case of a geometrically thin disc (optically thin as well as optically thick).

An optically thick disc (optical depth $\tau \gg 1$) may be studied in the 'diffusion approximation'. Let us consider the first moment of the radiative transfer equation (1.85). We assume that the change in $Q_{\rm rad}$ is insignificant there, and the left-hand side of (1.85) is zero. Thus, the radiation field spectrum is close to that of a blackbody: $J(z) = B(T)$. It follows from relation (1.87) that $\varepsilon_{\rm rad} = 4\pi B(T)/c \equiv aT^4$, and taking into account the isotropy of the radiation field, integrating the second moment of the radiative transfer equation (1.86) over frequency, we obtain:

$$\frac{c}{3\varkappa_{\rm R}\rho}\frac{{\rm d}(aT^4)}{{\rm d}z} = -Q_{\rm rad}\,, \tag{1.89}$$

where the Rosseland opacity $\varkappa_{\rm R}(z)$ is introduced

$$\frac{1}{\varkappa_{\rm R}} \equiv \frac{\int_0^\infty \frac{1}{\varkappa_{\rm a}(\nu)+\kappa_{\rm s}(\nu)}\frac{\partial B_\nu(T)}{\partial T}\,{\rm d}\nu}{\int_0^\infty \frac{\partial B_\nu(T)}{\partial T}\,{\rm d}\nu}\,. \tag{1.90}$$

If we consider quantities averaged over z, we obtain:

$$Q_{\rm rad} = \frac{1}{3}\frac{c}{\varkappa_{\rm R}\,\rho\,z_0}\,\varepsilon_{\rm rad}\,. \tag{1.91}$$

With allowance for convection, the vertical structure of discs was studied by Meyer and Meyer-Hofmeister (1982) for two variants of viscosity: proportional to the gas pressure and to the total pressure.

1.5.1.4 Dependence of the Surface Density on z

We introduce the quantity $\Sigma(z)$ for the surface density of the disc 'gathered' up to a certain height z, and with the help of this quantity we rewrite (1.41):

$$\frac{{\rm d}\Sigma}{{\rm d}z} = \rho\,. \tag{1.92}$$

1.5.2 Solution for the Vertical Structure

This section describes an approach to the solution of the disc vertical structure equations, proposed and implemented by Ketsaris and Shakura (1998). The method consists in finding similar solutions to the system of equations converted to a dimensionless form. The opacity coefficient and the rate of energy release are expressed as power-law functions of ρ and T. The obtained solution is compared to the numerical results of Suleimanov et al. (2007), and the agreement of the two methods is shown.

For sufficiently high temperatures ($> 10^6$ K), Thomson scattering off free electrons plays the most important part. The corresponding region of the disc, in which gas pressure dominates at the same time, is called zone B. Further out from the centre, where photo-ionization of ions from heavy elements and free-free transitions dominate, we have zone C. The contribution of radiation pressure to the total pressure in these two zones is neglected. In zone B, this assumption significantly limits the accuracy of the solution if $P_{\rm rad} \gtrsim (0.2\text{–}0.3)\, P_{\rm gas}$.

When calculating the disc vertical structure, we will assume that all heat from the work of viscous forces at given r and z is transformed to radiative energy. In particular, local energy balance (1.78) will apply. We replace everywhere $Q_{\rm rad}(z) = Q_{\rm vis}(z) = Q(z)$.

We list together the equations of the vertical structure of the disc (1.83), (1.84), (1.89), and (1.92):

$$
\begin{aligned}
\frac{1}{\rho}\frac{dP}{dz} &= -\,\omega_{\rm K}^2\, z\,, \\
\frac{d\Sigma}{dz} &= \rho\,, \\
\frac{dQ}{dz} &= \frac{3}{2}\,\omega_{\rm K}\, w_{r\varphi}^{\rm t}\,, \\
\frac{c}{3\varkappa_{\rm R}\rho}\,\frac{d(aT^4)}{dz} &= -Q\,.
\end{aligned}
\tag{1.93}
$$

The rate of energy release ε in α-discs is proportional to the pressure. The opacity coefficient is written as follows:

$$
\varkappa_{\rm R} = \varkappa_0\,\frac{\rho^{\varsigma}}{T^{\Upsilon}}\,. \tag{1.94}
$$

For hydrogen discs:

$$
\varsigma = \Upsilon = 0, \quad \varkappa_0 = 0.4\,{\rm cm^2/g}\,, \quad \text{if} \quad \varkappa_{\rm T} \gg \varkappa_{\rm ff}\,, \tag{1.95}
$$

$$
\varsigma = 1,\ \Upsilon = 7/2, \quad \varkappa_0 = 6.45\times10^{22}\,{\rm cm^5\,K^{7/2}/g^2}\,, \quad \text{if} \quad \varkappa_{\rm ff} \gg \varkappa_{\rm T}\,, \tag{1.96}
$$

and for discs with solar chemical abundances (Frank et al. 2002; Kurucz 1970, 1993):

$$
\begin{aligned}
\varsigma = \Upsilon = 0, &\quad \varkappa_0 = 0.335\,{\rm cm^2/g}\,, \quad \text{if} \quad \varkappa_{\rm T} \gg \varkappa_{\rm ff}\,, \\
\varsigma = 1,\ \Upsilon = 7/2, &\quad \varkappa_0 \approx 5\times10^{24}\,{\rm cm^5\,K^{7/2}/g^2}\,, \quad \text{if} \quad \varkappa_{\rm ff} \gg \varkappa_{\rm T}\,,
\end{aligned}
\tag{1.97}
$$

Calculations of absorption in the plasma, including collective and quantum effects, electron degeneracy, etc., performed by the OPAL project at Livermore laboratory

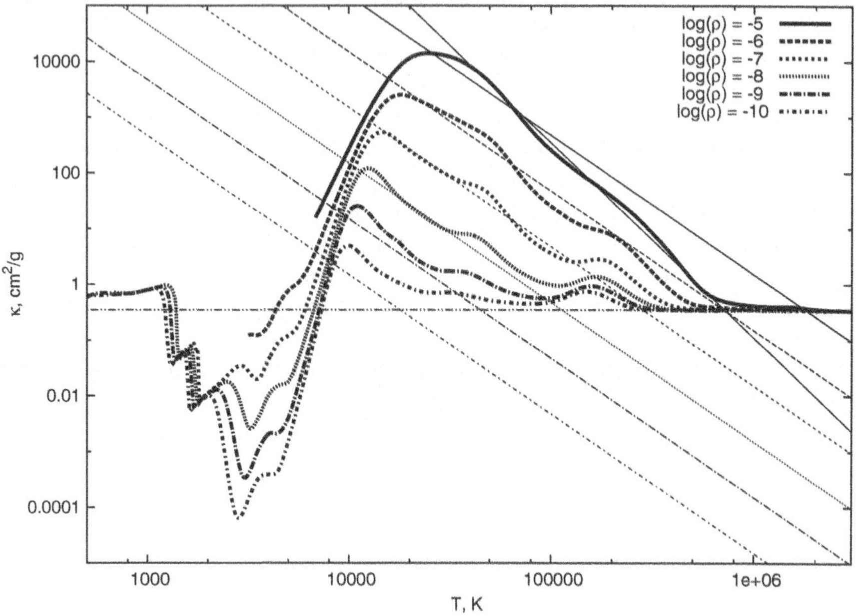

Fig. 1.5 Dependence of the opacity coefficient on density and temperature according to the OPAL project (Iglesias and Rogers 1996) and calculations for the low-temperature region in a medium with solar composition (Ferguson et al. 2005). The *horizontal line* corresponds to the value of the scattering coefficient off free electrons $\varkappa_0 \simeq 0.34\,\mathrm{cm}^2/\mathrm{g}$. Two fits are shown for a density of $\rho = 10^{-5}\,\mathrm{g/cm}^3$, namely, (1.98) and the dependence $\varkappa_\mathrm{R} = 1.2 \times 10^{25}\,\rho\,T^{-7/2}\,\mathrm{cm}^5\,\mathrm{K}^{7/2}/\mathrm{g}^2$ that gives a better fit in the high-temperature region (*solid lines*)

(Iglesias and Rogers 1996) (see Fig. 1.5) better fit another law in the absorption-dominated region:

$$\varsigma \approx 1,\ \Upsilon \approx 5/2, \quad \varkappa_0 \approx 1.5 \times 10^{20}\,\mathrm{cm}^5\,\mathrm{K}^{5/2}/\mathrm{g}^2\,, \quad \text{if} \quad \varkappa_\mathrm{ff} \gg \varkappa_\mathrm{T}\,. \tag{1.98}$$

For convenience, we introduce the dimensionless variable[3]

$$\sigma = \frac{2\,\Sigma(z)}{\Sigma_0}\,,$$

and in addition the dimensionless functions of this variable:

$$p = P(z)/P_\mathrm{c}\,, \quad \theta = T(z)/T_\mathrm{c}\,, \quad z' = z/z_0\,, \quad j = \rho(z)/\rho_\mathrm{c} \quad \text{and}\; q = Q(z)/Q_0\,.$$

[3]In the original paper by Ketsaris and Shakura (1998), the parameter Σ_0 was defined as half the total surface density of the disc. Due to this, there is a difference in the numerical coefficients in some of the formulas given below compared to the formulas in Ketsaris and Shakura (1998).

The symbols P_c, T_c, and ρ_c represent physical quantities in the equatorial plane of the disc and $Q_0 = (ac/4)\,T_{\mathrm{eff}}^4$ is the blackbody flux from one surface of the disc. We rewrite the system of Eq. (1.93) in the following form:

$$
\begin{aligned}
\frac{\mathrm{d}p}{\mathrm{d}\sigma} &= -\Pi_1\,\Pi_2\,z'\,; \quad \Pi_1 = \frac{\omega_K^2\,z_0^2\,\mu}{\Re\,T_c}\,;\\
\frac{\mathrm{d}z'}{\mathrm{d}\sigma} &= \Pi_2\,\frac{\theta}{p}\,; \qquad \Pi_2 = \frac{\Sigma_0}{2\,z_0\,\rho_c}\,;\\
\frac{\mathrm{d}q}{\mathrm{d}\sigma} &= \Pi_3\,\theta\,; \qquad \Pi_3 = \frac{3}{4}\,\frac{\alpha\,\omega_K\,\Re\,T_c\,\Sigma_0}{Q_0\,\mu} \equiv \frac{\alpha\,\Re\,T_c\,\Sigma_0}{W_{r\varphi}\,\mu}\,;\\
\frac{\mathrm{d}\theta}{\mathrm{d}\sigma} &= -\Pi_4\,\frac{q\,j^{\varsigma}}{\theta^{\Upsilon+3}}\,; \quad \Pi_4 = \frac{3}{32}\left(\frac{T_{\mathrm{ef}}}{T_c}\right)^4 \frac{\Sigma_0\,\varkappa_0\,\rho_c^{\varsigma}}{T_c^{\Upsilon}}\,.
\end{aligned}
\tag{1.99}
$$

The heating per gram $\varepsilon/\rho = \partial Q/\partial\Sigma$ determines the dependence of the temperature on z. In principle, the intensive mixing in the disc can lead to a situation where the energy output per unit mass is not dependent on the height z. The quantity ε depends in this case only on the density. The temperature dependence disappears from the equation describing the energy release (the third line in (1.99)), and Π_3 becomes equal to 1. A solution for such a case was also obtained by Ketsaris and Shakura (1998).

To find a solution to (1.99), i.e. to find the four functions $p(\sigma)$, $z'(\sigma)$, $q(\sigma)$, $\theta(\sigma)$ and the four unknown parameters, it is necessary to set eight boundary conditions—four at the surface of the disc and four in its symmetry plane. Ketsaris and Shakura (1998) performed a numerical integration of the equations and tabulated values $\Pi_{1..4}$. These values are given in Tables 1.1 and 1.2. Figure 1.7 shows functions $z'(\sigma)$, $p(\sigma)$, $\theta(\sigma)$, and $q(\sigma)$ in the Kramer opacity regime. Plots for other cases can be found in the work by Ketsaris and Shakura (1998).

In the symmetry plane of the disc for $\sigma = 0$, we have the obvious conditions:

$$p(0) = 1; \quad z'(0) = 0; \quad q(0) = 0; \quad \theta(0) = 1\,.$$

The first two boundary conditions at the disc surface can also be straightforwardly determined as:

$$z'(1) = 1; \qquad q(1) = 1\,.$$

The surface of the disc is defined as the level at which thermalization of radiation occurs. We may find boundary conditions for the pressure and temperature from approximate solutions to the equations of radiative transfer and hydrostatic balance close to the disc surface. Note that there is a difference in boundary conditions for different opacity regimes (see Fig. 1.6). In zone B, where absorption dominates, the disc surface is defined as the level in the photosphere where the optical depth,

Table 1.1 Dimensionless parameters of the solution to the equations of vertical structure for Thomson opacity versus the decimal logarithm of the free parameter δ

$\log\delta$	Π_1	Π_2	Π_3	Π_4
6.00	6.99	0.492	1.150	0.460
5.80	6.96	0.493	1.150	0.460
5.60	6.92	0.495	1.150	0.460
5.40	6.87	0.496	1.150	0.460
5.20	6.82	0.498	1.150	0.460
5.00	6.77	0.500	1.150	0.460
4.80	6.70	0.503	1.150	0.460
4.60	6.63	0.505	1.150	0.460
4.40	6.55	0.508	1.150	0.460
4.20	6.47	0.512	1.150	0.460
4.00	6.37	0.516	1.150	0.460
3.80	6.26	0.520	1.149	0.460
3.60	6.13	0.525	1.149	0.460
3.40	5.99	0.531	1.149	0.460
3.20	5.84	0.538	1.149	0.460
3.00	5.67	0.546	1.149	0.459
2.80	5.48	0.555	1.148	0.459
2.60	5.26	0.566	1.147	0.458
2.40	5.02	0.578	1.146	0.458
2.20	4.76	0.593	1.145	0.456
2.00	4.47	0.610	1.142	0.454
1.80	4.15	0.629	1.138	0.450
1.60	3.81	0.652	1.133	0.444
1.40	3.43	0.678	1.126	0.435
1.20	3.03	0.707	1.117	0.420
1.00	2.61	0.740	1.105	0.398
0.80	2.19	0.776	1.091	0.366
0.60	1.77	0.813	1.075	0.324
0.40	1.38	0.849	1.059	0.274
0.20	1.03	0.884	1.044	0.219
0.00	0.74	0.914	1.032	0.166

calculated from the outside inwards, is equal to 2/3. In the zone with predominant Thomson scattering, the disc surface is taken as the level where the effective optical depth, calculated including scattering, is equal to 1.

Let us derive the remaining boundary conditions in two opacity regimes.

1.5.2.1 Kramers Opacity

We will measure the optical depth τ from the surface of the disc in the direction of its symmetry plane, i.e. in the direction of decreasing height z. Deep inside the photosphere, where $\tau \sim 1$, we will use the solution to the equations of radiative

Table 1.2 Dimensionless parameters of the solution to the equations of vertical structure for Kramers opacity versus the decimal logarithm of the free parameter τ_0

$\log \tau_0$	Π_1	Π_2	Π_3	Π_4	$\log \tau$
6.00	7.75	0.465	1.131	0.399	6.046
5.80	7.71	0.466	1.131	0.399	5.847
5.60	7.67	0.468	1.131	0.399	5.646
5.40	7.62	0.469	1.131	0.399	5.445
5.20	7.56	0.471	1.131	0.399	5.245
5.00	7.50	0.473	1.131	0.399	5.045
4.80	7.44	0.475	1.131	0.399	4.845
4.60	7.36	0.477	1.131	0.399	4.644
4.40	7.27	0.480	1.131	0.399	4.444
4.20	7.18	0.483	1.131	0.399	4.244
4.00	7.07	0.487	1.131	0.399	4.043
3.80	6.95	0.491	1.131	0.399	3.843
3.60	6.82	0.496	1.131	0.399	3.643
3.40	6.67	0.501	1.131	0.399	3.443
3.20	6.50	0.508	1.131	0.398	3.243
3.00	6.31	0.515	1.131	0.398	3.043
2.80	6.10	0.524	1.130	0.398	2.842
2.60	5.87	0.534	1.130	0.398	2.642
2.40	5.60	0.546	1.129	0.397	2.442
2.20	5.31	0.560	1.128	0.397	2.241
2.00	4.98	0.576	1.126	0.395	2.040
1.80	4.62	0.596	1.124	0.393	1.839
1.60	4.23	0.619	1.120	0.389	1.638
1.40	3.79	0.647	1.114	0.383	1.434
1.20	3.33	0.679	1.106	0.371	1.232
1.00	2.83	0.716	1.095	0.354	1.025
0.80	2.34	0.756	1.081	0.326	0.819
0.60	1.86	0.798	1.065	0.286	0.613
0.40	1.42	0.838	1.050	0.237	0.406
0.20	1.05	0.876	1.036	0.185	0.202
0.00	0.75	0.908	1.025	0.136	−0.001

The rightmost column shows the decimal logarithm of the disc optical depth (1.114)

transfer and radiation balance for the case of LTE and for a frequency-independent absorption coefficient in the Eddington approximation (Sobolev 1969):

$$\frac{T}{T_{\rm eff}} = \left(\frac{1 + \frac{3}{2}\tau}{2} \right)^{1/4} . \tag{1.100}$$

Let the dimensionless variable $\sigma = 1$ at the level where $\tau = 2/3$ and $T = T_{\rm eff}$. Using the definition of the parameter Π_4, we obtain the boundary condition for the

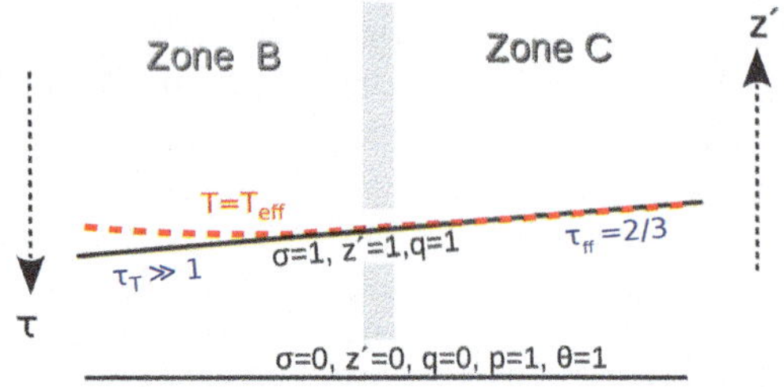

Fig. 1.6 The surfaces at which the boundary conditions are set, the upper surface of the disc and its equatorial plane (*solid lines*). Values of the dimensionless coordinate σ and functions at these surfaces are shown. The *arrows* indicate directions of increasing height $z' = z/z_0$ and optical depth τ, calculated from the exterior towards the equatorial plane. The two disc zones with different opacity regimes are separated nominally by the grey bar. In zone B (on the left), the optical depth at the disc surface $\tau_T(\tau^* = 1) \gg 1$. In zone C (on the right), $\tau_{ff} = 2/3$. The *dashed line* is the level where the disc temperature equals the effective temperature of the outgoing radiation

dimensionless temperature θ:

$$\theta\,(\sigma = 1) = \left[\frac{16}{3}\,\frac{\Pi_4}{\tau_0}\right]^{1/4},$$

where we have introduced the dimensionless parameter τ_0, proportional to the total optical depth of the accretion disc (see (1.96)):

$$\tau_0 = \frac{\Sigma_0\,\varkappa_0\,\rho_c}{2\,T_c^{7/2}}\,.$$

This quantity is a free parameter of the problem and varies widely (from a few to $\sim 10^6$).

To determine the boundary condition for the dimensionless pressure, we use the equation of hydrostatic balance (the first in system (1.93)). We divide both parts of this equation by the opacity coefficient $\varkappa_R$ and replace variables using the formula

$$d\tau = -\varkappa_R\,\rho dz$$

and making use of (1.96), arrive at:

$$\frac{1}{2}\,\frac{dP^2}{d\tau} = \frac{\omega_K^2\,z_0\,\mathfrak{R}\,T^{9/2}}{\varkappa_0\,\mu}\,.$$

Close to the photosphere, the z coordinate practically does not change and is equal to z_0. Integrating the last equation from $\tau = 0$ to $\tau = 2/3$, we get as a result the

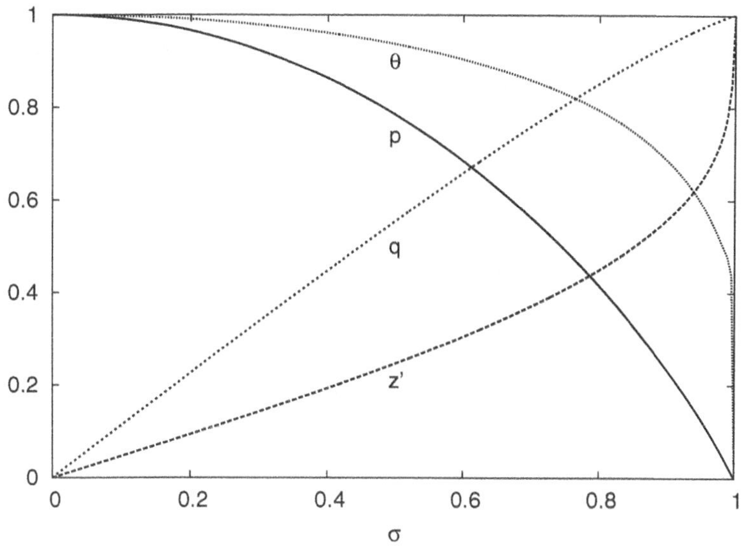

Fig. 1.7 Solution to the system of Eq. (1.99) in the form of dimensionless functions of the dimensionless variable σ, proportional to the column density: temperature $\theta(\sigma)$, pressure $p(\sigma)$, radiative flux $q(\sigma)$, and height from the equatorial plane $z'(\sigma)$. In the equatorial plane $\sigma = 0$, at the disc surface $\sigma = 1$

boundary condition for the dimensionless pressure:

$$p\,(\sigma = 1) = \left[\frac{3}{16 \times 2^{1/8}} \frac{\Pi_1\,\Pi_2}{\Pi_4} \left(\frac{16}{3} \frac{\Pi_4}{\tau_0}\right)^{17/8} f(\tau = 2/3)\right]^{1/2},$$

where

$$f(\tau) = \int_0^{\tau} (1 + \frac{3}{2}\tilde{\tau})^{9/8}\,\mathrm{d}\tilde{\tau}\,, \qquad f(\tau = 2/3) \approx 1.05\,.$$

Figure 1.7 shows the solution to the system of equations for the given case.

1.5.2.2 Thomson Scattering

If scattering processes are of high importance in the photosphere, thermalization occurs at the depth where the so-termed effective optical depth is of the order of 1:

$$\tau^* = -\int_{z_0}^{\infty} (\varkappa_{\rm ff}\,\varkappa_{\rm T})^{1/2}\,\rho\,\mathrm{d}z \approx 1\,.$$

The effective optical depth is accumulated as $\sqrt{\varkappa_{\rm ff}(\varkappa_{\rm T}+\varkappa_{\rm ff})}\,\rho\,{\rm d}z$ (see for example Zel'dovich and Shakura 1969, Mihalas 1978), which approximately gives the above condition. At this level, the optical depth due to scattering is much larger than 1:

$$\tau_{\rm T}(\tau^*=1) = -\int\limits_{z_0}^{\infty} \varkappa_{\rm T}\,\rho\,{\rm d}z \gg 1$$

and $T \simeq T_{\rm eff}\,(3\,\tau_{\rm T}/4)^{1/4}$ from (1.100). Thus, the boundary condition for the dimensionless temperature has the following form:

$$\theta\,(\sigma=1) \simeq \left[\frac{8\,\Pi_4\,\tau_{\rm T}(\tau^*=1)}{\varkappa_{\rm T}\,\Sigma_0}\right]^{1/4}.$$

For the pressure, we have:

$$p\,(\sigma=1) = 2\,\Pi_1\,\Pi_2\,\frac{\tau_{\rm T}(\tau^*=1)}{\varkappa_{\rm T}\,\Sigma_0}.$$

A convenient free parameter turns out to be the quantity

$$\delta = \frac{\varkappa_{\rm T}\,\Sigma_0/2}{\tau_{\rm T}(\tau^*=1)}. \qquad (1.101)$$

This parameter is the ratio of half the total optical depth due to scattering to the optical depth due to scattering at the thermalization depth.

1.5.3 Radial Dependence of Physical Parameters in Stationary α-Discs

In order to explain observations of sources with accretion discs as extended objects, whose properties vary significantly from the centre to the periphery, we have to calculate radial dependencies of the disc physical parameters. For this it is necessary to solve the equation of angular momentum transfer, which was done for the case of a stationary disc in Sect. 1.4, and also to solve the equations of vertical structure (see the previous section). Analytical approximations for radial dependencies of the disc parameters were given in the work by Suleimanov et al. (2007). We will describe these analytical approximations below.

We consider the following physical parameters: surface density $\Sigma(r)$, disc half-thickness $z_o(r)$, density $\rho_{\rm c}(r)$ and temperature $T_{\rm c}(r)$ at the symmetry plane of the disc for $z=0$. It is necessary to define what we consider to be the surface of the disc. When studying observed spectra it turns out to be convenient to assume that the disc surface corresponds to the level where the Rosseland optical depth $\tau_R = 2/3$.

The vertical structure of the disc is determined by Eq. (1.99) for known values of the dimensionless parameters $\Pi_{1..4}$. We express the quantities z_0, Σ, ρ_c, and T_c from (1.99). The resulting expressions contain the basic given parameters of the disc (accretion rate, mass of the central object, the turbulent α-parameter) as well as the radial structure defined by $\omega_K(r)$ and $W_{r\varphi}(r)$. We take the radial dependence of the vertically integrated component of the viscous stress tensor $W_{r\varphi}(r)$ for the case of a stationary disc (1.74), and the angular velocity of rotation we set equal to the Keplerian angular velocity $\omega_K = \sqrt{GM/R^3}$. The radial distribution of the radiative flux from the disc surface is determined by viscous stresses $W_{r\varphi}(r)$. We recall that the function $f(R)$, which describes the influence of the boundary conditions on the surface tension $W_{r\varphi}(r)$, is written as (cf. (1.49)):

$$f(r) = \frac{2\pi\, W_{r\varphi}(r)}{\dot{M}\,\omega} = \frac{F}{\dot{M}\,h}$$

in a disc with constant accretion rate. For a thin disc with a stress-free inner radius, we have

$$f(r) = \frac{8\pi}{3}\,\frac{Q_{\rm vis}}{\dot{M}\,\omega^2} = 1 - \frac{h_{\rm in}}{h}\,.$$

For the case $\dot{M} = \dot{M}(r,t) \neq const$, it is necessary to use the function $f(r)$ in its general form

$$f(r) = \frac{F(h,t)}{\dot{M}_{\rm in}(t)\,h} = \frac{F(h,t)/h}{\partial F(h,t)/\partial h\big|_{h=h_{\rm in}}}\,. \tag{1.102}$$

We normalise the accretion rate at the inner boundary of the disc and other parameters to their characteristic values in binary systems with stellar mass components:

$$\begin{array}{ll} M = m_x\, M_\odot, & \dot{M} = \dot{M}_{17} \times 10^{17}\text{g/s}, \\ r = R_7 \times 10^7\text{cm (zone B)} & \text{or } r = R_{10} \times 10^{10}\text{cm (zone C)}. \end{array} \tag{1.103}$$

As a characteristic value for the coefficient $\varkappa_0$ from expression (1.94) we use the quantity $\varkappa_T^* = 0.335\,\text{cm}^2/\text{g}$ in zone B, taken from an approximation to the tabulated values (Kurucz 1970, 1993), for a medium with mass fraction of hydrogen $X = 0.69$ and helium $Y = 0.27$ and $\varkappa_0^* = 5 \times 10^{24}\,\text{cm}^5\,\text{K}^{7/2}/\text{g}^2$ in zone C (see Frank et al. 2002, their chapter 5). The corresponding molecular weight μ=0.62. In a medium with such chemical composition, absorption of the radiation is mainly due to photoionization of ions of heavy elements. If we assume that all parameters $\Pi_{1..4}$ are equal to 1, $\varkappa_T = 0.4\,\text{cm}^2/\text{g}$, $\varkappa_0 = 6.4 \times 10^{22}\,\text{cm}^5\,\text{K}^{7/2}/\text{g}^2$, and $\mu = 0.5$, then the expressions for the radial dependencies of the physical parameters become identical to the expressions by Kato et al. (1998, their chapter 3) derived for hydrogen discs.

1.5.3.1 Zone B

In this zone, the main contribution to the optical depth comes from scattering off free electrons, and gas pressure dominates over radiation pressure. If we use expression (1.74) for the heat dissipated in the disc due to viscosity, normalising the parameters according to (1.103), we can solve the system of algebraic equations for $\Pi_{1..4}$ (the right part of the system (1.99)) and obtain:

$$\begin{aligned}
z_0/r &= 0.0092\, m_x^{-7/20}\, \dot{M}_{17}^{1/5}\, \alpha^{-1/10}\, R_7^{1/20}\, f(r)^{1/5} \left(\frac{\mu}{0.6}\right)^{-2/5} \left(\frac{\varkappa_\mathrm{T}}{\varkappa_\mathrm{T}^*}\right)^{1/10} \Pi_z\,, \\
\Sigma_0 &= 5.1 \times 10^3\, m_x^{1/5}\, \dot{M}_{17}^{3/5}\, \alpha^{-4/5} R_7^{-3/5}\, f(r)^{3/5} \left(\frac{\mu}{0.6}\right)^{4/5} \left(\frac{\varkappa_\mathrm{T}}{\varkappa_\mathrm{T}^*}\right)^{-1/5} \Pi_\Sigma\ [\mathrm{g/cm^2}], \\
\rho_\mathrm{c} &= 2.8 \times 10^{-2}\, m_x^{11/20}\, \dot{M}_{17}^{2/5}\, \alpha^{-7/10}\, R_7^{-33/20}\, f(r)^{2/5} \left(\frac{\mu}{0.6}\right)^{6/5} \times \\
&\qquad \times \left(\frac{\varkappa_\mathrm{T}}{\varkappa_\mathrm{T}^*}\right)^{-3/10} \Pi_\rho\ [\mathrm{g/cm^3}], \\
T_\mathrm{c} &= 8.2 \times 10^6\, m_x^{3/10}\, \dot{M}_{17}^{2/5}\, \alpha^{-1/5}\, R_7^{-9/10}\, f(r)^{2/5} \left(\frac{\mu}{0.6}\right)^{1/5} \left(\frac{\varkappa_\mathrm{T}}{\varkappa_\mathrm{T}^*}\right)^{1/5} \Pi_\mathrm{T}\ [\mathrm{K}].
\end{aligned} \tag{1.104}$$

The combinations of the dimensionless parameters Π_z, Π_Σ, Π_ρ, and Π_T are related in the following way to the parameters $\Pi_{1..4}$:

$$\begin{aligned}
\Pi_z &= \Pi_1^{1/2}\, \Pi_3^{1/10}\, \Pi_4^{-1/10} \approx 2.6\,, \\
\Pi_\Sigma &= \Pi_3^{4/5}\, \Pi_4^{1/5} \approx 0.96\,, \\
\Pi_\rho &= \Pi_1^{-1/2}\, \Pi_2^{-1}\, \Pi_3^{7/10}\, \Pi_4^{3/10} \approx 0.67\,, \\
\Pi_\mathrm{T} &= \Pi_3^{1/5}\, \Pi_4^{-1/5} \approx 1.2\,.
\end{aligned} \tag{1.105}$$

Their values versus the free parameter δ are shown in Fig. 1.8, left panel. The free parameter δ is derived from the expression (1.101) and may be estimated from the total optical depth of the disc τ and other disc parameters in the following way:

$$\delta = \sqrt{\frac{\kappa_0 \rho_\mathrm{c} T_\mathrm{c}^{-7/2}}{\varkappa_\mathrm{T}}}\, \tau\, X(\delta)\,, \tag{1.106}$$

where $\tau = \varkappa_\mathrm{T}\, \Sigma_0/2$. The numerical factor

$$X(\delta) = \delta \int_{1-1/\delta}^{1} (P/P_\mathrm{c})^{1/2}\, (T/T_\mathrm{c})^{-9/4} \mathrm{d}\sigma \sim 2\,, \tag{1.107}$$

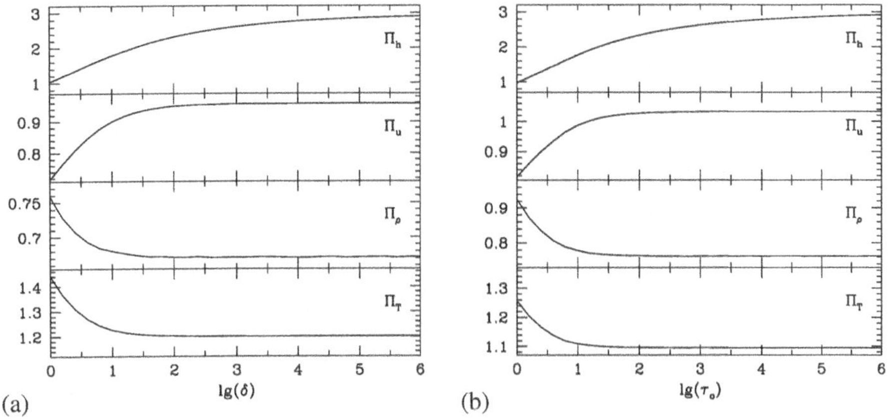

Fig. 1.8 (**a**) Left: The dependence of the dimensionless factors Π_z, Π_Σ, Π_ρ and Π_T for zone B (formulas (1.105)). (**b**) Right: The same factors for zone C (formulas (1.111)). The logarithms of the dimensionless parameters, characterizing the optical depth in each zone, are displayed along the horizontal axes. The dependencies are borrowed from Suleimanov et al. (2007) and constructed for values from Table 1.1 (graph to the left) and Table 1.2 (graph to the right)

which is independent of the absolute values of the disc parameters, is determined through integration of the equations of vertical structure. The value of δ may be found recursively with any desired precision, but this approach will be redundant in the sense of astronomical application of the obtained radial dependencies. It is sufficient to use the following estimate:

$$\delta = 440\, m_x^{-1/20}\, \dot{M}_{17}^{1/10}\, \alpha^{-4/5}\, R_7^{3/20}\, f(R)^{1/10} \left(\frac{\mu}{0.6}\right)^{21/20} \left(\frac{\varkappa_T}{\varkappa_T^*}\right)^{-1/5} \left(\frac{\varkappa_0}{\varkappa_0^*}\right)^{1/2}. \tag{1.108}$$

At high accretion rates, there is a zone in the disc where radiation pressure dominates (zone A). The radius at which the radiation pressure $a\,T_c^4/3$ is comparable to the gas pressure $\rho_c \Re T_c/\mu$ in the symmetry plane of the disc (the boundary between zones A and B, see Shakura and Sunyaev (1973)) may be approximately estimated as

$$R_{AB} \sim 10^7\, m_x^{1/3}\, \dot{M}_{17}^{16/21}\, \alpha^{2/21} \left(\frac{\mu}{0.6}\right)^{8/21} \left(\frac{\varkappa_T}{\varkappa_T^*}\right)^{6/7} \text{ cm.} \tag{1.109}$$

Here, we used characteristic values (1.105) for the dimensionless parameters $\Pi_{1..4}$ and $f(r) = 1$.

When the accretion rate decreases, zone B shifts radially towards the centre of the disc, giving way to zone C.

1.5.3.2 Zone C

The main contribution to the opacity in zone C comes from absorption processes in the form of free-free and bound-free transitions, and the gas pressure is much higher than the radiation pressure. As before, from the right-hand part of the system of Eq. (1.99) and from the expressions (1.74) and (1.103), we may find the radial dependencies of the parameters of the disc:

$$
\begin{aligned}
z_0/r &= 0.020\, m_x^{-3/8}\, \dot{M}_{17}^{3/20}\, \alpha^{-1/10}\, R_{10}^{1/8}\, f(r)^{3/20} \left(\frac{\mu}{0.6}\right)^{-3/8} \left(\frac{\varkappa_0}{\varkappa_0^*}\right)^{1/20} \Pi_z\,,\\
\Sigma_0 &= 33\, m_x^{1/4}\, \dot{M}_{17}^{7/10}\, \alpha^{-4/5}\, R_{10}^{-3/4}\, f(r)^{7/10} \left(\frac{\mu}{0.6}\right)^{3/4} \left(\frac{\varkappa_0}{\varkappa_0^*}\right)^{-1/10} \Pi_\Sigma \text{ [g/cm}^2\text{]},\\
\rho_c &= 8.0\times 10^{-8}\, m_x^{5/8}\, \dot{M}_{17}^{11/20}\, \alpha^{-7/10}\, R_{10}^{-15/8}\, f(r)^{11/20} \left(\frac{\mu}{0.6}\right)^{9/8} \times\\
&\quad \times \left(\frac{\varkappa_0}{\varkappa_0^*}\right)^{-3/20} \Pi_\rho \text{ [g/cm}^3\text{]},\\
T_c &= 4.0\times 10^{4}\, m_x^{1/4}\, \dot{M}_{17}^{3/10}\, \alpha^{-1/5}\, R_{10}^{-3/4}\, f(r)^{3/10} \left(\frac{\mu}{0.6}\right)^{1/4} \left(\frac{\varkappa_0}{\varkappa_0^*}\right)^{1/10} \Pi_T \text{ [K]},
\end{aligned}
\tag{1.110}
$$

We recall that $\dot{M}_{17}$ is the normalised accretion rate at the inner disc boundary. Note that if $\dot{M}(r,t) \neq const$, we need to substitute the value of the accretion rate at the inner boundary when using (1.104) and (1.110). This is convenient since in most cases this value determines the energetics of observed accreting systems.

The combinations of dimensionless parameters are related to the parameters $\Pi_{1..4}$ in the following way:

$$
\begin{aligned}
\Pi_z &= \Pi_1^{19/40}\,\Pi_2^{-1/20}\,\Pi_3^{1/10}\,\Pi_4^{-1/20} \approx 2.6\,,\\
\Pi_\Sigma &= \Pi_1^{1/20}\,\Pi_2^{1/10}\,\Pi_3^{4/5}\,\Pi_4^{1/10} \approx 1.03\,,\\
\Pi_\rho &= \Pi_1^{-17/40}\,\Pi_2^{-17/20}\,\Pi_3^{7/10}\,\Pi_4^{3/20} \approx 0.76\,,\\
\Pi_T &= \Pi_1^{-1/20}\,\Pi_2^{-1/10}\,\Pi_3^{1/5}\,\Pi_4^{-1/10} \approx 1.09\,,
\end{aligned}
\tag{1.111}
$$

and are shown in Fig. 1.8b as a function of the free parameter τ_0,

$$
\tau_0 = \frac{\varkappa_0 \rho_c}{T_c^{7/2}}\frac{\Sigma_0}{2} = 500\,\frac{\dot{M}_{17}^{1/5}\, f(r)^{1/5}}{\alpha^{4/5}} \left(\frac{\mu}{0.6}\right) \left(\frac{\varkappa_0}{\varkappa_0^*}\right)^{2/5} \frac{\Pi_3^{4/5}\,\Pi_4^{3/5}}{\Pi_1^{1/5}\,\Pi_2^{2/5}}\,, \tag{1.112}
$$

approximately equal to

$$
\tau_0 \sim 300\, \dot{M}_{17}^{1/5}\, \alpha^{-4/5} \left(\frac{\varkappa_0}{\varkappa_0^*}\right)^{2/5}. \tag{1.113}
$$

The full optical depth of the disc

$$\tau = \int_0^h \varkappa_0 \, \rho^2 \, T^{-7/2} \mathrm{d}z \tag{1.114}$$

is determined in the process of numerical solution of the vertical structure and is uniquely dependent on τ_0 (see Table 1.2). We also give the following formula, approximating the tabulated values to an error of less than 1% for $\tau_0 > 6$:

$$\tau \approx 1.042 \, \tau_0^{1.006} \, . \tag{1.115}$$

The dependencies of the parameters in zones B and C are depicted in Figs. 1.9 and 1.10. The boundary between zones B and C is approximately determined from

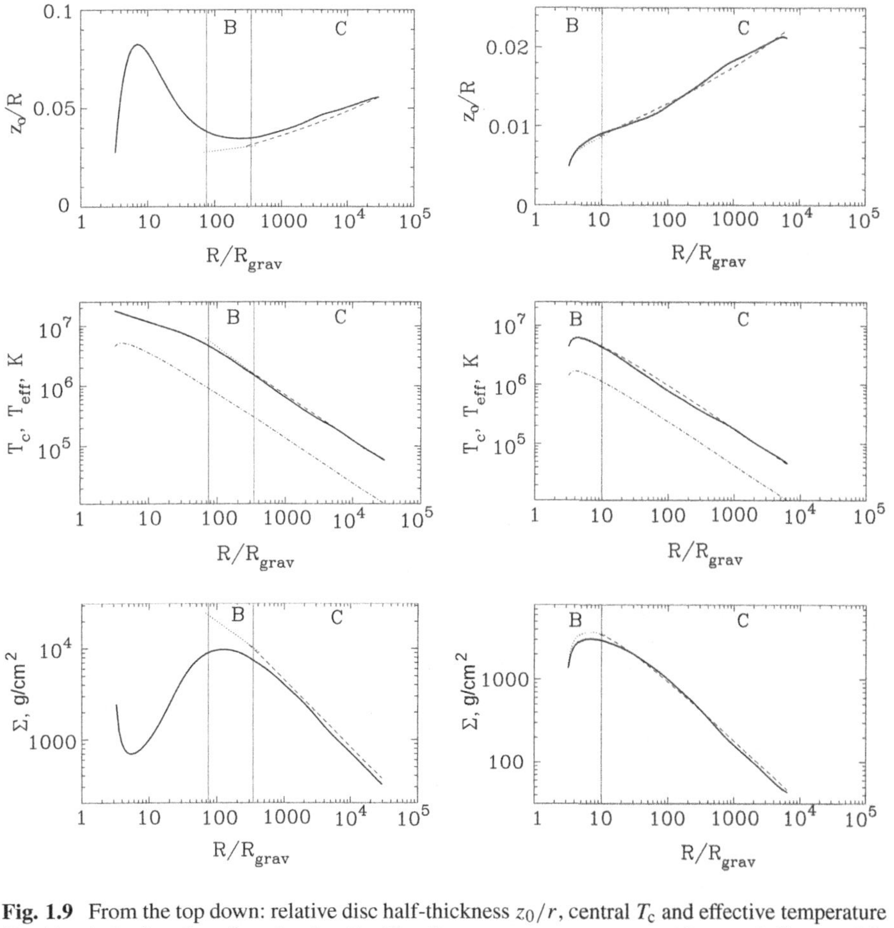

Fig. 1.9 From the top down: relative disc half-thickness z_0/r, central T_c and effective temperature $T_{\rm eff}$ (dot-dashes) and surface density Σ_0. The disc parameters are $m_x = 10$, $\mu = 0.62$, $\alpha = 0.3$, left: $\dot{M}_{17} = 33.6$ or $L_{\rm bol} = 0.2\,L_{\rm Edd}$, right: $\dot{M}_{17} = 0.336$ or $L_{\rm bol} = 0.002\,L_{\rm Edd}$. The *solid line* shows the result from the numerical calculation in Suleimanov et al. (2007). The *dotted line* shows the formulas (1.104) in zone B and the *dashed line* the formulas (1.110) in zone C. Figures from Suleimanov et al. (2007)

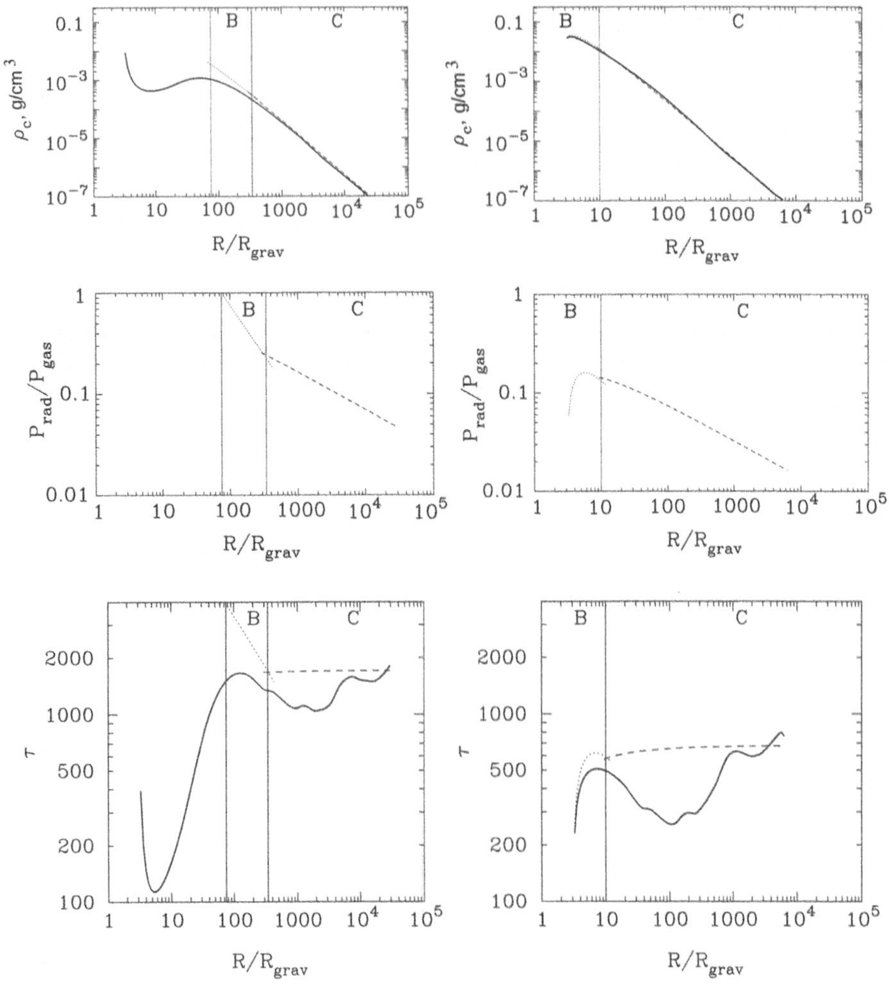

Fig. 1.10 From the top down: density in the disc symmetry plane ρ_c, ratio between radiation and gas pressure, and optical depth τ. Disc parameters: $m_x = 10$, $\mu = 0.62$, $\alpha = 0.3$, left: $\dot{M}_{17} = 33.6$ or $L_{bol} = 0.2\,L_{Edd}$, right: $\dot{M}_{17} = 0.336$ or $L_{bol} = 0.002\,L_{Edd}$. Notations as in Fig. 1.9. Figures from Suleimanov et al. (2007)

equating $\varkappa_T$ and $\varkappa_0 \rho T^{-7/2}$ in the equatorial plane of the disc

$$R_{BC} \sim 5 \times 10^7\, m_x^{1/3}\, \dot{M}_{17}^{2/3} \left(\frac{\mu}{0.6}\right)^{-1/3} \left(\frac{\varkappa_0}{\varkappa_0^*}\right)^{-2/3} \left(\frac{\varkappa_T}{\varkappa_T^*}\right)^{4/3} \text{ cm}$$

for characteristic values of the dimensionless parameters $\Pi_{1..4}$ and $f(r) = 1$.

As outer boundary of zone C we take the radius where recombination of hydrogen atoms sets in (at $T_{eff} \sim 10^4$ K). When this happens, thermal instabilities in the disc start developing, and due to a significant increase in the opacity coefficient of the matter, convection starts playing a role in the transfer of energy to the surface (Meyer and Meyer-Hofmeister 1981, 1982). In such regions, it is no longer correct to approximate the opacity coefficient $\varkappa_\mathrm{R}$ using Kramers law. Equating the right-hand side of (1.74) and $\sigma_\mathrm{SB}\, T_\mathrm{eff}^4$, we get:

$$R_\mathrm{C} \approx 3.5 \times 10^{10}\, m_x^{1/3}\, \dot{M}_{17}^{1/3} \left(\frac{T_\mathrm{eff}}{5000\ \mathrm{K}}\right)^{-4/3} \ \mathrm{cm}. \tag{1.116}$$

Due to irradiation of the outer parts of the disc by the central source, the boundary R_C can be further from the centre. This happens if the radiative X-ray flux, falling on the surface of the disc, thermalizes in its outer layers and heats them up so that the effective temperature of the disc surface does not drop below $\sim 10^4$ K (Dubus et al. 1999).

1.5.3.3 Thickness of the Disc

For the discs in binary systems with stellar-mass components during outbursts, the quantities $\lg(\delta)$ and $\lg(\tau_0)$ lie in the range of 2–4. For these values, the considered combinations of the quantities $\Pi_{1,2,3,4}$ practically do not change with radius, and inside each zone we may use the following characteristic values:

$$\text{(zone B)} \quad \Pi_z \approx 2.6, \quad \Pi_\Sigma = 0.96, \quad \Pi_\rho = 0.67, \quad \Pi_\mathrm{T} = 1.2, \tag{1.117}$$

$$\text{(zone C)} \quad \Pi_z \approx 2.6, \quad \Pi_\Sigma = 1.03, \quad \Pi_\rho = 0.76, \quad \Pi_\mathrm{T} = 1.09. \tag{1.118}$$

Let us consider a disc with matter consisting solely of hydrogen plasma ($\mu = 0.5$), choosing for the opacity a value $\varkappa_\mathrm{R} = 6.4 \times 10^{22}\, \rho\, T^{-7/2}\ \mathrm{cm}^2/\mathrm{g}$ (Kato et al. (1998); in the work by Shakura and Sunyaev (1973) a similar value was used), which is determined only by free-free electron transitions in the plasma. This value is two orders of magnitude less than the value of the opacity due to bound-free transitions $\varkappa_0^*$. However, the physical parameters depend only weakly on the opacity coefficient (1.110). For example, the half-thickness of the disc changes due to a direct decrease of $\varkappa_0$, μ, and also Π_z, since τ_0 decreases almost by a factor of 10 (see (1.112) and Fig. 1.8b). Thus, the disc half-thickness z_0 is $\sim 25\%$ less for $\mu = 0.5$ than for $\mu = 0.62$.

The numerical solution to the equations of vertical structure as described in this section gives a larger disc thickness compared to that of a vertically homogeneous disc, namely, the 'characteristic hydrostatic scale'. The latter is estimated as v_s/ω, where v_s is the sound speed in the disc symmetry plane. The presence of the factor two was indicated by Shakura and Sunyaev (1973). It is explained by the inhomogeneity of the distribution of density and temperature over the thickness of the disc. More exactly, this factor $\Pi_z \sim \sqrt{\Pi_1} \sim 2.5$, as can be seen from the first line in the system of Eq. (1.99):

$$z_0 = \sqrt{\Pi_1}\sqrt{\frac{\Re\, T_c}{\mu}}\,\frac{1}{\omega_K}\,. \tag{1.119}$$

1.5.3.4 'Dead' Discs

The formulas (1.104) and (1.110), describing radial dependencies in a disc, may be applied also for 'dead' discs or disc reservoirs (Syunyaev and Shakura 1977), i.e. discs in which transfer of matter through the inner boundary is not possible and thus $\dot{M}_{\rm in} = 0$. Since the inner accretion rate and $f(r)$ always show up as multiplicative factors in (1.104) and (1.110), the formulas could be converted using $\dot{M}_{\rm in}(t)\, f(r) = F(h,t)/h$ (cf. (1.102)).

1.6 Non-stationary Disc Accretion

Outbursts in accreting sources, for example in binary systems and active galactic nuclei, are of special interest. Bright events can be observed by instruments operating in different ranges of the electromagnetic spectrum, supplying a wealth of data about the physics of distant stars. Recently, due to the boom in studies of exoplanets, the subject of disc evolution in protoplanetary systems has become topical in astrophysics.

Transient phenomena in discs may be caused by different kinds of instabilities, which in general develop on different time scales. In this section, we will address the set up of and solution to the problem of non-stationary accretion in a viscous disc. The problem corresponds to the disc evolution that takes place on viscous time scales due to redistribution of angular momentum of matter in the disc.

1.6.1 Basic Equation of Non-stationary Accretion

In Sect. 1.4.1 we introduced the following quantities, integrated along the disc thickness: the surface density Σ_0 (1.41) and the integrated component of the turbulent viscosity tensor $W_{r\varphi}$ (1.45). We write down again the obtained equations for conservation of mass and angular momentum (1.42) and (1.44):

$$\frac{\partial \Sigma_0}{\partial t} = -\frac{1}{r}\frac{\partial}{\partial r}(\Sigma_0\, v_r\, r)\,,$$

$$\Sigma_0\, v_r\, r\,\frac{\partial\,(\omega r^2)}{\partial r} = -\frac{\partial}{\partial r}\,(W_{r\varphi} r^2)\,.$$

Substituting the combination $\Sigma_0 \, v_r \, r$ from the second line into the first, we obtain the basic equation for non-stationary accretion:

$$\frac{\partial \Sigma_0}{\partial t} = \frac{1}{r}\frac{\partial}{\partial r}\left[\frac{1}{\partial(\omega r^2)/\partial r}\frac{\partial}{\partial r}(W_{r\varphi}\, r^2)\right] . \tag{1.120}$$

This is an equation of diffusion type, a parabolic equation of the second order in partial derivatives.

The tensor component, integrated over the full thickness of the disc, is written in the framework of the gradient hypothesis of transfer of angular momentum by turbulent motions (1.38) in the following way:

$$W_{r\varphi}(r,t) = 2\int_0^{Z_0} w^{\mathrm{t}}_{r\varphi}\, \mathrm{d}Z \;= 3\,\omega_{\mathrm{K}}\int_0^{Z_0} \nu_{\mathrm{t}}\, \rho\, \mathrm{d}Z \, . \tag{1.121}$$

If the kinematic coefficient of the turbulent viscosity ν_{t} is independent of z, we get:

$$W_{r\varphi}(r,t) = \frac{3}{2}\,\omega_{\mathrm{K}}\, \nu_{\mathrm{t}}\, \Sigma_0 \, . \tag{1.122}$$

We introduce as a new independent parameter the specific angular momentum $h(r) = v_\varphi(r)\, r = \omega\, r^2$. We further define the specific angular momentum of a free particle, rotating in a Newtonian potential, as the quantity $h_{\mathrm{K}} \equiv \sqrt{G\,M\,r}$. Herewith, $\mathrm{d}r = 2\,h_{\mathrm{K}}\,\mathrm{d}h_{\mathrm{K}}/(G\,M)$.

In the case of Keplerian orbits, Eq. (1.120) taken together with (1.122) is written in the following form:

$$\frac{\partial \Sigma_0}{\partial t} = \frac{3}{4}\frac{(G\,M)^2}{h^3}\frac{\partial^2(\Sigma_0\, \nu_{\mathrm{t}}\, h)}{\partial h^2} , \qquad h \equiv h_{\mathrm{K}} \, . \tag{1.123}$$

We also consider an alternative version of this equation, convenient from the point of view of establishing boundary conditions in an evolving disc. It is, in addition, more appropriate for α-discs in models where the viscosity is parametrized using the turbulent α-parameter considered as a constant value, rather than using the kinematic viscosity coefficient ν_{t}.

We introduce the quantity $F = 2\pi\, W_{r\varphi} r^2$, which is equal to the total viscous torque, acting between neighbouring rings in the disc. At constant accretion rate in the disc, and using for $W_{r\varphi}$ a notation of the form (1.49), for a stress free inner boundary $W_{r\varphi}(r = r_{\mathrm{in}}) = 0$, we may write the quantity of the total viscous torque in the following way:

$$F = \dot{M}\sqrt{G\,M\,r}\left(1 - \sqrt{\frac{r_{\mathrm{in}}}{r}}\right) , \qquad \dot{M} = const \, . \tag{1.124}$$

As we can see, F is linearly proportional to the specific angular momentum $h = \sqrt{GMr}$ at large distances.

In the new variables, the equation of transfer of angular momentum (1.44) takes the form (note that v_r has a negative value):

$$-2\pi\, \Sigma_0\, v_r\, r \;=\dot{M}(r,t) = \left[\frac{\partial h}{\partial h_{\rm K}}\right]^{-1} \frac{\partial F}{\partial h_{\rm K}}\,, \tag{1.125}$$

and Eq. (1.120):

$$\frac{\partial \Sigma_0}{\partial t} = \frac{1}{4\pi} \frac{(GM)^2}{h_{\rm K}^3} \frac{\partial}{\partial h_{\rm K}} \left(\left[\frac{\partial h}{\partial h_{\rm K}}\right]^{-1} \frac{\partial F}{\partial h_{\rm K}}\right). \tag{1.126}$$

For a Keplerian disc, by definition, $\partial h/\partial h_{\rm K} \equiv 1$.

Which method to use for solving the equation of non-stationary accretion (1.123), depends on the form of the turbulent viscosity coefficient $\nu_{\rm t} = \nu_{\rm t}(r, \Sigma_0)$. In the framework of the model for α-turbulence, when the turbulent viscosity tensor is proportional to the pressure in the disc, the form of $\nu_{\rm t}(r, \Sigma_0)$, or in other words, the relationship between F and Σ_0, necessary for solving (1.126), may be derived from the equations of vertical structure.

1.6.2 Solutions to the Linear Equation of Viscous Evolution in the Disc

If F is linearly dependent on the surface density Σ_0, in other words, if $\nu_{\rm t}$ is a function only of radius and does not depend on the surface density, then (1.123) becomes a linear differential equation of diffusion type. In 1952, Lüst found particular solutions to the equation of viscous accretion, proposed by his teacher Weizsäcker (1948), and described the principles of constructing a general solution to both infinite and finite problems.

For a disc of infinite extension, Lynden-Bell and Pringle (1974) used a method of superposition of particular solutions to the equation of viscous evolution and, in particular, found Green's functions for two types of boundary conditions at the inner boundary. With the help of Green's functions it is possible to find F or Σ at any moment in time and at any point for arbitrary initial conditions. The inner radius of the disc in their solution is equal to zero. On long time scales, the dependencies in the disc are self-similar and the accretion rate through the inner boundary declines as a power law $\dot{M} \propto t^{-(1+l)}$, where the parameter $l < 1$. Pringle (1991) examined, with the help of Green's functions, an infinite disc with central inflow of angular momentum. This problem describes the evolution of a disc surrounding a binary system. A similar problem was solved by Tanaka (2011), with the difference that the inner boundary of the disc was considered to be located at a finite, non-zero inner radius. King and Ritter (1998) studied the evolution of a disc with finite radius and constant ν_t, and found that the accretion rate declines exponentially with time. The problem of a finite disc was also studied numerically in Zdziarski et al. (2009). The special case of Green's function for a finite disc was constructed in Wood et al. (2001) for a zero inner boundary. The full Green's function, which can be used together with an arbitrary initial distribution for two types of boundary conditions, was found by Lipunova (2015). This work also described the procedure of constructing a solution with non-zero and variable accretion rate at the outer boundary.

Note that in all these cases, the characteristic viscous time scale $\tau_{\rm vis} \sim r^2/\nu_t$ is constant in time.

1.6.3 Evolution of an Infinite Viscous Disc

Let us recall the solution obtained by Lynden-Bell and Pringle (1974). We write the kinematic viscosity coefficient in the form

$$\nu_t = \nu_0\, r^b\,.$$

Then the relation $F = 3\pi\, h\, \nu_t \Sigma_0$ (cf. (1.122)) may be written in the following way:

$$F = 3\pi\, h\, \nu_0\, \Sigma_0\, r^b\,. \tag{1.127}$$

For a Keplerian disc ($h \equiv h_*$), the equation of viscous torque (1.123) takes the following form:

$$\frac{\partial F}{\partial t} = \frac{3}{4}\, \nu_0\, h^{2b-2}\, (G\, M)^{2-b}\, \frac{\partial^2 F}{\partial h^2}\,, \tag{1.128}$$

or in a way similar to the notation in Lynden-Bell and Pringle,

$$\frac{\partial^2 F}{\partial h^2} = \frac{1}{4}\left(\frac{\kappa}{l}\right)^2 h^{1/l-2}\, \frac{\partial F}{\partial t}\,, \tag{1.129}$$

where the constant parameters are related in the following way:

$$\frac{1}{2l} = 2 - b\,, \qquad \kappa^2 = \frac{16\, l^2}{3\nu_0\, (G\, M)^{1/2l}}\,. \tag{1.130}$$

The general solution to the linear equation (1.129) may be found by expansion in eigenfunctions and superposition of particular solutions. The method of superposition allows for a general solution, satisfying the given initial or boundary conditions.

In the case of a linear equation, the method of separation of variables may also be used.

We will search for a particular solution of the form $F(h,t) = f(h_c\,\xi) \times \exp(-s\,t)$, where s is some constant of the same dimension as that of the inverse time, $\xi = h/h_c$, and h_c is some characteristic value of the specific angular momentum of the matter in the disc. Substituting such a function $F(h,t)$ into (1.129), we obtain a Lommel's transformation of the Bessel equation (see Sect. 4.31 in Watson 1944):

$$\frac{d^2 f}{dh^2} + \frac{s}{4}\left(\frac{\kappa}{l}\right)^2 h^{1/l-2} f = 0\,,$$

with the particular solution

$$f(x) = (k\,x)^l\,[A(k)\,J_l(k\,x) + B(k)\,J_{-l}(k\,x)]\,,$$

where J_l and J_{-l} are Bessel functions of non-integer order, $k^2 = s\,\kappa^2\,h_c^{1/l}$ and l are constants and $x = \xi^{1/2l} = (h/h_c)^{1/2l}$, where ξ is the normalised specific angular momentum. The general solution is equal to the superposition of particular solutions with all values of the parameters k, $A(k)$, $B(k)$ such that the specific boundary and initial conditions are satisfied:

$$F(h,t) = \int_0^\infty \exp\left(-\frac{k^2\,t}{\kappa^2\,h_c^{1/l}}\right)\,(kx)^l[A(k)\,J_l(k\,x)+B(k)\,J_{-l}(k\,x)]\,dk\,. \quad (1.131)$$

For example, the condition $F(h) = 0$ for $h = 0$ leads to the vanishing of all coefficients for Bessel functions with negative index: $B(k) \equiv 0$.

The following method was used to determine the coefficients $A(k)$ and $B(k)$. Let us choose a solution at $t = 0$, with the condition that all viscous stresses at the centre are equal to zero $F(h=0) = 0$, and write it using (1.131) in the form

$$F(h,t=0) = \int_0^\infty (kx)^l\,A(k)\,J_l(k\,x)\,dk\,.$$

We now use the Hankel inversion theorem (chapter II, theorem 19 in Sneddon (1951), see also Watson (1944) and MacRobert (1932)) for continuous functions $f(k)$ in the form

$$f(k') = \int_0^\infty x\,J_l(k'x)\left[\int_0^\infty k\,f(k)\,J_l(k\,x)dk\right]dx \quad \text{for } l \geq -1\,.$$

Substituting $f(k) = k^{l-1} A(k)$, we see that the integral within square brackets is equal to $F(h, t = 0)/x^l$. It follows that

$$(k')^{l-1} A(k') = \int_0^\infty F(h)\, J_l(k' x)\, x^{1-l}\, \mathrm{d}x \;, \tag{1.132}$$

where $F_0(h) \equiv F(h, t = 0)$. From here we can determine the coefficients $A(k')$.

If the initial distribution $F_0(h)$ is given, then the solution to the linear differential equation (1.129) has the form

$$F(h,t) = \int_0^\infty G(h, h_1, t)\, F_0(h_1)\, \mathrm{d}h_1 \;,$$

where G is the Green's function that is the solution to (1.129) at all points for $h \neq h_1$ and $t \neq 0$, and for which it is true that $G = 0$ for $t < 0$ in physical systems. It is possible to consider Eq. (1.129) as a linear system with input signal $F_0(h_1)$ and output signal $F(h, t)$, in which the Green's function has the role of a 'weighting function'. As is well known, the Green's function itself is a 'response' of the system to a delta impulse input signal, that is, it is a solution to (1.129), if the initial condition is a Dirac δ-function:

$$F_0 = \delta(x - x_1); \qquad F(h,t) = G(x, x_1, t) \;.$$

Substituting this initial distribution into (1.132), we find an expression for $A(k)$:

$$A(k) = (k\, x_1)^{1-l}\, J_l(k x_1) \;.$$

To obtain the Green's function we substitute $A(k)$ in expression (1.131):

$$G(x, x_1, t) = x^l\, x_1^{1-l} \int_0^\infty \exp\left(-\frac{k^2\, t}{\kappa^2\, h_\mathrm{c}^{1/l}} \right) k\, J_l(k x_1)\, J_l(k x)\, \mathrm{d}k \;.$$

The integral is found using Hankel's tables for integral transforms:

$$G(x, x_1, t) = \frac{\kappa^2\, h_\mathrm{c}^{1/l}\, x^l\, x_1^{1-l}}{2t} \exp\left(-\frac{x_1^2 + x^2}{4t}\, \kappa^2\, h_\mathrm{c}^{1/l} \right) I_l \left(\frac{x\, x_1}{2t}\, \kappa^2\, h_\mathrm{c}^{1/l} \right) , \tag{1.133}$$

where I_l is a modified Bessel function of the first kind (an Infeld function). Figure 1.11 shows the Green's function at four moments in time.

Let us choose the initial distribution F_0 in the form of a Dirac delta function with a physically motivated normalisation. We assume that the initial configuration is a

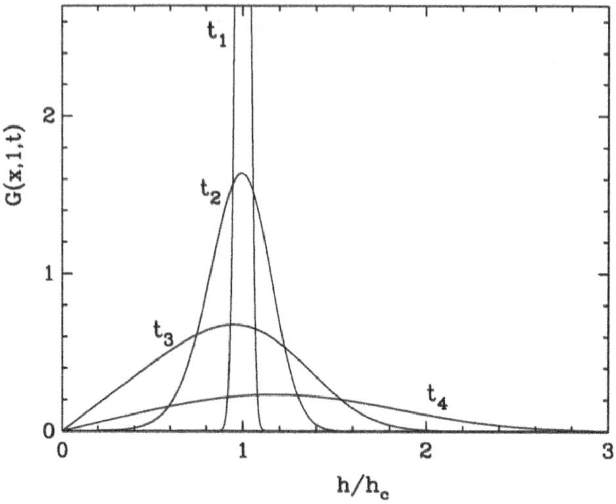

Fig. 1.11 The Green's function (1.133), found by Lynden-Bell and Pringle (1974), at four moments in time: $t_1 = 0.001$, $t_2 = 0.03$, $t_3 = t_{\max} = 0.1875$ and $t_4 = 1$. The parameters of the solution are $\kappa = 1$, $h_c = 1$, $l = 1/3$, $x_1 = 1$

narrow ring at radius r_s with total mass M_0. The specific angular momentum at this radius is equal to $h_s = x_s^{2l}\,h_c$. We write down the surface density for $t = 0$ as $\Sigma_0(h, t = 0) = M_0\delta(r - r_s)/2\pi r_s$. Using (1.127) and (1.130) we obtain for the earlier introduced variable $x = (h/h_c)^{1/2l}$:

$$F_0(x) = 2l\,M_0\,h_c^{1-1/l}\,\kappa^{-2}\,x_s^{2l-1}\,\delta(x - x_s)\,.$$

Here we used the equality $\delta(x - x_s)\,dx = \delta(r - r_s)\,ds$. The evolution of this narrow ring is determined with the help of the obtained Green's function:

$$F(x, t) = \int_0^\infty F_0(x_1)\,G(x, x_1, t)\,dx_1$$

and has the explicit form:

$$F(x, t) = \frac{M_0\,h_c\,l\,(x\,x_s)^l}{t}\,\exp\left(-\frac{x_s^2 + x^2}{4t}\,\kappa^2\,h_c^{1/l}\right)\,I_l\left(\frac{x\,x_s}{2t}\,\kappa^2\,h_c^{1/l}\right)\,. \tag{1.134}$$

We now consider the accretion rate at the inner boundary $\dot{M}_{\rm in} = (\partial F/\partial h)|_{h\to 0}$:

$$\dot{M}_{\rm in}(t) = \frac{x^{1-2l}}{2l\,h_c}\,\frac{\partial F(x, t)}{\partial x}\bigg|_{x\to 0} = \frac{M_0\,\tau_e^l}{\Gamma(l)}\,\frac{e^{-\tau_e/t}}{t^{1+l}}\,.$$

It is possible to rewrite the accretion rate using its peak value

$$\dot{M}_{\rm in}(t) = \dot{M}_{{\rm in},max} \left(\frac{\tau_{\rm pl}}{t}\right)^{1+l} e^{-\tau_e/t} ,$$

where we have introduced the characteristic time scale for exponential growth τ_e and power-law decline $\tau_{\rm pl}$:

$$\tau_e = \frac{\kappa^2 h_{\rm s}^{1/l}}{4} = \frac{1+l}{e} \tau_{\rm pl} .$$

The accretion rate reaches its peak value

$$\dot{M}_{{\rm in},max} = \frac{M_{\rm disc}}{t_{\rm max}} \frac{(1+l)^l}{e^{1+l}\, \Gamma(l)} \tag{1.135}$$

at time

$$t_{\rm max} = \frac{\kappa^2 h_{\rm s}^{1/l}}{4(1+l)} = \frac{\tau_{\rm pl}}{e} . \tag{1.136}$$

1.6.4 Solution for a Disc with a Fixed Outer Radius

The boundary conditions are of high importance for the type of solution to Eq. (1.128). Above, we considered a solution in which the disc increases in size without limitation. A part of the matter in the disc will with time acquire very high values of the specific angular momentum. In a number of astrophysical situations, it is clear that it is necessary to set conditions at a finite radius from the centre. This concerns generally discs in binary systems. The torque of tidal forces, appearing due to gravitational influence of the companion star and acting predominantly in the narrow area inside the Roche lobe, leads to the disc being truncated at a certain radius (Papaloizou and Pringle 1977; Paczynski 1977; Ichikawa and Osaki 1994; Hameury and Lasota 2005). Near the truncation radius, angular momentum is transferred from the disc to orbital motion of the binary system.

Thus, the problem now needs to be solved for a finite interval. The method of superposition of partial solutions is modified, and the general solution is found not as an integral (1.131), but as a sum of all the partial solutions that fulfill the specific boundary conditions (Lüst 1952):

$$F(x,t) = \sum_{i=1}^{\infty} e^{-t\, k_i^2 \kappa^{-2} h_{\rm out}^{-1/l}} (k_i\, x)^l \left[A_i\, J_l(k_i\, x) + B_i\, J_{-l}(k_i\, x)\right] , \tag{1.137}$$

Here we have also changed the characteristic value of the specific angular momentum to the value at the outer boundary $h_{\rm out}$, where the dimensionless parameter $x = 1$.

Let us set the boundary conditions at the outer radius of the disc:

$$\frac{\partial F}{\partial h} = \dot{M}_{\rm out}(t) \ \text{ at } \ h = h_{\rm out} \, . \tag{1.138}$$

In the simplest case, if $\dot{M}_{\rm out}(t) = 0$, this will be a homogeneous Dirichlet boundary condition. At the inner radius, we consider the same condition as earlier: $F(h) = 0$ for $h = 0$. The use of these two conditions gives an equation that every particular solution has to satisfy, that is, for any k

$$l\, J_l(k_i) + k_i\, J_l'(k_i) = 0 \, . \tag{1.139}$$

Since there in the series, representing the general solution, remain only terms with Bessel functions of positive order, the general solution at the starting point $t = 0$ is:

$$F(x, 0) = \sum_{i=1}^{\infty} (k_i\, x)^l\, A_i\, J_l(k_i\, x) \, . \tag{1.140}$$

Series of the form $\sum_{i=1}^{\infty} k_i^l\, A_i\, J_l(k_i\, x)$ with the condition (1.139) are called Dini series (see Watson 1944, Sect. 18.11). The function $f(x) = F(x, 0)\, x^{-l}$ can be expanded in Dini series if it satisfies the Dirichlet conditions at the given interval, and the coefficients of the expansion can be found as $k_i^l\, A_i = 2\, \bar{f}_J(k_i)\, J_l^{-2}(k_i)$ (Watson 1944; Sneddon 1951), where we have used the finite Hankel transform

$$\bar{f}_J(k_i) = \int_0^1 x\, f(x)\, J_l(k_i\, x)\, {\rm d}x \, .$$

To find the Green's function, we search for a solution to an initial condition of the form of a δ-function: $F(x, 0) = \delta(x - x_1)$. Using its properties, substituting $f(x)$ into the last expression, we get:

$$k_i^l\, A_i = 2\, x_1^{1-l}\, \frac{J_l(k_i\, x_1)}{J_l^2(k_i)} \, . \tag{1.141}$$

In this way we obtain the Green's function for a finite disc (Lipunova 2015):

$$G(x, x_1, t) = 2\, x^l\, x_1^{1-l} \sum_i e^{-t\, k_i^2\, \kappa^{-2}\, h_{\rm out}^{-1/l}}\, \frac{J_l(k_i\, x_1)\, J_l(k_i\, x)}{J_l^2(k_i)} \, , \tag{1.142}$$

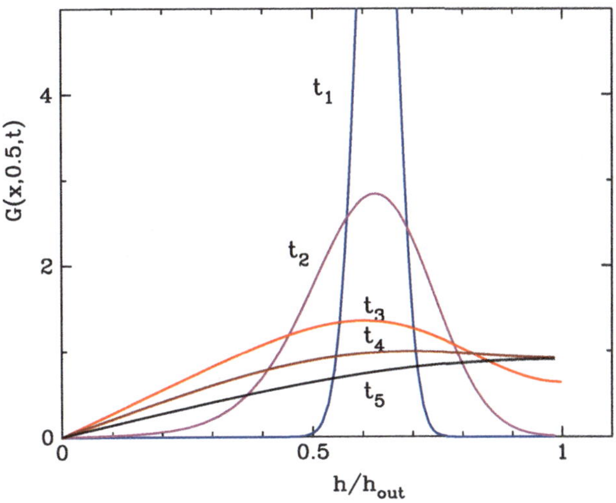

Fig. 1.12 Green's function of a finite disc with a zero torque at the centre at times $t_1 = 0.001$, $t_2 = 0.01$, $t_3 = t^{\infty}_{max} = 3/64$, $t_4 = 0.1$, $t_5 = 0.3$. The ring of matter was located at $x_s = (h/h_{\rm out})^{1/2l} = 0.5$ at time $t = 0$. The parameters are $\kappa = 1$ and $l = 1/3$

where k_i are the positive roots of the transcendental equation (1.139) and $x = (h/h_{\rm out})^{1/2l}$. The Green's function is depicted in Fig. 1.12 for a few moments in time. The curve at $t_3 = t_{\rm max}$ (see (1.136)) corresponds to the maximum accretion rate through the inner boundary of the disc.

For a specific initial distribution $F(x, 0)$, the distribution at any point in time $t > 0$ can be found as

$$F(x,t) = \int_0^1 F(x_1,0)\, G(x,x_1,t)\, \mathrm{d}x_1 \,. \tag{1.143}$$

The accretion rate at any point in time $t > 0$ is

$$\dot{M}(x,t) = \int_0^1 F(x_1,0)\, G_{\dot{M}}(x,x_1,t)\, \mathrm{d}x_1 \Big/ h_{\rm out} \,, \tag{1.144}$$

where the Green function for the accretion rate is

$$\begin{aligned} G_{\dot{M}}(x,x_1,t) &\equiv \frac{\partial G(x,x_1,t)}{\partial x^{2l}} = \\ &= \frac{(x\, x_1)^{1-l}}{l} \sum_i e^{-t\, k_i^2\, \kappa^{-2}\, h_{\rm out}^{-1/l}}\, k_i\, \frac{J_l(k_i\, x_1)\, J_{l-1}(k_i\, x)}{J_l^2(k_i)} \,. \end{aligned} \tag{1.145}$$

The functions G and $G_{\dot{M}}$ in the particular case of $x_1 = 1$ are found in the form of analytical asymptotics by Wood et al. (2001).

The initial distribution F can be expressed through the distribution of surface density, using (1.127) and (1.130):

$$F(x,0) = \frac{16\pi\, l^2}{\kappa^2\, h^{1/l}}\, r^2\, \Sigma(r)\, h\,, \tag{1.146}$$

where $r = h^2/GM$ and $h = h_{\rm out}\, x^{2l}$.

For large times t, the first term in the sum (1.145) dominates and the time dependence can be expressed as a simple exponential:

$$G_{\dot{M}}(0, x_1, t)\Big|_{t>t_{\rm vis}} = \frac{k_1^l\, x_1^{1-l}}{2\,l\,\Gamma(l)}\, \frac{J_l(k_1\, x_1)}{J_l^2(k_1)}\, \exp\left(-\frac{t\, k_1^2}{2\,l\, t_{\rm vis}}\right).$$

The characteristic time scale for exponential decrease of the accretion rate is equal to:

$$t_{\rm exp} = h_{\rm out}^{1/l}\, \frac{\kappa^2}{k_1^2} = \frac{16\, l^2}{3\, k_1^2}\, \frac{r_{\rm out}^2}{\nu_{\rm out}}\,, \tag{1.147}$$

where we have taken into account that $\nu_{\rm out} = \nu_0\, r_{\rm out}^b$. In Table 1.3, the first zero k_1 of the equation is shown for typical values of l. The table also provides the coefficients for calculating characteristic time scales for the growth (1.136) and the exponential decay (1.147) of the solution.

The disc becomes quasi stationary (i.e. the accretion rate practically does not change with radius) in regions where $r/r_{\rm out} < (t/t_{\rm exp})^{2l}$. The establishment of quasi stationarity in the central regions of the disc on viscous time scales is a common property for discs with any type of viscosity.

Table 1.3 Parameters of the Green function for a non-stationary disc

b	l	k_1	$t_{\rm max}(r_s^2/\nu_s)^{-1}$	$t_{\rm exp}(r_{\rm out}^2/\nu_{\rm out})^{-1}$	a_0	Comments
0	1/4	1.0585	1/15	0.298	1.267	$\nu = const$
1/2	1/3	1.2430	1/9	0.383	1.363	α-disc with $h/r = const$
3/5	5/14	1.2927	0.125	0.407	1.392	α-disc, $\tau_{\rm T} \gg \tau_{\rm ff}$
3/4	2/5	1.3793	0.152	0.449	1.444	α-disc, $\tau_{\rm ff} \gg \tau_{\rm T}$
1	1/2	1.5708	2/9	0.540	1.571	$F(h) \propto \sin((\pi/2)\, h/h_{\rm out})$
2	∞	–	–	–	–	$t_{\rm vis}$ independent of r

The columns are: Exponent in the power law $\nu \propto r^b$; l from expression (1.130); the first zero of Eq. (1.139); the numerical factor from (1.136) the numerical factor from (1.147); the parameter describing the radial profile, $a_0 = \dot{M}_{\rm in}\, h_{\rm out}/F_{\rm out}$. The solution to the linear equation may apply to α-discs on timescales of the order of or shorter than the viscous timescale. For α-discs the type of opacity is shown

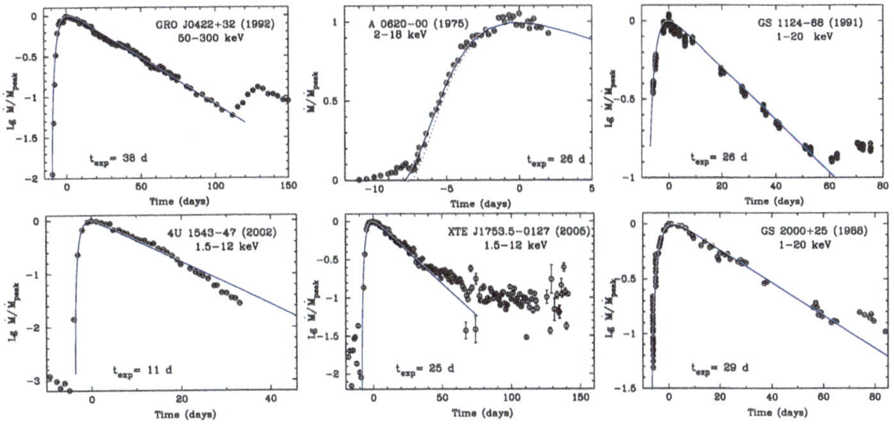

Fig. 1.13 Normalised lightcurves of the X-ray novae GRO J0422+32 (1992), A 0620-00 (1975), GS 1124-68 (1991), GS 2000+25 (1998) from Chen et al. (1997), 4U 1543-47 (2002) and XTE J1753.5-0127 (2005) (results from ASM/*RXTE*). The X-ray energy range for each flare is indicated in the plots. The solid curves show the peak-normalised accretion rates through the inner boundary calculated according to (1.144) for $l = 2/5$ and t_{exp} as indicated for each flare. The initial distribution of surface density in the disc is $\Sigma \propto r$ and the initial inner radius of the hot zone is $0.01 \times r_{out}$. For A 0620-00, two model lightcurves are shown, for inner radii, at $t = 0, 0.001 \times r_{out}$ (*solid line*) and $0.3 \times r_{out}$ (*dotted line*), respectively. Figure from Lipunova (2015)

Bright X-ray flares, known as outbursts of X-ray novae, are observed in binary systems consisting of a compact object and a low-mass normal star. It is well known that in the 'simplest' cases, outbursts in X-ray novae show lightcurves with a fast rise and an exponential decay, which are called FRED profiles (Chen et al. 1997). Such lightcurves are nicely produced within the framework of the model for viscous discs with a viscosity coefficient constant on time scales of the order of t_{vis} (Fig. 1.13). This is explained by the fact that on time scales of the order of one to two t_{vis}, a non-stationary α-disc and a disc such as considered in this section show similar evolution.

In order to fit the constant-viscosity solution to the evolution of a viscous α-disc, it is necessary to estimate the most appropriate value of parameter b in (1.128). This can be done using the relation (1.40) between the kinematic viscosity and the turbulence parameter:

$$\nu_t = \nu_0 \, r^b \simeq \frac{2}{3} \alpha \, \omega_K \, r^2 \left(\frac{z_0}{r}\right)^2 \frac{1}{\Pi_1}, \tag{1.148}$$

where the parameter Π_1 shows up from a consideration of the vertical structure, see (1.99). The solution for a stationary disc with dominant gas pressure and Kramers opacity gives $z_0/r \propto r^{1/8}$ (see (1.110)), thus $b \simeq 3/4$ or, equally, $l \simeq 2/5$, if we neglect the dependence of the disc thickness on the accretion rate.

One can estimate α for an X-ray nova using (1.147) and (1.148) (Lipunova and Malanchev 2017):

$$\alpha \sim 0.15 \left(\frac{r_{\rm out}}{2\,R_\odot}\right)^{3/2} \left(\frac{z_0/r_{\rm out}}{0.05}\right)^{-2} \left(\frac{M}{10\,M_\odot}\right)^{-1/2} \left(\frac{t_{\rm exp}}{30^{\rm d}}\right)^{-1} \times \Pi_1\,. \qquad (1.149)$$

Here, one should substitute z_0 corresponding to the peak of an X-ray nova outburst. The main uncertainty in the above formula is the radius of the disk. In addition, the evolution of the thickness of the α-disk leads to a variation of the numerical factor in (1.149). However, a numerical modelling of the disk evolution can provide a self-consistent value of α (see Sect. 1.7.3).

1.6.5 Solution to the Non-linear Equation for the Evolution of a Viscous α-Disc

Earlier we considered the case when the coefficient of kinematic viscosity depends solely on the radial coordinate in the disc. In the more general case, we may represent $\nu_{\rm t}$ as a power-law function of Σ and r. Such a dependence arises in particular if we consider discs with α-viscosity. In this case, (1.123) becomes a non-linear differential equation in partial derivatives. To search for a solution to such an equation, similarity methods can be used in many cases. A self-similar solution to a non-linear differential equation accurately describes the evolution if enough time has passed since the initial moment.

As we have seen in the previous section, self-similar solutions to a linear differential equation are characterized by the possibility to completely separate the time and coordinate parts of the solution. A particular solution is thus a product of functions of different variables. In the case of a nonlinear differential equation, such a simple separation is in general not possible. To approach the problem, we may use the method of introducing new dimensionless variables (parameters), which contain combinations of the dimensional parameters (for example time and coordinates) raised to various powers.

Self-similar solutions to non-linear differential equations can be divided into two kinds (Barenblatt 1996, 2003). The self-similar solution is of the first kind if the self-similar function, as well as the new dimensionless parameter, can be derived from dimensional analysis. This case is also called a complete self-similarity. The second kind, or incomplete self-similarity, is the more general case. Here the self-similar function is a particular solution to the problem itself (a non-linear eigenvalue problem, see Zeldovich and Raizer (1967)). A dimensional analysis does not allow us to determine the self-similar function, and in particular, find to which powers the dimensional parameters should be raised to produce a self-similarity dimensionless variable. For incomplete self-similarity, the type of solution depends on the value of the self-similar variable.

If the constant coefficients in a self-similar function can be found from conservation laws, then the self-similar solution will be of the first kind (for example energy conservation in J. I. Taylor's blast wave (Barenblatt 2003)) and conservation of the total angular momentum in an accretion disc (see further Sect. 1.6.6.3)). Self-similar solutions of the first kind were found for accretion discs with a non-linear viscous diffusion equation in the stage of evolution when the accretion rate is decaying (Pringle 1974, 1991). Solutions of the second kind have also been constructed (Lyubarskij and Shakura 1987). These solutions apply to an earlier evolutionary stage, that is, the spreading of an original ring of matter into a disc around the gravitating centre.

The form of the turbulence parameter ν_t is determined by the physical structure of the disc, which is dependent on the astrophysical conditions. For an α-disc with two variants of opacity (Kramers' law and Thomson scattering), within the framework of self-similar solutions of the first kind, it was found that the accretion rate declines as $\propto t^{-19/16}$ and $\propto t^{-5/4}$, respectively (Pringle 1974; Filipov 1984; Lyubarskij and Shakura 1987; Cannizzo et al. 1990; Pringle 1991). Lin and Pringle (1987) considered a molecular disc with a gravitational instability generating an effective viscosity $\nu_t \propto \Sigma^2 r^{9/2}$, and found that $\dot{M} \propto t^{-6/5}$. Lin and Bodenheimer (1982) studied the evolution of a protoplanetary disc under the influence of convective turbulent viscosity ($\nu_t \propto \Sigma^2$), for which $\dot{M} \propto t^{-15/14}$. Ogilvie (1999) investigated an advective accretion flow, the structure of which considerably differs from that of a thin viscous disc, and, using similarity methods, found a solution in the case of conserved total angular momentum.

The type of solution also depends on the boundary conditions. Pringle (1991) in addition considered the general case of an infinite cold protostellar disc with $\nu_t \propto \Sigma^3$ and a central source of angular momentum. Such a formulation of the problem corresponds to the evolution of a disc around a young binary system (see also Ivanov et al. 1999). In Rafikov (2013), a detailed consideration of the evolution of discs around binary black holes was presented, and self-similar solutions were found with different conditions at the inner boundary, suggesting a certain mass transfer through the inner boundary. Rafikov (2016) built self-similar solutions for a 'decretion' disc (disc with mass ejection from the centre).

For a disc with a zero (or very small) viscous stress at the inner boundary and with a limited outer radius, a solution was found by Lipunova and Shakura (2000). According to them, $\dot{M} \propto t^{-10/3}$ for Kramers opacity and $\dot{M} \propto t^{-5/2}$ for Thomson scattering (see Sect. 1.6.7 below).

If $\nu_t = \nu_0 \Sigma^a r^b$, the kinematic viscosity coefficient is not constant in time since the surface density varies. The relation $F = 3\pi h \nu_t \Sigma_0$ (cf. (1.122)) can be presented in the following way:

$$F = 3\pi h \nu_0 \Sigma_0^{a+1} r^b . \tag{1.150}$$

Then, for a Keplerian disc ($h \equiv h_*$), Eq. (1.123) takes the following form:

$$\frac{\partial F}{\partial t} = D \frac{F^m}{h^n} \frac{\partial^2 F}{\partial h^2}, \tag{1.151}$$

where D is a dimensional constant,

$$D = \frac{a+1}{2} (G\,M)^2 \left(\frac{3}{2} \frac{\nu_0}{(2\,\pi)^a \,(G\,M)^b} \right)^{1/(a+1)}, \tag{1.152}$$

and m and n are dimensionless constants,

$$m = \frac{a}{a+1}, \qquad n = \frac{3a+2-2b}{a+1}.$$

The values of the parameters D, m and n may be determined from the equations of vertical structure. The parameter D in (1.151) can be seen as a sort of 'diffusion coefficient'. It may be found from the relation between Σ_0, F and h (Filipov 1984; Lyubarskij and Shakura 1987):

$$\Sigma_0 = \frac{(GM)^2\, F^{1-m}}{4\,\pi\,(1-m)\, D\, h^{3-n}}. \tag{1.153}$$

Comparing the equation of disc evolution in the linear and non-linear cases, (1.129) and (1.151), we find that $D = 4\,(l/\kappa)^2$ for $m = 0$.

The non-linear problem of non-stationary accretion has the following distinctive features. Firstly, the self-similar solutions of the second kind exist only for $m \neq 0$. Secondly, self-similar solutions of the first kind in the third stage, while they exist for $m = 0$, have an exponential profile for $r \to \infty$, characteristic for linear problems (see for example Lynden-Bell and Pringle 1974). For $m \neq 0$, the boundary of the disc is fully determined.[4]

1.6.5.1 The α-Discs

Lyubarskij and Shakura (1987) give the equations of vertical structure in a form similar to (1.99). The opacity is given as:

$$\varkappa = \varkappa_0 \frac{\rho^\varsigma}{T^\gamma}.$$

[4]This property is similar to the one that arises in problems of thermal conductivity, when, due to the non-linearity, the heatwave boundary sharply separates the heated zone from the rest of the region (Zeldovich and Raizer 1967).

Table 1.4 Dimensionless parameters in the equations of non-stationary accretion for different forms of ν_t

	m	n	a	b	ς	Υ	α_{pl}
$\varkappa_T \gg \varkappa_{ff}$ and (1.94)	2/5	6/5	2/3	1	0	0	−19/16
$\varkappa_{ff} \gg \varkappa_T$ and (1.94)	3/10	4/5	3/7	15/14	1	7/2	−5/4
OPAL (Iglesias and Rogers 1996), full ionization of H and He	1/3	1	1/2	1	1	5/2	−11/9
Convective turbulence (Lin and Bodenheimer 1982)	2/3	8/3	2	0	–	–	−15/14
Molecular disc with gravitational instability (Lin and Pringle 1987)	2/3	−1/3	2	9/2	–	–	−6/5

The parameter α_{pl} is the power-law index of the time-dependence during the stage of declining accretion in an infinite disc: $\dot{M} \propto t^{\alpha_{pl}}$

After a few algebraic manipulations of the equations in the right column of (1.99), we may find the relation between Σ_0, $W_{r\varphi}\, r^2$ and $\omega\, r^2$, which together with (1.153) gives[5]:

$$D = \frac{1}{4(1-m)(2\pi)^m}\left\{\frac{2^{6+\varsigma+2\Upsilon}\alpha^{8+\varsigma+2\Upsilon}}{\Pi_1^{\varsigma}\,\Pi_2^{2\varsigma}\,\Pi_3^{8+\varsigma+2\Upsilon}\,\Pi_4^{2}}\left(\frac{\mathfrak{R}}{\mu}\right)^{8+2\Upsilon}\right.$$
$$\left.\times\left(\frac{9\varkappa_0}{8\,a\,c}\right)^2(GM)^{12+8\varsigma}\right\}^{\frac{1}{10+3\varsigma+2\Upsilon}}, \qquad (1.154)$$

where

$$\varsigma = -\frac{11m-2n-2}{7m-n-1}, \qquad \Upsilon = -\frac{1}{2}\frac{37m-4n-10}{7m-n-1} \qquad (1.155)$$

or

$$m = \frac{4+2\varsigma}{10+3\varsigma+2\Upsilon}, \qquad n = \frac{12+11\varsigma-2\Upsilon}{10+3\varsigma+2\Upsilon} \qquad (1.156)$$

(see Table 1.4).

It is important to note that the 'diffusion coefficient' D is only weakly dependent on the opacity coefficient: as a power function of $\varkappa_0$ with an index of $1/5$ or $1/10$. This reduces the impact of the uncertainty due to the dependence of the real opacity on the disc parameters. The combination of parameters $\Pi_{1,2,3,4}$ in (1.154) depends only weakly on the optical depth τ, i.e. on the radius in the disc (see Tables 1.1

[5]Note that here F is a factor of 2π larger than in the paper by Lyubarskij and Shakura (1987), and our quantity D is smaller by a factor of $(2\,\pi)^m$.

and 1.2). Thus, D may be considered a constant in the basic equation of non-stationary accretion (1.151).

1.6.6 Evolution of α-Disc from a Ring of Matter

It turns out that the global evolution of the disc can in general be divided into three stages: (1) the stage of formation of the disc from an initial ring made up of matter at some radius, (2) the establishment of a quasi-stationary distribution of parameters in the disc, and as a special case, increasing accretion rate onto the central body, and (3) 'spreading' of the disc away from the centre, accompanied by a decrease of the accretion rate.

The ring of matter around a star may be formed as a result of a mass-transfer from the neighbouring component in a binary system. In the presence of effective mechanisms of viscosity, the differentially rotating ring starts to smear out into a disc.

At the first stage, material from the inner edge of the ring, losing angular momentum to the outer layers, starts to move towards the centre. In the region $r \ll R_{\rm out}$, the flow evolves into some self-similar regime whose characteristics are independent of the initial mass distribution profile. The inner edge of the disc, which has the form of a stretched-out 'tounge', reaches the accreting centre in a finite time (Fig. 1.14a). The self-similar solution breaks down close to the radius of the innermost stable orbit around the black hole, or close to the magnetosphere of the neutron star. Nevertheless, after some transition period, accretion again evolves into another self-similar solution, the regime of quasi-stationary accretion (the second stage).

At the second stage, a practically radially constant distribution is rapidly established in the inner regions of the disc, by virtue of the small viscous time scales at small radii. The region of the quasi-stationary solution gradually expands outwards (Fig. 1.14b), while the accretion rate gradually increases in time. Meanwhile, in the outer region, conditions remain close to the original.

Further, the disc gradually evolves into the third final stage (the decay stage, Fig. 1.14c) at which the details of the initial distribution are 'forgotten', and only some integral quantities conserved during the accretion are important in finding the self-similar solution. This final stage is described by a self-similar solution of type I, whereas the two preceding cases are described by self-similar solutions of type II, i.e. solutions in which the self-similarity index is found not from dimensionality arguments but in the process of integrating the ordinary differential equation for the representative function (Lyubarskij and Shakura 1987).

Thus, each stage is characterized by the motions whose distinctive property is a similarity that is conserved in the motion itself. This means that the distribution of any quantity, for example, the viscous torque, may be represented in the form:

$$F(h,t) = h^{A_1}\, t^{A_2}\, f(\xi)\,, \tag{1.157}$$

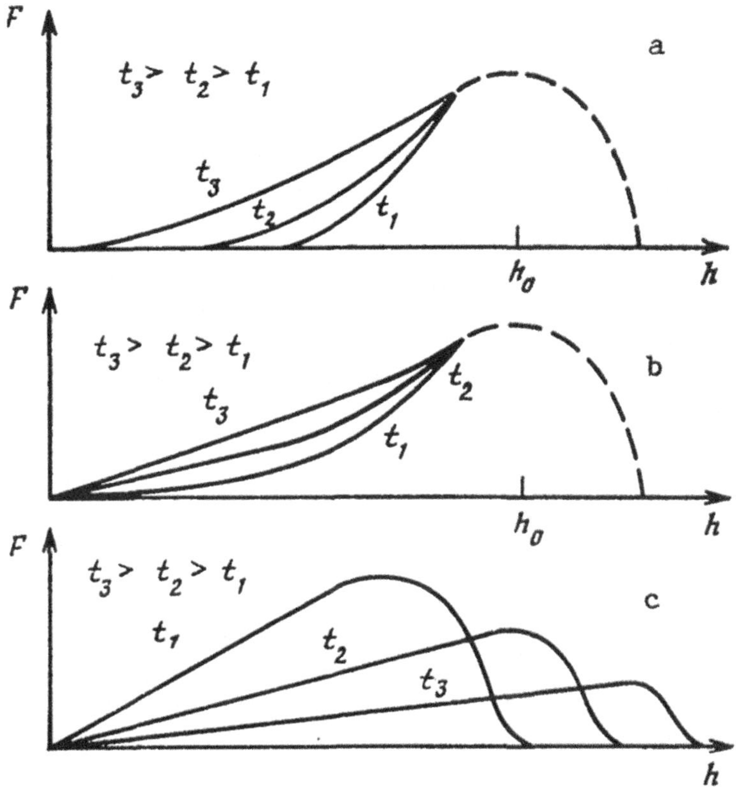

Fig. 1.14 Illustration by Lyubarskij and Shakura (1987) of the process of evolution of non-stationary disc accretion in the form of the dependence of viscous torques acting between adjacent rings of the disc, as a function of the specific angular momentum: (**a**) stage of formation and inward motion of self-similar 'tounge', (**b**) stage of formation of quasi-stationary regime, (**c**) stage of accretion decay. Dashes denote the regions into which the material was ejected and in which the solution is non self-similar. Each figure shows the distribution for three consecutive moments in time t_1, t_2, t_3. The calculated dependencies are shown in Figs. 1.16 and 1.17

where f is a function of a single self-similar variable $\xi = C\,h^{A_3}\,t^{A_4}$. For completely self-similar solutions, the parameters C and $A_{1..4}$ may be determined from dimensional arguments or from conservation laws. To determine the parameters for non completely self-similar solutions, a non-linear problem should be solved; in addition, the obtained parameters will depend on h and t.

1.6.6.1 'Tounge'-Formation Stage

Let us assume that the radius of the inner edge of the disc $r_{\rm in}$ or the equivalent value $h_{\rm in}$ decreases as a power law $h_{\rm in} \propto (-t)^{\gamma}$ ($t = 0$ when the centre is reached, thus the minus sign). As seen from Eq. (1.151), the combination $D\,F^m\,t/h^{n+2}$ is dimensionless, which permits the solution to be represented as

$$F(h,t) = \frac{h^{(n+2)/m}}{(-D\,t)^{1/m}}\,y(\xi)\,; \qquad 1 \le \xi = \frac{h}{A\,(-t)^{\gamma}} \le \infty\,, \tag{1.158}$$

where $y(\xi)$ is the representative function of the single self-similar variable ξ. It is not possible for the dimensionless variable ξ to be a combination of h, t and D, so we have to introduce the additional constant A of dimensionality [$\mathrm{cm}^2\mathrm{s}^{-(1+\gamma)}$], where the previously unknown exponent γ must be determined in the course of solving the problem. We thereby arrive at a self-similar problem of the second kind, similar to the problem of a converging shock wave (Zeldovich and Raizer 1967; Barenblatt 1996).

Substituting (1.158) into (1.151), we obtain an ordinary differential second-order equation for the representative

$$y^m\left[\xi^2 y'' + \frac{2(n+2)}{m}\xi y' + \frac{(n+2)(n+2-m)}{m^2}y\right] - \gamma\xi y' - \frac{y}{m} = 0$$

which can be characterised as an equation for a non-linear oscillator with dissipation (if γ is positive).

The boundary conditions are determined in the following manner. It is evident that the accretion rate through the inner edge of the ring can be considered to equal zero. Thus, at the inner boundary h_{in} (corresponding to $\xi = 1$), both the function $F(h_{\mathrm{in}},t)$ and its derivative $\partial F(h_{\mathrm{in}},t)/\partial h$ must vanish (cf. (1.125)). Otherwise, a δ-source (sink) appears with the substitution into Eq. (1.151). Consequently, we have two conditions:

$$y(1) = y'(1) = 0\,.$$

Another condition follows from the requirement that all physical quantities remain finite at time $t = 0$ (when the 'tounge' reaches the accreting centre), at any finite radius. It follows from (1.158) that $F(h,t)$ does not diverge as $t \to 0$ and $h \neq 0$ only if

$$y(\xi = \infty) = 0\,.$$

Thus, the solution of the second-order equation must satisfy three conditions, which is possible only for a specific value of γ.

Let us investigate qualitatively the equation for the representative function. For this we turn to the variable $x = \ln\xi + C$ (substituting C will not affect the resulting system of Eq. (1.159), but is important for adjustment of the solutions). The derivative with respect to x will be denoted by a dot. We write the resulting system of two equations of the first order:

$$\dot{y} = p\,,$$

$$\dot{p} = \frac{y^{1-m}}{m} + \gamma y^{-m} p - \frac{(n+2)(n+2-m)}{m^2}y - \frac{2n+4-m}{m}p\,. \tag{1.159}$$

We are interested in the solution which leaves the origin of the plane (p, y) at $\xi = 1$ and returns there at $\xi = \infty$. For $y \ll 1$ and $p \ll 1$, the system (1.159) has asymptotic solutions of the form

$$p = \frac{\gamma}{1-m} y^{1-m}; \qquad y = \left(\frac{\gamma m}{1-m} \ln \xi\right)^{1/m}, \tag{1.160}$$

$$p = -\frac{y}{\gamma m}; \qquad y = \xi^{-\frac{1}{\gamma m}}. \tag{1.161}$$

The functions (1.160) give asymptotics when $\xi \to 1$ and (1.161) when $\xi \to \infty$, respectively. The phase trajectories of the solutions to the equations are shown in Fig. 1.15 for four values of γ.

Each point for which $\dot{y} = 0$ and $\dot{p} = 0$ is a singular point. There is a stable focus in the phase plane with coordinates

$$p = 0, \qquad y_0 = \left[\frac{m}{(n+2)(n+2-m)}\right]^{1/m}.$$

For a certain $\gamma_{\rm cr}$, there exists a closed solution (Fig. 1.15b). Numerical investigation shows that for $m = 2/5$, $n = 6/5$ (the case $\varkappa_{\rm T} \gg \varkappa_{\rm ff}$), the sought after value is $\gamma_{\rm cr} \approx 0.595$, and for $m = 3/10$, $n = 4/5$ ($\varkappa_{\rm ff} \gg \varkappa_{\rm T}$), it is $\gamma_{\rm cr} \approx 0.696$. The phase trajectories are rearranged for some γ_+ and the stability of the focus changes (Fig. 1.15d).

Thus, the inner boundary of the disc moves towards the centre according to the law: $h_{\rm in} = A(-t)^{\gamma_{\rm cr}}$ (see Figs. 1.16 and 1.17). As follows from (1.161), the asymptotic solution of the initial equation (1.151) for $\xi \to \infty$ (i.e. for $t \to 0$, when the 'tongue' reaches the accreting centre), has the form

$$F = \frac{h^{\frac{n+2}{m}}}{(-Dt)^{\frac{1}{m}}} \left[\frac{A(-t)^{\gamma}}{h}\right]^{\frac{1}{\gamma m}} = \frac{A^{\frac{1}{\gamma_{\rm cr} m}} h^{\frac{n+2}{m} - \frac{1}{\gamma_{\rm cr} m}}}{D^{1/m}}. \tag{1.162}$$

We note that for large h, the profile $F(h, t)$ does not change with time during the 'tongue' formation stage. By 'sewing' the obtained self-similar solution and the initial profile $F_0(h)$ near the radius where the material was ejected at time $(-t_0)$, we may also determine the constant A. Within a dimensionless factor, we have from (1.162)

$$A = F_0^{\gamma_{\rm cr} m} D^{\gamma_{\rm cr}} / h_0^{\gamma_{\rm cr}(n+2)-1},$$

where $h_0 = \sqrt{G M r_0}$ is determined by the initial radius of the ring r_0.

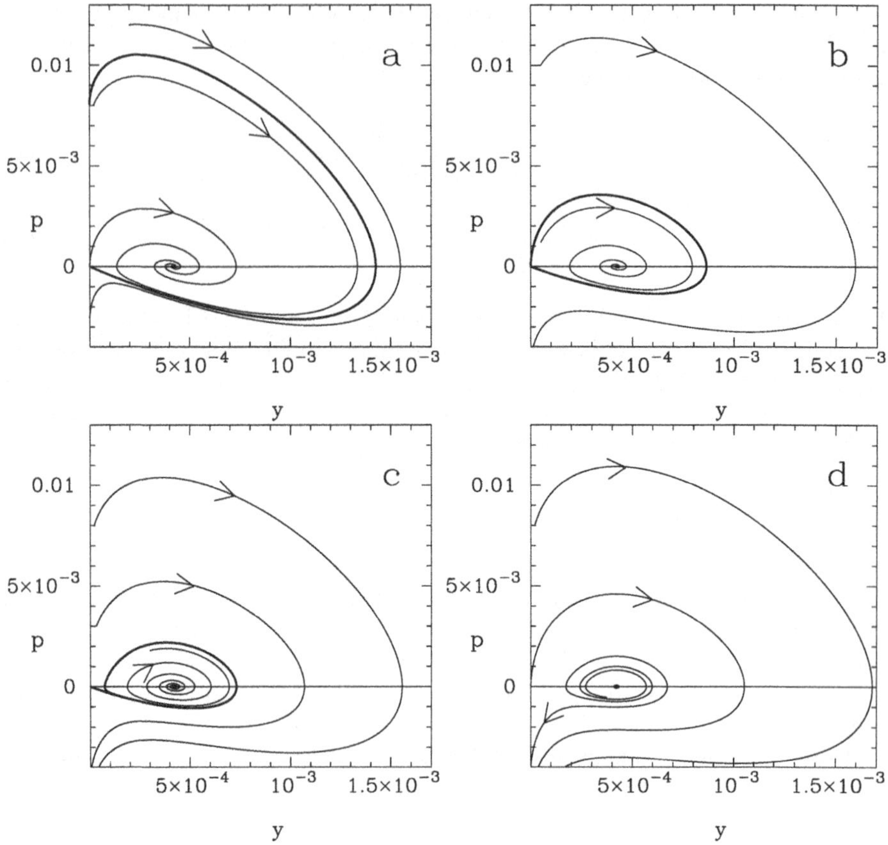

Fig. 1.15 Phase portrait of the system of Eq. (1.165) for different values γ. The *arrows* indicate the direction of change in ξ from 1 to ∞ (x from C to ∞). (**a**) For $\gamma < \gamma_{cr}$ the solution inside the separatrix, shown by the *bold curve*, reaches the stationary point (focus) on the horizontal axis $(0, y_0)$. (**b**) A closed solution is found for $\gamma = \gamma_{cr}$ and coincides with the separatrix. The separatrix at the same time forms a limit cycle of solutions in the region bounded by it, for $x \to -\infty$. (**c**) For $\gamma_{cr} < \gamma < \gamma_+$, the separatrix (*bold curve*) is gradually compressed. (**d**) For $\gamma = \gamma_+$ it is moving towards the point $(0, y_0)$

1.6.6.2 Quasi-Stationary Stage with Increasing Accretion

We seek a solution to (1.151) in the form:

$$F = \frac{h^{\frac{n+2}{m}}}{(D\,t)^{1/m}}\,y(\xi)\,, \qquad 0 \le \xi = \frac{h}{A\,t^{\beta}} \le \infty. \tag{1.163}$$

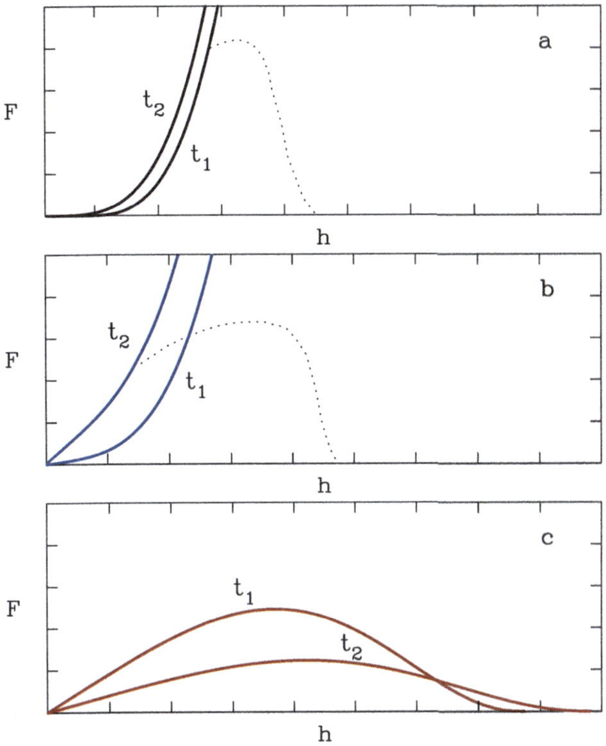

Fig. 1.16 Calculated profile $F(h)$ in the Lyubarski–Shakura solution at three stages of self-similar evolution: (**a**) formation of the 'tongue', $t_2/t_1 = 1/4$ (t is negative and approaches zero); (**b**) quasi-stationary accretion, $t_2/t_1 = 3$ (t is now positive); (**c**) accretion decay, $t_2/t_1 = 2$. The quantities F and h are normalised to arbitrary values. The *dotted lines* in the two upper panels give the symbolic dependence of $F(h)$ for regions where the (unknown) solution is non self-similar. The calculation is performed for opacity parameters $m = 2/5$, $n = 6/5$

The time t is now positive. Substituting (1.163) into (1.151), we obtain the equation for the representative function:

$$y^m \left[\xi^2 y'' + \frac{2(n+2)}{m}\xi y' + \frac{(n+2)(n+2-m)}{m^2} y\right] + \beta \xi y' + \frac{y}{m} = 0 \tag{1.164}$$

or a system of two equations

$$\dot{y} = p\,;$$

$$\dot{p} = -\frac{y^{1-m}}{m} - \beta\, y^{-m}\, p - \frac{(n+2)\,(n+2-m)}{m^2}\, y - \frac{2n+4-m}{m}\, p \tag{1.165}$$

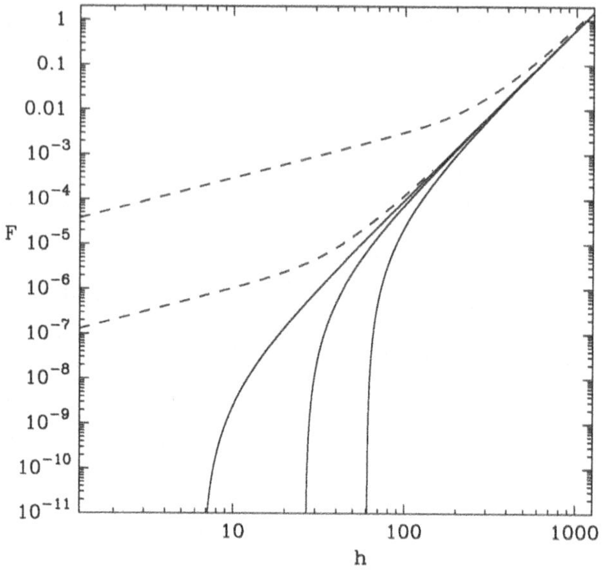

Fig. 1.17 Calculated profiles $F(h)$ in the Lyubarski–Shakura solution at the stages of 'tongue' formation (*solid line*) and quasi-stationary accretion (*dashed line*). At the first stage, we see the movement of the inner edge of the 'tongue' towards the centre. In the second stage, we can see how the zone of quasi-stationary accretion expands with time ($F \propto h$). The accretion rate increases with time from the lowest curve to the top. The quantities F and h are normalised to characteristic values

As $\xi \to \infty$, the asymptotic solution of this system has the form

$$p = -\frac{y}{\beta m}; \qquad y = \xi^{-\frac{1}{\beta m}}; \tag{1.166}$$

we notice that the main contribution comes from the two last terms of (1.164). Hence it follows that only if $\beta = \gamma_{\rm cr}$, the distribution $F(h, t)$ is the same as that at the preceding stage (1.162). Thus, the self-similarity index remains as before. For $\xi \to 0$ (at very large times t or at the accreting centre), there are two asymptotic solutions. (Now the main contribution comes from the terms in square brackets in (1.164)).

$$p = -\frac{n+2}{m} y; \qquad y = \xi^{-\frac{n+2}{m}}, \tag{1.167}$$

$$p = -\frac{n+2-m}{m} y; \qquad y = \xi^{-\frac{n+2-m}{m}}. \tag{1.168}$$

The first corresponds to $(\partial F/\partial h)_{h\to 0} = 0$, i.e. the solution without a material sink ($\dot{M}_{h\to 0} \to 0$), while the second corresponds to an accretion rate, radially constant at small h. Near the gravitating centre, the accretion rate depends on time according to

$$|\dot{M}| = \left|\frac{\partial F}{\partial h}\right|_{h\to 0} \approx \frac{(At^{\beta})^{\frac{n+2-m}{m}}}{(Dt)^{1/m}} = \frac{A^{\frac{n+2-m}{m}}\, t^{\frac{\gamma_{\rm cr}(n+2-m)-1}{m}}}{D^{1/m}} . \tag{1.169}$$

It is this solution with a material sink that describes the second accretion stage in our case (see Fig. 1.16b). We have $|\dot{M}| \propto t^{1.67}$ for $\varkappa_{\rm T} \gg \varkappa_{\rm ff}$ while the accretion rate increases as $|\dot{M}| \propto t^{2.47}$ for $\varkappa_{\rm ff} \gg \varkappa_{\rm T}$.

If we introduce the notation

$$\dot{M}_0 = \frac{F_0}{h_0}, \qquad \tau = \frac{h_0^{n+2}}{F_0^m\, D} \qquad \text{or} \qquad \tau = \frac{h_0^{n+2-m}}{\dot{M}_0^m\, D} ,$$

the accretion rate $\dot{M}(t)$ onto the gravitating centre during the quasi-stationary stage can be expressed in terms of the accretion rate $\dot{M}_0$, determined by the initial value of the viscous torque F_0 acting on the ring of matter with specific angular momentum h_0:

$$\dot{M} = \dot{M}_0 \left(\frac{t}{\tau}\right)^{\frac{\gamma_{\rm cr}(n+2-m)-1}{m}} \qquad \text{at} \quad t > \tau .$$

1.6.6.3 Accretion Decay Stage: Spreading of the Disc

We again seek a solution to (1.151) in the form (1.163), but now the variable ξ varies within the limits $0 \le \xi \le 1$ ($\xi = 1$ corresponds to the outer radius of the disc $r_{\rm out}$ or the specific angular momentum $h_{\rm out} = \sqrt{G\,M r_{\rm out}(t)}$). Thus, the solution for this stage is described, as before, by (1.164) or the equivalent system (1.165) with the boundary conditions $y(0) = y(1) = y'(1) = 0$. The value of the self-similar index β is now found from the law of conservation of the total angular momentum of the material in the disc. Indeed, if the ring was initially located at a radius r_0, much greater than the radius of the innermost stable orbit, then the quantity

$$K = 2\pi \int_0^{r_{\rm out}} \Sigma\, h\, r\, {\rm d}r = const . \tag{1.170}$$

is conserved during the accretion process. Substituting (1.163) into (1.170), with the use of the relationship (1.153) between Σ_0 and F, we obtain

$$K = \frac{1}{(1-m)\,D} \int_0^{h_{\rm out}} F^{1-m}\, h^{n+1} {\rm d}h = \frac{A^{\frac{n+2}{m}}\, t^{\frac{\beta(n+2)}{m}}}{(1-m)\,D^{1/m}\, t^{\frac{1-m}{m}}} \int_0^1 y^{1-m}(\xi)\, \xi^{\frac{n+2-m}{m}}\, {\rm d}\xi . \tag{1.171}$$

From the condition $\partial K/\partial t = 0$, we obtain $\beta = (1-m)/(n+2)$. Moreover, the expression (1.171) gives the exact relation for the constant A. For this β, the required solution to the equation for the representative function (1.164) can be found in explicit form. The method for solution of the non-linear ordinary second order differential equation (1.164), or the equivalent system of first order (1.165), is analogous to the solution of similar equations arising in heat propagation problems with temperature dependent thermal conductivity (Zeldovich and Kompaneets 1950). Since Eqs. (1.165) contain the variable x only as a differential, the order is lowered by introducing $p(y) = \mathrm{d}y/\mathrm{d}x$ as a new unknown function of the variable y

$$y^m \left[p\,\frac{\mathrm{d}p}{\mathrm{d}y} + \frac{2n+4-m}{m}\,p + \frac{(n+2)(n+2-m)}{m^2}\,y \right] + \beta\, p + \frac{y}{m} = 0\,.$$

It is then convenient to introduce function $Z(y) = p(y) + y\,(n+2-m)/m$:

$$y^m \left[\left(Z - \frac{n+2-m}{m}\,y \right) \frac{\mathrm{d}Z}{\mathrm{d}y} + \frac{n+2}{m}\,Z \right] + \beta \left(Z - \frac{n+2-m}{m}\,y \right) + \frac{y}{m} = 0\,.$$

We seek a solution of the form $Z(y) = B\,y^{1-m}$. Collecting the coefficients of powers of y^{1-m} in the last equation, we obtain $B = -\beta/(1-m)$. After substitution of $Z(y) = -\dfrac{\beta}{1-m}\,y^{1-m}$, the equation becomes a linear algebraic equation with respect to y. The left part of the equation vanishes for $\beta = (1-m)/(n+2)$. On the other hand, the equality of β to this value is a necessary condition for the existence of a self-similar solution at the stage of accretion decay (which follows from the condition $\partial K/\partial t = 0$). Thus,

$$p = -\frac{y^{1-m}}{n+2} - \frac{n+2-m}{m}\,y$$

is a particular solution satisfying the boundary condition $p(y=0) = \left.\dfrac{\mathrm{d}y}{\mathrm{d}x}\right|_{x=0} = 0$. Integrating this expression is elementary, and with the boundary condition $y(\xi = 1) = 0$, the solution can be written as

$$y(\xi) = \left[\frac{m}{(n+2)\,(n+2-m)} \right]^{1/m} \left(\frac{1}{\xi^{n+2-m}} - 1 \right)^{1/m}. \tag{1.172}$$

This solution implies that the integral on the right-hand side of (1.171), which is an Euler integral of the first kind, is reduced to the beta-function $B\left(\dfrac{n+3-m}{n+2-m}, \dfrac{1}{m}\right)$ with some coefficient, and the solution of the key equation (1.151) at the decay stage finally takes the form:

$$F = \frac{A^{\frac{n+2}{m}}}{D^{1/m}} \left[\frac{m}{(n+2)(n+2-m)}\right]^{1/m} \frac{\xi\,(1-\xi^{n+2-m})^{1/m}}{t} =$$

$$= \frac{K\,m\,(1-m)}{(n+2)\,B\left(\frac{n+3-m}{n+2-m},\,\frac{1}{m}\right)}\;\frac{\xi\,(1-\xi^{n+2-m})^{1/m}}{t}\,.$$

The accretion rate decays according to:

$$\dot{M} = \left|\frac{\partial F}{\partial h}\right|_{h\to 0} = \frac{A^{\frac{n+2-m}{m}}}{D^{1/m}} \left[\frac{m}{(n+2)(n+2-m)}\right]^{1/m} t^{-\frac{n+3-m}{n+2}}\,.$$

The exponent in the time dependence of $\dot{M}(t)$ can also be expressed through the exponents a and b, appearing in the expression $\nu_{\rm t} \propto \Sigma^a\, r^b$; it equals then $1 + 1/(5\,a - 2\,b + 4)$. For $\varkappa_{\rm T} \gg \varkappa_{\rm ff}$ we have $\dot{M} \propto t^{-19/16}$ and $\dot{M} \propto t^{-5/4}$ for $\varkappa_{\rm ff} \gg \varkappa_{\rm T}$.

At both the quasi-stationary stage and the decay stage, the total energy release in the disc is determined primarily by the release of gravitational energy in the inner regions of the disc. The disc luminosity is equal to $\eta_{\rm accr}\,\dot{M}_{\rm in}\,c^2$, where $\eta_{\rm accr}$ is the efficiency of energy release. During the 'tongue' stage, the energy release depends largely on the initial distribution $F(h)$ since the heat flux from a unit area of the disc surface $\propto F/h^7$ (cf. (1.75)).

The presented solutions describe processes in real accretion discs to some approximation. The assumption of constant opacity (or constant coefficients m and n) does not hold for the entire disc throughout its full evolution. To completely take into account changes in opacity, numerical calculations are required using tabulated values of the opacity coefficients as functions of temperature and density. On the other hand, the opacity coefficient has little effect on the presented solution, as it appears in D as a factor raised to a very small power. It should be noted that the opacity changes particularly strongly in regions with a variable degree of ionization.

1.6.7 Solution for α-Disc in a Binary System

As we have seen, the viscous evolution of a ring of matter eventually enters a stage of unconstrained spreading, when parts of the matter in the disc acquire a high angular momentum and reaches further and further from the centre. In binary systems, such spreading of the disc cannot continue indefinitely due to the gravitational effects of the companion star. Tidal forces from the companion star force the disc to be limited to within a certain radius from the centre inside the Roche lobe.

Lipunova and Shakura (2000) found a solution describing the evolution of an α-disc in a binary system. The obtained solution was used to model the optical and X-ray lightcurves of the X-ray novae A 0620-00 and GU Mus 1124-68 during the decline after the peak of their outbursts. As a result, new constrains on the turbulence parameter α were found (Lipunova and Shakura 2002; Suleimanov et al. 2008).

The angular momentum in the region of the outer radius is transferred from the matter in the disc into orbital motion of the binary system. Papaloizou and Pringle (1977) showed that the tidal truncation radius is on average ~ 0.9 times that of the Roche lobe. This radius is close to that of the last non-intersecting periodic orbit calculated for a three-body problem (Paczynski 1977). Numerical calculations have shown that the tidal stress tensor is important only in a rather narrow ring close to the outer radius. Significant perturbations in this region lead to the formation of strong spiral shock waves (Pringle 1991; Ichikawa and Osaki 1994; Hameury and Lasota 2005).

Since the outflow of angular momentum takes place in a narrow region close to the tidal truncation radius, we may choose not to examine this region in detail, considering it simply a δ-type channel. The function F grows as $r^{1/2}$ at radii much smaller than the tidal truncation radius. There, the stationary disc behaves according to the standard model, not 'noticing' the outer boundary conditions, and the dependence of the viscous torque on the radius is described by expression (1.124).

We also assume in the framework of the mathematical problem that the outer radius of the disc does not change, and that the rate of inflow of matter to the outer disc is negligible. The assumption that the outer radius remains unchanged is valid for transient activity phenomena during outbursts in some types of close binary systems. Numerical calculations, in which long-term evolution of non-stationary discs in binary systems (X-ray and dwarf novae) is modelled (DIM, Disc Instability Models), take into account the variability of the outer radius of the disc (Lasota 2001). During powerful flares in X-ray novae, when the brightness of the source may increase with up to a million times, the accretion rate inside the disc may be considered to be much higher than the rate of inflow of matter from the companion star. This corresponds to the vanishing of the derivative $F(h,t)$ with respect to h at the outer radius.

A solution to the basic equation of non-stationary accretion (1.151) for a disc with a constant outer radius can be found using the method of separation of variables:

$$F(h,t) = F(t) \times f_F(h/h_0)\,. \tag{1.173}$$

The quantity $h_0 = (G\,M\,r_{\rm out})^{1/2}$ equals the specific angular momentum at the outer edge of the disc. The above mentioned properties of the viscous torque lead to the following conditions at the outer radius:

$$f_F(1) = 1, \quad f_F'(1) = 0\,, \tag{1.174}$$

the first of which is normalising, and the other expresses the fact that the viscous torque has a maximum at the immediate vicinity of the disc outer radius (Fig. 1.18). This is equivalent to the condition of zero accretion rate at $r_{\rm out}$. It can also be said that the radial component of the velocity in the disc is zero at $r_{\rm out}$. A similar

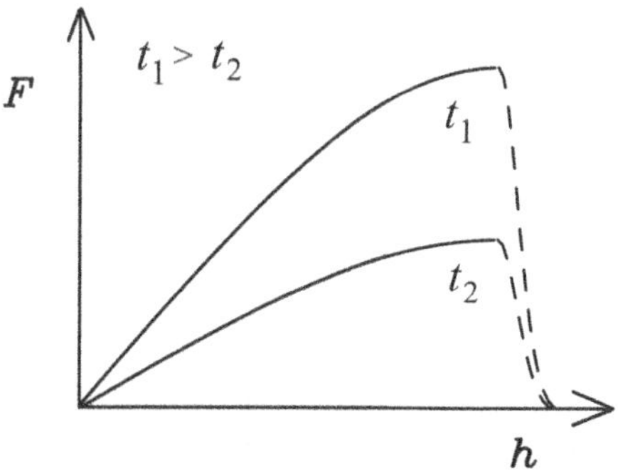

Fig. 1.18 Moment of viscous forces F as a function of the specific angular momentum h in an accretion disc in a binary system at two moments in time. Time t_2 is later than t_1. The accretion rate declines with time

approach was used, for example, by Pringle (1991) in a study of a disc surrounding a binary system. At the inner edge of such a disc, the angular momentum is transferred from the binary system into the disc, and the stars gradually move closer to each other.

Thus, tidal interactions determine the specific boundary conditions at the outer edge of the disc, thereby not changing the form of the equation that we solve (1.123) or (1.151).

Naturally, in reality the inner edge of the disc $r_{\rm in} \neq 0$. In many situations, however, $r_{\rm in}/r_{\rm out} \ll 1$.

Using $h \equiv h_{\rm K}$, we obtain from (1.125):

$$\dot{M}(h,t) = f_F'(h/h_0)F(t)/h_0 \,. \tag{1.175}$$

Substituting the product of the functions into the equation for non-stationary accretion (1.151), we obtain the time-dependent part of the solution, which gives the following asymptotic for the disc evolution after the peak of the outburst:

$$F(t) = \left(\frac{h_0^{n+2}}{\lambda\, m\, D\,(t + t_0)}\right)^{1/m} \,. \tag{1.176}$$

Here, D is a dimensional constant (1.154) that may be obtained by solving the equations of vertical structure, λ is a separation constant, which may be found from the boundary conditions imposed on $f_F(h/h_o)$, and t_0 is the integration constant in units of time.

The law of accretion rate change is written as:

$$\dot{M}(t) = \dot{M}(0)\,(1 + t/t_0)^{-1/m} \,, \tag{1.177}$$

where $\dot{M}(0)$ is the accretion rate at a certain moment in time $t = 0$, which can be chosen as any time at the stage of declining accretion. Then, parameter t_0 of the solution is

$$t_0 = \frac{h_0^{n+2}}{\lambda\, m\, D\, F^m(0)} \,,$$

where $F(0)$ is the value of $F(h, t = 0)$ at the outer radius $r_{\rm out}$. Substituting expression (1.152) for D and taking into account that $F_{\rm out} = 3\pi\, h_0\, \nu_{\rm out}\, \Sigma_0$, we get:

$$t_0 = \frac{4}{3\lambda\, a}\, \frac{r_{\rm out}^2}{\nu_{\rm out}(t=0)}\,, \tag{1.178}$$

where a is the power of Σ in the relation $\nu_{\rm t} \propto \Sigma^a\, r^b$.

After a separation of variables in the basic equation, we obtain a non-linear equation for $f_F(\xi)$. It constitutes a particular case of the general Emden-Fowler equation (Zaitsev and Polyanin 2012)

$$\frac{d^2 f_F}{d\xi^2} = -\lambda \xi^n\, f_F^{1-m}\,, \tag{1.179}$$

the solution to which we seek as a polynomial

$$f_F(\xi) = a_0\xi + a_1\xi^k + a_2\xi^l + \ldots\,. \tag{1.180}$$

Substituting $f_F(\xi)$ with the polynome into (1.179), we obtain for the second and the third term:

$$\begin{aligned} k &= 3 + n - m, \quad a_1 = \frac{-\lambda a_0^{1-m}}{k(k-1)}\,, \\ l &= 2k - 1, \quad a_2 = \frac{-\lambda a_0^{-m} a_1}{l(l-1)}(1-m)\,. \end{aligned} \tag{1.181}$$

Table 1.5 gives the values for the constants a_0 and λ, derived from the boundary conditions (1.174) on $f_F(\xi)$ in the opacity regimes of pure scattering and pure absorption as well as for an approximation based on the OPAL numerical calculations of opacity (Iglesias and Rogers 1996). The corresponding functions f_F are shown in Fig. 1.19. The OPAL case turns out to be effectively somewhere in the middle.

Table 1.5 Parameters of the analytical solution, presented by (1.176), (1.177), and (1.180), for the truncated α-disc decay

	m	n	λ	a_0	a_1	a_2	k	l
$\varkappa_{\rm T} \gg \varkappa_{\rm ff}$	2/5	6/5	3.482	1.376	−0.396	0.019	3.8	6.6
$\varkappa_{\rm ff} \gg \varkappa_{\rm T}$	3/10	4/5	3.137	1.430	−0.460	0.030	3.5	6.0
OPAL	1/3	1	3.319	1.400	−0.425	0.025	11/3	19/3

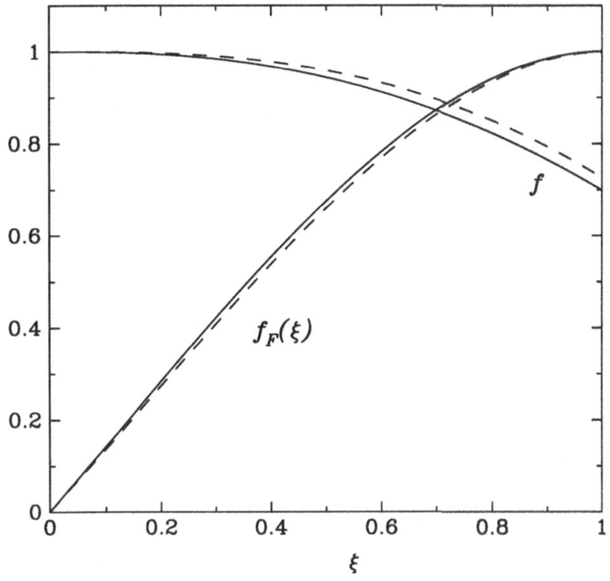

Fig. 1.19 The solution $f_F(\xi)$ for two cases: $\varkappa_{ff} \gg \varkappa_T$ (*solid line*) and $\varkappa_T \gg \varkappa_{ff}$ (dashes). The plot also shows the function $f(r)$, calculated using (1.102), that is included in the expression for calculating radial dependencies of physical parameters (Sect. 1.5.3). The accretion rate practically does not change with radius in the region, where $f \approx 1$. The variable $\xi = h/h_o$, where h_o is the specific angular momentum at the outer radius

The value a_0, included in the expression for the accretion rate

$$\dot{M}_{in} = \frac{F_{max}}{h_{max}} a_0 \ ,$$

can also be calculated for the self-similar solution by Lyubarski & Shakura during the concluding stage of disc decay (Sect. 1.6.6.3). Omitting the details, we only mention that in an unconstrained disc, h_{max} likewise corresponds to the maximum torque F_{max}. It is remarkable that the values of a_0 differ only by 2% between a constrained and an unconstrained disc. This means that the profile $F(h)$,in the region of the disc where $F(h)$ increases, is practically independent of the conditions outside this region.

1.6.7.1 Radial Dependencies for a Non-stationary Disc in a Binary System

Let us find the expressions for the evolution of physical parameters in the disc, using Eqs. (1.99), (1.153), (1.154), and (1.176).

Note that the relations (1.104) and (1.110) contain another function, $f(r)$ without an index. Function $f(r)$ is determined by relation (1.102). In the case of a stationary disc, we have $f_F = \xi\, f$. In the case of a disc with a radially variable accretion rate, for example a non-stationary disc, $f_F = a_0\, \xi\, f(r)$ (see Fig. 1.19).

Below, we derive expressions for the diffusion parameter D, surface density Σ_o, temperature in the central disc plane T_c, relative half-thickness z_0/r, and optical depth τ. We use for the mass of the compact object $m_x = M/M_\odot$. The values $\Pi_{1..4}$ should be chosen according to the appropriate opacity regime. The parameter t_0 depends on the accretion rate at $t = 0$:

$$t_0 = \frac{h_0^{n+2-m}\, a_0^m}{\lambda\, m\, D\, \dot{M}_{\rm in}^m(t=0)}\,. \tag{1.182}$$

It is important to remember that t_0 depends on the type of opacity.

Scattering-Dominated Opacity Regime ($\varkappa_{\rm T} \gg \varkappa_{\rm ff}$)

Substituting the numerical values of the constants into (1.154), we obtain:

$$D\ [{\rm cm}^{28/5}/{\rm g}^{2/5}/{\rm s}^{17/5}] = 1.40 \times 10^{38}\, \alpha^{4/5}\, m_{\rm x}^{6/5} \left(\frac{\mu}{0.5}\right)^{-4/5} \Pi_\Sigma^{-1}\, \varkappa_{\rm T}^{1/5}\,, \tag{1.183}$$

with the help of which we re-write (1.182):

$$t_0\ [{\rm days}] = 24.12\,\alpha^{-4/5} \left(\frac{r_{\rm out}}{R_\odot}\right)^{7/5} \left(\frac{\dot{M}_{\rm in}(t=0)}{10^{18}{\rm g/s}}\right)^{-2/5} m_{\rm x}^{1/5} \left(\frac{\mu}{0.5}\right)^{4/5} \Pi_\Sigma\, \varkappa_{\rm T}^{-1/5}\,, \tag{1.184}$$

where (1.95) determines $\varkappa_{\rm T}$. Substituting the combination $M_{\rm in}\, t_0^{1/m}$ from (1.184) into the expression for the declining accretion rate $\dot{M}(t) = \dot{M}(0)\,(1 + t/t_0)^{-1/m}$, and further the accretion rate and the function $f(r) = f_F/(a_0\sqrt{r/r_{\rm out}})$ into the radial dependencies (1.104) in zone B, we obtain the radial dependencies of the physical parameters in a non-stationary α-disc:

$$\Sigma_0\ [{\rm g/cm^2}] = 2.2 \times 10^2\ \alpha^{-2}\, m_{\rm x}^{1/2} \left(\frac{t+t_0}{10^{\rm d}}\right)^{-3/2} \left(\frac{r_{\rm out}}{R_\odot}\right)^{3/2} \left(\frac{r}{r_{\rm out}}\right)^{-9/10} f_F^{3/5} \times \\ \times \left(\frac{\mu}{0.5}\right)^2 \varkappa_{\rm T}^{-1/2}\, \Pi_\Sigma^{5/2}\,, \tag{1.185}$$

$$T_{\rm c}\ [{\rm K}] = 1.8 \times 10^4\,\alpha^{-1}\, m_{\rm x}^{1/2} \left(\frac{t+t_0}{10^{\rm d}}\right)^{-1} \left(\frac{r_{\rm out}}{R_\odot}\right)^{1/2} \left(\frac{r}{r_{\rm out}}\right)^{-11/10} f_F^{2/5}\, \frac{\mu}{0.5}\, \Pi_3\,, \tag{1.186}$$

$$\frac{z_0}{r} = 0.04\,\alpha^{-1/2}\, m_{\rm x}^{-1/4} \left(\frac{t+t_0}{10^{\rm d}}\right)^{-1/2} \left(\frac{r_{\rm out}}{R_\odot}\right)^{3/4} \left(\frac{r}{r_{\rm out}}\right)^{-1/20} f_F^{1/5}\, (\Pi_1\, \Pi_3)^{1/2}\,, \tag{1.187}$$

The dimensionless constants Π_Σ, $\Pi_{1..4}$ were introduced in Sect. 1.5.2 where we considered the vertical structure of the α-disc. Their interrelations are determined by expression (1.105), in particular $\Pi_3 = \Pi_T\,\Pi_\Sigma$ and $(\Pi_1\,\Pi_3)^{1/2} = \Pi_z\,\Pi_\Sigma^{1/2}$, and their values can be found in Table 1.1 and in Fig. 1.8. The effective optical depth of the disc can be estimated with the help of τ^*:

$$\tau^* = \left(\frac{\varkappa_{0,\mathrm{T}}\,\varkappa_{0,\mathrm{ff}}\,\rho_\mathrm{c}}{T_\mathrm{c}^{7/2}}\right)^{1/2} \Sigma_0 = 1.5\times10^2\,\alpha^{-1}\left(\frac{t+t_0}{10^\mathrm{d}}\right)^{-1/4}\left(\frac{r_\mathrm{out}}{R_\odot}\right)^{1/2}\times$$
$$\times\left(\frac{r}{r_\mathrm{out}}\right)^{1/10} f_F^{1/10}\left(\frac{\mu}{0.5}\right)^{5/4}\varkappa_\mathrm{T}^{-1/4}\left(\frac{\varkappa_{0,\mathrm{ff}}}{10^{22}}\right)^{1/2}\left(\frac{\Pi_3^4\,\Pi_4^3}{\Pi_1\,\Pi_2^2}\right)^{1/4},$$

where the units of $\varkappa_{0,\mathrm{ff}}$ are $[\mathrm{cm^2K^{7/2}/g^2}]$.

Absorption-Dominated Opacity Regime ($\varkappa_\mathrm{ff} \gg \varkappa_\mathrm{T}$)

This regime is established at lower temperatures and densities. In a similar fashion, we obtain:

$$D\,[\mathrm{cm^5/g^{3/10}/s^{16/5}}] = 2.41\times10^{34}\alpha^{4/5}\,m_\mathrm{x}\left(\frac{\mu}{0.5}\right)^{-3/4}\Pi_\Sigma^{-1}\left(\frac{\varkappa_{0,\mathrm{ff}}}{10^{22}}\right)^{1/10}, \tag{1.188}$$

$$t_0\,[\mathrm{days}] = 36.41\,\alpha^{-4/5}\left(\frac{r_\mathrm{out}}{R_\odot}\right)^{5/4}\left(\frac{\dot M_\mathrm{in}(t=0)}{10^{18}\mathrm{g/s}}\right)^{-3/10} m_\mathrm{x}^{1/4}\times \tag{1.189}$$
$$\times\left(\frac{\mu}{0.5}\right)^{3/4}\Pi_\Sigma\left(\frac{\varkappa_{0,\mathrm{ff}}}{10^{22}}\right)^{-1/10}.$$

The value $\varkappa_{0,\mathrm{ff}}\,[\mathrm{cm^2\,K^{7/2}/g^2}]$ can be taken from (1.97) or (1.96).

$$\Sigma_0\,[\mathrm{g/cm^2}] = 9.9\times10^2\,\alpha^{-8/3}\,m_\mathrm{x}^{5/6}\left(\frac{t+t_0}{10^\mathrm{d}}\right)^{-7/3}\left(\frac{r_\mathrm{out}}{R_\odot}\right)^{13/6}\left(\frac{r}{r_\mathrm{out}}\right)^{-11/10}\times$$
$$\times f_F^{7/10}\left(\frac{\mu}{0.5}\right)^{5/2}\left(\frac{\varkappa_{0,\mathrm{ff}}}{10^{22}}\right)^{-1/3}\Pi_\Sigma^{10/3}, \tag{1.190}$$

$$T_\mathrm{c}\,[\mathrm{K}] = 3.1\times10^4\alpha^{-1}m_\mathrm{x}^{1/2}\left(\frac{t+t_0}{10^\mathrm{d}}\right)^{-1}\left(\frac{r_\mathrm{out}}{R_\odot}\right)^{1/2}\left(\frac{r}{r_\mathrm{out}}\right)^{-9/10} f_F^{3/10}\left(\frac{\mu}{0.5}\right)\Pi_3, \tag{1.191}$$

$$\frac{z_0}{r} = 0.05\,\alpha^{-1/2}\,m_\mathrm{x}^{-1/4}\left(\frac{t+t_0}{10^\mathrm{d}}\right)^{-1/2}\left(\frac{r_\mathrm{out}}{R_\odot}\right)^{3/4}\left(\frac{r}{r_\mathrm{out}}\right)^{1/20} f_F^{3/20}\,(\Pi_1\,\Pi_3)^{1/2}. \tag{1.192}$$

The dimensionless coefficients Π_Σ, $\Pi_{1..4}$ were introduced in Sect. 1.5.2, when we considered the vertical structure of the α-disc. Their values can be found in Table 1.2 and Fig. 1.8. We recall that the surface density Σ_o is calculated between the bottom and the top surface of the disc. The full optical depth (for which (1.112) applies in the stationary case) is equal to:

$$\tau = \varkappa_{0,\mathrm{ff}}\, \rho_\mathrm{c}\, T_\mathrm{c}^{-7/2}\, \Sigma_0 = 2.4 \times 10^2\, \alpha^{-4/3}\, m_\mathrm{x}^{1/6} \left(\frac{t+t_0}{10^\mathrm{d}}\right)^{-2/3} \left(\frac{r_\mathrm{out}}{R_\odot}\right)^{5/6} \times \\ \times \left(\frac{r}{r_\mathrm{out}}\right)^{-1/10} f_F^{1/5} \left(\frac{\mu}{0.5}\right)^{3/2} \left(\frac{\varkappa_{0,\mathrm{ff}}}{10^{22}}\right)^{1/3} \left(\frac{\Pi_3^4\, \Pi_4^2}{\Pi_1^{1/2}\, \Pi_2}\right)^{1/3}. \tag{1.193}$$

Luminosity Dependence in an α-Disc with a Constant Outer Radius

In order to calculate the bolometric luminosity of the disc, we assume a quasi-stationary accretion rate $\dot{M}(t) = \dot{M}(0, t)$ (1.175), since the main part of the energy is released at distances from the centre $r \ll r_\mathrm{out}$. The quasi-stationarity is provided by the fact that the characteristic time scale for evolution (viscous time scale) at small radii is much smaller than that at large radii. Figure 1.19 illustrates this behaviour by the fact that the function $f(r)$ becomes approximately constant close to the disc centre.

Substituting t_0 into (1.177), we obtain for the luminosity $L = \eta_\mathrm{accr}\, \dot{M}(t)\, c^2$, where η_accr is the efficiency of accretion:

$$L_\mathrm{T}(t)\ [\mathrm{erg/s}] = 8.1 \times 10^{38}\, \alpha^{-2}\, m_\mathrm{x}^{1/2} \left(\frac{t+t_0}{10^\mathrm{d}}\right)^{-5/2} \left(\frac{r_\mathrm{out}}{R_\odot}\right)^{7/2} \left(\frac{\eta}{0.1}\right) \times \\ \times \left(\frac{\mu}{0.5}\right)^2 \varkappa_\mathrm{T}^{-1/2}\, \Pi_\Sigma^{5/2}, \tag{1.194}$$

if Thomson scattering dominates in the outer parts of the disc, and

$$L_\mathrm{ff}(t)\ [\mathrm{erg/s}] = 6.7 \times 10^{39}\, \alpha^{-8/3}\, m_\mathrm{x}^{5/6} \left(\frac{t+t_0}{10^\mathrm{d}}\right)^{-10/3} \left(\frac{r_\mathrm{out}}{R_\odot}\right)^{25/6} \left(\frac{\eta}{0.1}\right) \times \\ \times \left(\frac{\mu}{0.5}\right)^{5/2} \left(\frac{\varkappa_{0,\mathrm{ff}}}{10^{22}}\right)^{-1/3} \Pi_\Sigma^{10/3}, \tag{1.195}$$

if Kramer's opacity dominates. The quantities t_0 differ between expressions (1.194) and (1.195) and are determined using formulae (1.184) and (1.189), respectively.

Note that the quantities $t_0(\mathrm{T})$ and $t_0(\mathrm{ff})$ in the two regimes are not independent of each other. In a physically consistent model with a transition between the opacity regimes, it is necessary to find an intersection between the two solutions. This may be done by equating the torques F and the surface densities Σ_0 in the two regimes

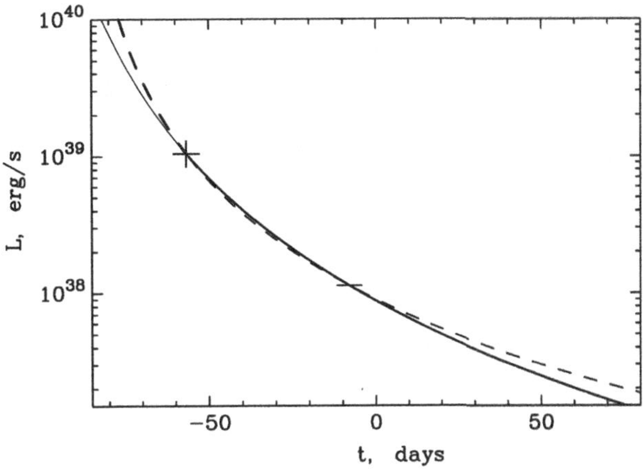

Fig. 1.20 Bolometric disc luminosity L_T and L_{ff} for the parameters $m_x = 3$, $\alpha = 0.3$, $\mu = 0.5$, $r_{out} = R_\odot$. The *dashed line* shows the scattering dominated opacity regime and the *solid line* shows the absorption dominated regime. The transition from the solution in the scattering regime to the solution in the absorption regime is marked with a *cross*. The second intersection of the curves is marked with a *bar*

at radius $r = 0.5\,r_{out}$. These two conditions specify the intersection time itself and the difference between times t_0 in the two regimes. The value of t_0 in one of the regimes is a free parameter, and can be chosen so that $t = 0$ corresponds to a certain accretion rate.

Figure 1.20 shows the bolometric lightcurves for the parameters $\alpha = 0.3$, $m_x = 3$ and $\varkappa_{0,ff} = 6.45 \times 10^{22}\ \mathrm{cm^5\,K^{7/2}/g^2}$ and $\varkappa_T = 0.4\,\mathrm{cm^2/g}$ in the two opacity regimes. Typical values for $\Pi_{1,2,3,4}$ are used. The normalised time in the absorption regime (1.189) $t_0(\mathrm{ff}) \approx 107$ days is obtained from the condition that the accretion rate is $\dot{M} = 10^{18}$ g/s at $t = 0$. Equality between F and Σ in the two different regimes occurs at radius $r/r_{out} = 0.5$ when

$$t + t_0(\mathrm{ff}) = t_{tr} \approx 48^{\mathrm{d}}(m_x/3)^{2/5}\,(\alpha/0.3)^{-4/5}\,(\mu/0.5)^{3/5}\,(r_{out}/R_\odot)^{4/5}\,.$$

The normalised time in the scattering regime can be uniquely determined: $t_0(\mathrm{T}) \approx 90$ days. The intersection of the lightcurves at time $t = t_{tr} - t_0(\mathrm{ff}) \approx -59^{\mathrm{d}}$ is marked with a cross in Fig. 1.20. We can see that there is a smooth transition between the solutions in the two regimes at this time. There is another intersection of the lightcurves at $t \approx -3^{\mathrm{d}}$, which represent a second point where the two functions $F_T(\xi, t + t_0(\mathrm{T})) = F_{ff}(\xi, t + t_0(\mathrm{ff}))$ take on equal values. This intersection exists only in a mathematical sense. The physical conditions in the disc at this moment are such that absorption dominates the opacity, and the values of the physical parameters in the disc, calculated according to (1.183)–(1.193), differ.

Let us not forget that we are working within the framework of the model for a geometrically thin disc with sub-critical accretion. Therefore, the solution considered is applicable only for luminosity below the Eddington value $L_{\rm Edd} \approx 1.4 \times 10^{38}\, m_{\rm x}$ erg/s. Figure 1.20 shows that the evolution of the disc with $L < L_{\rm Edd}$ proceeds almost entirely in the absorption-dominated opacity regime.

When the temperature in the equatorial disc plane $T_{\rm c}$ drops at large radii down to a value of $\sim 3 \times 10^4$ K, the opacity increases strongly due to the onset of recombination in the plasma. The coefficient D significantly changes, and the given solution is no longer applicable. As the mechanism of heat transfer to the surface changes, the vertical structure of the disc readjusts on the characteristic thermal time scale, and conditions arise for the onset of convection. This happens at $t \approx 80^{\rm d}$ for the disc parameters $m_{\rm x} = 3$ and $\alpha = 0.3$ (Fig. 1.20).

Figure 1.21 shows the bolometric lightcurve together with the lightcurves in two X-ray bands from a disc perpendicular to the line of sight at a distance of 1 kpc. The vertical line shows the moment in time after which the bolometric luminosity of the disc becomes lower than $L_{\rm Edd}$. The shape of the lightcurves describes well the exponential decay of the luminosity observed in outbursts of X-ray novae.

Suleimanov et al. (2008) modelled two outbursts of X-ray novae and compared them with observed lightcurves in the X-ray and optical bands (Fig. 1.22). The model included the illumination of the outer parts of the disc by the X-ray flux and its conversion to optical emission. The model also included the effect of distortion of the photon trajectories in the Kerr metric around the black hole (see Fig. 1.23) as well as the presence of an extended disc atmosphere, capable of scattering the X-ray emission at altitudes higher than the hydrodynamic thickness of the disc. As a result, limits on the parameters of discs and binary systems were found. If we know dynamical parameters of binaries from observations (their periods and companion

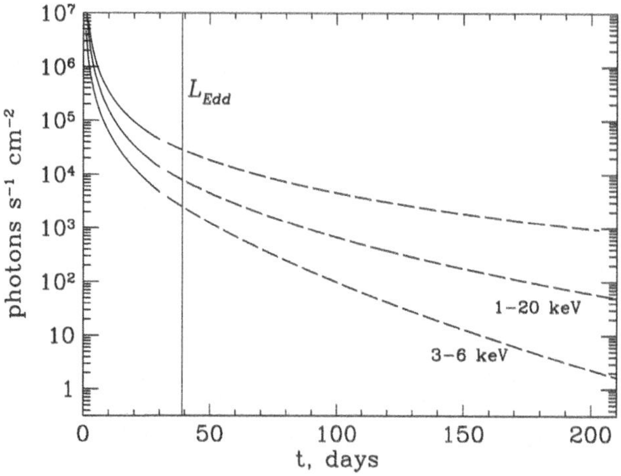

Fig. 1.21 Luminosity of the disc observed from a distance of 1 kpc, for parameters $m_{\rm x} = 3$, $\alpha = 0.3$, $\mu = 0.5$, and $r_{\rm out} = R_{\odot}$. The bolometric lightcurve (top) is shown together with the lightcurves in two X-ray bands, 1–20 keV and 3–6 keV

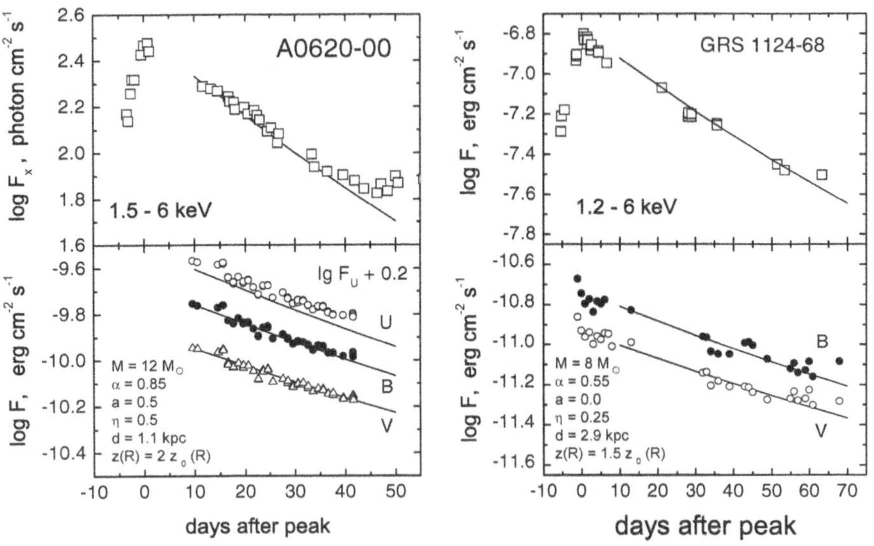

Fig. 1.22 Modelling of outbursts in X-ray novae A 06020-00 (1975) and GU Mus 1124-68 (1991) from Suleimanov et al. (2008). The parameters of the models are shown in the figure. In addition to the notations introduced in the text, we have the following parameters: the dimensionless Kerr parameter a of the black hole, the factor η of conversion of X-rays into optical emission, and the height of the scattering atmosphere $z(r)$

masses), we may find an interval of possible values for the turbulence parameter α. Figure 1.22 shows an example of the modelled lightcurves together with the corresponding parameters of the model.

1.7 Numerical Modelling of Non-stationary Disc Accretion

A numerical scheme, which is described in this section, is implemented in the `FREDDI`[6] code. `FREDDI` is intended for modelling the lightcurves of X-ray novae with fast rise and exponential decay (Lipunova and Malanchev 2017). With the help of `FREDDI` it is possible to describe the time-dependence of the accretion rate onto the black hole $\dot{M}(t)$ and to obtain lightcurves in various energy bands.

[6] http://xray.sai.msu.ru/~malanchev/freddi/.

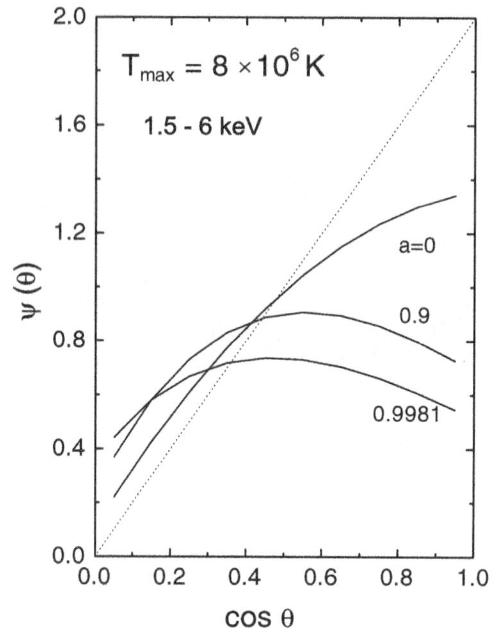

Fig. 1.23 Angular distribution $\Psi(\theta)$ of the intensity from a standard accretion disc around a Kerr black hole, integrated over 1.5–6 keV (from Suleimanov et al. 2008). The angle θ is the angle between the normal to the disc plane and the line of sight. $T_{\max}$ is the maximum effective temperature of the disc. The values of the dimensionless Kerr parameter a are indicated for the curves. The *dotted line* shows the angular distribution for a thin disc in the Newtonian approximation: $\Psi(\theta) = 2\cos(\theta)$. The observed flux can be found as $F = L\,\Psi(\theta)/(4\pi d)^2$, where L is the bolometric luminosity, d is the distance to the disc. The function $\Psi(\theta)$ is calculated using the code of Speith et al. (1995). The effects of limb-darkening are ignored here but are illustrated in Figure 9 in Suleimanov et al. (2007)

1.7.1 Solution to the Equations of Viscous Evolution

Let us examine the equation of viscous evolution of an accretion disc (1.123) obtained earlier in this chapter:

$$\frac{\partial \Sigma_0(h_{\rm K}, t)}{\partial t} = \frac{1}{4\pi}\frac{(GM)^2}{h_{\rm K}^3}\frac{\partial}{\partial h_{\rm K}}\left(\left[\frac{\partial h}{\partial h_{\rm K}}\right]^{-1}\frac{\partial F(h_{\rm K}, t)}{\partial h_{\rm K}}\right), \qquad (1.196)$$

where t is the time, $h(r) = \omega(r)\,r$ is the specific angular momentum in the disc, $h_{\rm K} = \sqrt{GMr}$ is the Keplerian angular momentum, and $\Sigma_0(h, t)$ is the surface density of the disc and $F(h, t)$ the viscous torque, acting on a layer of the disc.

We will consider the case of Keplerian rotation, when $h_{\rm K} = h$. Rotation in a relativistic potential will complicate the computations and place restrictions on the choice of nodes for the radial coordinate h. For a Schwarzschild potential, in the innermost regions of the disc, each following node must be located not further than twice as far from the centre as the previous one.

For a full set-up of the problem of viscous disc evolution, we need to give initial and boundary conditions. In the case of accretion onto a black hole, the boundary condition at the inner disc radius $R_{\rm in}$, corresponding to the innermost stable orbit (3.22), is given as the viscous torque F being equal to zero. If the accretion disc is limited by the magnetosphere of a neutron star or a young star, the inner boundary condition on the value of F is set by the conditions at the magnetospheric boundary. Thus, for a number of cases the inner boundary condition of the problem is a first (Dirichlet) type condition.

The type of outer boundary condition also depends on the astrophysical situation. In a binary system we may assume that angular momentum is removed only by tidal forces from the outer edge of the disc, corresponding to $h_{\rm out}$. Then, together with the assumption that matter flows into the accretion disc only through its outer boundary, we obtain a boundary condition of the second (Neumann) type: $\partial F/\partial h = \dot{M}_{\rm out}(t)$. In the more general case, if we take into account the radial distribution of tidal forces, removal of angular momentum from the disc surface through disc winds, capture of matter at a wide range of radii in the disc, etc., it becomes necessary to include additional terms in the original equation (1.196). If we consider the evolution of an infinite disc, for example a protoplanetary disc or a disc around a supermassive black hole in an active galactic nucleus, then from a mathematical point of view, a boundary condition at infinity is equivalent to the value and the derivative of the torque being equal to zero. However, from the point of view of numerical modelling, we cannot operate with infinite quantities of the specific angular momentum h. We may solve this problem in two ways. Firstly, we may limit the region of study to some value $h_{\rm out}$, to which, during the studied time-interval, no significant amount of matter will be able to reach, and there establish a boundary condition of the torque F being equal to zero. Secondly, we may replace the radial coordinate h with another coordinate, so that the infinite value h equals a finite value of the new coordinate, for example: $1/h$, $1-e^{-h}$ or $\mathrm{arcctg}\, h$. A change of the radial coordinate, however, complicates the original equation, and, as a consequence, place restrictions on the steps between the nodes for the new radial coordinate.

Equation (1.196) is written with respect to two unknown but related functions: $\Sigma_0(h,t)$ and $F(h,t)$. One of these quantities can be obtained for any value of the specific angular momentum h, and for any moment in time t, if the other quantity is known. Earlier in this chapter we studied the cases of linear and power-law relationships between $\Sigma_0(h)$ and $F(h)$, for which analytical solutions to Eq. (1.196) are possible. However, in the general case, the problem (1.196) has to be solved numerically. The problem is more convenient to solve with respect to the function $F(h,t)$, since the boundary conditions are set relative to this function. As we show below, using $F(h,t)$ as the unknown function is more convenient if we find the relationship between Σ_0 and F numerically from the equations of vertical structure. Thus we will express the surface density as a function of the radial coordinate and the torque: $\Sigma_0(F(h,t),\ h)$.

Note that the problem at hand is a specific case of the non-linear diffusion equation. Most often in physics, diffusion equations in which the non-linear diffusion coefficient is contained in the spatial derivative, are studied. As mentioned above, however, in our case it is more convenient to consider the problem with regard to the function $F(h,t)$. Then, the non-linear function $\Sigma_0(F(h,t),\ h)$ stands in the left part of Eq. (1.196). Below we will present a method of solving the equation, in which the non-linearity is included in the time-derivative. This method has a lot in common with the method studied in detail in the classical books on numerical methods, e.g., Press et al. (2002), used in the solution to diffusion equations with the non-linearity in the right part of the equation.

Let us consider the problem of evolution of an accretion disc in a binary system in the Newtonian potential, assuming that the removal of angular momentum is due to tidal forces from the outer edge of the disc only:

$$\begin{cases} \dfrac{\partial \Sigma_0(F(h,t),\ h)}{\partial t} = \dfrac{1}{4\pi}\dfrac{(GM)^2}{h^3}\dfrac{\partial^2 F(h,t)}{\partial h^2}, \\ F(h_{\rm in}, t) = F_{\rm in}(t), \\ \left.\dfrac{\partial F}{\partial h}\right|_{\rm out} = \dot{M}_{\rm out}(t), \\ F(h,0) = F_0(h), \\ h \in [h_{\rm in}, h_{\rm out}], \\ t \in [0, t_{\rm fin}], \end{cases} \tag{1.197}$$

where $F_0(h)$ is the initial condition satisfying the boundary conditions and $t_{\rm fin}$ is the time interval for which the calculation is performed.

To construct a finite difference scheme we introduce an arbitrary collection of nodes h_n:

$$\begin{aligned} & h_1 < h_2 < \cdots < h_n < \cdots < h_{N-1} < h_N, \\ & \Delta h_n \equiv h_n - h_{n-1}, \\ & n = 1 \ldots N, \end{aligned} \tag{1.198}$$

where h_1 and h_N correspond to the values of the specific Keplerian angular momentum at the inner and outer radius, respectively. We will consider a solution to the equation at the time-interval between t_0, with already known values of the desired function, and $t_0 + \Delta t$, for which we need to determine these values.

We substitute the two functions with their corresponding grid functions and introduce the following designations:

$$\begin{array}{llll} F(h_n, t_0) & \Rightarrow F_n, & F(h_n, t_0 + \Delta t) & \Rightarrow \tilde{F}_n, \\ \Sigma_0(F(h_n, t_0)\,, h_n) & \Rightarrow \Sigma_n, & \Sigma_0(F(h_n, t_0 + \Delta t)\,, h_n) & \Rightarrow \tilde{\Sigma}_n, \\ F_{\rm in}(t_0) & \Rightarrow F_{\rm in}, & F_{\rm in}(t_0 + \Delta t) & \Rightarrow \tilde{F}_{\rm in}, \\ \dot{M}_{\rm out}(t) & \Rightarrow F'_{\rm out}, & \dot{M}_{\rm out}(t_0 + \Delta t) & \Rightarrow \tilde{F}'_{\rm out}. \end{array} \quad (1.199)$$

Let us start constructing the finite difference scheme. To begin with, we write down the difference equations for the boundary conditions. The inner boundary condition of the first kind is written in exact form as:

$$F_1 = F_{\rm in}. \quad (1.200)$$

To write down the outer boundary condition of the second kind, we expand $\tilde{F}_{N-1}$ in Taylor series around the point h_N:

$$\tilde{F}_{N-1} = \tilde{F}_N - \Delta h_N \left.\frac{\partial F}{\partial h}\right|_{h_N} + \frac{\Delta h_N^2}{2} \left.\frac{\partial^2 F}{\partial h^2}\right|_{h_N} + o(\Delta h_N^2). \quad (1.201)$$

Note that in all the expressions considered here and below for the derivatives with respect to h, we use the value of the torque at time $t_0 + \Delta t$. Thus constructed numerical scheme is called implicit. It is numerically stable. As opposed to an explicit scheme, in which the derivatives with respect to h would be written using the known value F_n at time t_0, an implicit scheme guarantees that the errors introduced in this step will not grow in the next steps.

Without going into details, we note that, in addition to the explicit and implicit methods, there is also a mixed (Crank–Nicolson) method in which the values for the function at t_0 and $t_0 + \Delta t$ are both used to calculate the derivative with respect to the spatial coordinate. In some cases, the Crank–Nicolson method gives a higher accuracy of the solution. The node stencils used in the various methods are shown in Fig. 1.24.

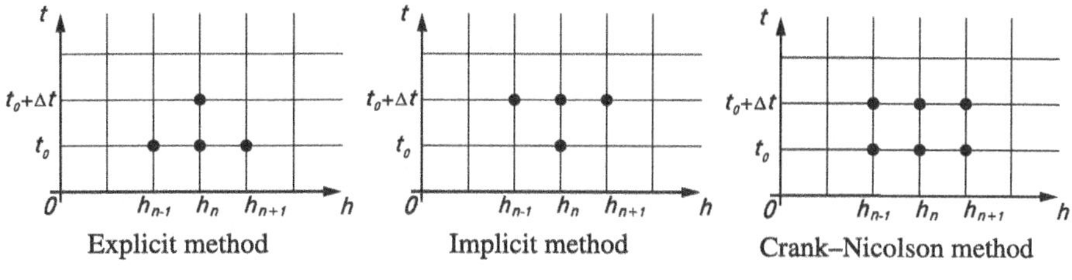

Fig. 1.24 Node stencils that are used for the n-th equation in the system (1.209) in different schemes. We use the implicit method stencil

Discarding the last term in (1.201), we obtain a simple expression for the numerical value of the first derivative of F with respect to h, with accuracy up to the first order of the expansion interval Δh_N:

$$\left.\frac{\partial F}{\partial h}\right|_{h_N} = \frac{\tilde{F}_N - \tilde{F}_{N-1}}{\Delta h_N} + o(\Delta h_N). \tag{1.202}$$

If we, however, in (1.201), substitute the value of the second derivative of F with respect to h, expressed from the original equation (1.197), we may increase the accuracy to the second order of Δh_N:

$$\left.\frac{\partial F}{\partial h}\right|_{h_N} = \frac{\tilde{F}_N - \tilde{F}_{N-1}}{\Delta h_N} + \Delta h_N \frac{2\pi h_N^3}{(GM)^2}\frac{\partial \Sigma_0(h_N)}{\partial t} + o(\Delta h_N^2), \tag{1.203}$$

where the expression for the derivative Σ_0 with respect to t by analogy with (1.202) takes the form:

$$\frac{\partial \Sigma_0(h_N)}{\partial t} = \frac{\tilde{\Sigma}_N - \Sigma_N}{\Delta t} + o(\Delta t). \tag{1.204}$$

In this way we obtain a final expression for the outer boundary condition:

$$\frac{\tilde{F}_N - \tilde{F}_{N-1}}{\Delta h_N} + \frac{\Delta h_N}{\Delta t}\frac{2\pi h_N^3}{(GM)^2}(\tilde{\Sigma}_N - \Sigma_N) + o(\Delta h_N^2) + o(\Delta t) = \tilde{F}'_{\rm out}. \tag{1.205}$$

Now that we have equations for the values of the function at both ends of the interval over h, we obtain the difference form of the differential equation itself from (1.197). Let us write down the Taylor expansion for $\tilde{F}_{n-1}$ and $\tilde{F}_{n+1}$ around the point h_n:

$$\begin{aligned}
\tilde{F}_{n-1} &= \tilde{F}_n - \Delta h_n \left.\frac{\partial F}{\partial h}\right|_{h_n} + \frac{\Delta h_n^2}{2}\left.\frac{\partial^2 F}{\partial h^2}\right|_{h_n} + o(\Delta h_n^2),\\
\tilde{F}_{n+1} &= \tilde{F}_n + \Delta h_{n+1} \left.\frac{\partial F}{\partial h}\right|_{h_n} + \frac{\Delta h_{n+1}^2}{2}\left.\frac{\partial^2 F}{\partial h^2}\right|_{h_n} + o(\Delta h_{n+1}^2),
\end{aligned} \tag{1.206}$$

where $n = 2 \ldots N-1$.

For convenience we introduce the notation $\Delta h = \max(h_n)$, where $n = 2 \ldots N$. Then, we may change $o(\Delta h_n)$ to $o(\Delta h)$ everywhere.

The second derivative of F with respect to h may be expressed from (1.206):

$$\left.\frac{\partial^2 F}{\partial h^2}\right|_{h_n} = 2\frac{\tilde{F}_{n-1}\frac{\Delta h_{n+1}}{\Delta h_n + \Delta h_{n+1}} - \tilde{F}_n + \tilde{F}_{n+1}\frac{\Delta h_n}{\Delta h_n + \Delta h_{n+1}}}{\Delta h_n \Delta h_{n+1}} + o(\Delta h^2). \tag{1.207}$$

Note that when using a homogeneous grid with respect to h, that is for $\Delta h_n = \Delta h_{n+1} = \Delta h$, the last expression takes a simpler form:

$$\left.\frac{\partial^2 F}{\partial h^2}\right|_{h_n} = \frac{\tilde{F}_{n-1} - 2\tilde{F}_n + \tilde{F}_{n+1}}{\Delta h^2}. \tag{1.208}$$

Substituting the values of the derivatives (1.204) and (1.207) into the differential equation from (1.197) and replacing the boundary conditions in (1.197) by their difference analogues (1.200) and (1.205), we obtain a finite difference scheme for the problem:

$$\begin{cases} \dfrac{4\pi h_n^3}{(GM)^2}\dfrac{\tilde{\Sigma}_n - \Sigma_n}{\Delta t} = 2\dfrac{\tilde{F}_{n-1}\frac{\Delta h_{n+1}}{\Delta h_n + \Delta h_{n+1}} - \tilde{F}_n + \tilde{F}_{n+1}\frac{\Delta h_n}{\Delta h_n + \Delta h_{n+1}}}{\Delta h_n \Delta h_{n+1}}, \\ \tilde{F}_1 = \tilde{F}_{\rm in}, \\ \dfrac{\tilde{F}_N - \tilde{F}_{N-1}}{\Delta h_N} + \dfrac{\Delta h_N}{\Delta t}\dfrac{2\pi h_N^3}{(GM)^2}(\tilde{\Sigma}_N - \Sigma_N) = \tilde{F}'_{\rm out}, \\ n = 2\ldots N-1. \end{cases} \tag{1.209}$$

Note that the level of accuracy in the obtained system is $o(\Delta h^2) + o(\Delta t)$.

As a result, we have reduced the solution of the differential equations with boundary conditions (1.197) to a subsequent solution of the system of N algebraic equations (1.209) at each time-step between $t = 0$ and $t = t_{\rm fin}$. This system is not linear, since Σ_n and F_n are related by the non-linear expression $\Sigma_n = \Sigma_0(F_n, h_n)$. One way to solve this system is to use the iterative root-finding algorithm for the value $\tilde{\Sigma}_n$. For this, some approximation to the value $\tilde{\Sigma}_n^{(1)}$ must first be chosen (the simplest variant is the value at the present time step Σ_n), and the system of linear algebraic equations is solved to find the intermediate value of $\tilde{\Sigma}_n^{(2)} = \Sigma_0(\tilde{F}_n^{(1)}, h_n)$ and then the system of linear algebraic equations is solved again. This simple iterative algorithm can be improved at the expense of extra memory usage; see Anderson (1965) for details.

One may think of a number of criteria to stop the integration. We will use one of them—the condition of small changes in the value for $\tilde{\Sigma}_n^{(s)}$ between two sequential iterations. We formalise this criterion:

$$\max_{n=2\ldots N}\left|\frac{\tilde{\Sigma}_n^{(s+1)} + \tilde{\Sigma}_n^{(s)}}{\tilde{\Sigma}_n^{(s+1)}}\right| < \epsilon, \tag{1.210}$$

where the top index in brackets refers to the number of performed iterations and ϵ is the dimensionless accuracy in the search for the value of $\tilde{\Sigma}_n$.

Note that in each iteration, the solution to the system of linear algebraic equations may be found by the tridiagonal matrix algorithm. The details of this algorithm can be found in textbooks on numerical methods, for example Press et al. (2002).

The described scheme (1.209) is implemented in the `FREDDI`[7] code. `FREDDI` is intended for modelling the lightcurves of X-ray novae with fast rise and exponential decay (Lipunova and Malanchev 2017). As initial conditions, we may choose either a quasi-stationary distribution (see Sect. 1.6.7), describing the radial structure of the disc after the peak in luminosity of the source, or the distribution corresponding to a dense torus far away from the central black hole.

1.7.2 Solving the Equations of Vertical Structure

In Sect. 1.5.2 we derived the equations for the vertical structure (1.93):

$$\begin{aligned}\frac{1}{\rho}\frac{\mathrm{d}P}{\mathrm{d}z} &= -\,\omega_{\mathrm{K}}^2\,z,\\ \frac{\mathrm{d}\Sigma}{\mathrm{d}z} &= \rho,\\ \frac{\mathrm{d}Q}{\mathrm{d}z} &= \frac{3}{2}\,\omega_{\mathrm{K}}\,w_{r\varphi},\\ \frac{c}{3\varkappa_{\mathrm{R}}\rho}\frac{\mathrm{d}(aT^4)}{\mathrm{d}z} &= -Q.\end{aligned}$$

To solve these equations, we have to choose suitable boundary conditions. If we consider the surface density at a given radius as known, we have only three boundary conditions: $\Sigma(z=0)=0$, $\Sigma(z=z_0)=\Sigma_0/2$, and $Q(z=z_0)=0$. On the other hand, if we consider the torque at a given radius as known, we may find the necessary number of boundary conditions to solve the system (1.211).

By analogy with the arguments in Sect. 1.5.2, we obtain the boundary condition for the pressure at the photosphere:

$$P(z=z_0)=\frac{2}{3}\frac{\omega_{\mathrm{K}}^2 z_0}{\varkappa_{\mathrm{R}}}. \tag{1.211}$$

If we assume that energy is released only in layers below the photosphere, the flux at the photosphere is determined by Eq. (1.73):

$$Q(z=z_0)=\frac{3}{8\pi}\frac{F\omega_{\mathrm{K}}}{r^2}. \tag{1.212}$$

[7] http://xray.sai.msu.ru/~malanchev/freddi/.

Due to symmetry, the flux is equal to zero in the plane of the disc:

$$Q(z = 0) = 0. \tag{1.213}$$

We consider the emitted spectrum to be that of a blackbody, so we may take the temperature in the photosphere to be equal to the effective temperature:

$$T(z = z_0) = \left(\frac{Q(z = z_0)}{\sigma_{\rm SB}} \right)^{1/4}. \tag{1.214}$$

The boundary condition $\Sigma = 0$ may be set at the surface of the disc as well as in its symmetric plane. It turns out that in the symmetry plane there are only two boundary conditions, on the flux Q (1.186) and on the surface density. However, if we set Σ equal to zero at the disc surface and integrate the system towards the central plane, we can find the boundary values of all four unknown functions: the pressure P (1.211), the surface density Σ, the flux Q (1.213), and the temperature T (1.214). Thus, in what follows, we shall consider integration along the direction from the disc surface towards its symmetry plane.

While all four boundary conditions at the photosphere are known, we still do not know the disc half-thickness z_0. For convenience in integrating the system (1.211) from the photosphere to the symmetry plane, we rewrite it with aspect to the alternative vertical parameter $\hat{z} \equiv z_0 - z$:

$$\frac{1}{\rho} \frac{{\rm d}P}{{\rm d}\hat{z}} = \omega_{\rm K}^2 (z_0 - \hat{z}), \tag{1.215}$$

$$\frac{{\rm d}\Sigma}{{\rm d}\hat{z}} = \rho, \tag{1.216}$$

$$\frac{{\rm d}Q}{{\rm d}\hat{z}} = -\frac{3}{2} \omega_{\rm K} w_{r\varphi}, \tag{1.217}$$

$$\frac{c}{3\varkappa_{\rm R}\rho} \frac{{\rm d}(aT^4)}{{\rm d}\hat{z}} = Q, \tag{1.218}$$

$$P(\hat{z} = 0) = \frac{2}{3} \frac{\omega_{\rm K}^2 z_0}{\varkappa_{\rm R}}, \tag{1.219}$$

$$\hat{\Sigma}(\hat{z} = 0) = 0, \tag{1.220}$$

$$Q(\hat{z} = 0) = \frac{3}{8\pi} \frac{F\omega_{\rm K}}{r^2}, \tag{1.221}$$

$$Q(\hat{z} = z_0) = 0, \tag{1.222}$$

$$T(\hat{z} = 0) = \left(\frac{Q(\hat{z} = 0)}{\sigma_{\rm SB}} \right)^{1/4}, \tag{1.223}$$

where $\hat{\Sigma}(\hat{z}) = \Sigma_0/2 - \Sigma(\hat{z})$ is calculated in the direction from the disc surface.

This system consists of four equations, five boundary conditions and one unknown—z_0. We need to choose a value of z_0 such that when integrating the system (1.223) from $\hat{z} = 0$ to $\hat{z} = z_0$, the boundary condition $Q(\hat{z} = z_0) = 0$ is fulfilled. As an initial approximation, we can use values obtained analytically (see Sect. 1.5.3), and then search for z_0 using any method of root-finding.

1.7.2.1 Irradiation of the Accretion Disc

In X-ray binaries the outer parts of the disc with photospheric temperature of the order of 10^4 K may be irradiated by photons from the inner parts of the disc, direct or scattered in the corona, with temperatures of the order of 10^7 K. The surface of a neutron star may serve as an additional source of hard photons. The surface of protoplanetary discs, with temperatures of the order of 10^2 K, is irradiated by ultra violet radiation from a newly formed star.

Let us consider the case with irradiation by hard radiation incident on the disc surface at an angle $\arccos \zeta$. If the disc is illuminated by a point source located in its centre, and if the disc itself can be considered as thin, we may use the relation

$$\zeta = \frac{dz_0}{dr} - \frac{z_0}{r}. \qquad (1.224)$$

Then the illuminating flux incident on the disc surface at radius r equals $\zeta L_\mathrm{X}/(4\pi r^2)$. If the source of the hard radiation is the disc itself, then the radiation pattern is not isotropic. Assuming that the central source is point-like, the flux may be written as $\zeta L_\mathrm{X}/(4\pi r^2) \times \Psi(\theta)$. The function $\Psi(\theta)$ is shown in Fig. 1.23, and θ is the angle measured from the vertical axis.

A detailed calculation of the effect of irradiation on the vertical structure of the disc is rather complicated, and was presented, e.g., in the work by Mescheryakov et al. (2011b). In a first approximation, we may limit ourselves to changing the boundary condition on the flux originating from the surface of the disc:

$$Q(\hat{z} = 0) = \frac{3}{8\pi}\frac{F\omega_\mathrm{K}}{r^2} + \zeta \frac{L_\mathrm{X}}{4\pi r^2}\Psi(\theta).$$

In order to explain the observed optical lightcurves from X-ray novae, the effective thickness of the disc for radiation interception in formula (1.224) needs to be twice as large as z_0 (Suleimanov et al. 2008). It is assumed in their calculations that the lower layers of the disc atmosphere above the photosphere are opaque to soft X-rays from the central parts of the disc. Furthermore, it was shown by Mescheryakov et al. (2011a), from modelling lightcurves of the illuminated stellar companion in the burster GS 1826-238, a low-mass X-ray binary with a neutron star, that the effective thickness of the disc for interception of X-rays is approximately twice as large as z_0.

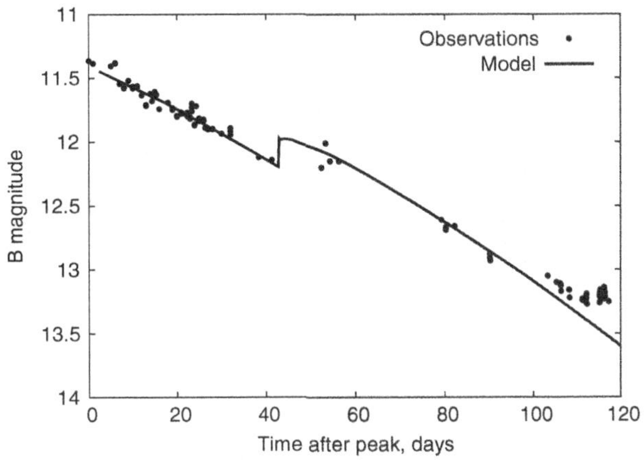

Fig. 1.25 Lightcurve of the X-ray nova A 0620-00 in the photometric B-band. Data from Duerbeck and Walter (1976), Lloyd et al. (1977) are shown with *filled circles*. The *solid line* shows our modelling of the lightcurve

1.7.3 Example Numerical Modelling of a FRED Lightcurve of an X-Ray Nova

Let us now turn to the numerical modelling of an outburst of X-ray nova A 0620-00. The following parameters of the binary system are used: mass of the compact object (a black hole) 6.6 $M_{\odot}$, mass of the optical companion 0.5 $M_{\odot}$, orbital period 0.323 days, inclination of the orbital plane to the line of sight 53.5°, and distance to the system 1.1 kpc. These parameters are observational results from analyses of the lightcurves of the system in quiescence (Cantrell et al. 2010; Gou et al. 2010).

In Figs. 1.25 and 1.26, lightcurves of the source after the peak of the outburst in 1975 in soft X-rays and in the B-band are shown. The lightcurve of this outburst is an example of a FRED-type lightcurve,[8] in which a fast rise in luminosity is followed by a quasi-exponential decay.

An interesting feature in most FRED-type lightcurves is the existence of a secondary peak. The nature of this secondary peak is currently not understood.

To reproduce the secondary peak, it has been suggested that a significant amount of matter was supplied to the disc by the donor star on the 43rd day after the peak. Within the framework of this model, this matter instantaneously increases the surface density of the disc in its outer parts, which leads to a jump in optical luminosity (Fig. 1.25).

Due to the increase in surface density in the outer regions of the disc, a gradual increase of the accretion rate takes place in the central regions of the disc. This leads to an increase in temperature and thereby X-ray luminosity of the disc. In this way, a local maximum shows up in the lightcurve (Fig. 1.26).

The maximal accretion rate $\dot{M}_{\rm max} \approx 0.2\,\dot{M}_{\rm Edd}$, and the α-parameter, ≈ 0.3, are determined from the part of the X-ray lightcurve before the secondary peak.

[8]Fast-rise exponential-decay.

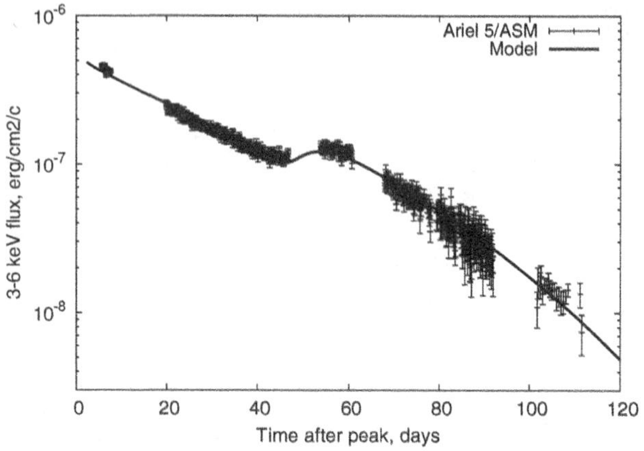

Fig. 1.26 Lightcurve of the X-ray nova A 0620-00 at 3–6 keV. The *vertical bars* show data with errors from Ariel 5 (Kaluzienski et al. 1977), and the *solid line* shows the model lightcurve

From the part of the optical lightcurve before the secondary peak, we estimate the effective thickness of the disc for X-ray radiation interception, which turns out to be $\approx 2\,z_0$ (Malanchev and Shakura 2015).

To model the lightcurve of an X-ray nova, one has to keep in mind that in general the disc is not physically uniform, but has a hot inner part with ionized matter (zones A, B and C; see Sect. 1.5) and a colder outer part with lower accretion rate. When the disc cools down to temperatures at which hydrogen recombines, the α-parameter decreases by approximately an order of magnitude (Smak 1984). In a first approximation, we may assume that accretion in the cold outer parts ceases. The boundary between the hot and cold parts gradually moves towards the centre following the hydrogen recombination front.

The open code `FREDDI` is provided by the authors to model FRED-type lightcurves of X-ray novae. This code calculates the disc evolution for a fully ionized disc, as well as for a disc with a cold front propagating inwards. Using this code, the outburst of the X-ray nova 4U 1543-47 in 2002, hosting a black hole, was modelled by Lipunova and Malanchev (2017).

Using `FREDDI`, estimates of α can be derived, which are more accurate than (1.149):

$$\alpha \approx 0.21 \left(\frac{R_{\rm hot}}{R_\odot}\right)^{25/16} \left(\frac{t_{\rm exp}}{30^{\rm d}}\right)^{-5/4} \left(\frac{\dot{M}_{\rm max}}{10^{18}\,{\rm g/s}}\right)^{-3/8} m_x^{5/16}, \tag{1.225}$$

for the Kramers opacity, and

$$\alpha \approx 0.20 \left(\frac{R_{\rm hot}}{R_\odot}\right)^{12/7} \left(\frac{t_{\rm exp}}{30^{\rm d}}\right)^{-9/7} \left(\frac{\dot{M}_{\rm max}}{10^{18}\,{\rm g/s}}\right)^{-3/7} m_x^{2/7} \tag{1.226}$$

for the OPAL approximation (Lipunova and Malanchev 2017). Here, R_{hot} is the radius of the hot zone of the disc at the peak of an outburst. Power indexes in the above expressions are obtained when substituting the thickness of the disk in (1.149) by its analytic expression from (1.104) or (1.110). The numerical factors in the expressions for α are found by fitting `FREDDI` results; their accuracy is around 5%.

References

Abramowicz MA (2016) Velocity, acceleration and gravity in Einstein's relativity. ArXiv e-prints 1608.07136

Abramowicz MA, Fragile PC (2013) Foundations of black hole accretion disk theory. Living Rev Relativ 16:1. https://doi.org/10.12942/lrr-2013-1. ArXiv 1104.5499

Anderson DG (1965) Iterative procedures for nonlinear integral equations. J ACM 12(4):547–560. https://doi.org/10.1145/321296.321305. http://doi.acm.org/10.1145/321296.321305

Balbus SA, Hawley JF (1991) A powerful local shear instability in weakly magnetized disks. I - Linear analysis. II - Nonlinear evolution. Astrophys J 376:214–233. https://doi.org/10.1086/170270

Balbus SA, Hawley JF (1998) Instability, turbulence, and enhanced transport in accretion disks. Rev Mod Phys 70:1–53. https://doi.org/10.1103/RevModPhys.70.1

Barenblatt GI (1996) Scaling, Self-similarity, and intermediate asymptotics: dimensional analysis and intermediate asymptotics. Cambridge texts in applied mathematics. Cambridge University Press, Cambridge

Barenblatt G (2003) Scaling. Cambridge texts in applied mathematics. Cambridge University Press, Cambridge. https://books.google.ru/books?id=05zBYET6tR0C

Bisikalo DV, Zhilkin AG, Boyarchuk AA (2013) Gaseous dynamic of close binary stars. FIZMATLIT (in Russian), Moscow

Bisnovatyi-Kogan GS, Blinnikov SI (1976) A hot corona around a black-hole accretion disk as a model for Cygnus X-1. Sov Astron Lett 2:191–193. ArXiv astro-ph/0003275

Bisnovatyi-Kogan GS, Lovelace RVE (2001) Advective accretion disks and related problems including magnetic fields. New Astron Rev 45:663–742. https://doi.org/10.1016/S1387-6473(01)00146-4. ArXiv astro-ph/0207625

Brandenburg A, Nordlund A, Stein RF, Torkelsson U (1996) The disk accretion rate for dynamo-generated turbulence. Astrophys J 458:L45. https://doi.org/10.1086/309913

Cannizzo JK (1992) Accretion disks in active galactic nuclei - vertically explicit models. Astrophys J 385:94–107. https://doi.org/10.1086/170918

Cannizzo JK (1998) The accretion disk limit cycle mechanism in the black hole x-ray binaries: toward an understanding of the systematic effects. Astrophys J 494:366. https://doi.org/10.1086/305210

Cannizzo JK, Lee HM, Goodman J (1990) The disk accretion of a tidally disrupted star onto a massive black hole. Astrophys J 351:38–46. https://doi.org/10.1086/168442

Cantrell AG, Bailyn CD, Orosz JA, McClintock JE, Remillard RA, Froning CS, Neilsen J, Gelino DM, Gou L (2010) The inclination of the soft x-ray transient A0620-00 and the mass of its black hole. Astrophys J 710:1127–1141. https://doi.org/10.1088/0004-637X/710/2/1127. ArXiv 1001.0261

Chandrasekhar S (1960) The stability of non-dissipative Couette flow in hydromagnetics. Proc Natl Acad Sci 46:253–257. https://doi.org/10.1073/pnas.46.2.253

Chen W, Shrader CR, Livio M (1997) The properties of x-ray and optical light curves of x-ray novae. Astrophys J 491:312. https://doi.org/10.1086/304921

Coroniti FV (1981) On the magnetic viscosity in Keplerian accretion disks. Astrophys J 244:587–599. https://doi.org/10.1086/158739
Deardorff JW (1970) A numerical study of three-dimensional turbulent channel flow at large Reynolds numbers. J Fluid Mech 41:453–480. https://doi.org/10.1017/S0022112070000691
Dubrulle B (1993) Differential rotation as a source of angular momentum transfer in the solar nebula. Icarus 106:59. https://doi.org/10.1006/icar.1993.1158
Dubus G, Lasota JP, Hameury JM, Charles P (1999) X-ray irradiation in low-mass binary systems. Mon Not R Astron Soc 303:139–147
Duerbeck HW, Walter K (1976) A periodic brightness variation of the optical counterpart of A 0620-00. NASA Spec Publ 389:343–345
Eardley DM, Lightman AP (1975) Magnetic viscosity in relativistic accretion disks. Astrophys J 200:187–203. https://doi.org/10.1086/153777
Favre AJ (1969) Statistical equations of turbulents gases. 231–267
Felten JE, Rees MJ (1972) Continuum radiative transfer in a hot plasma, with application to Scorpius X-l. Astron Astrophys 17:226
Ferguson JW, Alexander DR, Allard F, Barman T, Bodnarik JG, Hauschildt PH, Heffner-Wong A, Tamanai A (2005) Low-temperature opacities. Astrophys J 623:585–596. https://doi.org/10.1086/428642. ArXiv astro-ph/0502045
Filipov LG (1984) Self-similar problems of the time-dependant discs accretion and the nature of the temporary X-ray sources. Adv Space Res 3:305–313. https://doi.org/10.1016/0273-1177(84)90107-8
Frank J, King A, Raine DJ (2002) Accretion power in astrophysics, 3rd edn. Cambridge University Press, Cambridge
Fridman AM (1989) On the dynamics of a viscous differentially rotating gravitating medium. Sov Astron Lett 15:487
Galeev AA, Rosner R, Vaiana GS (1979) Structured coronae of accretion disks. Astrophys J 229:318–326. https://doi.org/10.1086/156957
Gorbatskii VG (1965) Disk-like envelopes in close binary systems and their effect on stellar spectra. Sov. Astron. 8:680
Gou L, McClintock JE, Steiner JF, Narayan R, Cantrell AG, Bailyn CD, Orosz JA (2010) The spin of the black hole in the soft x-ray transient A0620-00. Astrophys J 718:L122–L126. https://doi.org/10.1088/2041-8205/718/2/L122. ArXiv 1002.2211
Hameury JM, Lasota JP (2005) Tidal torques, disc radius variations, and instabilities in dwarf novae and soft X-ray transients. Astron Astrophys 443:283–289. https://doi.org/10.1051/0004-6361:20053691. ArXiv arXiv:astro-ph/0508509
Hameury JM, Menou K, Dubus G, Lasota JP, Hure JM (1998) Accretion disc outbursts: a new version of an old model. Mon Not R Astron Soc 298:1048–1060
Ichikawa S, Osaki Y (1994) Tidal torques on accretion disks in close binary systems. Publ Astron Soc Jpn 46:621–628
Iglesias CA, Rogers FJ (1996) Updated opal opacities. Astrophys J 464:943. https://doi.org/10.1086/177381
Ivanov PB, Papaloizou JCB, Polnarev AG (1999) The evolution of a supermassive binary caused by an accretion disc. Mon Not R Astron Soc 307:79–90. https://doi.org/10.1046/j.1365-8711.1999.02623.x. ArXiv astro-ph/9812198
Kaluzienski LJ, Holt SS, Boldt EA, Serlemitsos PJ (1977) All-Sky Monitor observations of the decay of A0620-00 /Nova Monocerotis 1975/. Astrophys J 212:203–210. https://doi.org/10.1086/155036
Kato S, Fukue J, Mineshige S (1998) Black-hole accretion disks. Kyoto University Press, Kyoto
Kato S, Fukue J, Mineshige S (2008) Black-hole accretion disks — towards a new paradigm —, 549 pp
Ketsaris NA, Shakura NI (1998) On the calculation of the vertical structure of accretion discs. Astron Astrophys Trans 15:193. https://doi.org/10.1080/10556799808201769
King AR, Ritter H (1998) The light curves of soft X-ray transients. Mon Not R Astron Soc 293:L42–L48

Kotko I, Lasota JP (2012) The viscosity parameter α and the properties of accretion disc outbursts in close binaries. Astron Astrophys 545:A115. https://doi.org/10.1051/0004-6361/201219618. ArXiv 1209.0017

Kurucz RL (1970) Atlas: a computer program for calculating model stellar atmospheres. SAO Special Report. Smithsonian Astrophysical Observatory, Cambridge

Kurucz R (1993) Kurucz CD-ROMs, Smithsonian Astrophysical Observatory, Cambridge

Landau LD, Lifshitz EM (1959) Fluid mechanics, Course of Theoretical Physics, vol 6

Landau LD, Lifshitz EM (1973) The classical theory of fields. Course of theoretical physics, vol 2, Sect. 88. Pergamon, Oxford

Lasota JP (2001) The disc instability model of dwarf novae and low-mass X-ray binary transients. New Astron Rev 45:449–508

Lasota JP (2015) Black hole accretion discs. ArXiv e-prints 1505.02172

Lightman AP, Eardley DM (1974) Black holes in binary systems: instability of disk accretion. Astrophys J 187:L1+

Lin DNC, Bodenheimer P (1982) On the evolution of convective accretion disk models of the primordial solar nebula. Astrophys J 262:768–779. https://doi.org/10.1086/160472

Lin DNC, Pringle JE (1987) A viscosity prescription for a self-gravitating accretion disc. Mon Not R Astron Soc 225:607–613

Lipunova GV (2015) Evolution of finite viscous disks with time-independent viscosity. Astrophys J 804:87. https://doi.org/10.1088/0004-637X/804/2/87. ArXiv 1503.09093

Lipunova GV, Malanchev KL (2017) Determination of the turbulent parameter in accretion discs: effects of self-irradiation in 4U 1543-47 during the 2002 outburst. Mon Not R Astron Soc 468:4735–4747. https://doi.org/10.1093/mnras/stx768. ArXiv 1610.01399

Lipunova GV, Shakura NI (2000) New solution to viscous evolution of accretion disks in binary systems. Astron Astrophys 356:363–372

Lipunova GV, Shakura NI (2002) Non-steady-state accretion disks in x-ray novae: outburst models for Nova Monocerotis 1975 and Nova Muscae 1991. Astron Rep 46:366–379

Lloyd C, Noble R, Penston MV (1977) The light curve of V616 Mon = A0620-00. Mon Not R Astron Soc 179:675–681. https://doi.org/10.1093/mnras/179.4.675

Lüst RZ (1952) Die Entwicklung einer um einen Zentralkörper rotierenden Gasmasse. I. Lösungen der hydrodynamischen Gleichungen mit turbulenter Reibung. Zeitschrift Naturforschung Teil A 7:87

Lynden-Bell D (1969) Galactic nuclei as collapsed old quasars. Nature 223:690

Lynden-Bell D, Pringle JE (1974) The evolution of viscous discs and the origin of the nebular variables. Mon Not R Astron Soc 168:603–637

Lyubarskij YE, Shakura NI (1987) Nonlinear self-similar problems of nonstationary disk accretion. Sov Astron Lett 13:386

MacRobert TM (1932) Fourier integrals. Proc R Soc Edinb 51:116–126

Malanchev KL, Shakura NI (2015) Vertical convection in turbulent accretion disks and light curves of the X-ray nova A0620-00 1975 outburst. Astron Lett 41:797–808. https://doi.org/10.1134/S1063773715120087. ArXiv 1511.02356

Marov M, Kolesnichenko A (2011) Turbulence and self-organization: modeling astrophysical objects. Astrophysics and Space Science Library, Springer, New York. https://books.google.ru/books?id=y1r2sgEACAAJ

Mescheryakov AV, Revnivtsev MG, Filippova EV (2011a) Parameters of irradiated accretion disks from optical and X-ray observations of GS 1826-238. Astron Lett 37:826–844. https://doi.org/10.1134/S1063773711120073

Mescheryakov AV, Shakura NI, Suleimanov VF (2011b) Vertical structure of the outer accretion disk in persistent low-mass X-ray binaries. Astron Lett 37:311–331. https://doi.org/10.1134/S1063773711050045. ArXiv 1108.4222

Meyer F, Meyer-Hofmeister E (1981) On the elusive cause of cataclysmic variable outbursts. Astron Astrophys 104:L10

Meyer F, Meyer-Hofmeister E (1982) Vertical structure of accretion disks. Astron Astrophys 106:34–42

Meyer F, Meyer-Hofmeister E (1984) HZ Her/Her X-1 - an alternative model for the 35d cycle? Astron Astrophys 140:L35–L38

Mihalas D (1978) Stellar atmospheres, 2nd edn. W. H. Freeman and Co., San Francisco

Mihalas D, Mihalas BW (1984) Foundations of radiation hydrodynamics. Courier Corporation, North Chelmsford

Monin A, Yaglom A (1971) Statistical fluid mechanics: mechanics of turbulence. vol 1. Peace Corps. https://books.google.ru/books?id=7BvYQwAACAAJ

Morozov AG, Khoperskov AV (2005) Physics of discs. Volgograd University Press, Volgograd (in Russian)

Nakao Y, Kato S (1995) Vertical dependence of the viscous heating in accretion disks. Publ Astron Soc Jpn 47:451–461

Novikov ID, Thorne KS (1973) Astrophysics of black holes. In: DeWitt C, DeWitt BS (eds) Black holes (Les Astres Occlus), Gordon and Breach, New York, pp 343–450

Ogilvie GI (1999) Time-dependent quasi-spherical accretion. Mon Not R Astron Soc 306:L9–L13

Paczynski B (1977) A model of accretion disks in close binaries. Astrophys J 216:822–826

Paczynski B, Bisnovatyi-Kogan G (1981) A model of a thin accretion disk around a Black Hole. Acta Astron. 31:283

Paczynsky B, Wiita PJ (1980) Thick accretion disks and supercritical luminosities. Astron Astrophys 88:23–31

Papaloizou J, Pringle JE (1977) Tidal torques on accretion discs in close binary systems. Mon Not R Astron Soc 181:441–454

Pletcher R, Tannehill J, Anderson D (1997) Computational fluid mechanics and heat transfer, 2nd edn. Series in computational and physical processes in mechanics and thermal sciences, Taylor & Francis. http://books.google.ru/books?id=ZJPbtHeilCgC

Prandtl L (1925) Bericht uber Untersuchungen zur ausgebildeten Turbulenz. Z Angew Math Mech 5:136–139. http://naca.central.cranfield.ac.uk/reports/1949/naca-tm-1231.pdf

Press WH, Teukolsky SA, Vetterling WT, Flannery BP (2002) Numerical recipes in C: the art of scientific computing, 2nd edn. Cambridge University Press, Cambridge

Pringle JE (1974) PhD thesis, University of Cambridge, 1974

Pringle JE (1991) The properties of external accretion discs. Mon Not R Astron Soc 248:754–759

Pringle JE, Rees MJ (1972) Accretion disc models for compact x-ray sources. Astron Astrophys 21:1

Rafikov RR (2013) Structure and evolution of circumbinary disks around supermassive black hole binaries. Astrophys J 774:144. https://doi.org/10.1088/0004-637X/774/2/144. ArXiv 1205.5017

Rafikov RR (2016) Generalized similarity for accretion/decretion disks. Astrophys J 830:7. https://doi.org/10.3847/0004-637X/830/1/7. ArXiv 1604.07439

Rayleigh L (1917) On the dynamics of revolving fluids. Proc R Soc Lond Ser A 93:148–154. https://doi.org/10.1098/rspa.1917.0010

Richard D, Zahn JP (1999) Turbulence in differentially rotating flows. What can be learned from the Couette-Taylor experiment. Astron Astrophys 347:734–738. ArXiv astro-ph/9903374

Shafee R, Narayan R, McClintock JE (2008) Viscous torque and dissipation in the inner regions of a thin accretion disk: implications for measuring black hole spin. Astrophys J 676:549–561. https://doi.org/10.1086/527346. ArXiv 0705.2244

Shakura NI (1973) Disk model of gas accretion on a relativistic star in a close binary system. Sov Astron 16:756

Shakura NI, Sunyaev RA (1973) Black holes in binary systems. Observational appearance. Astron Astrophys 24:337–355

Shakura NI, Sunyaev RA (1976) A theory of the instability of disk accretion on to black holes and the variability of binary X-ray sources, galactic nuclei and quasars. Mon Not R Astron Soc 175:613–632

Shakura NI, Sunyaev RA, Zilitinkevich SS (1978) On the turbulent energy transport in accretion discs. Astron Astrophys 62:179–187

Shapiro SL, Teukolsky SA (1983) Black holes, white dwarfs, and neutron stars: The physics of compact objects. Research supported by the National Science Foundation. Wiley-Interscience, New York, 663 pp
Shaviv G, Wehrse R (1986) The vertical temperature stratification and corona formation of accretion disc atmospheres. Astron Astrophys 159:L5–L7
Shibazaki N, Hōshi R (1975) Structure and stability of accretion-disk around a black-hole. Prog Theor Phys 54:706–718
Smak J (1984) Accretion in cataclysmic binaries. IV - Accretion disks in dwarf novae. Acta Astron 34:161–189
Sneddon IN (1951) Fourier transforms. International series in pure and applied mathematics. McGraw-Hill. http://books.google.ru/books?id=HRAJAQAAIAAJ
Sobolev VV (1969) Course in theoretical astrophysics, 1, vol. F-531. NASA, nasa technical translation 1
Speith R, Riffert H, Ruder H (1995) The photon transfer function for accretion disks around a Kerr black hole. Comput Phys Commun 88:109–120. https://doi.org/10.1016/0010-4655(95)00067-P
Suleimanov VF (1992) Modeling the accretion disks and spectra of cataclysmic variables - Part One - V603-AQUILAE. Sov Astron Lett 18:104–+
Suleimanov VF, Lipunova GV, Shakura NI (2007) The thickness of accretion α-disks: theory and observations. Astron Rep 51:549–562. https://doi.org/10.1134/S1063772907070049
Suleimanov VF, Lipunova GV, Shakura NI (2008) Modeling of non-stationary accretion disks in X-ray novae A 0620-00 and GRS 1124-68 during outburst. Astron Astrophys 491:267–277. https://doi.org/10.1051/0004-6361:200810155. ArXiv 0805.1001
Syunyaev RA, Shakura NI (1977) Disk reservoirs in binary systems and prospects for observing them. Sov Astron Lett 3:138–141
Tanaka T (2011) Exact time-dependent solutions for the thin accretion disc equation: boundary conditions at finite radius. Mon Not R Astron Soc 410:1007–1017. https://doi.org/10.1111/j.1365-2966.2010.17496.x. ArXiv 1007.4474
Tayler RJ (1980) Vertical energy transport in optically thick steady accretion discs. Mon Not R Astron Soc 191:135–150
Thorne KS, Price RH, MacDonald DA (1986) Black holes: the membrane paradigm. Yale University Press, New Haven
Tout CA, Pringle JE (1992) Accretion disc viscosity - a simple model for a magnetic dynamo. Mon Not R Astron Soc 259:604–612
Velikhov EP (1959) Stability of an ideally conducting liquid flowing between cylinders rotating in a magnetic field. Sov J Exp Theor Phys 9:995–998
Watson G (1944) A treatise on the theory of Bessel functions. Cambridge University Press. http://books.google.ru/books?id=WNoIAQAAIAAJ
Weizsäcker CFV (1948) Die Rotation kosmischer Gasmassen. Zeitschrift Naturforschung Teil A 3:524
Wood KS, Titarchuk L, Ray PS, Wolff MT, Lovellette MN, Bandyopadhyay RM (2001) Disk diffusion propagation model for the outburst of XTE J1118+480. Astrophys J 563:246–254. https://doi.org/10.1086/323768. ArXiv astro-ph/0108189
Zaitsev V, Polyanin A (2012) Handbook of exact solutions for ordinary differential equations. Taylor & Francis. http://books.google.ru/books?id=JjPDfRwOmAIC
Zdziarski AA, Kawabata R, Mineshige S (2009) Viscous propagation of mass flow variability in accretion discs. Mon Not R Astron Soc 399:1633–1640. https://doi.org/10.1111/j.1365-2966.2009.15386.x. ArXiv 0902.4530
Zeldovich YB (1981) On the friction of fluids between rotating cylinders. R Soc Lond Proc Ser 374:299–312. https://doi.org/10.1098/rspa.1981.0024
Zeldovich YB, Kompaneets AS (1950) Sbornik posvyashchennyy 70-leityu A. F. Ioffe (in: Collection of papers celebrating the seventieth birthday of A. F. Ioffe). AN SSSR, Moscow, in Russian

Zeldovich YB, Raizer YP (1967) Physics of shock waves and high-temperature hydrodynamic phenomena. Dover, New York
Zel'dovich YB, Shakura NI (1969) X-ray emission accompanying the accretion of gas by a neutron star. Sov Astron 13:175

Chapter 2
The Properties of Spherical Geodesics in the Kerr Metric

Nikolay Shakura

Abstract This small methodological chapter is devoted to considering the motion of particles along spherical geodesical trajectories around rotating black holes. The study of this motion is necessary for understanding the inner structure of the disc tilted to the equatorial plane of the rotating black hole. Moreover, this chapter uses a special approach to find out how the values that are measured in a local Lorentz frame of observers falling freely in an axially symmetric gravitational field are related to each other. This approach allows us to understand better the basic principles of measuring physical values in general relativity. These basic principles, which are systematically presented in the next chapter, are required for a more comprehensive understanding the structure of relativistic tilted accretion discs.

In the Kerr metric, a squared interval may be expressed in the Boyer–Lindquist coordinates as follows (see, for example, Misner et al. 1977):

$$ds^2 = -\left(1 - \tfrac{2r}{\rho^2}\right) dt^2 - \tfrac{4ar\sin^2\theta}{\rho^2} dt d\phi + \tfrac{\rho^2}{\Delta} dr^2 + \rho^2 d\theta^2 + \left(r^2 + a^2 + \tfrac{2a^2 r \sin^2\theta}{\rho^2}\right) \sin^2\theta d\phi^2 , \quad (2.1)$$

where

$$\rho^2 = r^2 + a^2\cos^2\theta , \quad \Delta = r^2 - 2r + a^2 , \quad -1 \le a \le 1 . \quad (2.2)$$

In these coordinates, $GM/c^2 = 1$ and $c = 1$, i.e. the distances are measured in units of half a Schwarzschild radius of the black hole (BH). Thus an unit time is GM/c^3. The Kerr metric describes the structure of spacetime around a rotating BH with mass M and specific dimensionless angular momentum $a = Jc/(GM^2)$. There exist four

N. Shakura (✉)
Sternberg Astronomical Institute, Lomonosov Moscow State University, Moscow, Russia

Kazan Federal University, Kazan, Russia

N. Shakura (ed.), *Accretion Flows in Astrophysics*, Astrophysics and Space Science Library 454, https://doi.org/10.1007/978-3-319-93009-1_2

integrals of motion in this metric: (a) the energy of a test particle $E = -p_t$; (b) the projection of angular momentum of the particle onto the BH rotation axis L_z; (c) the rest mass of the particle m; and (d) the so-called Carter integral of motion

$$Q = p_\phi^2 + \cos^2\theta \left[a^2(m^2 - E^2) + \frac{L_z^2}{\sin^2\theta} \right], \tag{2.3}$$

which is related to the square of the total angular momentum of the particle.

The motion of test particles in the Kerr metric reduces to solving the following system of ordinary differential equations of the first order (Carter 1968):

$$\rho^2 \frac{dr}{d\lambda} = \pm\sqrt{R(r)} \tag{2.4}$$

$$\rho^2 \frac{d\theta}{d\lambda} = \pm\sqrt{\Theta(\theta)} \tag{2.5}$$

$$\rho^2 \frac{d\phi}{d\lambda} = -\left(aE - \frac{L_z}{\sin^2\theta} \right) + \frac{aP}{\Delta} \tag{2.6}$$

$$\rho^2 \frac{dt}{d\lambda} = -a\left(aE\sin^2\theta - L_z \right) + \frac{(r^2 + a^2)P}{\Delta}, \tag{2.7}$$

where differentiation is performed with respect to the affine parameter λ related to the proper time of the particle $\tau = m\lambda$, and

$$R(r) = P^2 - \Delta\left[m^2 r^2 + (L_z - aE)^2 + Q \right], \tag{2.8}$$

$$\Theta(\theta) = Q - \cos^2\theta \left[a^2(m^2 - E^2) + \frac{L_z^2}{\sin^2\theta} \right], \tag{2.9}$$

$$P = E(r^2 + a^2) - L_z a\,. \tag{2.10}$$

The general solution of the system of equations (2.4)–(2.7) may be expressed through elliptical integrals (Bardeen et al. 1972). In spite of the apparent complexity of the right-hand parts of these equations, the motion of particles in the Kerr metric turns out fairly simple from a qualitative point of view. Indeed, let a particle of mass m move with a constant velocity along a straight line in empty space with respect to a spherical system of coordinates. Equating $M = a = 0$ in the system (2.4)–(2.10), we obtain

$$\frac{dr}{d\lambda} = \pm\sqrt{E^2 - m^2 - \frac{L_z^2 + Q}{r^2}}\,, \tag{2.11}$$

$$r^2\frac{d\theta}{d\lambda} = \pm\sqrt{Q + L_z^2 - \frac{L_z^2}{\sin^2\theta}}\,, \tag{2.12}$$

$$r^2\frac{d\phi}{d\lambda} = \frac{L_z}{\sin^2\theta}\,, \tag{2.13}$$

$$\frac{dt}{d\lambda} = E\,. \tag{2.14}$$

It can be seen that the "effective repulsive potential" $\frac{L_z^2+Q}{r^2}$ arises in this system of coordinates, where $L_z^2 + Q$ has the meaning of the total angular momentum relative to the origin of the coordinate system. The rate of change of the radial component $dr/d\lambda$ becomes zero at the minimal travelling path $r_{min} = \sqrt{(L_z^2 + Q)/(E^2 - m^2)}$. In the general case, the equatorial plane of the coordinate system is oriented arbitrarily with respect to the plane passing through the velocity vector and the origin. The joint solution to Eqs. (2.12)–(2.13) makes it possible to find the trajectory trace in the angular coordinates θ, ϕ. Note that if these planes are oriented orthogonally, the angular velocity of the particle $d\theta/d\lambda$ changes the sign at the moment the particle passes through the poles $\theta_p = 0, \pi$ of the coordinate system. Naturally, the actual velocity of the particle does not change at all. Equation (2.14) illustrates the well-known relativistic effect of time dilation in a moving coordinate system.

Further, let us place a mass with a spherically symmetric gravitational field (for example, a Schwarzschild BH) in the origin of the coordinate system. A comparison of (2.5) and (2.6) to (2.12) and (2.13) shows that if $M \neq 0$ and $a = 0$, the right-hand parts of the equations for the angular variables θ and ϕ remain the same as for the free particle, implying that the general solution $\theta(\phi)$ remains the same for a spherically symmetric mass. This, in particular, means that the particle always remains in the same plane when moving in a spherically symmetric gravitational field. The gravitating mass in the origin of coordinate system effectively changes the motion along the r coordinate (giving rise to gravitationally bound orbits with energy $E < m$). In addition, the time dilation effect becomes more appreciable in a gravitational field.

When examining a Kerr black hole ($a \neq 0$), it is necessary to consider a qualitatively new effect related to the particle's being dragged additionally along the ϕ coordinate (the Lense-Thirring effect) due to the vortical component of the stationary gravitational field of the rotating BH. The part of the field due to gravitational potential,[1] although remaining axially symmetric, loses its spherical symmetry, providing the rotating BH with an effective quadrupole moment proportional to Ma^2, which complicates quantitatively the form of the right-hand parts of Eqs. (2.4)–(2.7).

In many astrophysical applications (especially, accretion discs around rotating BHs; see, for example, Bardeen and Petterson (1975) and the following chapters), the so-called spherical orbits with $r = const$ are of special interest.

In order for particles to move along spherical orbits, the following conditions should be fulfilled (Wilkins 1972):

$$R(r) = E^2(r^4 + 2a^2 r + a^2 r^2) - 4EL_z ar - L_z^2(r^2 - 2r) - \Delta(m^2 r^2 + Q) = 0\,, \tag{2.15}$$

$$\frac{\partial R}{\partial r} = E^2(2r^3 + a^2 + a^2 r) - 2EL_z a - (r-1)(L_z^2 + Q + m^2 r^2) - \Delta m^2 r = 0\,. \tag{2.16}$$

From these equations, the functions $E(Q, r)$ and $L_z(Q, r)$ may be expressed in an explicit form. However, according to Bardeen et al. (1972), who was the first to note this circumstance, it is more natural and convenient to perform calculations in the reference frame of a fiducial observer (FIDO), who does not possess any angular momentum with respect to the BH, although rotating with the Lense-Thirring frequency relative to an infinitely remote observer. Any test particle with zero projection of angular momentum onto the BH rotation axis ($L_z = 0$), will have a zero ϕ component of velocity travelling near such a local inertial observer (see further below).

In the general case, the squared interval in the stationary axially symmetric metric is

$$ds^2 = -e^{2\nu}dt^2 + e^{2\psi}(d\phi - \omega dt)^2 + e^{2\lambda}dr^2 + e^{2\mu}d\theta^2\,. \tag{2.17}$$

Since the non-diagonal metric coefficient $g_{t\phi} = -\omega g_{\phi\phi}$ is not equal to zero, it turns out that a FIDO moves with the Lense-Thirring angular velocity:

$$\omega = -\frac{g_{t\phi}}{g_{\phi\phi}}\,. \tag{2.18}$$

[1]The part that is responsible for the gravitational acceleration of a reference observer; see Thorne et al. (1986) for details.

The contravariant components of the metric tensor in Eq. (2.17) are:

$$g^{tt} = -e^{-2\nu}, \quad g^{t\phi} = -\omega e^{-2\nu}, \quad g^{\phi\phi} = e^{-2\psi} - \omega^2 e^{-2\nu},$$
$$g^{rr} = e^{-2\lambda}, \quad g^{\theta\theta} = e^{-2\mu}. \tag{2.19}$$

Thus it follows that the general expression for the squared four-momentum p of a particle with mass m

$$p_\alpha p^\alpha = g^{\alpha\beta} p_\alpha p_\beta = -m^2 \tag{2.20}$$

can be written in the form

$$-m^2 = -e^{-2\nu} p_t^2 - 2\omega e^{-2\nu} p_t p_\phi + (e^{-2\psi} - \omega^2 e^{-2\nu}) p_\phi^2 + e^{-2\lambda} p_r^2 + e^{-2\mu} p_\theta^2 =$$
$$-e^{-2\nu}(p_t + \omega p_\phi)^2 + e^{-2\psi} p_\phi^2 + e^{-2\lambda} p_r^2 + e^{-2\mu} p_\theta^2 . \tag{2.21}$$

In the reference frame of a FIDO, spacetime is locally flat (pseudo-Euclidean) and can be described by the Minkowski metric $g_{\alpha\beta} = \eta_{(\alpha)(\beta)} = diag(-1, 1, 1, 1)$.[2] The general formula for the four-momentum squared (2.20) is expressed using physical values that can be measured by a FIDO, namely, the particle's energy $p^{(t)} = m/\sqrt{1-\beta^2} \equiv m\gamma$ and three spatial components of the momentum $p^{(i)} = m\gamma\beta^{(i)}$ with $i = r, \theta, \phi$ in the standard Lorentz-invariant manner:

$$-(p^{(t)})^2 + (p^{(r)})^2 + (p^{(\theta)})^2 + (p^{(\phi)})^2 = -m^2 , \tag{2.22}$$

where the spatial velocity components are related through the expression $\beta^2 = (\beta^{(r)})^2 + (\beta^{(\theta)})^2 + (\beta^{(\phi)})^2$.

In the general case, the energy $E = -p_t$ of a particle moving along some path in a stationary axially symmetric metric (2.17) is conserved with respect to an infinitely remote observer, and so is the particle's angular momentum projected onto the symmetry axis of the BH, $L_z = p_\phi$ (Bardeen et al. 1972). Equations (2.21) and (2.22) may be used to find the relationship between the energy and the spatial components of the particle's momentum (locally measured by a FIDO) through these constants and metric coefficients:

$$p^{(t)} = m\gamma = e^{-\nu}(E - \omega L_z), \quad p^{(\phi)} = m\gamma\beta^{(\phi)} = e^{-\psi} L_z,$$
$$p^{(r)} = m\gamma\beta^{(r)} = e^{-\lambda} p_r, \quad p^{(\theta)} = m\gamma\beta^{(\theta)} = e^{-\mu} p_\theta \tag{2.23}$$

(note that these quantities may also be directly calculated through the FIDO tetrad components in the given metric; see the next chapter for details).

[2] More details on reference frames and the tetrad representation may be found in Sect. 3.2.

Let us now examine the particular case of spherical orbits, for which $p^{(r)} = 0$. In this case, expressions for locally measurable velocity components can be derived from Eqs. (2.22) and (2.23):

$$\beta^{(\phi)} = \frac{\tilde{L}_z}{\tilde{E} - \omega\tilde{L}_z} e^{\nu-\psi}, \tag{2.24}$$

$$\beta^{(\theta)} = \left[1 - \frac{e^{2\nu}(1 + e^{-2\psi})\tilde{L}_z^2}{(\tilde{E} - \omega\tilde{L}_z)^2}\right]^{1/2}, \tag{2.25}$$

where $\tilde{E} = E/m$ and $\tilde{L}_z = L_z/m$.

For further consideration, it is convenient to introduce the angular velocity of the particle with respect to a remote observer:

$$\Omega(r,\theta) \equiv \frac{d\phi}{dt} = \frac{d\phi/d\tau}{dt/d\tau} = \frac{p^\phi}{p^t} = \frac{g^{\phi\phi}p_\phi + g^{\phi t}p_t}{g^{tt}p_t + g^{\phi t}p_\phi} = \omega + \frac{\tilde{L}_z}{\tilde{E} - \omega\tilde{L}_z} e^{2\nu-2\psi}. \tag{2.26}$$

Substituting (2.26) into (2.24), we obtain

$$\beta^{(\phi)} = (\Omega - \omega)e^{\psi-\nu}. \tag{2.27}$$

As a matter of fact, we have derived the well-known property that the particle, which rotates with an angular velocity $\Omega = \omega$ due to the Lense-Thirring frame-dragging effect with respect to a remote observer, does not rotate in the reference frame of a local FIDO: $\beta^{(\phi)} = 0$.

The motion of particles along spherical orbits crossing the BH equatorial plane is confined in latitude (except for the degenerate case of $L_z = 0$), implying that there exist angle coordinate values θ_* and $\pi - \theta_*$ (the turning points) above and below the equatorial plane, respectively, where $\beta^{(\theta)}(\theta_*) = 0$. This allows us to find the energy for the spherical orbits using Eqs. (2.24), (2.25) and (2.27):

$$\tilde{E} = \frac{e^\nu[1 + \omega^*(\Omega^* - \omega^*)e^{2(\psi-\nu)}]}{[1 - (\Omega^* - \omega^*)^2 e^{2(\psi-\nu)}]^{1/2}} \tag{2.28}$$

and the angular momentum projected onto the BH rotation axis:

$$\tilde{L}_z = \gamma\beta^{(\phi)}e^\psi = \frac{(\Omega^* - \omega^*)e^{2\psi-\nu}}{[1 - (\Omega^* - \omega^*)^2 e^{2(\psi-\nu)}]^{1/2}}, \tag{2.29}$$

where the angular velocities Ω^* and ω^*, along with the corresponding metric coefficients, are to be calculated at the turning points θ_* or $\pi - \theta_*$. At these points, the angular velocity of the particle can be derived from the geodesic equation

$$\frac{d^2x^\mu}{d\lambda^2} + \Gamma^\mu_{\alpha\beta} p^\alpha p^\beta = 0 \,. \tag{2.30}$$

In case of spherical orbits, the first term vanishes for the radial component of Eq. (2.30). Furthermore, $p^{(\theta)}(\theta_*) = 0$ at the turning points, which yields

$$\frac{\Gamma^*_{rtt} dt^2 + 2\Gamma^*_{rt\phi} dt d\phi + \Gamma^*_{r\phi\phi} d\phi^2}{d\lambda^2} = 0 \,, \tag{2.31}$$

where in view of the axial symmetry of the problem, the Christoffel symbols depend only on partial derivatives of the metric tensor over the r coordinate:

$$\Gamma^*_{rtt} = -\frac{1}{2}\left(\frac{\partial g_{tt}}{\partial r}\right)_{\theta=\theta_*} = -\frac{1}{2} g^*_{tt,r}; \quad \Gamma^*_{rt\phi} = -\frac{1}{2} g^*_{t\phi,r}, \quad \Gamma^*_{r\phi\phi} = -\frac{1}{2} g^*_{\phi\phi,r} \,. \tag{2.32}$$

Taking into account the definition of the angular velocity $\Omega = d\phi/dt$, we see that Eq. (2.31) becomes an algebraic equation relative to Ω; its two roots

$$\Omega^*_{1,2} = \frac{-g^*_{t\phi,r} \pm \sqrt{g^{*2}_{t\phi,r} - g^*_{\phi\phi,r} g^*_{tt,r}}}{g^*_{\phi\phi,r}} \tag{2.33}$$

correspond to the angular velocities at the turning points in direct spherical orbits (coincident with the direction of BH rotation) and retrograde spherical orbits, respectively. Using the explicit form of the Kerr metric (2.1), we find

$$\Omega^*_{1,2} = \frac{q_*}{\sin\theta_* (\pm \rho_*^2 \sqrt{r} + a q_* \sin\theta_*)} \,, \tag{2.34}$$

where $q_*^2 = r^2 - a^2\cos^2\theta_*$ and $\rho_*^2 = r^2 + a^2\cos^2\theta_*$. Substituting (2.34) and the values of the corresponding metric coefficients from the metric (2.1) into (2.28) and (2.29) and cancelling out the common multiple $\rho_*^2\sqrt{r}/(\rho_*^2\sqrt{r} \pm a q_* \sin\theta_*)$ in the numerators and denominators of (2.28) and (2.29), we finally obtain:

$$\tilde{E} = \frac{1 - \frac{2r}{\rho_*^2} \pm \frac{a q_*}{\rho_*^2}\sqrt{\frac{1}{r}}\sin\theta_*}{\left[1 - \frac{3r}{\rho_*^2} \pm \frac{2a q_*}{\rho_*^2}\sqrt{\frac{1}{r}}\sin\theta_* + \frac{a^2}{\rho_*^2 r}\cos^2\theta_*\right]^{1/2}} \,, \tag{2.35}$$

$$\tilde{L}_z = \frac{\pm\frac{q_*}{\rho_*^2}\sqrt{\frac{1}{r}}(r^2+a^2)\sin\theta_* - \frac{2ar}{\rho_*^2}\sin^2\theta_*}{\left[1-\frac{3r}{\rho_*^2}\pm\frac{2aq_*}{\rho_*^2}\sqrt{\frac{1}{r}}\sin\theta_* + \frac{a^2}{\rho_*^2 r}\cos^2\theta_*\right]^{1/2}}. \tag{2.36}$$

The direct substitution shows that (2.35) and (2.36) take the familiar form for circular equatorial orbits with $\theta_* = \pi/2$ (Bardeen 1973), whereas for spherical polar orbits with $L_z = 0$ and $\theta_* = 0$, the expression (2.35) coincides with the result provided in Lightman et al. (1975).

In the end, using (2.9) (taking into account that $d\theta/d\lambda = 0$, we have $\Theta(\theta_*) = 0$ at the points θ_*) along with (2.35) and (2.36), we find the Carter integral:

$$\tilde{Q} = \cos^2\theta_* r \times$$
$$\frac{1 \mp \frac{4aq_* r^2}{\rho_*^4}\sqrt{\frac{1}{r}}\sin\theta_* - \frac{4a^2 r}{\rho_*^4}\cos^2\theta_* + \frac{a^2(3r^2+a^2\cos^2\theta_*)}{\rho_*^4}\sin^2\theta_*}{\left[1-\frac{3r}{\rho_*^2}\pm\frac{2aq_*}{\rho_*^2}\sqrt{\frac{1}{r}}\sin\theta_* + \frac{a^2}{\rho_*^2 r}\cos^2\theta_*\right]^{1/2}}. \tag{2.37}$$

In disc accretion onto BHs, the angle θ_* determines, at large distances, the orientation of the disc plane relative to the BH equatorial plane. The structure of tilted relativistic discs around rotating BHs is discussed in the next two chapters in detail.

References

Bardeen JM (1973) Timelike and null geodesics in the Kerr metric. In: Dewitt C, Dewitt BS (eds) Black holes (Les Astres Occlus). Gordon and Breach, New York, pp 215–239

Bardeen JM, Petterson JA (1975) The Lense-Thirring effect and accretion disks around Kerr black holes. Astrophys J 195:L65. https://doi.org/10.1086/181711

Bardeen JM, Press WH, Teukolsky SA (1972) Rotating black holes: locally nonrotating frames, energy extraction, and scalar synchrotron radiation. Astrophys J 178:347–370. https://doi.org/10.1086/151796

Carter B (1968) Global structure of the Kerr family of gravitational fields. Phys Rev 174:1559–1571. https://doi.org/10.1103/PhysRev.174.1559

Lightman AP, Press WH, Price RH, Teukolsky SA (1975) Problem book in relativity and gravitation. Princeton University Press, Princeton

Misner CW, Thorne KS, Wheeler JA (1977) Gravitation, vol 3. Princeton University Press, Princeton

Thorne KS, Price RH, MacDonald DA (1986) Black holes: the membrane paradigm. Yale University Press, New Haven

Wilkins DC (1972) Bound geodesics in the Kerr metric. Phys Rev D 5:814–822. https://doi.org/10.1103/PhysRevD.5.814

Chapter 3
Relativistic Standard Accretion Disc

Viacheslav Zhuravlev

Abstract In this chapter we present a model of a standard accretion disc around a rotating black hole taking general relativity effects into full account. This model was first described in the paper by Novikov and Thorne (Black holes (Les Astres Occlus). Gordon and Breach, New York, 1973) and has since then been used in many studies to obtain convincing evidence of the existence of black holes, in both stellar binary systems and active galactic nuclei. It remains topical since a full account of the general relativistic properties of the motion of matter in the disc, and the generation of disc emission, allows the position of the inner disc radius and hence the black hole spin to be inferred from observations. In addition, the standard accretion disc is the basis for more complicated theories of warped (twisted) accretion discs, which are formed when the accreting matter moves outside the equatorial plane of a rotating black hole.

Everywhere below, natural units, $G = c = 1$, are used. If the mass is measured in units of the black hole mass, M, the unit of length is half the Schwarzschild gravitational radius, $R_g/2$, such that $R_g/2 = GM/c^2 = 1$ and the unit of time is the light crossing time for a unit length.

In addition, Latin indices $i, j, k \ldots$, taking values from 0 to 3, are used to denote components of vectors, with the zero component standing for the time coordinate. Also, wherever needed the Einstein summation convention is used.

V. Zhuravlev (✉)
Sternberg Astronomical Institute, Lomonosov Moscow State University, Moscow, Russia
e-mail: zhuravlev@sai.msu.ru

N. Shakura (ed.), *Accretion Flows in Astrophysics*, Astrophysics and Space Science Library 454, https://doi.org/10.1007/978-3-319-93009-1_3

3.1 Space-Time Near Rotating Black Holes

3.1.1 The Kerr Metric

The properties of spacetime near a rotating black hole are described by an axially symmetric and stationary metric of the form

$$ds^2 = \left(1 - \tfrac{2R}{\rho^2}\right) dt^2 - \tfrac{4aR\sin^2\theta}{\rho^2} dt d\phi - \tfrac{\rho^2}{\Delta} dR^2 - \rho^2 d\theta^2 - \left(R^2 + a^2 + \tfrac{2a^2 R\sin^2\theta}{\rho^2}\right) \sin^2\theta d\phi^2 \,, \tag{3.1}$$

where the signature $(1, -1, -1, -1)$ is chosen and the following notations are used:

$$\rho^2 = R^2 + a^2\cos^2\theta \,, \quad \Delta = R^2 - 2R + a^2 \,, \quad -1 \le a \le 1 \,. \tag{3.2}$$

The coordinates $\{t, \phi, R, \theta\}$ are called the Boyer-Lindquist coordinates.[1] Far away from the gravitating body, the spatial part of these coordinates, in the limit of a zero black hole spin parameter, a, transforms into the usual spherical coordinates, where ϕ is the azimuthal angle. For a non-zero a, it transforms to generalized spherical coordinates in which the surfaces of constant radial distance, $R = const$, represent spheroids with the aspect ratio $R/(R^2 + a^2)^{1/2}$.

The space-time described by (3.1) is axially symmetric with respect to the line $\theta = 0$, called the black hole rotation axis. The plane corresponding to $\theta = \pi/2$, is called the black hole equatorial plane.

In (3.1) an important quantity appears:

$$\omega = \frac{2aR}{\Sigma^2} \,, \tag{3.3}$$

which has the dimension of frequency. This is the angular velocity that any freely moving observer acquires in the direction of the black hole rotation.

As described in the literature on the structure of rotating black holes (see, e.g., Chandrasekhar (1992), paragraph 58), the metric (3.1) has several special hypersurfaces, including the event horizon and the ergosphere. However, as we show below, for the astrophysical problem under consideration, of most importance is the dynamics of free circular motion of particles in the equatorial plane of the gravitating body. This motion has additional features in comparison to the corresponding Newtonian problem. Note that in the next chapter we also discuss weakly elliptical orbits, slightly inclined to the equatorial plane.

[1] Unlike in the previous chapter, in this chapter the spherical radial coordinate in the Kerr metric is denoted by R. The lowercase letter, r, is reserved for the cylindrical coordinates.

We consider a standard, and hence geometrically thin, accretion disc. In this chapter we discuss the basic case of a flat disc. By definition, this is a stationary flow of matter with mirror symmetry with respect to its midplane and axial symmetry with respect to the line perpendicular to this plane. Clearly, such a model flow can be described by dynamical equations in an axially symmetric metric only if the disc symmetry plane coincides with the equatorial plane of the black hole. To tackle the problem, it is sufficient to know the form of the metric close to the plane $\theta = \pi/2$. Changing to cylindrical coordinates using the standard transformation

$$r = R\sin\theta, \quad z = R\cos\theta,$$

all metric coefficients g_{ik} in (3.1) can be expanded in power series of the small ratio $z/r \ll 1$. For geometrically thin discs, corrections to g_{ik} due to non-equatorial motion up to $(z/r)^2$ are sufficient. Indeed, one of the basic equations describing the disc, namely, the projection of the relativistic analogue of the Euler equation onto the direction normal to the disc plane, must be odd with respect to the coordinate reflection $z \to -z$ due to the mirror symmetry of the disc. This means that in its expansion in (z/r) only odd powers of (z/r) must be present. According to the main assumption of the smallness of (z/r), only the first term in this expansion should be kept. This, in turn, corresponds to an expansion in series of g_{ik} up to quadratic terms, since only first derivatives of g_{ik}, characterizing the 'strength' of the gravitational field, enter the dynamical equations.

Note, however, that hydrodynamic equations also contain a second covariant derivative of the velocity field (see below), and hence the final expressions can involve second derivatives of g_{ik} with respect to z, which may seem to require that we keep terms of the order of $(z/r)^3$ in g_{ik}. But this is not required, since, as follows from the explicit form of the stress-energy tensor, such terms can appear only when multiplied by some of the viscous coefficients, which in turn cannot be greater than of the order of (z/r) being proportional to the characteristic mixing length in the fluid. The latter is initially assumed to be less than the disc thickness.

As regards the other equations, namely (see below): the two projections of the relativistic analogue of the Euler equation onto the disc plane, the energy balance equation and the rest energy conservation law—the same symmetry considerations imply that they are even under the coordinate reflection $z \to -z$. Therefore, the leading term is of the zeroth order in (z/r) in the metric expansion.

Using these expansions and expressions for the coordinate differentials

$$dR = \left(1 - \frac{1}{2}\frac{z^2}{r^2}\right) dr + \frac{z}{r} dz,$$

$$d\theta = \frac{z}{r}\frac{dr}{r} - \left(1 - \frac{z^2}{r^2}\right)\frac{dz}{r},$$

we find the metric in the following form (see also Riffert and Herold 1995):

$$ds^2 = \left[1 - \frac{2}{r} + \frac{z^2}{r^3}\left(1 + \frac{2a^2}{r^2}\right)\right] dt^2 - \left[r^2 + a^2 + \frac{2a^2}{r}\right.$$
$$\left. - \frac{a^2 z^2}{r^2}\left(1 + \frac{5}{r} + \frac{2a^2}{r^3}\right)\right] d\phi^2 + \frac{2a}{r}\left[2 - \frac{z^2}{r^2}\left(3 + \frac{2a^2}{r^2}\right)\right] dt d\phi -$$
$$\left\{1 - \frac{z^2}{r^2 D}\left[\frac{3}{r} - \frac{4}{r^2} - \frac{a^2}{r^2}\left(3 - \frac{6}{r} + \frac{2a^2}{r^2}\right)\right]\right\} \frac{dr^2}{D} - \frac{2z}{rD}\left(\frac{2}{r} - \frac{a^2}{r^2}\right) dr dz -$$
$$\left[1 + \frac{z^2}{r^2 D}\left(\frac{2}{r} - \frac{2a^2}{r^3} + \frac{a^4}{r^4}\right)\right] dz^2, \tag{3.4}$$

where the notation

$$D = 1 - \frac{2}{r} + \frac{a^2}{r^2}$$

is introduced. Below, we also use (with a few exceptions) the notations introduced in the original paper by Novikov and Thorne (1973) for the relativistic correction coefficients.

Finally, the inverse of the matrix g^{ik} corresponding to a double-contravariant tensor has the form:

$$g^{ik} = \begin{Vmatrix} (g_{tt} g_{\phi\phi} - g_{t\phi}^2)^{-1} \times \begin{vmatrix} g_{\phi\phi} & -g_{t\phi} \\ -g_{t\phi} & -g_{tt} \end{vmatrix} & 0 \\ 0 & (g_{rr} g_{zz} - g_{rz}^2)^{-1} \times \begin{vmatrix} g_{zz} & -g_{rz} \\ -g_{rz} & -g_{rr} \end{vmatrix} \end{Vmatrix} \tag{3.5}$$

3.1.2 Circular Equatorial Geodesics

The expression for circular equatorial geodesics can be conveniently found from the extremum condition for the distance along them. Here we follow the exposition from Hobson et al. (2006), (their paragraphs 13.10 and 13.13). Indeed, for time-like trajectories the functional

$$S = \int L ds = \int g_{ik} \frac{dx^i}{ds} \frac{dx^k}{ds} ds,$$

should be minimal, which is equivalent to the Euler-Lagrange equations for L:

$$\frac{d}{ds}\left(\frac{\partial L}{\partial \dot{x}^i}\right) - \frac{\partial L}{\partial x^i} = 0, \tag{3.6}$$

where $U_g^i \equiv dx^i/ds \equiv \dot{x}^i$ is the four-velocity in Boyer-Lindquist coordinates. As L does not explicitly depend on t and ϕ, the following quantities are conserved:

$$g_{ti}U_g^i = k,$$

$$g_{\phi i}U_g^i = -h,$$

where k and h have the meaning of the time and azimuthal covariant velocity components, respectively.

In explicit form, using the components g_{ik} from (3.4) at $z = 0$, we find:

$$\left(1 - \frac{2}{r}\right)\dot{t} + \frac{2a}{r}\dot{\phi} = k, \tag{3.7}$$

$$\frac{2a}{r}\dot{t} - \left(r^2 + a^2 + \frac{2a^2}{r}\right)\dot{\phi} = -h. \tag{3.8}$$

We temporarily assume that the motion is not necessarily circular and $U_g^r \neq 0$. Instead of the r-component of the Euler-Lagrange equations, it is more convenient to use the condition of normalization of the four-velocity of particles with non-zero mass:

$$g^{tt}k^2 - 2g^{t\phi}kh + g^{\phi\phi}h^2 + g^{rr}(U_r)^2 = 1. \tag{3.9}$$

This yields the following equation for k and h:

$$\frac{\dot{r}^2}{2} + V_{eff}(r) = \frac{k^2 - 1}{2}, \tag{3.10}$$

where we introduce the effective potential

$$V_{eff} = -\frac{1}{r} + \frac{h^2 - a^2(k^2 - 1)}{2r^2} - \frac{(h - ak)^2}{r^3}. \tag{3.11}$$

The conditions for circular motion include, first, $\dot{r} = 0$ and, second, $\ddot{r} = 0$ (for the particle to stay in a circular orbit). The latter condition is equivalent to the vanishing of the derivative of V_{eff} with respect to r:

$$1 + \frac{a^2(k^2 - 1) - h^2}{r} + \frac{3(h - ak)^2}{r^2} = 0. \tag{3.12}$$

Equation (3.10) with $\dot{r} = 0$ and Eq. (3.12) allow us to determine k and h as functions of r and then, using (3.7) and (3.8), to find U_g^t and U_g^ϕ.

To solve the first problem, let us introduce the new variable $\mu \equiv h - ak$ and, to facilitate manipulations, make the change $u \equiv 1/r$. Then Eq. (3.10), with $\dot{r} = 0$

together with Eq. (3.12), yield the following equation for μ:

$$u^2[(3u-1)^2-4a^2u^3]\mu^4-2u[(3u-1)(a^2u-1)-2ua^2(u-1)]\mu^2+(au-1)^2=0. \tag{3.13}$$

The solution of (3.13) for a stable circular prograde orbit has the form:

$$\mu=-\frac{a\sqrt{u}-1}{[u(1-3u+2au^{3/2})]^{1/2}}. \tag{3.14}$$

Using (3.14) and (3.10) taken at $\dot{r}=0$, we find the constants h and k, as well as the components U_g^i:

$$U_g{}^t=C^{-1/2}B, \quad U_g{}^\phi=(r^3C)^{-1/2}, \quad U_g{}^r=U_g{}^z=0, \tag{3.15}$$

where

$$B=1+\frac{a}{r^{3/2}}, \quad C=1-\frac{3}{r}+\frac{2a}{r^{3/2}}. \tag{3.16}$$

It is easy to verify that the modulus of this vector is equal to unity:

$$g_{ik}\,U_g{}^iU_g{}^k=1$$

The angular velocity as measured by the clock of an infinite observer (who measures the coordinate time t), corresponding to such motion, is

$$\Omega=\frac{d\phi}{dt}=r^{-3/2}B^{-1}. \tag{3.17}$$

It follows that in the Schwarzschild case this value exactly coincides with the Keplerian angular velocity.

3.1.3 Radius of the Innermost (Marginally) Stable Orbit

Stable circular motion is no longer possible when the minimum of the function $V_{eff}(r,h(r_c),k(r_c))$ disappears at $r=r_c$, where r_c is the radius of a circular orbit. This is equivalent to the condition

$$\left.\frac{d^2V_{eff}}{dr^2}\right|_{r=r_c}=0,$$

which leads to the quartic equation

$$z^4 - 6z^2 + 8az - 3a^2 = 0, \tag{3.18}$$

where $z \equiv r^{1/2}$.

Using the Ferrari method (see, e.g., Korn and Korn 2000), we write the corresponding auxiliary cubic equation:

$$y^3 - 12y^2 + 12(3 + a^2)y - 64a^2 = 0. \tag{3.19}$$

The real root of Eq. (3.19) is related to the Cardano solution to the corresponding incomplete cubic equation and is given by

$$y_1 = -2(1 - a^2)^{1/3}[(1 + a)^{1/3} + (1 - a)^{1/3}] + 4. \tag{3.20}$$

Next, having obtained y_1, it is possible to use the Ferrari solution to write the quadratic equation that gives two real roots of (3.18):

$$p^2 + \sqrt{y_1}\,p + \frac{1}{2}\left(-6 + y_1 - \frac{8a}{\sqrt{y_1}}\right) = 0\,. \tag{3.21}$$

The larger root of (3.21), p_1, determines the boundary of the stable circular motion of a test particle in the equatorial plane, which we denote as $r = r_{ms}$. Thus,

$$r_{ms} = p_1^2 = 3 + \frac{4a}{\sqrt{y_1}} - (-y_1^2/4 + 4a\sqrt{y_1} + 3y_1)^{1/2}. \tag{3.22}$$

It is easy to verify that the result (3.22) coincides with the expression presented in Page and Thorne (1974), (see formula (15k) therein), taking into account that the auxiliary values $Z_{1,2}$ there take the form $Z_1 \equiv 3 - y_1/2$ and $Z_2 \equiv 4a/\sqrt{y_1}$ for $a \geq 0$ in our notation.

In the Schwarzschild metric, $a = 0$, we recover the well-known result that the circular motion becomes unstable for $r < 6$, i.e. at distances smaller than three gravitational radii from the black hole. For slow rotation, $1 \gg a > 0$, we have $r_{ms} \approx 6 - 4\sqrt{6}a/3$, and hence the zone of stable motion moves closer to the event horizon. In the limit case $a = 1$ we find $r_{ms} = 1$, i.e. the marginally stable circular orbit coincides with the gravitational radius of a black hole of extreme spin.

During accretion, gas elements in the disc slowly approach r_{ms} by losing their angular momentum due to the action of viscous forces. Once the gas elements fall into the region with $r < r_{ms}$, due to instability of the circular motion, they no longer need to lose angular momentum to approach the black hole. This means that the matter falls freely inside r_{ms}, and the standard accretion disc model assumes that r_{ms} is the inner disc radius.

3.2 Choice of Reference Frame

3.2.1 Bases in General Relativity

Mechanical laws, formulated in the form of vector equations, can be written in symbolic form irrespective of observer or reference frame. But to represent some physical quantity describing a natural phenomenon in the form of a set of numerical values, the measurement procedure should be specified. In Newtonian mechanics, this means that the observer introduces a coordinate system, and then at each point in space he/she arbitrarily constructs three basis vectors. The coordinate system and the basis vectors can evolve over time, which is equal at all points. A tool measuring time, together with the coordinate grid and a vector basis, form a reference frame, in which any physical value (scalar, vector or tensor) can be measured, i.e. it can be represented by a number or a set of numbers.

The situation in relativistic mechanics is different: since it is not possible to consider time independently, it becomes the fourth component of the space-time continuum. Therefore, the choice of reference frame reduces to the construction of a coordinate system and four basis vectors determined at each point in space-time. In general, this set of basis vectors is usually referred to as a *tetrad*. Here, there is no universal observer any more; instead, a set of observers moving along a certain family of world lines is considered. If one of the tetrad orts, conventionally corresponding to the time direction, is tangent at each point to these world lines, the tetrad is said to be 'transported' by the observers. The last statement can easily be understood, since in such a basis, the four-velocity of each observer at any time has a non-zero projection only on the 'time' ort. In other words, the observers are at rest in this basis, i.e. transport it with them.

3.2.1.1 Coordinate Representation

Thus, the choice of coordinate system and the choice of the tetrad are independent procedures. Nevertheless, if there is a coordinate system, x^i, the tetrad is frequently chosen in such a way that each basis vector, $\mathbf{e}_i$, is tangent to the corresponding coordinate line. Here the moduli of orts of this so-called coordinate basis are chosen such that their pairwise scalar products are equal to the corresponding metric coefficients:

$$(\mathbf{e}_i \cdot \mathbf{e}_k) = g_{ik}. \tag{3.23}$$

We recall that in differential geometry (see paragraphs 3.1–3.4 in Hobson et al. 2006), such coordinate orts are introduced as objects isomorphic to the partial derivatives of an arbitrary scalar function on the manifold with respect to the coordinates,

$$\mathbf{e}_i \equiv \frac{\partial}{\partial x^i}. \tag{3.24}$$

Any (tangent) vector is a linear combination of the coordinate orts, and the components of this linear combination are called contravariant components of the vector.

In addition to $\mathbf{e_i}$ the so-called dual basis, $\mathbf{e^i}$, is introduced with the orts defined as

$$(\mathbf{e_i} \cdot \mathbf{e^j}) = \delta_i{}^j, \tag{3.25}$$

where $\delta_i{}^j$ is the Kronecker symbol. The condition (3.25) implies that each ort of the dual basis has a unit projection on the corresponding ort of the coordinate basis and is orthogonal to all other orts of the coordinate basis.

The dual coordinate orts, in turn, are introduced as objects isomorphic to the coordinate differentials,

$$\mathbf{e}^j \equiv dx^i. \tag{3.26}$$

Next, if we use the fact that any tangent vector **A** can be alternatively presented as a linear combination of dual coordinate orts, whose coefficients are referred to as covariant vector components, we obtain the well-known rule of lowering vector indices:

$$A_k = A_i(\mathbf{e}^i \cdot \mathbf{e}_k) = (A_i\mathbf{e}^i \cdot \mathbf{e}_k) = (A^i\mathbf{e}_i \cdot \mathbf{e}_k) = A^i(\mathbf{e}_i \cdot \mathbf{e}_k) = A^i g_{ik}. \tag{3.27}$$

In a similar way it is easy to show that if we introduce the notation $g^{ik} \equiv (\mathbf{e}^i \cdot \mathbf{e}^k)$, then due to the duality of bases, the matrix g^{ik} is inverse to the matrix g_{ik}, and the rule of raising of vector indices holds. A similar representation in coordinate bases can be extended to the more general case of tensors.

3.2.1.2 Tetrad Representation

In this and the subsequent sections we mostly follow the exposition given in paragraph 7 of Chandrasekhar (1992). Assume that now we project the same vectors and tensors on an arbitrary tetrad defined by the relations

$$\mathbf{e}_{(a)} = e_{(a)}{}^i \frac{\partial}{\partial x^i}, \tag{3.28}$$

where $e_{(a)}{}^i$ are some functions of coordinates, and the indices labelling the tetrad orts are in parentheses.

From the duality condition (3.25) we can introduce the dual tetrad:

$$\mathbf{e}^{(a)} = e^{(a)}{}_i\, dx^i, \tag{3.29}$$

where $e^{(a)}{}_i$ is the matrix inverse to $e_{(a)}{}^i$.

In these matrices, there are two kinds of indices: coordinate and tetrad ones. The coordinate indices can be lowered or raised using the metric (3.1). We can impose an additional constraint on the tetrad:

$$e_{(a)}{}^i e_{(b)i} = \eta_{(a)(b)}, \quad e^{(a)i} e^{(b)}{}_i = \eta^{(a)(b)}, \tag{3.30}$$

where

$$\eta_{(a)(c)}\, \eta^{(c)(b)} = \delta_{(a)}{}^{(b)} \tag{3.31}$$

are mutually inverse matrices and $\eta_{(c)(b)} = diag(1, -1, -1, -1)$ is the Minkowski matrix. In other words, we require that the original and dual tetrads be orthonormal in four-dimensional pseudo-Euclidean space.

Using the above relations, it is straightforward to show that

$$e^{(a)}{}_i\, e_{(a)j} = g_{ij}, \tag{3.32}$$

and therefore the following alternative expression for the interval squared holds:

$$ds^2 = \eta_{(a)(b)}(e^{(a)}{}_i\, dx^i)\,(e^{(b)}{}_k\, dx^k) = \eta_{(a)(b)}\, \mathbf{e}^{(a)}\, \mathbf{e}^{(b)}, \tag{3.33}$$

which is useful below.

Note that the values in parentheses on the right-hand side of (3.33) can be considered as infinitesimal shifts along the corresponding orts of the tetrad. Therefore, in the introduced tetrad representation with an orthonormal tetrad, the square of an interval takes exactly the same form as in the Minkowski space-time of special relativity. Similarly, the expressions (3.28) can be thought of as directional derivatives along the tetrad's orts, and these have exactly the form that the usual partial derivatives, with respect to coordinates in the coordinate basis, take when changing from the coordinate basis to the tetrad one.

Using the definitions and relations given above, it is easy to see how the tetrad components of vectors are expressed through their coordinate components. Tetrad components of a vector are written as

$$A_{(a)} = e^i_{(a)} A_i, \quad A^{(a)} = e^{(a)}_i A^i = \eta^{ab} A_{(b)}. \tag{3.34}$$

Conversely,

$$A_i = e^{(a)}_i A_{(a)}, \quad A^i = e^i_{(a)} A^{(a)}.$$

Similar expressions can be written for a tensor of any valence. For example, for a two-covariant tensor, we have

$$T_{(a)(b)} = e^i_{(a)} e^j_{(b)} T_{ij} = e^i_{(a)} T_{i(b)},$$

and conversely,

$$T_{ij} = e^{(a)}_i e^{(b)}_j T_{(a)(b)} = e^{(a)}_i T_{(a)j}. \tag{3.35}$$

Note in conclusion that the relations (3.34) and isomorphism (3.26) can be used to find contravariant components of the four-velocity in the tetrad representation:

$$U^{(a)} = \frac{\mathbf{e}^{(a)}}{ds}. \tag{3.36}$$

That is, it is again a unit tangent vector along the world line, but now its components are given by small shifts along the corresponding orts of the dual basis. Using (3.34) it is easy to find the relation between the conventional coordinate components of four-velocity, $U^i = dx^i/ds$, and its tetrad components. Covariant tetrad components are derived from contravariant ones using the standard rule in special relativity: lowering a spatial index is equivalent to changing the sign of the corresponding component.

3.2.1.3 Covariant Derivative in Tetrad Representation

Let us calculate the directional derivative along a tetrad ort from a contravariant component of a vector:

$$A_{(a),(b)} = e^i_{(b)} \frac{\partial}{\partial x^i} A_{(a)} = e^i_{(b)} \frac{\partial}{\partial x^i} e^j_{(a)} A_j = e^i_{(b)} [e^j_{(a)} A_{j;i} + A_k e^k_{(a);i}], \tag{3.37}$$

where the semicolon denotes the usual covariant derivative in the coordinate basis.

Expression (3.37) can be recast to the form:

$$A_{(a),(b)} = e^j_{(a)} A_{j;i} e^i_{(b)} + e_{(a)k;i} e^i_{(b)} e^k_{(c)} A^{(c)}, \tag{3.38}$$

whence

$$e^j_{(a)} A_{j;i} e^i_{(b)} = A_{(a),(b)} - \gamma_{(c)(a)(b)} A^{(c)}, \tag{3.39}$$

where $\gamma_{(c)(a)(b)}$ are the so-called *Ricci rotation coefficients*,

$$\gamma_{(a)(b)(c)} = e_{(b)k;i} e^i_{(c)} e^k_{(a)}. \tag{3.40}$$

An important point is that for orthonormal bases satisfying (3.30), the coefficients $\gamma_{(a)(b)(c)}$ are antisymmetric in the first two indices. Indeed,

$$0 = (\eta_{(b)(a)})_{,i} = (e_{(b)k}e^k_{(a)})_{;i} = e_{(b)k;i}e^k_{(a)} + e_{(b)k}e^k_{(a);i} = e_{(b)k;i}e^k_{(a)} + e^k_{(b)}e_{(a)k;i}.$$

Comparing this relation with (3.40) proves the stated property of the Ricci coefficients.

Finally, let us discuss one more useful property of coefficients (3.40): to calculate these coefficients only partial derivatives of the components of the tetrad basis orts are needed, and therefore the Christoffel symbols are not required. Indeed, we consider auxiliary combinations

$$\lambda_{(a)(b)(c)} = e_{(b)i,j}[e^i_{(a)}e^j_{(c)} - e^j_{(a)}e^i_{(c)}], \tag{3.41}$$

and rewrite them in the form

$$\lambda_{(a)(b)(c)} = e^i_{(a)}e^j_{(c)}[e_{(b)i,j} - e_{(b)j,i}]. \tag{3.42}$$

In the last expression, the ordinary partial derivatives can be substituted by covariant ones, since the additional terms with Christoffel symbols are symmetric in i, j. Then expression (3.42) is equal to the difference $\gamma_{(a)(b)(c)} - \gamma_{(c)(b)(a)}$.

But in this case,

$$\gamma_{(a)(b)(c)} = 1/2[\lambda_{(a)(b)(c)} + \lambda_{(c)(a)(b)} - \lambda_{(b)(c)(a)}] \tag{3.43}$$

and, using (3.41), it is possible to calculate the Ricci rotation coefficients by taking the partial derivatives of the components of the tetrad basis orts.

We now consider formula (3.39). The left-hand side represents simply the projection on the tetrad basis of a rank-2 covariant tensor obtained by taking the derivative of some vector field. Therefore, this combination has the meaning of the covariant derivative of a vector in a non-coordinate basis.

Next, the right-hand side of (3.39) has exactly the same form as the covariant derivative in a coordinate basis, with the only difference that it involves tetrad indices (which can be raised or lowered, including for $\gamma_{(a)(b)(c)}$, using the Minkowski metric). It can be shown that the same holds for contravariant components of a vector field and for tensor fields in general.

Thus, as the components of a covariant derivative in a tetrad basis has the same form as in a coordinate basis, it is convenient to use the same notations and terms that are used in the coordinate basis. In particular, the Ricci rotation coefficients are simply referred to as connection symbols in a given basis. We emphasize once again that they should not be confused with the Christoffel symbols, which represent another limit case of connection coefficients in a coordinate basis and have different index symmetry.

3.2.2 A Tetrad Transported by Rotating Observers

We construct a tetrad basis related at each point in space-time to observers moving around a black hole in equatorial circular orbits with angular velocity Ω.

Strictly at $z = 0$, this is the free motion along geodesics found in Sect. 3.1.2. However, for a small deviation from the equatorial plane, such motion, corresponding to constant z, is possible only if there is some external supporting force. In the case of a gas disc, for example, this force is due to the pressure gradient.

To start the construction, we direct the time ort of the tetrad along the world line under discussion. Using the four-vector of the geodesic found in Sect. 3.1.2, we write it in the form

$$\mathbf{e}_{(t)} = ({U_g}^t + Z_0)\,\frac{\partial}{\partial t} + {U_g}^\phi\,\frac{\partial}{\partial \phi},$$

where we add the correction factor $Z_0(z/r)$ to the time coordinate component, since the modulus of the vector $\mathbf{e}_{(t)}$ should be equal to unity away from the equatorial plane as well, whereas the vector $\mathbf{U}_g$ itself is unitary only at $z = 0$. Clearly, with account for this correction, $\mathbf{e}_{(t)}$ would correspond to the four-velocity of the real motion. A calculation of the modulus of the vector $\mathbf{e}_{(t)}$ in metric (3.4) shows that it is equal to unity under the following condition:

$$Z_0 = -\left(\frac{z}{r}\right)^2 \frac{H}{2rGC^{1/2}},$$

where we introduce the relativistic correction coefficients

$$G = 1 - \frac{2}{r} + \frac{a}{r^{3/2}} \tag{3.44}$$

and

$$H = 1 - \frac{4a}{r^{3/2}} + \frac{3a^2}{r^2}. \tag{3.45}$$

Thus, the ort $\mathbf{e}_{(t)}$ is transported by the observer rotating around the black hole with a frequency equal to the ϕ-component of $\mathbf{e}_{(t)}$, which is independent of z. This frequency corresponds to the free circular motion in the equatorial plane of the black hole, and rotation occurs in planes of constant z.

We now calculate the time ort of the dual basis. According to the convention rule for raising and lowering coordinate indices, we have

$$\mathbf{e}^{(t)} = ({U_g}^t g_{tt} + {U_g}^\phi g_{t\phi})\, dt + ({U_g}^t g_{\phi t} + {U_g}^\phi g_{\phi\phi})\, d\phi.$$

Next, we consider the part of metric (3.4) containing the differentials dr and dz. It can be rewritten in the form (see the result (3.33) in the previous section):

$$ds_{rz}^2 = -\left[\mathbf{e}^{(r)}\right]^2 - \left[\mathbf{e}^{(z)}\right]^2,$$

where

$$\mathbf{e}^{(r)} = |g_{rr}|^{1/2}\,dr - \frac{g_{rz}}{|g_{rr}|^{1/2}}\,dz,$$

$$\mathbf{e}^{(z)} = \left(|g_{zz}| - \frac{{g_{rz}}^2}{|g_{rr}|}\right)^{1/2} dz$$

are the radial and vertical orts of the dual basis, respectively. The coordinate components of the vectors $\mathbf{e}^{(t)}$, $\mathbf{e}^{(r)}$ and $\mathbf{e}^{(z)}$ satisfy the orthonormality condition (3.30), as can be easily verified by direct substitution.

The orthonormality condition for a tetrad can now be used to determine the fourth ort corresponding to the azimuthal direction.

From three orthonormality conditions for three already known vectors, we obtain that for these conditions to be consistent, the following relation should hold:

$$e^{(\phi)}{}_r = e^{(\phi)}{}_z = 0,$$

and the time and azimuthal components should be related as

$$e^{(\phi)}{}_\phi = -\,e^{(\phi)}{}_t\,\frac{{e_{(t)}}^t}{{e_{(t)}}^\phi}.$$

Finally, the normalization condition for $\mathbf{e}^{(\phi)}$ yields a quadratic equation for $e^{(\phi)}{}_t$, and the sign of the solution is dictated by the additional requirement of the choice of a right-hand triple of space orts of the tetrad.

We thus obtain the dual tetrad basis with the leading corrections in (z/r) due to out-of-equatorial-plane motion in the form

$$\mathbf{e}^{(t)} = C^{-1/2}\left\{G + \left(\frac{z}{r}\right)^2 \frac{1}{2rG}\left(D + \frac{2a}{r^{3/2}}\left(F - \frac{a}{r^{3/2}} + \frac{a^2}{r^2}\right)\right)\right\} dt - C^{-1/2}\left\{r^{1/2}F + \left(\frac{z}{r}\right)^2 \frac{a}{rG} Z_1\right\} d\phi, \qquad (3.46)$$

$$\mathbf{e}^{(\phi)} = -\left\{\left(\frac{D}{rC}\right)^{1/2} + \frac{1}{2}\left(\frac{z}{r}\right)^2 \frac{1-a/r}{r^{3/2}}(DC)^{-1/2}\right\} dt +$$

$$\left\{ rB\left(\frac{D}{C}\right)^{1/2} + \frac{1}{2}\left(\frac{z}{r}\right)^2 \left[(1-a/r)\frac{B}{(DC)^{1/2}} - \frac{H}{G}\left(\frac{D}{C}\right)^{1/2} \right] \right\} d\phi, \tag{3.47}$$

$$\mathbf{e}^{(r)} = D^{-1/2}\left\{1 - \frac{1}{2D}\left(\frac{z}{r}\right)^2 Z_2\right\} dr + \frac{z}{r} D^{-1/2}\left(\frac{2}{r} - \frac{a^2}{r^2}\right) dz, \tag{3.48}$$

$$\mathbf{e}^{(z)} = \left(1 + \frac{z^2}{r^3}\right) dz. \tag{3.49}$$

To obtain the original basis, which we use to write the equations of motion, it suffices to calculate the inverse of the matrix ${e^{(a)}}_i$, which yields

$$\mathbf{e}_{(t)} = C^{-1/2}\left[B - \left(\frac{z}{r}\right)^2 \frac{H}{2rG}\right] \frac{\partial}{\partial t} + (r^3 C)^{-1/2} \frac{\partial}{\partial \phi}, \tag{3.50}$$

$$\mathbf{e}_{(\phi)} = \left\{\frac{F}{(rCD)^{1/2}} + O\left(\frac{z^2}{r^2}\right)\right\} \frac{\partial}{\partial t} + \left\{\frac{G}{r(DC)^{1/2}} + O\left(\frac{z^2}{r^2}\right)\right\} \frac{\partial}{\partial \phi}, \tag{3.51}$$

$$\mathbf{e}_{(r)} = \left\{D^{1/2} + \frac{1}{2}\left(\frac{z}{r}\right)^2 \frac{Z_2}{D^{1/2}}\right\} \frac{\partial}{\partial r}, \tag{3.52}$$

$$\mathbf{e}_{(z)} = -\frac{z}{r^2}(2 - a^2/r) \frac{\partial}{\partial r} + \left(1 - \frac{z^2}{r^3}\right) \frac{\partial}{\partial z}. \tag{3.53}$$

The following notations for the relativistic correction coefficients are introduced in the expressions for the original and dual bases:

$$F = 1 - \frac{2a}{r^{3/2}} + \frac{a^2}{r^2}, \tag{3.54}$$

$$Z_1 = 3 - \frac{5}{r} - \frac{a}{r^{1/2}} + \frac{3a}{r^{3/2}} - \frac{3a^2}{r^3} + \frac{a^2}{r^2} + \frac{2a^3}{r^{7/2}}, \tag{3.55}$$

$$Z_2 = \frac{3}{r} - \frac{4}{r^2} - \frac{a^2}{r^2}\left(3 - \frac{6}{r} + \frac{2a^2}{r^2}\right). \tag{3.56}$$

Here, we omit the terms $\sim O(z^2/r^2)$ in the expression for the azimuthal ort of the original basis due to their complexity. In addition, as we will see below, these terms are not required in the standard accretion disc model.

For the reader's convenience, we here preserve the notations introduced in Novikov and Thorne (1973) for the coefficients B, C, D, F, G, but use the standard style of Latin letters, which is more familiar to the reader. In addition, the coefficient H is equivalent to the coefficient C introduced in Riffert and Herold (1995). We would also like to point out that two other coefficients, A and B, introduced in the same paper, are equivalent to our coefficients D and C, respectively. It can be verified that the original and dual bases presented in Novikov and Thorne (1973) coincide with the bases derived here at $z = 0$.

Using formulas (3.46–3.49) and (3.36), it is easy to deduce that solution (3.15) indeed yields $U^{(a)} = (1, 0, 0, 0)$ in the equatorial plane.

3.2.2.1 The Connection Coefficients

Using (3.41) and then (3.43) and knowing the matrices of the original and dual bases given above, we can calculate the connection coefficients $\gamma_{(a)(b)(c)}$.

Of the 64 coefficients, 16 are equal to zero due to anti-symmetry of $\gamma_{(a)(b)(c)}$ in the first two indices. For the same reason, only half of the other coefficients (i.e. 24) have to be found. Since we are interested in the region near the equatorial plane of the black hole, it makes sense to separate these coefficients into two groups: those that are $\sim (z/r)^0$ in the leading order, and those that are proportional to the first power of (z/r). As mentioned in Sect. 3.1.1, the latter coefficients must appear in the vertical projection of the relativistic Euler equation, while the former emerge in other equations.

It can be shown that

(1) if there is no index (z) among the indices of $\gamma_{(a)(b)(c)}$, then $\gamma_{(a)(b)(c)} \sim (z/r)^0$,
(2) if only one such index is present, then $\gamma_{(a)(b)(c)} \sim (z/r)$, and, finally,
(3) if two indices (z) appear in $\gamma_{(a)(b)(c)}$, then this coefficient is of the second order in (z/r).

Indeed, we examine formula (3.41). Here the brackets contain the original basis components, which are summed with the coordinate derivatives of the dual basis components (the raising of a tetrad index can only change the sign of the component).
In case (1) $(a), (b), (c) \neq (z)$. As the (t)-, (ϕ)- and (r)-orts of the original basis have no z-component, only terms which do not contain derivatives with respect to z of the dual basis components, and have no z-component of the dual (r)-ort, make a non-zero contribution to $\gamma_{(a)(b)(c)}$. Only in these two cases can the contribution $\sim (z/r)$ appear, and hence we prove statement (1).

Now, in (3.41) let $(b) = (z)$ and $(a), (c) \neq (z)$. Then the non-zero contribution can only be due to terms containing the z-component of the (t)-, (ϕ)- and (r)-orts of the original basis, which are absent. Therefore, to check case (2) we should consider only the variant when in (3.41) $(a) = (z)$ or $(c) = (z)$. Here, the terms containing separately either r- or z-components of the (z)-ort of the original basis contribute. In the first variant, the proportionality to $\sim (z/r)$ is due to exactly the component $e^r_{(z)}$, while in the second, it is due to the derivative with respect to z of one of the dual basis components that is always an even function of z, as can be easily verified.

We leave it to the reader to prove statement (3).

Counting shows that there must be 9 connection coefficients without the index (z), and hence an even function of z, and 12 coefficients with the index (z) and hence odd functions of z. The calculation indicates that only 4 coefficients of the first type are non-zero, namely:

$$\gamma_{(t)(\phi)(r)} = -\frac{1}{2}\frac{H}{r^{3/2}C} \qquad \gamma_{(t)(r)(\phi)} = -r^{-3/2} \tag{3.57}$$

$$\gamma_{(\phi)(r)(t)} = -r^{-3/2} \qquad \gamma_{(\phi)(r)(\phi)} = -\frac{1}{r}\frac{d}{dr}\left(rD^{1/2}\right). \tag{3.58}$$

To compute coefficients (3.57) and (3.58) it is sufficient to use bases taken without corrections in z. When constructing the standard disc model, the following facts are also important. First, a direct calculation shows that another 5 connection coefficients of this type are zero through corrections of the order of $\sim\ (z/r)^2$ inclusively. This is a rigorous result, since the coefficients $\gamma_{(a)(b)(c)}$ under discussion have no derivatives of the basis components with respect to z, and therefore any possible unaccounted for corrections due to the terms $\sim\ (z/r)^3$ in $e_z^{(r)}$ and $e^r_{(z)}$ cannot contribute. Second, a direct calculation similarly shows that $\gamma_{(t)(z)(z)} = 0$ through the order $\sim (z/r)^2$.

A calculation of all non-zero coefficients of the second type is a much more cumbersome task. But as we will see below, the only coefficient of this type that is needed has the form

$$\gamma_{(z)(t)(t)} = \frac{z}{r^3}\frac{H}{C}.$$

We note that all connection coefficients of the type $\gamma_{(a)(t)(t)}$ vanish in the equatorial plane $z = 0$. This is consistent with the requirement that the four-velocity $U^{(a)} = (1, 0, 0, 0)$ must satisfy the geodesic equation at $z = 0$:

$$\frac{DU^a}{Ds} = U^b\,\mathbf{e}_{(b)}\left(U^{(a)}\right) + \eta^{(a)(c)}\gamma_{(c)(b)(d)}U^{(b)}U^{(d)} = \gamma_{(a)(t)(t)} = 0. \tag{3.59}$$

3.2.3 Relativistic Hydrodynamic Equations

Everywhere below, we only use the tetrad components of vectors, tensors and covariant derivatives. Therefore, starting from this section, we will substitute the tetrad notation by the standard one, which is familiar when using the coordinate basis. This means that from now on we do not put tetrad indices in parentheses but denote them by the Latin letters i, j, k.[2] In addition, we denote the connection coefficients by Γ.

The stress-energy tensor of a viscous fluid with energy flux has the form (see, e.g., paragraph 4.3 in Mihalas and Weibel Mihalas 1984 or paragraph 22.6 in Misner et al. 1973)

$$T^{ik} = (\rho + \epsilon + p)U^i U^k - p\eta^{ik} + 2\eta\sigma^{ik} + \zeta\Theta P^{ik} - U^i q^k - U^k q^i, \qquad (3.60)$$

where ρ, ϵ, p, η and ζ are the rest-energy density, internal energy density, pressure and two viscosity coefficients, respectively, as measured in the local comoving fluid volume, and $\mathbf{q}$ is the energy flux inside the fluid as measured by a local comoving observer.

The shear tensor is

$$\sigma^{ik} = \frac{1}{2}\left(U^i{}_{;j} P^{jk} + U^k{}_{;j} P^{ji}\right) - \frac{1}{3} U^j{}_{;j} P^{ik}, \qquad (3.61)$$

with the projection operator

$$P^{ik} = \eta^{ik} - U^i U^k. \qquad (3.62)$$

The divergence of four-velocity is

$$\Theta = U^i{}_{;i}. \qquad (3.63)$$

The relativistic Euler equation is written as

$$P_{is}\, T^{sk}{}_{;k} = 0. \qquad (3.64)$$

The energy conservation law has the form

$$U_s\, T^{sk}{}_{;k} = 0. \qquad (3.65)$$

The rest-energy conservation law reads

$$(\rho\, U^k)_{;k} = 0. \qquad (3.66)$$

The covariant derivative in a non-coordinate basis is

$$A^i{}_{;j} = \mathbf{e}_j(A^i) + \Gamma^i{}_{kj} A^k,$$

[2]If one of the symbols t, ϕ, r, z, appears among the indices, it means that the corresponding index takes this value.

while the divergence of a rank-2 contravariant tensor is

$$A^{ij}{}_{;j} = \mathbf{e}_j(A^{ij}) + \Gamma^i{}_{kj}A^{kj} + \Gamma^j{}_{kj}A^{ik}.$$

The energy flux vector and the shear tensor (the deformation tensor free from pure scaling) are purely space-like objects:

$$U_i q^i = 0, \quad U_i \sigma^{ik} = 0, \quad \sigma^i{}_i = 0. \tag{3.67}$$

3.3 Construction of the Standard Accretion Disc Model

3.3.1 Basic Assumptions and the Vertical Balance Equation

Thus, we consider a disc from the standpoint of local observers rotating around a black hole near its equatorial plane with a relativistic Keplerian velocity. Before writing down the dynamic equations in the projection onto tetrad ((3.50))–((3.53)), we discuss the basic assumptions of the model and their consequences. In addition to obvious assumptions about axial symmetry and stationarity of the flow (meaning that the derivatives ∂_t and ∂_ϕ are zero) the main assumption, which we have already used, is that of a small disc thickness, $\delta = h(r)/r \ll 1$, where $h(r)$ is the characteristic height of the disc along the z-axis (more precisely, the disc half-thickness).

The disc symmetry with respect to the plane $z = 0$ implies that U^t, U^ϕ, U^r, q^t, q^ϕ, q^r, ρ, p, η, ζ, and ϵ are even functions of z, and U^z and q^z are odd functions of z.

We also assume that the characteristic scale of variations of these quantities in the radial direction is much larger than that in the vertical direction, that is, their ratio is greater than $\sim \delta^{-1}$.[3]

Next, kinematic arguments suggest that

$$U^z \sim \delta U^r. \tag{3.68}$$

If the energy flux determined by the vector $\mathbf{q}$ is proportional to the internal energy gradient ϵ, then, for the local comoving observer, $q^t_{loc} = 0$ and $q^\phi_{loc}, q^r_{loc} \sim \delta q^z_{loc}$. Taking (3.68) into account implies that the projection of $\mathbf{q}$ onto the four-velocity of circular equatorial motion is also small, i.e. of the order of $\sim \delta q^z_{loc}$. From the

[3]We note that we also need to make the assumption that the velocity components in the disc plane, U^r, U^ϕ, may only change substantially in the vertical direction on scales $\sim r$. Otherwise, the terms in the shear tensor could arise that strongly contribute dynamically to the vertical balance condition, which would lead to a disc totally different from the basic case of interest here.

standard Lorentz transformations, we obtain that $q^t, q^\phi, q^r \sim \delta q^z$, i.e. the energy flux with respect to the tetrad should be directed almost normally to the disc plane.

Now, taking all of the above into account, we consider the projection of the relativistic analog of the Euler equation (3.64) onto the ort $\mathbf{e}_z$ in more detail:

$$T^{zk}{}_{;k} + U_z U_s T^{sk}{}_{;k} = 0. \tag{3.69}$$

Using the symmetry of the physical quantities discussed above, and symmetry properties of tetrad orts and connection coefficients (which become odd functions of z if they have at least one index z), as discussed in Sect. 3.2.2, it is easy to check that Eq. (3.69) is indeed an odd function of z. Further, we see that the first term in (3.69) separately yields the term $\partial_z p$, and other terms containing p are smaller due to the smallness of U^z. All other terms together can always be written as $\sim z\rho f(r)(1 + g(r, z))$ with the function $g(r, z) \sim O(\delta^0)$.

Thus, we arrive at the important conclusion that necessarily

$$\frac{1}{\rho}\frac{\partial p}{\partial z} \sim \delta \ll 1. \tag{3.70}$$

This means that in a thin disc the variables p, $\partial_r p \sim \delta^2$, are small relative to the dominant action of the gravitational force in this direction. Therefore, particles of the disc must move in trajectories close to geodesic ones. Clearly, in a steady-state and axially symmetric flow, this can be realized only in two cases: when the matter moves almost radially towards the gravitating centre (and the specific angular momentum in the disc is close to zero everywhere) or when the matter moves in almost circular orbits (and the specific angular momentum, oppositely, is maximal). We note that both cases are consistent with the general assumptions discussed above and the result (3.70). However, in the last case, strict vertical hydrostatic equilibrium holds in the disc in the first order in δ. In other words, (3.69) can be rewritten in the form

$$\frac{1}{\rho}\frac{\partial p}{\partial z} \sim z f(r)(1 + \delta^2 \ldots). \tag{3.71}$$

When the flow is almost radial, the corrections in parentheses in (3.71) are not small, and their value is determined by the contribution from the prevailing radial motion, when, due to the change in the disc thickness at each radius, the particles are accelerated in the z direction.

Thus, the standard disc model includes one more independent assumption regarding the closeness of the fluid particle trajectories to equatorial circular orbits around the central black hole. Therefore, we will additionally assume that in our reference frame $U^\phi, U^r \sim s U^t$ with $s \ll 1$ and later we will see how this second small parameter is related to δ.

Consequently, we write equations first not only in the leading order in δ but also by assuming $s = 0$, i.e. that the flow moves along geodesic orbits and $U^i = (1, 0, 0, 0)$. Wherever needed, we then additionally evaluate the contribution from the terms in the leading order in s.

3.3.1.1 Deformation of the Velocity Field

We first find the non-zero components of the shear tensor in the leading order. First, the velocity divergence vanishes:

$$\Theta = U^j{}_{;j} = \Gamma^j{}_{kj}\, U^k = \Gamma^j{}_{0j} = 0. \tag{3.72}$$

(Recall that $\Gamma^i{}_{jk}$ are the tetrad connection coefficients, rather than Christoffel symbols, which are connection coefficients in the coordinate basis.)

Next, we have

$$U^i_{;j} P^{jk} = \Gamma^i_{tk} \eta^{kk} - \Gamma^i_{tt},$$

and, in view of the symmetry in i and k, we see that the only non-zero components of the shear tensor σ^{ik} are

$$\sigma^{r\phi} = -\frac{1}{2}\left(\Gamma^{\phi}{}_{tr} + \Gamma^{r}{}_{t\phi}\right) = \frac{1}{2}\left(\frac{1}{2}\frac{H}{r^{3/2}C} + r^{3/2}\right) = \frac{3}{4}\frac{D}{r^{3/2}C}, \tag{3.73}$$

$$\sigma^{rz} = -\frac{1}{2}\Gamma^{z}{}_{t\phi} = O(z). \tag{3.74}$$

3.3.1.2 Equation of Hydrostatic Equilibrium

Substituting $U^i = (1, 0, 0, 0)$ in (3.69) and taking the smallness (due to the small sound velocity in the flow), of several non-zero terms containing η and components of $\mathbf{q}$, into account, we obtain

$$\frac{\partial p}{\partial z} = \rho\, \Gamma^z_{tt} = -\rho \frac{z}{r^3}\frac{H}{C}. \tag{3.75}$$

3.3.1.3 Radial Direction

The radial projection of the relativistic Euler equation for $s = 0$ reads

$$T^{rk}{}_{;k} = 0, \tag{3.76}$$

and excluding terms $\sim \delta^4$ containing the connection coefficients and components of **q**, we have only one non-zero term of the order of δ^2 which has the form[4]

$$-[p\eta^{rk}]_{;k} = D^{1/2}\frac{\partial p}{\partial r}$$

Clearly, this term should be balanced by the leading terms $\sim s$. Evidently, the contribution from

$$[\rho U^r U^k]_{;k}$$

should be considered first, and here it can only be due to terms containing one of the connection coefficients of zeroth order in z and the time velocity component. There is only one such term: $2\Gamma^r_{t\phi} U^t U^\phi = 2r^{-3/2}U^\phi$.

Hence, we reach the important conclusion that $s \sim \delta^2$, i.e. the velocity components in the disc plane are

$$U^r, U^\phi \sim \delta^2, \quad (3.77)$$

which is used when determining the force balance in the azimuthal direction.

3.3.2 *Azimuthal Direction*

We consider the last projection of the relativistic Euler equation, or more specifically, its component along the azimuthal ort. Let us proceed in the same way as above and first write down the terms that are present in the case $s = 0$. Again, we take $U^i = (1, 0, 0, 0)$ and see that

$$[(\rho + \epsilon + p)U^\phi U^k]_{;k} = 0,$$

since $\Gamma^\phi_{tt} = 0$ through the order $\sim \delta^2$ (see the discussion at the end of Sect. 3.2.2). Next, the term including the pressure is absent by virtue of the axial symmetry, and terms with q^i cannot contribute to any order higher than $\sim \delta^4$.

[4]The order of components q^i can be estimated as follows. In the stationary case, the divergence of the energy flux must be of the order of the power generated due to viscous dissipation, which is, in turn, proportional to some scalar characterizing the degree of the velocity shear and the viscosity coefficient η. In our case, the viscosity coefficient $\eta < \rho h c_s \sim \delta^2$. The divergence is mainly due to the term $\partial_z q^z$. This immediately implies that $q^z \sim \delta^3$ and $q^{t,\phi,r} \sim \delta^4$.

It thus remains to consider the contribution

$$[2\eta\sigma^{\phi k}]_{;k} = D^{1/2}(2\eta\sigma^{r\phi})_{,r} + (2\eta\sigma^{\phi z})_{,z} + 4\eta\Gamma^{\phi}_{r\phi}\sigma^{r\phi} + \eta\, O(\delta^2) =$$

$$= -\frac{3}{2}D^{1/2}\left(\eta\frac{D}{r^{3/2}C}\right)_{,r} + (\eta\Gamma^{t}_{z\phi})_{,z} + 3\eta\,(rD^{1/2})_{,r}\frac{D}{r^{5/2}C} + \eta\, O(\delta^2). \quad (3.78)$$

Here, we are also dealing with terms of the second order in δ^2. Therefore, it is necessary to find the leading contribution from terms $\sim s$. Again, we consider only the prevailing part due to the perfect fluid term:

$$(\eta_{\phi i} - U_\phi U_i)[\rho U^i U^k]_{;k}.$$

The second part, which is proportional to U_ϕ, can be neglected since the term in square brackets cannot contribute to the zeroth order in δ, as there are no connection coefficients of the form $\Gamma^i_{tt} \sim \delta^0$, as was discussed at the end of Sect. 3.2.2.

As a result, we obtain

$$[\rho U^\phi U^k]_{;k} = \rho\Gamma^{\phi}_{lk}U^l U^k + \rho\Gamma^{k}_{lk}U^\phi U^l = \rho(\Gamma^{\phi}_{tr} + \Gamma^{\phi}_{rt})U^r =$$

$$-\rho\frac{U^r}{r^{3/2}}\left(\frac{1}{2}\frac{H}{C} - 1\right) \equiv \rho\frac{U^r}{2r^{3/2}}\frac{E}{C}, \quad (3.79)$$

where

$$E = 1 - \frac{6}{r} + \frac{8a}{r^{3/2}} - \frac{3a^2}{r^2}. \quad (3.80)$$

We now introduce the notation

$$T_\nu \equiv \int_{-h}^{+h} T^{r\phi}_\nu = 2\sigma^{r\phi}\int_{-h}^{+h}\eta\, dz, \quad (3.81)$$

where T_ν is the vertically integrated density of the flux of the ϕ-component of momentum in the radial direction. Then, by integrating (3.78) and (3.79) over the disc thickness and combining them in one equation, we have

$$\frac{\partial T_\nu}{\partial r} + \frac{2T_\nu}{rD}\left(1 - \frac{1}{r}\right) + \frac{\Sigma U^r}{2r^{3/2}}\frac{E}{CD^{1/2}} = 0, \quad (3.82)$$

where the contribution from $\sigma^{\phi z}$ vanishes due to its being an odd function of z, and we neglect the dependence of U^r on z, which gives rise to a higher-order correction (see footnote 2). In formula (3.82) we have introduced the surface density of the disc

$$\Sigma \equiv \int_{-h}^{+h} \rho \, dz. \tag{3.83}$$

The important Eq. (3.82) together with known boundary conditions at the inner disc radius allows us to calculate the profile $T_\nu(r)$ for the disc, provided that the radial velocity distribution is known.

We note that the equation for T_ν can also be derived from the angular momentum conservation law, which was used in the original paper by Novikov and Thorne (1973) (see equations (5.6.3)–(5.6.6) therein).

3.3.3 *Rest Energy Conservation Law and Radial Momentum Transfer*

To solve Eq. (3.82), the radial velocity profile should be specified. It can be obtained from the rest energy conservation law (3.66):

$$\mathbf{e}_r(\rho U^r) + \mathbf{e}_z(\rho U^z) + \Gamma^i_{ki}\rho U^k = 0. \tag{3.84}$$

Clearly, the substitution $U^i = (1, 0, 0, 0)$ does not yield non-zero terms up to the order $\sim \delta^2$ (see the discussion at the end of Sect. 3.2.2). In our reference frame, this fact can be easily understood: the circular axially symmetric motion corresponds to zero velocity divergence. It is straightforward to check that the following terms $\sim s$ will appear in the continuity equation:

$$D^{1/2}(\rho U^r)_{,r} + (\rho U^z)_{,z} - \frac{(r\, D^{1/2})_{,r}}{r} \rho U^r = 0, \tag{3.85}$$

where the last term arises due to the contribution from $\Gamma^\phi_{r\phi}\rho U^r$, and similar terms with other velocity components, even if they appear, have an order higher than $\sim \delta^4$.

After integrating over z, the contribution from the second term in (3.85) vanishes since $\rho \to 0$ far from the equatorial disc plane, and we obtain

$$(\Sigma U^r r D^{1/2})_{,r} = 0. \tag{3.86}$$

The combination whose derivative is found in (3.86) is a constant, which must be identified as the radial flux of matter. After additional integration over ϕ we obtain that

$$2\pi \, \Sigma U^r r D^{1/2} = -\dot{M}, \tag{3.87}$$

where $\dot{M} > 0$ is the rate of the matter inflow into the disc at infinity, i.e. the mass accretion rate.

After substituting (3.87) into (3.82), we finally obtain

$$\frac{dT_\nu}{dr} + P_1 T_\nu + P_2 = 0, \tag{3.88}$$

where

$$P_1 = \frac{2}{rD}\left(1 - \frac{1}{r}\right),$$

$$P_2 = -\frac{\dot{M}}{4\pi}\frac{E}{r^{5/2}CD}.$$

The solution to (3.88) with the boundary condition $T|_{r_{ms}} = 0$ is written in the form

$$T_\nu = \frac{1}{F(r)}\int_{r_{ms}}^{r} P_2(x)F(x)\,dx, \tag{3.89}$$

$$F(r) = exp\left(\int_{r_{ms}}^{r} P_1(x)dx\right). \tag{3.90}$$

The integral (3.90) is elementary, and as a result we obtain

$$T_\nu = \frac{\dot{M}}{4\pi r^2 D}\int_{r_{ms}}^{r}\frac{E}{r^{1/2}C}\,dr. \tag{3.91}$$

3.3.4 Energy Balance

Here, we consider Eq. (3.65). As above, let us set $U^i = (1, 0, 0, 0)$ and find terms of the leading order in δ. As in the case of the azimuthal projection of the relativistic Euler equation, 'perfect' terms $[(\rho + \epsilon + p)U^t U^k]_{;k}$ and $p\eta^{0k}{}_{;k}$ do not contribute here. From the shear term, we have

$$[2\eta\sigma^{tk}]_{;k} = \Gamma^t_{lk}\sigma^{lk} = 2\eta[(\Gamma_{t\phi r} + \Gamma_{tr\phi})\sigma^{r\phi} + O(\delta^2)] = 2\eta[4\sigma^{r\phi} + O(\delta^2)].$$

Terms with q^i contribute due to rapid change in the energy flux component normal to the disc with z:

$$(U^t q^k)_{;k} = \frac{\partial q^z}{\partial z} + O(\delta^4).$$

Summing all terms, we obtain from the energy balance equation

$$\frac{\partial q^z}{\partial z} = 4\eta \left(\sigma^{r\phi}\right)^2 = \frac{3}{2} T_\nu^{r\phi} \frac{D}{r^{3/2} C}, \tag{3.92}$$

whence, after integrating over the disc thickness, we derive the important relation

$$Q = \frac{3}{4} \frac{D}{r^{3/2} C} T_\nu, \tag{3.93}$$

where $Q = q^z(z = h)$ is the vertical energy flux escaping from the disc. At this point it is appropriate to give the dimensional energy flux $\tilde{Q}$ [erg cm^{-2} c^{-1}]:

$$\tilde{Q} = \frac{3}{8\pi} \frac{\dot{M}_0 GM}{\tilde{r}^3} \Phi, \tag{3.94}$$

where the dimensionless function Φ was obtained in the remarkable paper by Page and Thorne (1974):

$$\Phi = \left[\frac{\chi - \chi_0 - \frac{3}{2} a \ln \frac{\chi}{\chi_0} - \mathscr{A} - \mathscr{B} - \mathscr{C}}{\chi(\chi^3 - 3\chi + 2a)} \right], \tag{3.95}$$

where

$$\chi \equiv \sqrt{\tilde{r}/R_0}, \quad R_0 \equiv \frac{GM}{c^2}, \ \chi_0 \equiv \sqrt{\tilde{r}_{ms}/R_0}$$

$$\mathscr{A} = \frac{3(\chi_1 - a)^2}{\chi_1(\chi_1 - \chi_2)(\chi_1 - \chi_3)} \ln \frac{\chi - \chi_1}{\chi_0 - \chi_1},$$

$$\mathscr{B} = \frac{3(\chi_2 - a)^2}{\chi_2(\chi_2 - \chi_1)(\chi_2 - \chi_3)} \ln \frac{\chi - \chi_2}{\chi_0 - \chi_2},$$

$$\mathscr{C} = \frac{3(\chi_3 - a)^2}{\chi_3(\chi_3 - \chi_1)(\chi_3 - \chi_2)} \ln \frac{\chi - \chi_3}{\chi_0 - \chi_3},$$

$$\chi_1 = 2\cos\left(\frac{\arccos a - \pi}{3}\right), \quad \chi_2 = 2\cos\left(\frac{\arccos a + \pi}{3}\right),$$

$$\chi_3 = -2\cos\left(\frac{\arccos a}{3}\right). \tag{3.96}$$

The quantities χ_1, χ_2 and χ_3 are the roots of the cubic equation

$$\chi^3 - 3\chi + 2a = 0.$$

$\Phi \to 1$ as $\tilde{r} \to \infty$, and $\Phi \to (\tilde{r} - \tilde{r}_{ms})^2$ at the inner edge of the disc $\tilde{r} \to \tilde{r}_{ms}$. Note that in the Newtonian model

$$\Phi_N(\tilde{r}) = 1 - \sqrt{\frac{\tilde{r}_{ms}}{\tilde{r}}}. \tag{3.97}$$

After specifying Q, we can calculate the radial profile of the effective temperature of the disc surface, since by definition $Q = \sigma T_{eff}^4$. This is the universal result of the standard accretion disc theory: T_{eff} does not depend on the specific nature of the dissipation of the kinetic energy of matter or on the mechanism of thermal energy transfer towards the disc surface, and is proportional to the value of $\dot{M}$, times some universal known function of r.

Thus, we have obtained the explicit form of the viscous stress integrated over the disc thickness, T_ν, and the explicit form of the radiation energy flux from its surface, Q. At the same time, we know only the combination ΣU^r, and not each of these variables separately. In addition, we should determine the disc half-thickness profile, $h(r)$, and the temperature, pressure and density distributions, $T(r, z)$, $p(r, z)$ and $\rho(r, z)$, inside it. To do this, the vertical disc structure should be calculated.

3.3.5 Energy Transfer Equation and the Vertical Disc Structure

The vertical disc structure is defined by three equations. Two of them have already been obtained above: the vertical hydrostatic balance equation (3.75) and the thermal energy generation equation (3.92).

The remaining equation is the transfer equation for energy dissipating in the disc. In the simplest case, the energy transfer is due to photon diffusion in the heated matter. Strictly speaking, we should write a relativistic analogue of the radiation heat conductivity equation, which is a variant of the kinetic Boltzmann equation for photons when their mean free path is much smaller than the characteristic spatial length of the problem. The Boltzmann equation is relativistically generalized in Section 2.6 in Novikov and Thorne (1973), in which the full transfer equation is given as Eq. 2.6.22 (where one must multiply the second term on the left hand side by the intensity I_ν due to a misprint). The standard transition to the diffusion approximation yields the following equation (see expression 2.6.43 in Novikov and Thorne 1973):

$$q^i = \frac{1}{\tilde{\kappa}\rho}\frac{4}{3}bT^3 P^{ik}(\mathbf{e}_k(T) + a_k T), \tag{3.98}$$

where $\tilde{\kappa}$ is the Rosseland mean opacity of matter, T is the temperature, b is the radiation constant and $a_k \equiv U_{k;j}U^j$ is the four-acceleration. A discussion of Eq. (3.98) can be also found on p. 165 of Mihalas and Weibel Mihalas (1984).

As regards (3.98), we first note that the four-acceleration never exceeds the order $\sim \delta^2$, since the four-velocity itself differs from the geodesic value (free circular equatorial motion) only in the second order in δ. In contrast, the derivative in the first term in parentheses on the right-hand side of (3.98) for $k = z$ raises the order in δ, since T, as well as ϵ, vary significantly across the disc thickness. As a result, as already discussed in Sect. 3.3.1, we see that q^z is the leading component of the vector **q** and is determined by the equation

$$q^z = -\frac{1}{3\tilde{\kappa}\rho}\frac{\partial(bT^4)}{\partial z}, \tag{3.99}$$

which is identical to the Newtonian form for a thin disc (see Chap. 1).

Equations (3.75), (3.92) and (3.99) must be supplemented with the equation of state of matter

$$p(\rho, T),$$

the opacity law

$$\tilde{\kappa}(\rho, T),$$

and the explicit form of

$$\eta(\rho, T), \text{or} \quad T_\nu^{r\phi}(\rho, T)$$

depending on the type of parametrization of the turbulent viscosity in the disc.

In addition, it is necessary to set boundary conditions at the integration interval $z \in [0, h]$. In the simplest case, we assume that the disc has no atmosphere and

$$\rho|_{z=h} = T|_{z=h} = 0.$$

Furthermore, the energy flux vanishes in the disc equatorial plane:

$$q^z|_{z=0} = 0.$$

Finally,

$$2\int_0^h T_\nu^{r\phi} dz = T_\nu.$$

Note that the above equations and boundary conditions for the vertical disc structure automatically guarantee the validity of Eqs. (3.87), (3.91) and (3.93) for the radial disc structure.

After calculating the vertical structure, we can specify the surface density distribution using (3.83) and then U^r using (3.87).

3.3.6 *Parametrization of Turbulent Viscosity and Explicit Disc Structure*

Estimates carried out in Shakura and Sunyaev (1973) and Novikov and Thorne (1973), according to the algorithm described in Sect. 3.3.5, show that at sufficiently high accretion rate $\dot{M}$, which is the free parameter of the problem, radiation energy becomes dominant in the inner parts of the disc. An estimate of the threshold value of $\dot{M}$ can be found, for example, in Shakura and Sunyaev (1973) (see formula 2.18 therein). It turns out that the disc thickness far away from its inner radius is independent of r and for $\dot{M}$ of the order of and above the critical value, $\dot{M}_{cr}$ (when the disc luminosity reaches the Eddington value in the inner parts of the disc), $\delta > 1$, corresponding to a spherization of the flow (see expression 7.1 and the discussion in Shakura and Sunyaev 1973). In addition, later studies showed that the radiation-dominated region is both thermally (Shakura and Sunyaev 1976) and convectively (Bisnovatyi-Kogan and Blinnikov 1977) unstable.

This implies that for a correct description of the inner parts of accretion discs at high accretion rates, when δ increases, terms of higher order in δ should be taken into account. These include the radial pressure gradient $\sim \delta^2$ in the radial force balance and the advection term, $U^r T \partial S/\partial r \sim \delta^4$, which arises in the energy balance and accounts for the radial heat transfer. The latter, in fact, implies that the heat diffusion time in the vertical direction is comparable to its radial advection due to radial transfer of matter. In other words, the main property of the standard accretion disc model considered here—the local energy balance in the disc—is violated, i.e. the heat generated due to turbulent energy dissipation is no longer released locally from the disc surface. It was found that taking the new terms into account also allows for a correct description in the region near r_{ms}, where in the standard model $U^r \to \infty$, and the construction of a stationary solution, with $\delta < 1$ for $\dot{M}$ of the order of and above $\dot{M}_{cr}$, which is stable against thermal perturbations (the so-called 'slim-disc', see Paczynski and Bisnovatyi-Kogan 1981 and Abramowicz et al. 1988 and their citations list, and, e.g., Klepnev and Bisnovatyi-Kogan 2010). Later, these results were confirmed by numerical simulations (see, e.g. Fujita and Okuda 1998 and Agol et al. 2001). We would like to add that a transition from a standard disc to a slim disc with increasing $\dot{M}$, in the relativistic model around a rotating black hole, should occur even earlier due to the higher accretion efficiency (which is, in turn, due to both decreasing r_{ms} and additional angular momentum loss from the disc surface by radiation).

Now, assuming that $\dot{M} \ll \dot{M}_{cr}$, let us estimate the disc semi-thickness profile, which will be useful in the next chapter, in the simplest case when the pressure is mainly determined by fully ionized hydrogen plasma, i.e.

$$p = 2\rho kT/m_p, \tag{3.100}$$

where m_p is the mass of a proton, k_B is the Boltzmann constant, and the opacity is determined by Thomson scattering, $\tilde{\kappa} = \kappa_T = 0.4\,\mathrm{cm}^2/\mathrm{g}$.

Let us also assume that the kinematic viscosity ν is independent of z and can be parametrized in the form

$$\nu = \alpha c_s h, \tag{3.101}$$

where $0 < \alpha < 1$ is the Shakura parameter determining the turbulent viscosity in the disc (see Shakura 1972 and Shakura and Sunyaev 1973), and c_s is the speed of sound in the equatorial disc plane. Herein, due to (3.100),

$$c_s^2 = 2kT_c/m_p, \tag{3.102}$$

where $T_c = T(z = 0)$.

Equation (3.99) yields

$$\int_0^h dz q^z \rho = -\frac{1}{3\kappa_T} bT^4 \bigg|_0^h = \frac{1}{3\kappa_T} bT_c^4.$$

On the other hand,

$$\int_0^h dz q^z \rho = C_q F \int_0^h \rho dz = \frac{1}{2} C_q \Sigma F,$$

where C_q is some correction factor of the order of unity corresponding to the difference between the escaping radiation flux, Q, and its mean value along the disc thickness. As a result, we obtain

$$T_c = \left(\frac{3\kappa_T}{2} \frac{C_q}{b} \Sigma F \right)^{1/4}, \tag{3.103}$$

Next, for simplicity we assume that the entropy is constant along z and, dividing the left-hand side of (3.75) by ρ, we introduce the enthalpy, $dw = dp/\rho$, integrate (3.75) over z and obtain the equatorial value of w, $w_c \equiv w(z = 0)$:

$$w_c = -\int_0^h dw = \int_0^h \frac{z}{r^3} \frac{H}{C} = \frac{h^2}{2r^3} \frac{H}{C},$$

Hence, using that $w_c = nc_s^2$, where n is the polytropic index, we obtain

$$c_s^2 = \frac{h^2}{2nr^3} \frac{H}{C}. \tag{3.104}$$

Finally, as a result of definition (3.81), parametrization (3.101) and Eq. (3.91) we find

$$T_\nu = \frac{3}{2} \frac{D}{r^{3/2} C} \alpha \Sigma c_s h = \frac{\dot{M}}{2\pi} \frac{Y}{r^{3/2} D}, \tag{3.105}$$

where in the second equality we introduce the new variable

$$Y \equiv (2r)^{-1/2} \int_{r_{ms}}^{r} \frac{E}{r^{1/2} C} \, dr, \tag{3.106}$$

which in the Newtonian limit, far away from the inner edge of the disc, tends to unity.

Equations (3.93), (3.102), (3.103) and (3.104) are sufficient to exclude all unknowns except Σ and the free parameters $\dot{M}$ and α from (3.105). We thus obtain the following surface density profile Σ:

$$\Sigma = \Sigma_0 \alpha^{-4/5} \dot{M}^{3/5} r^{-3/5} C^{3/5} D^{-8/5} H^{2/5} Y^{3/5}, \tag{3.107}$$

where the dimensional constant Σ_0 combines all relevant physical constants and numerical coefficients. Its explicit form and numerical value (which depends on the black hole mass to which we normalize all quantities) can be found by the reader.

Now, using formulas (3.104), (3.102), (3.103) and (3.107), it is possible to derive the profile $h(r)$. The resulting disc aspect ratio reads $\delta(r) = h(r)/r$:

$$\delta(r) = \delta_* r^{1/20} C^{9/20} D^{-1/5} H^{-9/20} Y^{1/5}, \tag{3.108}$$

where δ_* is a constant that determines the characteristic disc thickness δ (which is of the order of 0.001–0.01).

The observational appearance of relativistic standard accretion disc was modelled for the first time in Cunningham (1975) (for a comprehensive review, see Gierlinski et al. (2001) and Li et al. (2005), references therein and their citations list). The Novikov-Thorne solution is also used as a background solution in theory of relativistic twisted discs presented in Zhuravlev and Ivanov (2011) and discussed in more detail in Chap. 4.

References

Abramowicz MA, Czerny B, Lasota JP, Szuszkiewicz E (1988) Astrophys J 332:646
Agol E, Krolik J, Turner NJ, M SJ (2001) Astrophys J 558:543
Bisnovatyi-Kogan GS, Blinnikov SI (1977) Astrophys J 59:111
Chandrasekhar S (1992) The mathematical theory of black holes. Clarendon Press, Oxford
Cunningham CT (1975) Astrophys J 202:788
Fujita M, Okuda T (1998) Publ Astron Soc Jpn 50:639
Gierlinski M, Maciolek-Niedzwiecki A, Ebisawa K (2001) Mon Not R Astron Soc 325:1253
Hobson MP, Efstathiou GP, Lasenby AN (2006) General relativity: an introduction for physicists. Cambridge University Press, Cambridge. https://doi.org/10.2277/0521829518
Klepnev AS, Bisnovatyi-Kogan GS (2010) Astrophysics 53:409
Korn GA, Korn TM (2000) Mathematical handbook for scientists and engineers: definitions, theorems, and formulas for reference and review. Dover Publications, Mineola
Li LX, Zimmerman ER, Narayan R, McClintock JE (2005) Astrophys J Suppl Ser 157:335

Mihalas D, Weibel Mihalas B (1984) Foundations of radiation hydrodynamics. Oxford University Press, New York
Misner CW, Thorne KS, Wheeler JA (1973) Gravitation, vol 2. W.H. Freeman, San Francisco
Novikov ID, Thorne KS (1973) Astrophysics of black holes. In: Dewitt C, Dewitt BS (eds) Black holes (Les Astres Occlus). Gordon and Breach, New York, pp 343–450
Paczynski B, Bisnovatyi-Kogan G (1981) Acta Astronomica 31:283
Page DN, Thorne KS (1974) Astrophys J 191:499
Riffert H, Herold H (1995) Astrophys J 450:508
Shakura NI (1972) Astron Zh 49:921
Shakura NI, Sunyaev RA (1973) Astrophys J 24:337
Shakura NI, Sunyaev RA (1976) Mon Not R Astron Soc 175:613
Zhuravlev VV, Ivanov PB (2011) Mon Not R Astron Soc 415:2122

Chapter 4
Relativistic Twisted Accretion Disc

Viacheslav Zhuravlev

Abstract A twisted disc forms around a rotating black hole each time when the disc outskirts are not aligned with the black hole's equatorial plane. We derive equations describing the evolution of the shape of twisted discs and perturbations of density and velocity necessarily arising in such a disc. This is done under the following simplifying assumptions: a small aspect ratio of the disc, a slow rotation of the black hole, and a small tilt angle of the disc rings with respect to the black hole equatorial plane. Nevertheless, the GR effects are considered accurately. Additionally, an analysisă of particular regimes of non-stationary twist dynamics (the wave andă diffusion regimes) is presented both in the framework of the Newtonian dynamics and taking into account Einstein's relativistic precession. At the end of the chapter, a calculation of the shape of a stationary relativistic twisted accretion disc for different values of free parameters of the model is done.

4.1 Introductory Remarks

In the previous chapter we described a flat disc in the equatorial plane of a rotating black hole. The axially symmetric structure of a flat disc is evident and consistent with the symmetry of space near the black hole. If we now relax the main assumption that the flow of matter at all distances coincides with the equatorial plane, the question arises: what would be the dynamics of this more complicated, stationary or non-stationary, flow? Is this flow similar to a disc in any way? For thin discs as considered here, the answer to this question proves to be positive under certain conditions.

V. Zhuravlev (✉)
Sternberg Astronomical Institute, Lomonosov Moscow State University, Moscow, Russia
e-mail: zhuravlev@sai.msu.ru

N. Shakura (ed.), *Accretion Flows in Astrophysics*, Astrophysics and Space Science Library 454, https://doi.org/10.1007/978-3-319-93009-1_4

The Gravitomagnetic Force

The main reason for the deformation of (for example, an initially flat) disc is that the black hole spin gives rise to an additional off-centre gravitational interaction with the gas elements of the flow. It can be shown that far away from the event horizon, but close to the equatorial plane of the black hole, this interaction is represented by an axially symmetric force field directed towards the black hole spin axis in planes parallel to the equatorial (see Thorne et al. (1986), Chapter 3, paragraph A). The physical meaning of gravitomagnetic force becomes particularly obvious far away from the gravitating body at $r \gg R_g$. In this case the equation of motion for a particle of mass m near a body of mass M, and proper angular momentum J, reads

$$m_p \frac{d^2 \boldsymbol{r}}{d\tau^2} = m_p \left(\boldsymbol{g} + \frac{d\boldsymbol{r}}{d\tau} \times \boldsymbol{H} \right) \tag{4.1}$$

where

$$\boldsymbol{g} = -\frac{M}{r^2} \boldsymbol{e}_r \tag{4.2}$$

is the standard radial component of the gravitational force, and

$$\boldsymbol{H} = 2 \frac{\boldsymbol{J} - 3\boldsymbol{e}_r(\boldsymbol{J}\boldsymbol{e}_r)}{r^3} f \tag{4.3}$$

is an additional dipole field generated by the spinning body. By similarity of Eq. (4.1) to the equation of motion for a charged particle moving in an electromagnetic field, this additional field is referred to as gravitomagnetic, since it acts similar to the Lorentz force. The gravitomagnetic force is directed perpendicular to the spin of the central body. Clearly, this external force can change the proper angular momentum of disc elements moving outside the equatorial plane of the black hole (and hence deform the disc). Here, only the projection of the gravitomagnetic force onto the angular momentum direction matters, which is proportional to the sine of the angle between the angular momentum vector and the black hole spin axis. As we will see shortly, the restriction that allows us to treat the new configuration as a disc (both stationary and non-stationary) requires that the gravitomagnetic force is smaller than the central gravitational attraction force, i.e. requires the parameter $a \ll 1$ to be small. In addition, one more restriction can be formulated: the non-complanarity of the disc with the equatorial plane of the black hole, as well as the degree of its deviation from the planar form (i.e. twist, warp) should not exceed certain small values for the disc to be hydrodynamically stable (see Ivanov and Illarionov (1997) paragraph 7 and Zhuravlev and Ivanov (2011) paragraph 4.2.4).

Hydrostatic Equilibrium in a Disc Under the Action of a Gravitomagnetic Force

Let us split a thin planar disc into rings of narrow width. In each ring, the motion of gas elements is mainly due to the gravitational attraction force from the central body. The characteristic time of this motion is $t_d \sim \Omega^{-1}$. In addition, t_d determines the time it takes for the disc to restore hydrodynamic equilibrium across the ring, since the disc aspect ratio (the ratio of the disc thickness to the radial distance) is of the order of the ratio of the sound velocity to the orbital velocity. This conclusion can be also arrived at by noticing that the vertical pressure gradient is δ^{-1} times smaller than the unit mass gas element acceleration, i.e. exactly as small as the ratio of the radial size of the ring to its vertical scale. Thus, we can conclude that if other forces, acting on a given ring from the adjacent rings or from the black hole, lead to dynamics with characteristic time t_{ev} much greater than the dynamical one, i.e. $t_{ev} \gg t_d$, and hydrostatic equilibrium is maintained in the ring. In other words, the ring remains flat, and the entire flow preserves a disc-like form. This is undoubtedly so in a flat disc, since in this case equally oriented rings interact via the viscous force acting in the azimuthal direction and the angular momentum changes due to inflow and outflow of matter accreting through the ring, with both processes occurring on the diffusion time scale, $t_\nu \sim \Omega^{-1}\delta^{-2} \gg t_d$.

Now, let the disc be tilted with respect to the equatorial plane of the black hole by a small angle $\beta \ll 1$. In a flat disc the gravitomagnetic force contributes only to the modulus of acceleration of gas elements moving in circular orbits, but now, due to the non-zero projection of this force ($\propto \beta$) onto the angular momentum of the gas elements, this force makes the orbits precess around the black hole spin axis. For free particles, this effect is described in detail in the second part of the next paragraph in terms of the difference between the frequencies of circular and vertical motion. We also show that the precession frequency is much smaller than the circular frequency for $a \ll 1$ (see formula (4.14)), which is equivalent to the condition $t_{ev} \gg t_d$ for a ring composed of gas elements.

Differential Precession and the Twist of a Tilted Disc

Equation (4.14) suggests that the precession of the rings is differential, i.e. depends on the distance to the centre. As a result, the relative orientation of initially coaxial rings changes and the disc is no longer flat. However, we keep in mind that under the condition $t_{ev} \gg t_d$, each of the rings behaves 'rigidly' in its own vertical direction, which is now also a function of r. The new configuration is similar to a twisted (or warped) disc, i.e. a flow symmetric relative to some (now not planar) surface, which can be called the *equatorial surface* of the twisted disc. Here, the intersection of the equatorial surface with a plane passing through the centre makes up a circle—the instantaneous shape of orbits of gas elements rotating with a given radial distance r. The disc turns into a set of rings, tilted with respect to the black hole equatorial plane by a constant angle β, but with node lines (the line formed by the intersection of the ring planes with the black hole equatorial plane) depending on r, see Fig. 4.1. The node line is now determined by the position angle $\gamma(r)$, measured in the equatorial plane in the positive direction from some fixed direction to the ascending node of a given ring.

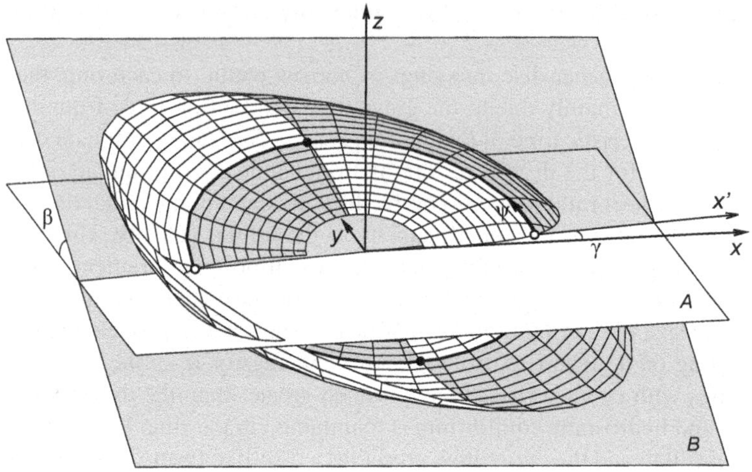

Fig. 4.1 The geometrical image of a twisted disc consisting of rings, each tilted with the same angle $\beta = const$ with respect to the plane A, but having a position angle γ depending on its radius. The axis x' is directed towards the ascending node of the particular ring (bold line). The disc is intersected by the semi-transparent plane B, which contains this ring. The azimuthal angle ψ is measured counterclockwise from the ascending node of the ring. It can be seen that the plane B is partially below the disc surface (plotted in light colour in this case) and partially above it (plotted in dark colour in this case). Additionally, two groups of points are plotted on the ring displayed by the bold line: the first of them are located at $\psi \pm \pi/2$ (filled circles), while the others are located at $\psi = 0, \pi$ (empty circles). The first group corresponds to the position on the ring, where the normals to the plane B and to the disc surface coincide with each other, while the second group corresponds to the position on the ring, where those normals are the most divergent

Radial Projection of the Pressure Gradient

The key point here is that the pressure gradient in a twisted disc, directed (as in any thin disc in general) almost normal to its warped surface, is not normal to the planes of the rings composing the disc. Therefore, we conclude that the pressure gradient acquires two projections, see Fig. 4.2. The main projection is coaxial with the rotational axis of each ring. Let us here denote it by $(\nabla p)_\xi$, where ξ is the distance from the equatorial surface of the twisted disc measured along the direction of rotation of the ring (ξ reduces to z in the case of a flat disc). We note from the beginning that $(\nabla p)_\xi \propto \xi$ due to hydrostatic equilibrium across the ring. The second projection of the pressure gradient, conventionally denoted as $(\nabla p)_r$, lies in the ring's plane along the radial direction connecting the disc centre and a given gas element of the ring. The ratio of these two projections is a small value proportional to the rate of change of orientations of the rings in the disc, which, in turn, depends on the radial direction chosen in the plane of the given ring. From purely geometrical considerations, we show in detail in what follows that for a disc with $\beta = const$, the ratio $(\nabla p)_r/(\nabla p)_\xi \propto \beta d\gamma/dr \cos\psi$, where ψ is the angle measured in the azimuthal direction for the given ring from its ascending node to the given gas element. Note that the normal to the twisted disc surface is orthogonal

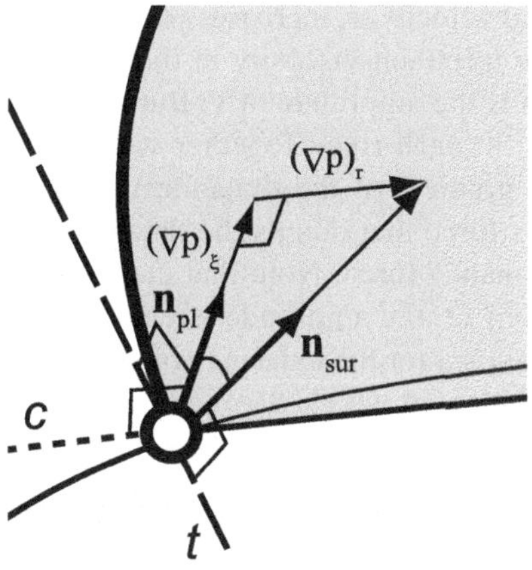

Fig. 4.2 A sketch of a zone of a twisted disc in the vicinity of an empty circle, located at the particular disc ring at $\psi = \pi$, see Fig. 4.1. The straight line t is tangent to the ring, while the straight line c is the intersection of the planes A and B in Fig. 4.1. The dashed part of the line indicate that it is obscured to the observer by the disc surface. There are two normals originating from the point displayed by an empty circle: one to the plane of the ring, $\mathbf{n}_{pl}$, and one to the disc surface, $\mathbf{n}_{sur}$

to the ring's plane only in two diametrically opposite points—where the plane of the given ring intersects with the planes of the adjacent rings. At $\beta = const$, these points are characterized by $\psi = \pm\pi/2$. At the same time, in other pair of points with $\psi = 0, \pi$, the value of $(\nabla p)_r$ reaches both positive and negative maxima.

Flow Lines, Keplerian Resonance, and Perturbation of Velocity

Thus, in a flat disc, in the leading order in $\delta \ll 1$, the dynamics in the radial direction is controlled by the gravitational force, and the corrections $\sim \delta^2$ are neglected, whereas in a twisted disc the radial projection of the pressure gradient starts additionally contributing to the radial balance. This addition, on the one hand, depends on the degree of the twist, and on the other hand, increases proportionally to the distance from the equatorial disc surface, ξ. Further, since it also depends harmonically on the azimuthal direction, the gas elements (for $\xi \neq 0$) are subjected to a periodic, with the orbital period, disturbance by this force, and their orbits become ellipses with low eccentricity. As is well known, the eigenfrequency of the small oscillations of free particles in eccentric orbits is equal to the epicyclic frequency, κ. As the pressure gradient projection considered here excites exactly such oscillations, the radial profile of the epicyclic frequency, $\kappa(r)$, is an important characteristic that determines the shape of both stationary and non-stationary twisted configurations. In the next section, we derive the required relativistic profile $\kappa(r)$ for equatorial circular orbits in the Kerr metric (see Eq. (4.10)). Note from the beginning that in the special case of Newtonian gravity $\kappa = \Omega$, and hence the action of the external exciting force on gas elements with the same frequency results in a resonance. This means that the amplitude of the perturbed motion, characterized by perturbation of the orbital velocity, $\mathbf{v}$, increases without limit. This growth, however, is always limited by the turbulent viscosity in the disc. Indeed, since the exciting force amplitude $\propto \xi$, so is the amplitude of $\mathbf{v}$. But this would suggest the presence of a vertical shear, $\partial_\xi \mathbf{v}$, in each ring. Together with the vertical density gradient (and hence the vertical gradient of the dynamic viscosity) in the disc, this gives rise to a volume viscous force that damps the driving of individual layers of each disc ring through a resonance force. Note that close to the black hole, where the frequency κ deviates from Ω, the amplitude $\mathbf{v}$ remains limited even in the absence of viscous forces. This allows for the existence of stationary twisted discs with low viscosity around black holes, in which $\beta(r)$ takes an oscillatory form (see Ivanov and Illarionov (1997)).

The Asymmetrical Density Distribution in a Twisted Disc and the Torque due to a Central Gravitational Force

Thus, we see that the twist of the disc caused by the gravitomagnetic force necessarily results in a perturbation of the circular motion of gas elements in the disc rings. The velocity field of this perturbation, $\mathbf{v}$, depends on r (in addition to being proportional to $\propto \xi$, as explained above) and is determined by the current shape of the disc. By virtue of the continuity of the flow, this gives rise to density inhomogeneities outside the disc equatorial surface, $\rho_1 \propto \xi$. Since $(\nabla p)_r \propto \cos\psi$, these inhomogeneities take opposite signs in the diametrically opposite points of any given ring. But this implies that the ring is subject to the total torque of the central gravitational force acting on the domains with enhanced density outside of the equatorial plane of the ring (i.e. outside $\xi = 0$). We denote this torque by $\mathbf{T}_g$. Since the disc is thin and the gravitational acceleration along the axis of a ring is itself $\propto \xi$, the corresponding component of the gravitational force, and $\mathbf{T}_g$ as well, are quadratic in ξ. In addition, recall that the torque $\mathbf{T}_g$ is proportional to the small warp magnitude, $\mathbf{T}_g \propto \beta d\gamma/dr$. Thus, we arrive at the conclusion that the dynamics of the twisted disc rings is controlled by $\mathbf{T}_g$, together with the torque due to the gravitomagnetic force discussed earlier in this introductory section. Note that in the case $\beta = const$ considered here, $(\nabla p)_r$ and, correspondingly, ρ_1 take their maximum absolute values (with opposite signs) at $\psi = 0, \pi$, i.e. at the node line of each ring,[1] see also Fig. 4.4. But this implies that $\mathbf{T}_g$ lies in the plane formed by the angular momentum of each ring and the black hole spin axis. By virtue of the symmetry of the problem, the total contribution to $\mathbf{T}_g$ from other azimuths does not alter its direction. Therefore, immediately after the gravitomagnetic force turns an imaginary tilted planar disc into a twisted configuration with $\beta = const$, the gravitational force acting on the asymmetrically located matter of the disc, relative to the surface $\xi = 0$, tends to change the tilt angles of the disc rings: either to align the rings with the equatorial plane of the black hole or, conversely, to remove them from it. On the other hand, once β becomes dependent on r, the maxima of absolute

[1]For the sake of making the description as rigorous as possible, it is important also to add that the coincidence of azimuthal location of maxima of $(\nabla p)_r$ and ρ_1 occurs only when the effect of viscosity on the gas elements of the ring is neglected.

values of $(\nabla p)_r$ are shifted from the node line of each ring to some new ψ, which gives rise to a component in $\mathbf{T}_g$ that also contributes to the precession motion of the disc rings, just like the gravitomagnetic torque.

Additional Influence of the Viscous Torque

The dynamics of twisted discs as sketched above is complicated due to the presence of non-zero viscosity in the disc. First of all, each ring of the disc is subjected to the action of the viscous force arising due to the difference between the direction of the tangential velocity of the ring and that of the adjacent rings. This difference is maximal in the directions where the ring planes intersect, i.e. exactly where $(\nabla p)_r$ vanishes, see Fig. 4.3. In the above example of a configuration with $\beta = const$, this corresponds to $\psi = \pm\pi/2$, i.e. perpendicular to the node line of the rings. The viscous force, being proportional to the difference in tangential velocities, is directed at these points perpendicular to the ring plane and has different signs on different sides of the node line. Therefore, the corresponding torque, $\mathbf{T}_\nu$, is perpendicular to the plane made up by the angular momentum of the ring and the black hole spin axis. In other words, the viscous interaction between the disc rings leads only to their precession around the black hole spin axis, see Fig. 4.4. Note also that the viscous torque $\mathbf{T}_\nu \propto \beta d\gamma/dr$, which appears due to the difference between the tangential velocities of adjacent rings, and $\mathbf{T}_\nu \propto \xi^2$ due to the viscosity coefficient. It is important to note that as soon as the profile $\beta(r)$ is formed due to the gravitational torque $\mathbf{T}_g$, the viscous torque $\mathbf{T}_\nu$ also starts causing alignment/misalignment of the

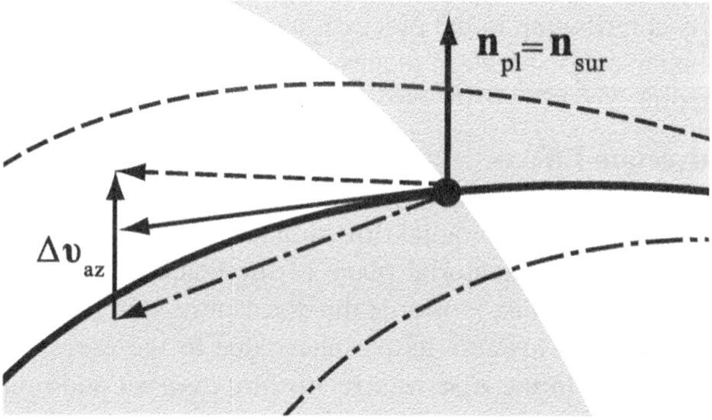

Fig. 4.3 A sketch of the zone of a twisted disc in the vicinity of the filled circle located at the particular ring at $\psi = \pi/2$, see Fig. 4.1. The visible part of the disc is transparent in the figure, while the part of it obscured by the semi-transparent plane B shaded, see Fig. 4.1. Dashed and dot-dashed lines represent the rings adjacent to the particular ring, with larger and smaller radii, respectively. There are two normals originating from the point indicated by the filled circle: to the plane of the ring, $\mathbf{n}_{pl}$, and to the disc surface, $\mathbf{n}_{sur}$. In addition, we have the velocities of the flow rotation corresponding to the displayed rings, evaluated at the same $\psi = \pi/2$ and transported to this point. Their difference is the relative velocity $\Delta \mathbf{v}_{az}$, which causes the emergence of a viscous force acting on the particular ring from the adjacent rings

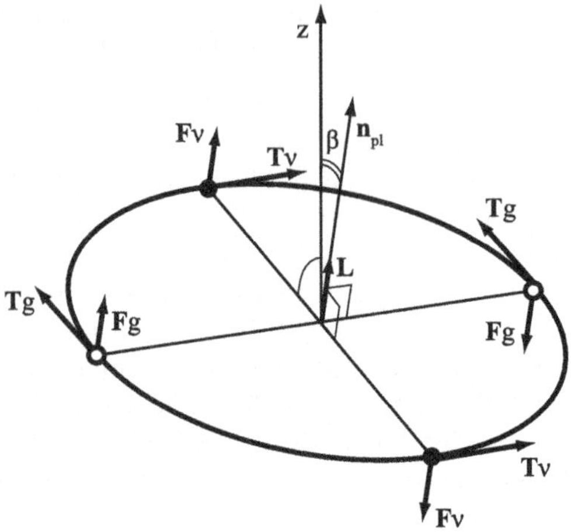

Fig. 4.4 A sketch of the forces acting on a particular ring in a twisted disc, see Fig. 4.1. Although both the central gravitational force, $\mathbf{F}_g$, and the viscous force, $\mathbf{F}_\nu$, are non-zero everywhere along the ring, their vectors are displayed here in the points of their maximum magnitude only, see Fig. 4.1 and also Figs. 4.2 and 4.3. The full averages of these forces vanish, whereas the corresponding full torques, $\mathbf{T}_g$ and $\mathbf{T}_\nu$, are non-zero. In this case (look at the direction of the angular momentum of the particular ring, $\mathbf{L}$) the viscous torque causes precession of the ring, whereas the central gravitational torque causes evolution of its inclination

ring with the equatorial plane of the black hole. This happens for the same reasons as $\mathbf{T}_g$ also starts contributing to the precession motion, and as discussed above, the location of the intersection between planes of adjacent rings becomes shifted in the azimuthal direction.

Additional Advection Effects

In addition to causing the appearance of $\mathbf{T}_\nu$, the viscosity in a twisted disc, like in a flat accretion disc, leads to radial diffusion transfer of the angular momentum component parallel to the equatorial plane of the black hole (which is non zero exactly for a tilted/twisted disc) towards the disc centre, due to simple transport of the accreting matter, and towards its periphery due to the corresponding outflow of angular momentum. In the case of a relativistic disc, an additional loss of this angular momentum component occurs due to thermal energy outflow in the form of radiation from the disc surface (see Eq. (C6) in Zhuravlev and Ivanov (2011)).

All forces participating in the dynamics of a twisted disc appear in the so-called 'twist' equation—the master equation of the twisted disc theory. This equation is derived and analyzed in the subsequent sections.

4.1.1 Weakly Perturbed Circular Equatorial Motion: Epicyclic Frequency and Frequency of Vertical Oscillations

In a twisted disc, we assume motion of matter outside the equatorial plane of the Kerr metric. This motion is not necessarily circular in the projection onto that plane. Therefore, we first analyze the properties of free particles moving in orbits slightly different from circular ones.

We first assume that the particles move exactly in the equatorial plane but in slightly non-circular orbits. This problem can be solved using relativistic hydrodynamic equations with zero pressure, and by assuming that there is a small deviation from the purely circular velocity. Also, instead of Eqs. (3.64) and (3.65), it is better to use the original equations in the form

$$T^{ik}{}_{;k} = 0, \tag{4.4}$$

where in the considered case of free motion, $T^{ik} = \rho U^i U^k$ and $\rho = const$. Under the last assumption, the velocity field, as follows from the rest-energy conservation law (3.66), is divergence-free, and (4.4) is equivalent to the following equation:

$$U^i{}_{;k} U^k = 0. \tag{4.5}$$

We now consider a small deviation from the four-velocity of the circular equatorial motion and denote it by v^i. Unperturbed motion corresponds to rest in the projection onto tetrad (3.50)–(3.53), used to construct the flat accretion disc model, i.e. is given by the four-velocity $U_0^i = \{1, 0, 0, 0\}$. Substituting the sum $U_0^i + v_i$ in (4.5), we obtain linear equations for small perturbations of the four-velocity v_i, which is assumed to be a function of t only:

$$v^i{}_{;k} U_0^k + U^i_{0;k} v^k = 0. \tag{4.6}$$

Taking into account that $U^i_{0;k} = \Gamma^i_{tk}$, for $i = 1, 2$ we obtain the set of equations

$$v^r{}_{;t} + \Gamma^r_{t\phi} v^\phi = C^{-1/2} B \frac{dv^r}{dt} - 2r^{-3/2} v^\phi = 0, \tag{4.7}$$

$$v^\phi{}_{;t} + \Gamma^\phi_{tr} v^r = C^{-1/2} B \frac{dv^\phi}{dt} + r^{-3/2} \left(1 - \frac{1}{2}\frac{H}{C}\right) v^r = 0. \tag{4.8}$$

It follows that small perturbations of the four-velocity components in the equatorial plane of the rotating black hole oscillate in time. For example, v^r satisfies the equation

$$\frac{d^2 v^r}{dt^2} + \frac{2C}{r^3 B^2} \left(1 - \frac{H}{2C}\right) v^r = 0, \tag{4.9}$$

which implies that the square of the frequency of these oscillations, epicyclic by definition, has the form

$$\kappa^2 = r^{-3}B^{-2}(2C - H) = r^{-3}\left(1 + \frac{a}{r^{3/2}}\right)^{-2}\left(1 - \frac{6}{r} + \frac{8a}{r^{3/2}} - \frac{3a^2}{r^2}\right). \tag{4.10}$$

A somewhat different derivation of κ can be found in the Appendix in Okazaki et al. (1987). It is important to note that (4.10) contains a derivative with respect to the coordinate time, and therefore the epicyclic frequency is determined by the clock of an infinitely remote observer, similarly to the circular frequency (3.17) introduced above. By comparing Eq. (3.18), which defines the location of the innermost stable circular equatorial orbit in the Kerr metric, r_{ms}, with (4.10), we infer that $\kappa^2(r_{ms}) = 0$. For $r < r_{ms}$ the epicyclic frequency becomes imaginary, and Eq. (4.9) has exponentially growing solutions. This must be true since, in this region, free circular motion around a rotating black hole becomes unstable. In Sect. 3.1.3 this result was obtained from an analysis of the form of the effective centrifugal potential, in which a test particle moves in an equatorial circular orbit. We see that r_{ms} can be determined alternatively from the calculation of the profile $\kappa^2(r)$ in the Kerr metric.

It is well known that for Newtonian motion, so-called Keplerian degeneration occurs when $\kappa = \Omega$ for non-circular motion, which causes the non-relativistic orbits to be closed. However, this symmetry is broken for relativistic free motion, and the epicyclic frequency κ differs from Ω already close to a non-rotating ($a = 0$) black hole, where its square is

$$\kappa^2 = r^{-3}\left(1 - \frac{6}{r}\right) = \Omega^2\left(1 - \frac{6}{r}\right) < \Omega^2. \tag{4.11}$$

The difference between the epicyclic and circular frequencies results in the well known effect of precession of the elliptical orbit. Far away from the horizon of a Schwarzschild black hole, i.e. for $r \gg 1$, the frequency of the orbits rotation, called the Einstein precession frequency, is $\Omega_p \approx 3/r^{5/2}$.

Suppose now that we rotate together with the test particle at some radius. When considering the problem in the projection onto tetrad (3.50)–(3.53), this particle remains at rest. We now impart to the particle a small velocity in the direction perpendicular to the equatorial plane. Equation (3.75) of hydrostatic equilibrium for a flat disc implies that in our reference frame the particle, being in free motion, is subjected to acceleration that is proportional to z and tends to return the particle to $z = 0$. As a result, the test particle oscillates harmonically with a frequency the square of which is

$$\Omega_v'^2 = \frac{H}{r^3 C}, \tag{4.12}$$

where the superscript ‘l’ serves as a remainder that the frequency is measured in the reference frame comoving with the particle in its main circular equatorial motion. To re-define this frequency as measured by the clock of an infinite observer, as was done for both circular and epicyclic frequencies, the frequency Ω_v^l must be divided by the time dilation factor (the difference between the proper time of the particle and the time at infinity), i.e. by the t-component of the four-velocity (3.15). Thus, the square of the frequency of vertical oscillations is

$$\Omega_v{}^2 = r^{-3}B^{-2}H = r^{-3}\left(1+\frac{a}{r^{3/2}}\right)^{-2}\left(1-\frac{4a}{r^{3/2}}+\frac{3a^2}{r^2}\right), \tag{4.13}$$

which coincides, for example, with the expression presented in Kato (1990) (see also Ipser (1996)). Equation (4.13) implies that around a non-rotating black hole $\Omega_v = \Omega$. This means that the vertical and circular motions have the same period, and the total motion of the particle is again a circular motion in a closed orbit whose plane, however, is now slightly tilted towards the initial equatorial plane. The situation changes for $a \neq 0$, since for $\Omega_v \neq \Omega$ the orbit is not closed any more, and the orbital plane starts precessing around the spin axis of the black hole. The frequency of the orbital precession is equal to the difference between the circular and vertical frequencies. For a slowly rotating black hole with $a \ll 1$ the precession frequency of a slightly tilted orbit is

$$\Omega_{LT} = \Omega - \Omega_v \approx r^{3/2}\left(1-\frac{a}{r^{3/2}}\right) - r^{3/2}\left(1-\frac{3a}{r^{3/2}}\right) = \frac{2a}{r^3} \ll \Omega. \tag{4.14}$$

This is simply the angular velocity of the frame dragging by the rotating black hole (see Eq. (3.3)) in the limit $a \ll 1$. The frequency Ω_{LT} is also referred to as the Lense-Thirring frequency.

In the most general case, where the test particle deviates from circular motion simultaneously in the vertical and horizontal directions, the particle's motion in space can be described by a slightly elliptical orbit, with both plane and apse line turning with an angular velocity proportional to the difference between the circular and vertical frequency and the difference between the circular and epicyclic frequency, respectively. For $a \ll 1$, the precession of the orbital plane occurs on a timescale much longer than the dynamical time, $t_{LT} \gg t_d$, where $t_{LT} \sim \Omega_{LT}^{-1}$ (see the discussion in the previous section).

4.2 Choice of Reference Frame

4.2.1 The Metric

Taking the general conclusions of Sect. 4.1 into account, we here consider slowly rotating black holes, $a \ll 1$. In this case, a linear expansion of the Kerr metric in the parameter a is sufficient. Then the formula (3.1) takes the form

$$ds^2 = (1-2/R)dt^2 - (1-2/R)^{-1}dR^2 - R^2(d\theta^2 + sin^2\theta d\phi^2) + 4\frac{a}{R}sin^2\theta\, d\phi\, dt. \tag{4.15}$$

Metric (4.15) is identical to that of a non-rotating black hole written in Schwarzschild coordinates, except for one non-diagonal term responsible for the Lense-Thirring precession.

Our main purpose in this section is to introduce the relativistic reference frame that follows the disc twist. The symmetry of the problem implies that the equations of motion should have the simplest form in such a frame. As for a flat disc, it is convenient to use some orthonormal non-coordinate basis. For this basis to follow the disc shape, its two spatial orts should be tangential to the disc symmetry plane. At each spatial point we take the orts of the 'flat' basis, which are determined, say, by the equatorial plane of the black hole, and rotate them by the angles β and γ, defining the disc shape. This is done in the simplest way by using a Cartesian coordinate system with the z-axis parallel to the black hole spin. However, we should first understand which four-dimensional basis (whose dual tetrad must transform the metric (4.15) into the Minkowski metric) in the flat-space limit would produce the spatial part described by the Cartesian reference frame.

This can be done by changing the radial variable in (4.15), namely, by passing from R to the so-called 'isotropic' radial coordinate, R_I:

$$R = R_I\left(1 + \frac{1}{2R_I}\right)^2. \tag{4.16}$$

Substituting (4.16) into (4.15) yields

$$ds^2 = \left(\frac{1-\frac{1}{2R_I}}{1+\frac{1}{2R_I}}\right)^2 dt^2 - \left(1 + \frac{1}{2R_I}\right)^4 (dR_I^2 + R_I^2 d\theta^2 + R_I^2 sin^2\theta d\phi^2)$$

$$+4\frac{a\, sin^2\theta}{R_I\left(1+\frac{1}{2R_I}\right)^2}dt\, d\phi, \tag{4.17}$$

where the second term represents an elementary spherical volume. Now, it is easy to transform to Cartesian coordinates via the change $\{x = R_I\cos\phi\sin\theta,\, y = R_I\sin\phi\sin\theta,\, z = R_I\cos\theta\}$. Using $R_I^2\sin^2\theta d\phi = xdy - ydx$ we have

$$ds^2 = K_1^2 dt^2 + 2aK_1K_3(xdy - ydx)dt - K_2^2(dx^2 + dy^2 + dz^2), \tag{4.18}$$

where

$$K_1 = \frac{1-\frac{1}{2R_I}}{1+\frac{1}{2R_I}}, \quad K_2 = \left(1+\frac{1}{2R_I}\right)^2, \quad K_3 = \frac{2}{R_I^3}\frac{1}{1-\left(\frac{1}{2R_I}\right)^2}, \tag{4.19}$$

are functions of $R_I = (x^2+y^2+z^2)^{1/2}$ only.

Metric (4.18) generates the following dual basis

$$\mathbf{e}^t = K_1 dt + aK_3(xdy - ydx), \quad \mathbf{e}^x = K_2 dx, \quad \mathbf{e}^y = K_2 dy, \quad \mathbf{e}^z = K_2 dz. \tag{4.20}$$

Note that basis (4.20) corresponds to observers at rest in Schwarzschild coordinates, since their world lines, defined by the condition $U^i = \mathbf{e}^i/ds = \{1, 0, 0, 0\}$, correspond to the equalities $dx = dy = dz = 0$. Their identical clocks are synchronized in such a way that in equal time intervals, determined by the ort $\mathbf{e}^t$, light travels an equal distance in any direction defined by the combination of $\mathbf{e}^{x,y,z}$. If the observers were to use the coordinate time t, they would discover, for example, that the light signal in the azimuthal direction, prograde with the black hole spin, travels a larger distance than in the opposite (retrograde) direction. This follows from the frame-dragging effect of a rotating black hole and is equivalent to the well-known tilt of light cones in the azimuthal direction. Finally, we note that another choice of orthonormal basis is possible in principle, which also compensates for the space-dragging effect. This basis is called the frame of locally non-rotating observers, and is moving with an azimuthal angular velocity equal to (3.3). Mathematically, this corresponds to a correction of the azimuthal ort instead of the time ort (see Bardeen et al. (1972)).

Below, we need to rotate the spatial part of (4.20), so as to obtain the dual twisted basis and then the original basis, which, as we recall, is needed to write down the projection of the hydrodynamic equations. For this, let us first introduce the twisted cylindrical coordinates.

4.2.2 *Twisted Coordinates*

We define the twisted cylindrical coordinates $\{\tau, r, \psi, \xi\}$ such that the condition $\xi = 0$ determines a coordinate surface coincident with the equatorial surface of a twisted disc. Here, τ, r, ψ and ξ are the new time variable and twisted analogues of the radial, azimuthal and vertical cylindrical coordinates, respectively.[2] These coordinates were first introduced in Petterson (1977, 1978). At each fixed $r = const$, the angle ψ is measured in the positive direction from the ascending

[2]Here and hereafter, r denotes the twisted radial coordinate.

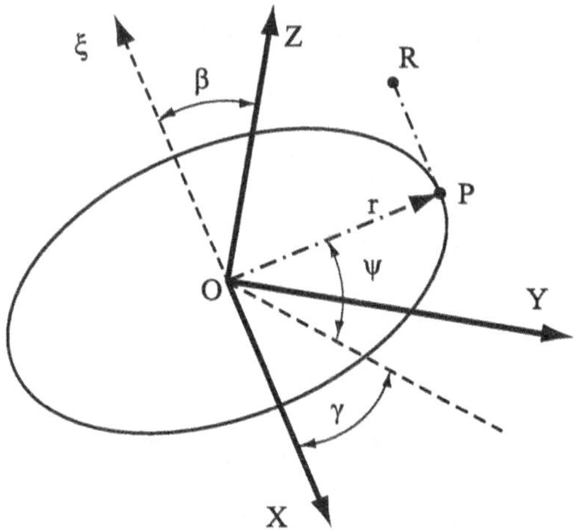

Fig. 4.5 Twisted cylindrical coordinates $\{r,\ \psi,\ \xi\}$

node of the circle $\xi = 0$ crossing the equatorial plane of the black hole. The relation between $\{\tau, r,\ \psi\ \xi\}$ and $\{t,\ x,\ y\ z\}$ can be obtained by a sequence of rotations at each radial distance by the angles $\beta(r, \tau)$ and $\gamma(r, \tau)$ (Fig. 4.5).

Let us take the radius vector with coordinates

$$\begin{bmatrix} \tau \\ r\cos\psi \\ r\sin\psi \\ \xi \end{bmatrix}, \tag{4.21}$$

where the three spatial Cartesian coordinates are defined in a frame with the z-axis tilted by the angle $\beta(r, \tau)$ towards the black hole spin, and the x-axis lying in the black hole equatorial plane and turned by the angle $\gamma(r)$ relative to some direction common for all r.

Next, we consecutively rotate this frame by the angle $\beta(r, \tau)$ about its x-axis in the negative direction and then by the angle $\gamma(r, \tau)$ about its z-axis in the negative direction. After these two rotations, this frame transforms into a 'flat' Cartesian frame common for all r, with the xy-plane coinciding with the equatorial plane of the black hole. Herewith, the new coordinates of the radius-vector are obtained by multiplying (4.21) first by the matrix

$$A_I(\beta) = \begin{bmatrix} 1 & 0 & 0 & 0 \\ 0 & 1 & 0 & 0 \\ 0 & 0 & \cos\beta & -\sin\beta \\ 0 & 0 & \sin\beta & \cos\beta \end{bmatrix}, \tag{4.22}$$

and then by the matrix

$$A_2(\gamma) = \begin{bmatrix} 1 & 0 & 0 & 0 \\ 0 & \cos\gamma & -\sin\gamma & 0 \\ 0 & \sin\gamma & \cos\gamma & 0 \\ 0 & 0 & 0 & 1 \end{bmatrix}. \tag{4.23}$$

As a result, we obtain the following relation between the twisted cylindrical and the 'flat' Cartesian coordinates in the linear approximation in small β:

$$\begin{aligned} t &= \tau \\ x &= r\cos\gamma\cos\psi - \sin\gamma\,(r\sin\psi - \xi\beta) \\ y &= r\sin\gamma\cos\psi + \cos\gamma\,(r\sin\psi - \xi\beta) \\ z &= r\beta\sin\psi + \xi. \end{aligned} \tag{4.24}$$

4.2.3 A Tetrad Transported by Observers Following the Twist

We now move from the 'flat' basis (4.20) to the twisted one by rotating its spatial orts by the twisting angles at each spatial point. First, we need to perform the rotation strictly opposite to what we did in the previous paragraph. This means that we should take basis (4.20) as a column and first multiply it by the matrix $A_2(-\gamma)$ and then by the matrix $A_1(-\beta)$. After that, since we wish to obtain the basis corresponding to the (twisted) cylindrical frame, it is necessary to additionally 'advance' the three spatial orts by the azimuthal angle ψ, which is achieved by additional multiplication of the basis by the matrix $A_2(-\psi)$.

As a result, we obtain a twisted dual basis, which contains some linear combinations of the 'flat' coordinate orts, $\{dt, dx, dy, dz\}$. It remains to express it as linear combinations of coordinate orts of the twisted coordinate frame, $\{d\tau, dr, d\psi, d\xi\}$. For this, it is sufficient to take differentials of the coordinate transformation (given by (4.24) in the linear approximation in β) and to substitute them in the twisted dual basis obtained after the rotations. It can be verified that in linear order in β and a, we have

$$\boldsymbol{e}^{\tau} = (K_1 - ar\xi K_3\partial_\varphi U)d\tau + a\xi K_3\partial_\varphi(Z - rW)dr + arK_3(r - \xi Z)d\varphi - arK_3\partial_\varphi Z d\xi, \tag{4.25}$$

$$\boldsymbol{e}^{r} = -\xi K_2 U d\tau + K_2(1 - \xi W)dr, \tag{4.26}$$

$$\boldsymbol{e}^{\varphi} = -\xi K_2\partial_\varphi U d\tau - \xi K_2\partial_\varphi W dr + rK_2 d\varphi, \tag{4.27}$$

$$\boldsymbol{e}^{\xi} = rK_2 U d\tau + rK_2 W dr + K_2 d\xi, \tag{4.28}$$

where we introduce the new azimuthal variable $\varphi = \psi + \gamma(r)$ and change to partial derivatives with respect to the corresponding new coordinates.

We also introduce new variables characterizing the disc geometry:

$$\Psi_1 = \beta \cos\gamma,\ \Psi_2 = \beta \sin\gamma \tag{4.29}$$

and from now on use them instead of the angles β and γ. Additionally,

$$Z = \beta \sin\psi = \Psi_1 \sin\varphi - \Psi_2 \cos\varphi, \quad U = \dot{Z}, \quad W = Z', \tag{4.30}$$

where partial derivatives with respect to τ and r are denoted by the dot and the prime.

It follows that for $\beta = \gamma = 0$ and with an additional transition to Cartesian coordinates, basis (4.25)–(4.28) is transformed into the 'flat' basis (4.20).

As discussed above, observers transporting basis (4.20) are at rest in the Schwarzschild coordinates. On the contrary, observers associated with basis (4.25)–(4.28) move in space by following the changing shape of the twisted disc (in non-stationary dynamics).

As we have seen in Chap. 3, the original basis, onto which the hydrodynamic equations are projected is obtained by inverting the dual basis matrix. Using (4.25)–(4.28) in an approximation linear in β and a, we have

$$\mathbf{e}_\tau = \frac{1}{K_1}\left(\partial_\tau + \xi U \partial_r + \frac{\xi}{r}\partial_\varphi U \partial_\varphi - rU\partial_\xi\right), \tag{4.31}$$

$$\mathbf{e}_r = \frac{1}{K_2}\left(-a\xi\frac{K_3}{K_1}\partial_\varphi Z \partial_\tau + (1+\xi W)\partial_r + \frac{\xi}{r}\partial_\varphi W \partial_\varphi - rW\partial_\xi\right), \tag{4.32}$$

$$\mathbf{e}_\varphi = \frac{1}{K_2}\left(-a\frac{K_3}{K_1}(r-\xi Z)\partial_\tau - a\xi\frac{K_3}{K_1}rU\partial_r + \left(\frac{1}{r} - a\xi\frac{K_3}{K_1}\partial_\varphi U\right)\partial_\varphi + ar\frac{K_3}{K_1}rU\partial_\xi\right), \tag{4.33}$$

$$\mathbf{e}_\xi = \frac{1}{K_2}\left(ar\frac{K_3}{K_1}\partial_\varphi Z \partial_\tau + \partial_\xi\right). \tag{4.34}$$

Using the original and dual bases, together with the algorithm presented in Sect. 3.2.1, we can now calculate the connection coefficients. This very cumbersome but straightforward procedure yields the following non-zero connection coefficients in a linear approximation in β and a:

$$\Gamma_{\tau r\tau} = \frac{K_1'}{K_1K_2}, \qquad \Gamma_{\tau r\varphi} = a\frac{K_3}{K_2^2}\left(1 - \tfrac{1}{2}(r - \xi Z)K_4\right),$$

$$\Gamma_{\tau r\xi} = -a\frac{K_3}{K_2^2}\partial_\varphi Z\left(1 - \tfrac{1}{2r}(r^2+\xi^2)K_4\right), \quad \Gamma_{\tau\varphi r} = -\Gamma_{\tau r\varphi},$$

$$\Gamma_{\tau\varphi\xi} = a\frac{K_3}{K_2^2}\left(Z + \tfrac{\xi}{2r}(r - \xi Z)K_4\right), \qquad \Gamma_{\tau\xi\tau} = \frac{\xi}{r}\frac{K_1'}{K_1K_2},$$

$$\Gamma_{\tau\xi r} = -\Gamma_{\tau r\xi}, \qquad \Gamma_{\tau\xi\varphi} = -\Gamma_{\tau\varphi\xi},$$

$$\Gamma_{r\varphi\tau} = \frac{\xi}{r}\frac{1}{K_1}\partial_\varphi U - \Gamma_{\tau r\varphi}, \qquad \Gamma_{r\varphi r} = \frac{\xi}{r}\frac{1}{K_2}\partial_\varphi W,$$

$$\Gamma_{r\varphi\varphi} = \frac{(rK_2)'}{rK_2^2} - a\xi\frac{K_3}{K_1K_2}\partial_\varphi U, \qquad \Gamma_{r\xi\tau} = \frac{U}{K_1} - \Gamma_{\tau r\xi},$$

$$\Gamma_{r\xi r} = \frac{W}{K_2} - \frac{\xi}{r}\frac{K_2'}{K_2^2}, \qquad \Gamma_{r\xi\varphi} = -ar\frac{K_3}{K_1K_2}U,$$

$$\Gamma_{r\xi\xi} = \frac{K_2'}{K_2^2}, \qquad \Gamma_{\varphi\xi\tau} = \frac{1}{K_1}\partial_\varphi U - \Gamma_{\tau\varphi\xi},$$

$$\Gamma_{\varphi\xi r} = \frac{1}{K_2}\partial_\varphi W, \qquad \Gamma_{\varphi\xi\varphi} = -\frac{\xi}{r}\frac{K_2'}{K_2^2} - ar\frac{K_3}{K_1K_2}\partial_\varphi U, \tag{4.35}$$

where $K_4 \equiv K_3/K_1(K_1/K_3)'$. The other non-zero Γ_{ijk}, as usual, can be obtained by taking the asymmetry in the first two indices into account.

Thus, the basis (4.31)–(4.34) together with the connection coefficients (4.35) are the sum of two parts: the main part that persists at $\beta = 0$ and a small additional part $\propto \beta$. In what follows, we denote these parts 'B_0' and "B_1', respectively.

4.3 The Set of Twist Equations

4.3.1 Projection of the Dynamical Equations onto the Twisted Basis for a Thin Disc

4.3.1.1 Separation of the Equations into Two Sets Describing a Flat and a Twisted Disc

Let us use the relativistic hydrodynamic equations in the original form:

$$T^{ik}_{;k} = 0, \tag{4.36}$$

where the stress-energy tensor and its components are presented in Sect. 3.2.3. Equations (4.36) should now be projected onto the twisted basis (4.31)–(4.34). To do this, we assume that $\beta \ll 1$. In other words, mathematically we consider the twist of the disc as a small perturbation to its 'ground' state, i.e. to the model of a flat disc, also referred to as the background. It is important to note that the appearance of a twist gives rise to new terms in the equations, not only due to the bending of the basis, but also due to the appearance of additional perturbations of the physical quantities themselves that enter the stress-energy tensor, including the density, pressure and four-velocity.

Thus for a twisted disc, instead of (4.36), we may write

$$((T_0{}^{ik} + T_1{}^{ik})_{;k})_0 + ((T_0{}^{ik} + T_1{}^{ik})_{;k})_1 = 0, \tag{4.37}$$

where $T_0{}^{ik}$ corresponds to the background state and $T_1{}^{ik}$ is a small *Eulerian* perturbation of the stress-energy tensor. The indices 0 and 1 that follow the notation of the covariant divergence mean that the divergence is taken in bases B_0 and B_1, respectively.

The action of the covariant divergence with index 0 on $T_0{}^{ik}$, evidently, yields 0, since these are equations for the background:

$$(T_0{}^{ik}{}_{;k})_0 = 0. \tag{4.38}$$

Then, in the approximation linear in β, we find the twist equations:

$$(T_1{}^{ik}{}_{;k})_0 + (T_0{}^{ik}{}_{;k})_1 = 0. \tag{4.39}$$

We assume that in a twisted disc the four-velocity, pressure, rest-mass energy density, internal energy, viscosity coefficient and energy flux density, as defined in their standard sense (see Sect. 3.2.3), are given by

$$U^i = U_0^i + v^i, \; p = p_0 + p_1, \; \rho = \rho_0 + \rho_1, \; \epsilon = \epsilon_0 + \epsilon_1,$$

$$\eta = \eta_0 + \eta_1, \; q^i = q_0^i + q_1^i,$$

respectively. Here, the indices 0 and 1 denote values related to the background and perturbations, respectively, and v^i are perturbations of the four-velocity.[3]

[3]To shorten the equations, we omit the term with the second viscosity ζ: as it can be shown using the analysis given below, this term does not contribute to the final equations in the leading order in the small parameters of the problem.

Thus, T_0^{ik} is a stress-energy tensor which contains only unperturbed quantities in accordance with definition (3.60) and its perturbation has the form

$$T_1^{ik} = w_1 U_0^i U_0^k + w_0 (v^i U_0^k + U_0^i v^k) - p_1 \eta^{ik} + 2\eta_1 \sigma_0^{ik} + 2\eta_0 \sigma_1^{ik} -$$

$$U_0^i q_1^k - U_0^k q_1^i - v^i q_0^k - v^k q_0^i, \tag{4.40}$$

where $w_0 = \rho_0 + \epsilon_0 + p_0$ is the background enthalpy and $w_1 = \rho_1 + p_1 + \epsilon_1$ its perturbation.

In addition, σ_0^{ik} is a shear tensor which contains only unperturbed quantities in accordance with definition (3.61), and σ_1^{ik} is its perturbed part of the form

$$\sigma_1^{ik} = \frac{1}{2}[(v^i_{;j})_0 P_0^{jk} + (v^k_{;j})_0 P_0^{ji}] - \frac{1}{3}(v^j_{;j})_0 P_0^{ik} +$$

$$\frac{1}{2}[(U_0{}^i_{;j})_0 P_1^{jk} + (U_0{}^k_{;j})_0 P_1^{ji}] - \frac{1}{3}(U_0{}^j_{;j})_0 P_1^{ik} +$$

$$\frac{1}{2}[(U_0{}^i_{;j})_1 P_0^{jk} + (U_0{}^k_{;j})_1 P_0^{ji}] - \frac{1}{3}(U_0{}^j_{;j})_1 P_0^{ik}, \tag{4.41}$$

where P_0^{ik} is the projection tensor that contains only unperturbed quantities in accordance with definition (3.62), and its perturbation is written as $P_1^{ik} = -U_0^i v^k - U_0^k v^i$.

Everywhere below we omit the index 0 for the unperturbed quantities. In addition, the viscous part of the stress-energy tensor in the disc is marked with "ν" wherever necessary: $T_\nu^{ik} \equiv 2\eta\sigma^{ik}$.

4.3.1.2 Additional Relations Used to Write the Equations

The relations given below are valid up to terms of the order of $\propto \delta^2$, which is sufficient for the theory of twisted discs in the leading order in the small parameter δ. In deriving these relations, this simplification enables us to consider that in the background solution, only U^τ and U^φ are non-zero, while $U^r \propto \delta^2$ and U^r can be temporarily set equal to zero.

We first note that the following relation between the components U^τ and U^φ is used below:

$$(U^\tau)^2 = (U^\varphi)^2 + 1, \tag{4.42}$$

which follows from the expression for the norm of the four-velocity in an orthonormal basis. Constraint (4.42) is also useful in differential form:

$$U^\tau dU^\tau = U^\varphi dU^\varphi. \tag{4.43}$$

Next, since the normalization of the four-velocity is also valid in the twisted disc, and the four-velocity perturbations are small, in the linear approximation we have

$$(U^\tau + v^\tau)^2 - (U^\varphi + v^\varphi)^2 = (U^\tau)^2 + 2U^\tau v^\tau - (U^\varphi)^2 - 2U^\varphi v^\varphi = 1,$$

and hence, with account for (4.42), v^i is 'orthogonal' to U^i:

$$U^\tau v^\tau = U^\varphi v^\varphi. \tag{4.44}$$

Finally, from the condition that σ^{ik} is space-like, we have

$$\sigma^{r\tau} U^\tau = \sigma^{r\varphi} U^\varphi,$$

and thus, in the basis B_0 used in this section, in the flat disc model, not only $T_\nu^{r\varphi}$, but also $T_\nu^{r\tau}$ is non-zero in the order of δ that is of interest to us here:

$$T_\nu^{r\tau} = \frac{U^\varphi}{U^\tau} T_\nu^{r\varphi}. \tag{4.45}$$

Note that in basis (3.50)–(3.53) co-moving with the azimuthal motion, only the component $T_\nu^{r\varphi}$ is non-zero (see (3.73)).

4.3.1.3 Equation of Free Azimuthal Motion

The quantities corresponding to the background model and entering the twist equations (4.39) should be obtained separately from Eq. (4.38). For this, it is sufficient to use the results from Chap. 3, taking only the transition from basis (3.50)–(3.53) to the basis B_0 into account.

Nevertheless, when deriving the twist equations, it is also necessary to use some of equations (4.38) written exactly in the basis B_0. This regards the r- and ξ-projections of these equations in the leading order in the small parameter of disc thickness which, as we know, describe its azimuthal rotation in the equatorial plane of the black hole and its vertical hydrostatic equilibrium. We emphasize that these relations are valid for both a stationary and a non-stationary accretion flow, for any viscosity parametrization, as well as for any specific vertical and radial structure of the flow. Only the condition $\delta \ll 1$ is important.

At the first stage of deriving the twist equations we will need only the r-projection of (4.38). Setting $T^{ik} = \rho U^i U^k$ we find that $T^{rk}{}_{;k} = 0$ yields

$$\frac{K_1'}{K_1}(U^\tau)^2 + a\frac{K_3}{K_2}(2 - rK_4)U^\tau U^\varphi - \frac{(rK_2)'}{rK_2}(U^\varphi)^2 = 0. \tag{4.46}$$

Exactly this combination (4.46) is used in the derivation; however, it can be checked that together with (4.42) in the approximation linear in a it gives the solution

$$U^{\varphi} = (r_S - 3)^{-1/2}\left(1 - a r_S^{-1/2}(r_S - 3)^{-1}\right), \tag{4.47}$$

where we have switched to the Schwarzschild radial coordinate r_S which is equivalent to r, which we used in Chap. 3 in the expression for U_g^{φ} (see formula (3.15)). It is easy to check that $U_g^{\varphi} = U^{\varphi}/r_S$, as must be the case with the transition from the coordinate basis to B_0 taken into account.

4.3.1.4 'Gauge' Condition of the Twisted Frame

The principal kinematic constraint for the twisted reference frame requires a constant vertical position of fluid particles:

$$\frac{d\xi}{d\tau} = 0, \tag{4.48}$$

which is provided by fast establishment of hydrostatic equilibrium across the disc compared to the dynamical time of the twist change, as discussed in Sect. 4.1. However, as has been already noted in Hatchett et al. (1981), an important point is that this does not mean that the projection of the four-velocity of the fluid onto $\mathbf{e}_{\xi}$ is also zero, because our basis is non-coordinate and its orts are not tangent to the coordinate lines.

By definition,

$$v^{\xi} = \frac{\mathbf{e}^{\xi}}{ds}.$$

Using (4.28) we have:

$$v^{\xi} = r K_2 U \frac{d\tau}{ds} + r K_2 W \frac{dr}{ds};$$

where we should substitute $d\tau/ds$ and dr/ds in this relation in the zeroth order in β, in other words, as values corresponding to the flat disc dynamics. Expressions for $\mathbf{e}^{\tau}$, $\mathbf{e}^{\varphi}$ and $\mathbf{e}^{r}$ at $\beta = 0$ give

$$\frac{d\tau}{ds} = \frac{1}{K_1}\left(U^{\tau} - a r^2 K_3 \frac{d\varphi}{ds}\right), \quad \frac{dr}{ds} = \frac{U^r}{K_2}, \frac{d\varphi}{ds} = \frac{U^{\varphi}}{r K_2}, \tag{4.49}$$

where by definition $U^i \equiv \mathbf{e}^i/ds$. As a result, we obtain

$$v^{\xi} = r U^{\tau} K \frac{K_2}{K_1} U + r U^r W, \tag{4.50}$$

where

$$K = \left(1 - ar\frac{K_3}{K_2}\frac{U^\varphi}{U^\tau}\right).$$

In (4.50) the velocity components U^τ and U^r should be taken from the corresponding background solution for a flat disc.

4.3.1.5 Explicit Form of the Set of Equations of a Twisted Disc

Now, using (4.42)–(4.50), we write Eq. (4.39) in explicit form by keeping only the terms in the leading order in the two small parameters δ and $u \equiv t_d/t_{ev}$.[4] Here, we take into account that quantities of 'thermal' origin in the background solution are small, i.e. $p,\ e,\ \eta \propto \delta^2 \rho$ and $q^\xi \propto \delta^3 \rho,\ q^{r,\varphi} \propto \delta^4$ (see Chap. 3).

We postpone discussing the effects of the fluid non-ideality for a while. Note that this assumption not only concerns the vanishing of terms including the viscosity coefficient and energy flux density, or their perturbations, but also implies the absence of contributions $\propto U^r$. To select the leading-order terms in the ideal fluid approximation, we start by considering second terms in the τ-, r- and φ-projections of (4.39). It turns out that such terms are proportional to $\delta\beta$ here, and in the r-projection of (4.39) this contribution is due to the projection of the vertical pressure gradient onto the orbital plane of motion of matter in the twisted disc (see the analysis in Sect. 4.1, where this quantity was denoted by $(\nabla p)_r$). In addition, the τ- and φ-projections of (4.39) involve terms $\propto \delta^{-1} u \beta$, which should be kept. On the other hand, the first terms in the τ-, r- and φ-projections of (4.39) give rise to terms containing Eulerian velocity perturbations, $v^{\tau,r,\varphi}$, as well as the Eulerian rest-mass energy density perturbation, ρ_1. Hence, we conclude that

$$v^{\tau,r,\varphi} \propto \max\{\delta, \delta^{-1} u\}\, \beta, \quad \text{and} \quad \rho_1 \propto \max\{\delta, \delta^{-1} u\}\, \rho\beta. \tag{4.51}$$

In addition, for reasons that become clear below, we temporarily keep partial derivatives of v^i and ρ_1 with respect to time, despite their being u^{-1} times smaller than the quantities themselves. Finally, the first terms of the τ- and φ-projections of (4.39) also contain terms with the combination $\partial_\xi \rho v^\xi$, whose amplitudes are restricted to the order $\propto \max\{\delta, \delta^{-1} u\}\, \beta$ by Eq. (4.50).

Now, using the result (4.51), it is easy to select the leading terms entering in the τ-, r- and φ-projections of (4.39) due to fluid non-ideality. The most troublesome here is the contribution due to the shear tensor perturbation, $2\eta\sigma_1^{ik}$, which appears

[4]As we discussed above, the smallness of t_d/t_{ev} is necessary to ensure that the accretion flow outside the equatorial plane of the black hole can be considered a 'disc'. In turn, this is jointly ensured by the smallness of both δ and $t_d/t_{LT} \ll 1$ (see Sect. 4.1.1).

in T_1^{ik} (see (4.40) and (4.41)). However, most of the terms from this contribution contain jointly $\eta \propto \delta^2$ and $v^i \propto \delta\beta$. Therefore, it is necessary to include only the terms in which the derivative with respect to ξ (lowering the order in δ) occurs twice. This fact strongly reduces the number of 'viscous' terms to be kept. By similar considerations, the final expressions will not contain terms with $\mathbf{q}$, $\mathbf{q}_1$ and η_1. Finally, we stress once again that in addition to the purely 'viscous' terms mentioned above, the contribution due to the radial advection that appears in the background solution with non-zero viscosity should not be forgotten. We are concerned with the terms that may appear in the 'non-viscous' part of the stress-energy tensor (see the first term in (3.60)) due to the non-zero value of $U^r \propto \delta^2$.

Taking all of the above into account and using the relations derived in the previous three sections, we obtain the τ-, r- and φ-projections of (4.39) in the form

$$K\frac{K_2}{K_1}(U^\tau)^2\dot{\rho}_1 + \left(2U^\varphi - ar\frac{K_3}{K_2}\frac{(U^\varphi)^2+(U^\tau)^2}{U^\tau}\right)\frac{K_2}{K_1}\rho\dot{v}^\varphi + \frac{1}{r}U^\tau U^\varphi \partial_\varphi \rho_1 +$$
$$\frac{1}{r}\frac{(U^\varphi)^2+(U^\tau)^2}{U^\tau}\rho\partial_\varphi v^\varphi + \partial_r(\rho U^\tau v^r) + \partial_\xi \rho U^\tau v^\xi + \frac{(rK_1^2K_2^2)'}{rK_1^2K_2^2}\rho U^\tau v^r + F_\nu^\tau =$$
$$r\partial_\xi \rho (U^\tau)^2 K\frac{K_2}{K_1}U + \frac{\xi}{r}\rho U^\tau U^\varphi \partial_\varphi W, \qquad (4.52)$$

$$K\frac{K_2}{K_1}U^\tau \dot{v}^r + \frac{U^\varphi}{r}\partial_\varphi v^r - \left[2\frac{K_1'}{K_1 U^\varphi} + a\frac{K_1}{rK_2U^\tau}\left(\frac{r^2K_3}{K_1}\right)'\right]v^\varphi + \frac{1}{\rho}F_\nu^r =$$
$$Wr\frac{\partial_\xi p}{\rho} - a\xi\frac{K_3^2}{K_1K_2}\left(\frac{K_1}{K_3}\right)' ZU^\tau U^\varphi, \qquad (4.53)$$

$$K\frac{K_2}{K_1}U^\tau U^\varphi \dot{\rho}_1 + \left(\frac{(U^\varphi)^2+(U^\tau)^2}{U^\tau} - 2ar\frac{K_3}{K_2}U^\varphi\right)\frac{K_2}{K_1}\rho\dot{v}^\varphi + \frac{(U^\varphi)^2}{r}\partial_\varphi \rho_1$$
$$+2\frac{U^\varphi}{r}\rho\partial_\varphi v^\varphi + \partial_r(\rho U^\varphi v^r) + \partial_\xi \rho U^\varphi v^\xi + \frac{(r^2K_1K_2^3)'}{r^2K_1K_2^3}U^\varphi \rho v^r$$
$$-a\frac{K_1}{rK_2}\left(\frac{r^2K_3}{K_1}\right)' U^\tau \rho v^r + F_\nu^\varphi = K\frac{K_2}{K_1}r\partial_\xi \rho U^\tau U^\varphi U$$
$$+\frac{\xi}{r}\rho(U^\varphi)^2\partial_\varphi W, \qquad (4.54)$$

where

$$K = \left(1 - ar\frac{K_3}{K_2}\frac{U^\varphi}{U^\tau}\right),$$

and $F_\nu^{\tau,r,\varphi}$ is the total contribution due to non-zero viscous forces and the radial advection of matter in the background solution $\propto U^r$.

Explicitly,

$$F_\nu^\tau = \frac{U^\varphi}{U^\tau}(\partial_\xi T_\nu^{\varphi\xi} - rW\partial_\xi T_\nu^{r\varphi}) - r\partial_\xi \rho U^\tau U^r W, \quad F_\nu^r = \partial_\xi T_\nu^{r\xi},$$

$$F_\nu^\varphi = (\partial_\xi T_\nu^{\varphi\xi} - rW\partial_\xi T_\nu^{r\varphi}) - r\partial_\xi \rho U^\varphi U^r W, \qquad (4.55)$$

where

$$T_\nu^{r\xi} = -\frac{\eta}{K_2}(\partial_\xi v^r + U^\varphi \partial_\varphi W), \quad T_\nu^{\varphi\xi} = -\frac{\eta}{K_2}\left(\partial_\xi v^\varphi - 2a\frac{K_3}{K_2}U^\tau (U^\varphi)^2 Z\right),$$

$$T_\nu^{r\varphi} = -\eta r\left(\frac{U^\varphi}{rK_2}\right)'. \qquad (4.56)$$

We note that $T_\nu^{r\xi}$ and $T_\nu^{\varphi\xi}$ have the meaning of perturbations of the viscous stress tensor. In these expressions, the terms $\propto \beta$ contributing to the shear tensor perturbations appear due to the twisted basis. Conversely, $T_\nu^{r\varphi}$ relates to the background. Nevertheless, for the sake of brevity, we use the same notation with the index ν for these three quantities.

Finally, we assume in (4.52)–(4.56) that in the relativistic coefficients K_1, K_2 and K_3, the argument R_I is replaced by r, since $R_I^2 = r^2 + \xi^2$ and accounting for the dependence on ξ here always gives rise to a small correction $\propto \delta^2$ only.

It remains to write the explicit form of the ξ-projection of (4.39). Similarly, we start with the contribution of terms in the ideal fluid approximation, and first rearrange the first term in (4.39). The leading-order terms in δ here are, in particular, ρv^φ and ρ_1, but additionally multiplied by ξ. This means that their amplitudes are restricted to the order $\max\{\delta^2, u\}\,\beta$. Furthermore, v^ξ now enters the term $U^\varphi \rho \partial_\varphi v^\xi$ which also implies the raising of the order of smallness by δ compared to (4.52)–(4.54) (it can be seen that in formulas (4.52) and (4.54) v^ξ enter in combination with $\partial_\xi \rho$). Besides, of all terms of a 'thermal' origin we must now keep the term with $\partial_\xi p_1$, since it also is of the order of δ^2 due to the fact that $p_1 \sim \delta^2 \rho_1 \propto \rho\delta^3\beta$.

Turning now to the second term in the ξ-projection of (4.39), we write all terms up to the order $\propto \max\{\delta^2, u\}\,\beta$. From similar considerations, the terms due to fluid non-ideality (including 'advective' terms proportional to $\propto U^r$) are also kept here, with their smallness increased by the coefficient δ compared to (4.52)–(4.54).

We thus obtain the following equation:

$$U^\varphi \partial_\varphi v^\xi + r\frac{\partial_\xi p_1}{\rho} + \xi\frac{(U^\varphi)^2}{r}\left(1 - 2ar\frac{K_3}{K_2}\frac{U^\tau}{U^\varphi}\right)\frac{\rho_1}{\rho} +$$
$$2\xi U^\varphi v^\varphi \left[\frac{K_1'}{K_1} - \frac{K_2'}{K_2} - \frac{ar}{2}\frac{K_3^2}{K_1 K_2}\left(\frac{K_1}{K_3}\right)'\left(\frac{U^\tau}{U^\varphi} + \frac{U^\varphi}{U^\tau}\right)\right] + \frac{r}{\rho}F_\nu^\xi =$$
$$-\left[\frac{K_2}{K_1}\partial_\varphi U - 2a\frac{K_3 Z}{K_2} + a\frac{\xi^2}{r}\frac{K_3^2 Z}{K_1 K_2}\left(\frac{K_1}{K_3}\right)'\right] rU^\tau U^\varphi + ar^2\frac{K_3}{K_1}(U^\varphi)^2\partial_\varphi U, \tag{4.57}$$

where

$$F_\nu^\xi = \frac{1}{rK_1K_2^3}\partial_r(rK_1K_2^3 T_\nu^{r\xi}) + \partial_\xi T_\nu^{\xi\xi} + \frac{1}{r}\partial_\varphi T_\nu^{\varphi\xi} +$$
$$\partial_\varphi W(T_\nu^{r\varphi} + T_{adv}^{r\varphi}) + a\frac{K_1}{rK_2}\left(\frac{r^2K_3}{K_1}\right)'\partial_\varphi Z\left(\frac{U^\varphi}{U^\tau}T_\nu^{r\varphi} + \frac{U^\tau}{U^\varphi}T_{adv}^{r\varphi}\right), \tag{4.58}$$

and $T_{adv}^{r\varphi} = \rho U^\varphi U^r$. We do not provide the explicit form of $T_\nu^{\xi\xi}$ here, as it is not required in the final form of the twist equations.

Everywhere in (4.57)–(4.58), except in the second term in square brackets on the right-hand side of (4.57), the argument R_I in the relativistic coefficients K_1, K_2 and K_3 is replaced by r. This term is an exception since it alone is of the zeroth order in small parameters δ and u in Eq. (4.57). But because we have kept the terms $\propto \max\{\delta^2, u\}$ in (4.57), in the term under discussion it should be necessary to take corrections $\propto \delta^2$ into account due to the dependence of the relativistic coefficients K_2 and K_3 on ξ. We did not do this for the reason discussed in the next paragraph.

4.3.2 Completing the Derivation of the Twist Equations

We have thus written the twist equations in the leading orders in the small parameters δ and u. All corrections linear in the Kerr parameter a have been taken into account. If we temporarily set $a = 0$ and consider Eq. (4.57), we see that, on the one hand, it contains terms proportional to the rate of change of the disc twist, U, and on the other hand, it has terms containing perturbations of physical quantities of the order $\propto \delta^2$. Thus, we can say that a thin twisted disc evolves on long timescales such that $u \sim \delta^2$, due to internal forces only. Then it becomes clear that Eqs. (4.52)–(4.54) are restricted to the order $\propto \delta$, and Eq. (4.57) is restricted to the order $\propto \delta^2$.

At the same time, when the parameter a is non-zero, a 'large' term of the zeroth order in δ and $\propto aZ$ arises on the right-hand side of Eq. (4.57). This term describes the gravitomagnetic interaction of the rotating black hole with the tilted/twisted disc. In order that all terms in (4.57) be balanced with each other, we must assume that $a \sim \delta^2$. But then it becomes clear that all additional corrections $\sim a$ in Eqs. (4.52)–(4.54) are of the next order in δ and can be omitted. The same applies to all terms $\propto a\delta^2$ in Eq. (4.57), including the correction $\propto \delta^2$ due to the dependence of the relativistic coefficients on R_I in the gravitomagnetic term itself.

In fact, this means that when considering the dynamics of a twisted thin accretion disc near a rotating black hole, it is sufficient to use the background model, i.e. the corresponding flat disc, in the Schwarzschild metric with $a = 0$. The assumption of slow black hole rotation was required as otherwise the accretion flow (including the non-stationary one) could not be regarded as a disc, since the vertical hydrostatic equilibrium there would be violated (see Sect. 4.1). Of course, these conclusions only apply to slightly tilted/twisted and geometrically thin discs with $\beta \ll 1, \delta \ll 1$.

In what follows, we therefore set $a = 0$ in all terms except the gravitomagnetic. This significantly simplifies further calculations required for obtaining the twist equations in the final form. Let us first analyze Eqs. (4.52) and (4.54). It is convenient to consider their combinations, which will contain neither $\dot{\rho}$ nor $\dot{v}^\varphi$.

Eliminating $\dot{v}^\varphi$ for $a = 0$ we obtain the equation

$$U^\varphi \partial_\varphi \rho_1 + \frac{1}{(U^\tau)^2}\rho \partial_\varphi v^\varphi + \frac{U^\tau}{K_2^2}\frac{\partial}{\partial r}\left(r K_2^2 \frac{\rho v^r}{U^\tau}\right) =$$
$$\xi U^\varphi \rho \partial_\varphi W + \frac{U^\varphi}{(U^\tau)^2}(\partial_\xi T_\nu^{\varphi\xi} - r W \partial_\xi T_\nu^{r\varphi}), \qquad (4.59)$$

where we have omitted the term $\dot{\rho}_1$, which is of the next order in δ, compared to the other terms. In the Newtonian limit, as $r \to \infty$, Eq. (4.59) reduces to the continuity equation for perturbations.

Next, eliminating $\dot{\rho}_1$ for $a = 0$,[5] we obtain the equation

$$\frac{K_2}{K_1}\dot{v}^\varphi + \frac{1}{r}\frac{U^\varphi}{U^\tau}\partial_\varphi v^\varphi + \left(\frac{\partial_r U^\varphi}{U^\tau} + \frac{K_1'}{K_1}\frac{U^\tau}{U^\varphi}\right) v^r + \frac{1}{\rho U^\tau}(\partial_\xi T_\nu^{\varphi\xi} - r W \partial_\xi T_\nu^{r\varphi}) = 0. \qquad (4.60)$$

In the Newtonian limit, (4.60) reduces to the azimuthal component of the Navier-Stokes equation for perturbations.

Finally, (4.53) with $a = 0$ takes the form

$$\frac{K_2}{K_1}U^\tau \dot{v}^r + \frac{U^\varphi}{r}\partial_\varphi v^r - 2\frac{K_1'}{K_1 U^\varphi}v^\varphi + \frac{1}{\rho}\partial_\xi T_\nu^{r\xi} = W r \frac{\partial_\xi p}{\rho}. \qquad (4.61)$$

[5] $a = 0$ also in the expression for $T_\nu^{\varphi\xi}$.

In the Newtonian limit, (4.61) reduces to the radial component of the Navier-Stokes equation for perturbations.

It is important to explain why we have retained terms with $\dot{v}^r$ and $\dot{v}^\varphi$ in Eqs. (4.60) and (4.61) although they are of the next order in δ. As mentioned in Sect. 4.1, in the Newtonian limit the epicyclic frequency becomes equal to the Keplerian circular frequency, which results in a resonance growth of the amplitude of velocity perturbations of gas elements in the disc, under the action of the radial projection of the vertical pressure gradient, $(\nabla p)_r$, which is limited only by the viscosity. Mathematically expressed, in the limit of an inviscid Keplerian disc, Eq. (4.60) yields, in the leading order in the parameter u (with the term with $\propto \dot{v}^\varphi$ omitted), a relation between v^r and v^φ such that the sum of the second and the third terms in (4.61) vanishes. But, as there is a term $\propto \delta\beta$ in the right-hand side of (4.61), it follows that $\dot{v}^r$ (and hence $\dot{v}^\varphi$ as well) acquires the first order in δ in the considered case. Either viscosity or relativistic corrections eliminate the Keplerian resonance, and the amplitudes of $\dot{v}^r$ and $\dot{v}^\varphi$ decrease again to the third order in δ.

Now, from Eq. (4.57) we need to derive the so-called twist equation that plays the principal role in the theory of twisted discs. For this, we need to explicitly determine the value $\partial_\xi p/\rho$, which is done in the next section. Although the Schwarzschild approximation is sufficient, we also take linear corrections in a into account. This is required below to obtain an additional expression for the Lense-Thirring frequency in terms of the relativistic coefficients used in the twisted basis.

4.3.2.1 Equation of Vertical Hydrostatic Equilibrium

Let us write down the ξ-projection of (4.38) in the basis B_0 to the leading order in δ, as we did in Chap. 3, employing basis (3.50)–(3.53) (see Eq. (3.75)). Taking into account that the four-velocity of the flow is $\{U^\tau, 0, U^\varphi, 0\}$ in the leading order in δ, we obtain the following equation

$$\frac{\partial_\xi p}{\rho} = \frac{\xi}{r}(U^\varphi)^2 \left[\frac{K_2'}{K_2} - \left(\frac{U^\tau}{U^\varphi}\right)^2 + ar\frac{K_3 K_4}{K_2}\frac{U^\tau}{U^\varphi} \right], \tag{4.62}$$

where U^τ and U^φ satisfy the normalization condition (4.42) and the geodesic Eq. (4.46). With this in mind, we arrive at the final form of the hydrostatic equilibrium equation

$$\frac{\partial_\xi p}{\rho} = -\frac{\xi}{r}\frac{(U^\varphi)^2}{r}\left(1 - 2ar\frac{K_3}{K_2}\frac{U^\tau}{U^\varphi}\right), \tag{4.63}$$

where the Schwarzschild profiles of U^τ and U^φ are used in the term including parameter a.

It can be verified that with the substitution $\xi \to z/K_2$, Eq. (4.63) is equivalent to (3.75) in a linear approximation in a. Here, we should only take into account that $r_S = K_2 r$, where r_S is the Schwarzschild coordinate equivalent to the coordinate r in (3.75).

4.3.2.2 Twist Equation

Our goal is to rewrite (4.57) in divergent form. Without accounting for the gravitomagnetic term, Eq. (4.57), in which we also set $a = 0$, must respect the conservation law of the angular momentum projection of the twisted disc onto the equatorial plane of the black hole (the conservation of the disc angular momentum projection onto the black hole spin in our problem, linear in β, follows from equations for the background, since the corrections due to the small tilt angle are proportional to $\propto 1 - \cos\beta \sim \beta^2$), which reflects spherical symmetry of the Schwarzschild metric.

It turns out that to do this it is necessary to eliminate v^φ and ρ_1, on the left-hand side of equation (4.57). Therefore, we will use Eqs. (4.59)–(4.61) with $\dot{v}^r = \dot{v}^\varphi = 0$ for our purposes, since we will not deal with resonance combinations of v^r and v^φ that vanishes in the main order in u in the Keplerian inviscid limit (see the comment to Eqs. (4.60) and (4.61) above).

First, on the right-hand side of (4.59) we rewrite the term with $\partial_\varphi W$ through v^r and v^φ using (4.61) and (4.63) with $a = 0$. In the resulting expression for ρ_1, we replace v^φ using (4.60). Here, the derivative with respect to φ can be eliminated using the harmonic dependence on φ (see (4.30)). In other words, $\partial_{\varphi\varphi} = -1$. Substituting the obtained expressions for ρ_1 and v^φ in (4.57), integrating over ξ, and performing integration by parts wherever necessary, using the fact that the corresponding surface terms vanish as $\rho \to 0$, we arrive at the compact equation

$$\Sigma U^\tau U^\varphi \left\{ \partial_\varphi U - a \frac{K_1 K_3}{K_2^2} Z \right\} + \partial_\varphi W \frac{K_1}{K_2} \{ \Sigma U^\varphi U^r + \bar{T}_\nu^{r\varphi} \} =$$

$$- \frac{1}{2r^2 K_2^4} \int d\xi \, \{ \partial_r (\xi r K_1 K_2^3 U^\varphi \rho \partial_\varphi v^r + r^2 K_1 K_2^3 T^{r\xi}), \tag{4.64}$$

where, as usual, $\Sigma = \int \rho \, d\xi$ is the surface density of the disc, and the bar over $T^{r\varphi}$ indicates that it is integrated over ξ. In the Appendix of Zhuravlev and Ivanov (2011), it is shown that (4.64) can be used to obtain the angular momentum conservation law for a twisted disc.

Equation (4.64) is a master equation governing the dynamics of twisted discs. It is usually referred to as the "twist equation". In Sect. 4.1 we gave a qualitative description of the dynamics of a twisted disc splitted into rings. At the quantitative level, this information is contained in (4.64). Let us explain the physical meaning of each term in the twist equation. The first term on the LHS of (4.64) $\propto \partial_\varphi U$ characterizes the rate of change of the horizontal component of angular momentum for a particular ring of the twisted disc. The second term on the LHS of (4.64) represents the gravitomagnetic torque, while the combination in front of Z gives the Lense-Thirring frequency, see (4.68) in the next paragraph. Finally, the combination $\propto \partial_\varphi W$ shows the radiative transfer of the horizontal component of the angular momentum for a particular ring, which arises due to accretion with speed U^r, as well as due to the action of viscous forces in the plane (r, φ). Further, the first term on the RHS of (4.64) provides torque due to the central gravitational force, which emerges due to the asymmetric density distribution perpendicular to the disc equatorial surface. This term (along with the gravitomagnetic torque) remains non-zero in the inviscid case. The last term on the LHS of (4.64) represents the viscous torque acting from the adjacent rings divergent from the chosen ring. The force associated with the latter torque is directed along the axis of rotation of the chosen ring, see the corresponding paragraph in Sect. 4.1. With regards to the last two terms, note that the central gravitational force *is always larger* than the viscous force. These forces become comparable to each other in the most viscous discs with a viscosity parameter $\alpha \sim 1$.

Equations (4.60), (4.61) and (4.64) represent a closed set of equations describing the dynamics of twisted configurations provided that the corresponding model of the background is specified. Unknown variables in this set include the velocity perturbations v^r and v^φ and the quantity Z characterizing the disc geometry. We emphasize that in deriving these equations we essentially used only three main assumptions: $a \ll 1$, $\delta \ll 1$ and $\beta \ll 1$. This means that the equations describe the dynamics of any geometrically thin accretion flow (disc) with any parametrization of viscosity and any radial or vertical structure, in both the stationary and the non-stationary case, (with non-stationary here, referring to a non-stationary background). Consequently, the equations determine not only the dynamics of twist perturbations propagating in a stationary flat disc, but also the dynamics of the twisted rings/tori, when evolution of the geometry is accompanied with their expansion in the radial direction due to turbulent viscosity, which means that the background itself is evolving.

4.3.2.3 Once Again About the Characteristic Frequencies of the Problem

In Sect. 4.1.1, we already obtained relativistic expressions for the characteristic frequencies of the problem. These include the circular and epicyclic frequencies of free equatorial motion, as well as the frequency of vertical oscillations and the precession frequency of tilted orbits. Here, we wish to obtain expressions for these frequencies, but now in terms of the values used above to construct the theory of twisted discs, i.e. in the basis B_0. These expressions are required to write the twist equations in a more compact form.

The circular frequency of free equatorial motion as measured by the clock of an infinitely remote observer, which we already presented in Eq. (3.17), can be obtained simply by dividing $d\varphi/ds$ by $d\tau/ds$ given in (4.49). We obtain

$$\Omega = \frac{K_1}{K K_2} \frac{U^\varphi}{r U^\tau}. \tag{4.65}$$

Using (4.47) and (4.42), and remembering that $r_S = r K_2$, we can check that (4.65) coincides with (3.17) in the linear approximation in a.

We now consider small vertical deviations from the circular equatorial motion. In Sect. 4.1.1, we discussed that the frequency of vertical oscillations as measured by an infinitely remote observer, Ω_v, is the locally measured frequency, Ω_l, divided by the t-component of the four-velocity of circular motion, U_g^t. The frequency Ω_l explicitly enters the equation of hydrostatic equilibrium (see Eq. (3.75) or equivalent, Eq. (4.63) with the substitution $\xi \rightarrow z/K_2$). Using relations (4.49), we express $U_g^t \equiv d\tau/ds$ in terms of U^τ:

$$U_g^t = K K_1^{-1} U^\tau,$$

whence

$$\Omega_l = \Omega_v \frac{K U^\tau}{K_1} = \frac{U^\varphi}{r K_2} \frac{\Omega_v}{\Omega}, \tag{4.66}$$

where the final expression was obtained using (4.65).

But then, from a comparison of (4.66) with (4.63), we see that

$$\Omega_v = \Omega \left(1 - a r \frac{K_3}{K_2} \frac{U^\tau}{U^\varphi} \right), \tag{4.67}$$

where the Schwarzschild profiles for U^τ and U^φ are used in the term with the parameter a.

Then, using (4.14), we obtain the Lense-Thirring frequency

$$\Omega_{LT} = a \frac{K_1 K_3}{K_2^2}. \tag{4.68}$$

It is sufficient for our purposes to know the epicyclic frequency in the Schwarzschild case with $a = 0$. This expression can be most easily derived directly from the twist equations, more precisely, from the part of these equations that describes the dynamics in the plane of the disc rings, i.e. from (4.60) and (4.61). Setting the 'viscous' terms and radial projection of the pressure gradient on the right-hand side of (4.61) equal to zero, as well as omitting the dependence of v^r and v^φ on φ, we obtain equations for the Eulerian perturbations which describe free motion of gas elements slightly deviating from circular motion. Clearly, these equations are equivalent to (4.7)–(4.8) which were written in the basis (3.50)–(3.53). From these equations, we obtain the following equation for v^r:

$$\ddot{v}^r + 2 \frac{K_1 K_1'}{K_2^2 (U^\tau)^2} \left(\frac{\partial_r U^\varphi}{U^\varphi} + \frac{K_1'}{K_1} \frac{(U^\tau)^2}{(U^\varphi)^2} \right) v^r = 0, \tag{4.69}$$

where the expression before v^r is equal to κ^2. It can be rewritten in a more compact form

$$\kappa^2 = 2\frac{K_1'(K_1 U^\tau)'}{K_2^2 U^\tau (U^\varphi)^2} \tag{4.70}$$

to ensure that it coincides with (4.11), considering that the radial Schwarzschild coordinate $r_S = r K_2$ enters the last equation.

Finally, for convenience, we introduce the following quantity with the dimension of frequency that appears in our problem. In the Schwarzschild case, $a = 0$,

$$\tilde{\Omega} = \frac{K_1'}{K_2}\frac{1}{U^\tau U^\varphi} = \frac{r_S - 3}{r_S^2 (r_S - 2)^{1/2}}, \tag{4.71}$$

which tends to the Keplerian value in the Newtonian limit.

Using (4.65), (4.70) and (4.71) allows us to write equations (4.60) and (4.61) in a more compact form. Lense-Thirring frequency (4.68), evidently, enters the gravitomagnetic term in (4.64). However, we address this rewriting in the next section when considering a specific background model.

4.3.3 Twist Equations in the Particular Case of a Stationary Vertically Isothermal α-Disc

We now consider the form which the twist equations take in the specific background of a stationary α-disc, which was discussed in Chap. 3. This does not mean, however, that only stationary twisted solutions are to be considered. In other words, the equations we obtain are also applicable to arbitrary non-stationary dynamics of the corresponding twisted disc. For example, they enable us to calculate the evolution of the shape of an (infinite) initially flat disc instantly tilted to the equatorial plane of a rotating black hole. The initial stage of the evolution of such a disc was qualitatively described in Sect. 4.1. In addition, these equations describe the wave-like (in the case of a disc with sufficiently small $\alpha < \delta$; see also Papaloizou and Lin (1995)) or diffusion-like (in the case of a disc with sufficiently large $\alpha > \delta$; see also Papaloizou and Pringle (1983)) dynamics of some twist perturbation imposed on the disc lying initially in the equatorial plane of the black hole.

4.3.3.1 Explicit Form of the Necessary Background Profiles

The twist equations contain the quantity $\bar{T}_\nu^{r\varphi}$ (as well as $\bar{\eta}$), related to the corresponding flat disc model. We could obtain the explicit form of these quantities by integrating the $\tau-$ and $\varphi-$projections of Eq. (4.38). However, it is simpler to use the results from Chap. 3, where we have already obtained this quantity, there denoted by T_ν (see Eq. (3.91)). We should only take into account that now we are working in another basis than that used for the flat disc, and therefore the transition from T_ν to $\bar{T}_\nu^{r\varphi}$ should be specified. First, using the orthogonality condition for the shear tensor and, hence, for the viscous stress tensor, (3.67), we see that only one component of the viscous stress tensor, $T_\nu^{r\varphi'}$, is non-zero in basis (3.50)–(3.53), as the four-velocity there has only non-zero time components up to terms $\propto \delta^2$. The prime here marks basis (3.50)–(3.53). Further, the (orthonormal) bases are different only in that an observer associated with basis (3.50)–(3.53) moves in the azimuthal direction with the velocity of the free equatorial circular motion, whereas the basis B_0 corresponds to an observer at rest. Therefore, the transformations of vectors and tensors must be equivalent to the usual Lorentz transformations. Using Landau and Lifshitz (2000) (see Exercise 1, Paragraph 6 therein) we see that $T_\nu^{r\varphi} = U^\tau T_\nu^{r\varphi'}$ where U^τ is the Lorentz factor of azimuthal motion. Finally, it should be taken into account that integration over ξ differs from that over z by the coefficient K_2. As a result, we obtain

$$\bar{T}_\nu^{r\varphi} = \frac{U^\tau}{K_2} T_\nu. \tag{4.72}$$

We note that it is possible to change from $T_\nu^{r\varphi'}$ to $T_\nu^{r\varphi}$, using the relation (3.35), by writing it for two bases, equating the right-hand sides and then multiplying one of the sides of the obtained equalities by matrices inverse to the basis matrices there. Here, we should only take into account that in the basis B_0 the radial coordinate was changed, (4.16), i.e. that $r_S = r K_2$ in the notation of this part of the chapter.

Next, for the case $a = 0$, which is sufficient here, it is easy to express T_ν in terms of elementary functions. Indeed, the integral in (3.91) can be used with the substitution $y \equiv \sqrt{r_S}$:

$$\int \frac{E}{r_S^{1/2} C} dr_S = \int \frac{y^2 - 6}{y^2 - 3} dy = y + \frac{\sqrt{3}}{2} \ln \frac{y + \sqrt{3}}{y - \sqrt{3}}.$$

For $\bar{T}^{r\varphi}$ with account for (4.72) we finally obtain

$$\bar{T}^{r\varphi} = \frac{\dot{M}}{2\pi} U^\tau r^{-3/2} \frac{L(r)}{K_2^{5/2} K_1^2}, \tag{4.73}$$

where

$$L = 1 - \frac{\sqrt{6}}{y} - \frac{\sqrt{3}}{2y} \ln \frac{(y - \sqrt{3})(3 + 2\sqrt{2})}{(y + \sqrt{3})}. \tag{4.74}$$

By necessity, $L = 0$ at $r_S = 6$. We note that $L = Y(a = 0)$ where Y was defined in (3.106).

On the other hand, the expression for $\sigma^{r\varphi}$ in (3.73), in our case $a = 0$ in the basis B_0, can be rewritten in the form

$$\sigma^{r\varphi} = \frac{3}{4}\frac{D}{r_S^{3/2}C}U^\tau = \frac{3}{4}K_1^2 U_g^\varphi U_g^\tau U^\tau = \frac{3}{4}\frac{K_1}{rK_2}U^\varphi (U^\tau)^2,$$

where, as usual, we use relations (4.49).

Then

$$\bar{T}^{r\varphi} = \frac{3}{2}\bar{\eta}\frac{K_1}{rK_2}(U^\tau)^2 U^\varphi. \tag{4.75}$$

As in Sect. 3.3.6, equating expressions (4.73) and (4.75) we obtain

$$\bar{\eta} = \frac{\dot{M}}{3\pi}\left(\frac{r^{-1/2}}{U^\tau U^\varphi}\frac{L}{K_1^3 K_2^{3/2}}\right). \tag{4.76}$$

In the Newtonian limit, far away from the inner edge of the disc, Eq. (4.76) gives the well-known result $\bar{\eta} = \dot{M}/(3\pi)$.

We assume that the kinematic viscosity is proportional to the characteristic disc half-thickness times the sound velocity in the disc:

$$\nu \sim \alpha c_s h_p, \tag{4.77}$$

where h_p is the proper characteristic disc half-thickness, which in our coordinate system is $h_{proper} = K_2 h$ and α is the Shakura parameter, which is assumed to be constant. Since (4.63) implies that $c_s \sim \sqrt{P/\rho} \sim U^\varphi h/r$, we finally define α by the equality

$$\nu = \alpha K_2 U^\varphi h^2/r. \tag{4.78}$$

Using (4.76) and (4.78), we obtain the relation

$$\Sigma h^2 = \frac{\dot{M}}{3\pi\alpha}\left(\frac{r^{1/2}}{U^\tau (U^\varphi)^2}\frac{L}{K_1^3 K_2^{5/2}}\right). \tag{4.79}$$

To find U^r in the advective term in (4.64), we use the rest-energy conservation law in the basis B_0 for the stationary disc. Again, we use result (3.87). Recalling the transition to the isotropic radial coordinate, the relation between the coordinate and physical velocities (3.48) and (4.49), and the difference in the definitions of Σ, we obtain

$$-\frac{\dot{M}}{2\pi} = \Sigma K_1 K_2^2 r U^r. \tag{4.80}$$

Then U^r can be derived from (4.80) and (4.79) as

$$U^r = -\frac{3\alpha}{2}\frac{\delta^2}{L} K_1^2 U^\tau (U^\varphi)^2 \sqrt{K_2 r}. \tag{4.81}$$

Finally, we need to know the profile $\delta(r)$. Note that this value is invariant under the transition between the bases (3.50)–(3.53) and B_0, since the change from h_p to h and from r_S to r is scaled with the same coefficient K_2.

In a gas-pressure-dominated disc with Thomson scattering opacity, it follows from (3.108) with $a = 0$ that

$$\delta(r) = \delta_* K_1^{1/2} K_2^{1/20} (U^\tau)^{-9/10} L^{1/5} r^{1/20}. \tag{4.82}$$

In order to derive a simpler form of the twist equations, we need to specify the vertical profile of the rest-energy density. Here we use it in its simplest form in an isothermal disc:

$$\rho = \rho_c \exp\left(-\frac{\xi^2}{2h^2}\right), \tag{4.83}$$

where $\rho_c(r)$ is the equatorial density.

4.3.3.2 Switching to Complex Amplitudes

In the case of an isothermal disc, the velocity perturbations v^r and v^φ in the form

$$v^\varphi = \xi(A_1 \sin\varphi + A_2 \cos\varphi) \quad v^r = \xi(B_1 \sin\varphi + B_2 \cos\varphi) \tag{4.84}$$

satisfy Eqs. (4.60) and (4.61), provided that ν does not change with the height, and that the amplitudes A_1, A_2, B_1 and B_2 are functions of r and τ. Indeed, in this case, all 'thermal' terms are $\propto \xi$, and the dependence on ξ with the ansatz (4.84) is identically satisfied in the considered equations.

Let us introduce the complex amplitudes

$$\mathbf{A} = A_2 + iA_1, \quad \mathbf{B} = B_2 + iB_1 \quad \text{and} \quad \mathbf{W} = \Psi_1 + i\Psi_2 = \beta e^{i\gamma} \tag{4.85}$$

By constructing two combinations, (4.60) + i ∂_φ (4.60) and (4.61) + i ∂_φ (4.61), we see that all terms in these combinations are $\propto$ $e^{-i\varphi}$. In particular, the terms containing W and $\partial_\varphi W$ transforms into terms containing $-i\mathbf{W}'e^{-i\varphi}$ and $\mathbf{W}'e^{-i\varphi}$, respectively).

As a result, we obtain the following complex equations

$$\dot{\mathbf{A}} - (i - \alpha)\Omega\mathbf{A} + \frac{\kappa^2}{2\tilde{\Omega}}\mathbf{B} = -\frac{3}{2}i\alpha K_1 (U^\tau)^2 U^\varphi \Omega \mathbf{W}', \tag{4.86}$$

$$\dot{\mathbf{B}} - (i - \alpha)\Omega\mathbf{B} - 2\tilde{\Omega}\mathbf{A} = -(i + \alpha) U^\varphi \Omega\, \mathbf{W}', \tag{4.87}$$

where we have used Eq. (4.78), as well as the expressions for the frequencies (4.65), (4.70) and (4.71) obtained in Sect. 4.3.2.

In a similar way, by using (4.84) and (4.85) and constructing the combination (4.64) + i ∂_φ (4.64), we derive an equation for complex amplitudes. On the right-hand side of this equation, an integration over ξ should be performed for the derivative with respect to r. For an isothermal disc with density distribution (4.83), the equality $\int \rho \xi^2 d\xi = \Sigma h^2$ holds. Thus, the derivative with respect to r acts on terms proportional to Σh^2 or $\bar{\eta}$. Instead of these combinations, we substitute Eqs. (4.79) and (4.76) into our equation, and group common constant factors before the derivative with respect to r. Additionally, instead of U^r and $\bar{T}^{r\varphi}$ we substitute expressions (4.81) and (4.73) into the left-hand side of our equation and then divide the whole equation by Σ. The obtained equation contains $\dot{M}$ and Σ only in the combination $\dot{M}/\Sigma$, which we express through δ^2 and other known quantities using (4.79). Also using the expression for Lense-Thirring frequency (4.68), we finally arrive at the following equation

$$\dot{\mathbf{W}} - i\Omega_{LT}\mathbf{W} + \frac{3}{2}\alpha\delta^2 \frac{K_1^2}{K_2} U^\varphi \left(U^\tau - K_1 (r K_2)^{1/2} \frac{U^\varphi}{L} \right) \mathbf{W}' =$$
$$\frac{\delta^2 K_1^3 U^\varphi}{2 r^{1/2} K_2^{3/2} L} \frac{\partial}{\partial r} \left\{ r^{3/2} K_2^{1/2} \frac{L}{K_1^2 U^\tau U^\varphi} \left((i + \alpha)\mathbf{B} + \alpha U^\varphi \mathbf{W}' \right) \right\}. \tag{4.88}$$

Equations (4.86)–(4.88) form a closed set of equations for the quantities **A**, **B** and **W** as functions of r and τ. In the weak gravity limit they reduce to equations (30), (31) and (33) in Demianski and Ivanov (1997).

Importantly, under the condition $u \equiv t_d/t_{ev} \ll 1$, the derived set of equations can be reduced to *two* equations for the variables **B** and **W**, since (4.86)–(4.87) prove to be equivalent to a single equation (4.105): see the next section, where this issue is considered using the language of frequencies. A derivation of (4.105) can be found later on in Sect. 4.4.2.

4.4 Non-stationary Twist Dynamics

4.4.1 The Local Dispersion Relation

As noted in the introductory remarks to Sect. 4.3.3, Eqs. (4.86)–(4.88) describe also non-stationary twist dynamics exhibited by an infinite stationary accretion α-disc. Let us consider this dynamics in the example of harmonic twist perturbation, which depends on τ and r as $\propto \exp(-\mathrm{i}\omega\tau + ikr)$. For simplicity, we assume that the perturbation is local, i.e. that $k \gg 1$. If so, we can neglect the radial variation of the coefficients in Eqs. (4.86)–(4.88), as far as the typical spatial scale of changes in these coefficients is much larger than the wavelength of the perturbation $\sim k^{-1}$.

Additionally, we assume that effects of viscous damping and General Relativity are weak, i.e. $\alpha \ll 1$ and $r^{-1} \ll 1$. The latter allows to shorten calculations, while not overlooking any of the regimes of twist dynamics. In the leading order in small parameters we have

$$\dot{\mathbf{A}} - (\mathrm{i} - \alpha)\Omega \mathbf{A} + \frac{\Omega}{2}\left(1 - \frac{4}{r}\right)\mathbf{B} = -\frac{3}{2}\mathrm{i}\alpha U^{\varphi}\Omega \mathbf{W}', \tag{4.89}$$

$$\dot{\mathbf{B}} - (\mathrm{i} - \alpha)\Omega \mathbf{B} - 2\Omega\left(1 - \frac{2}{r}\right)\mathbf{A} = -\left[\mathrm{i}\left(1 + \frac{V_1}{r}\right) + \alpha\right] U^{\varphi}\Omega \mathbf{W}', \tag{4.90}$$

$$\dot{\mathbf{W}} - \mathrm{i}\Omega_{LT}\mathbf{W} = \frac{\delta^2 r}{2}\left[\mathrm{i}\left(1 + \frac{V_2}{r}\right)\mathbf{B}' + \alpha(\mathbf{B}' + U^{\varphi}\mathbf{W}')\right], \tag{4.91}$$

where V_1 and V_2 are constants that appear in the expansion of the relativistic coefficients over small r^{-1}, which we do not need to find explicitly for our purposes. It is implied here that all known quantities entering Eqs. (4.89)–(4.91) take their Newtonian values, thus, $\Omega = r^{-3/2}$, $U^{\varphi} = r^{-1/2}$, whereas τ and r are the variables measured in absence of relativistic effects. In particular, this means that in this case $r = r_S$.

Changing the derivatives according to $\partial/\partial\tau \to -\mathrm{i}\omega$ and $\partial/\partial r \to ik$ and excluding the Fourier amplitude corresponding to $\mathbf{A}$, we obtain the set of algebraic equations

$$\left[\omega^2 + 2\omega(1 + \mathrm{i}\alpha)\Omega + (1 + \mathrm{i}\alpha)^2\Omega^2 - \Omega^2 + 2\Omega_p\Omega\right]\hat{\mathbf{B}} =$$
$$\left[\left(\mathrm{i}\left(1 + \frac{V_1}{r}\right) + \alpha\right)(\omega + (1 + \mathrm{i}\alpha)\Omega) - 3\alpha\Omega\right] U^{\varphi}\Omega k \hat{\mathbf{W}}, \tag{4.92}$$

$$\frac{\delta^2 r}{2}\left[\mathrm{i}\left(1 + \frac{V_2}{r}\right) + \alpha\right] ik\hat{\mathbf{B}} = \left[-\mathrm{i}(\omega + \Omega_{LT}) + \alpha\frac{\delta^2 r}{2}U^{\varphi}k^2\right]\hat{\mathbf{W}}, \tag{4.93}$$

where $\hat{\mathbf{B}}$ and $\hat{\mathbf{W}}$ are the Fourier amplitudes of the variables $\mathbf{B}$ and $\mathbf{W}$.

In (4.92)–(4.93) we used the Einstein frequency, $\Omega_p \equiv 3\Omega/r \ll \Omega$, introduced in Sect. 4.1.1, see the expression for epicyclic frequency squared (4.11), and the discussion in that section. Finally, we need to keep in mind that the solution we are looking for must obey the condition $\omega \ll \Omega$ by virtue of the smallness of $u = t_d/t_{ev}$, see the discussion in the introduction to this chapter and the introductory remarks to the derivation of the twist equations given on page 168.

First, let us pay attention to the terms in square brackets on the LHS of (4.92). We find that the leading terms, both equal to Ω^2, cancel each other out, leaving a set of only small terms. This is how the Keplerian resonance, or alternatively, the Keplerian degeneration (i.e. the coincidence of epicyclic and rotational frequencies), briefly discussed in the introduction to this chapter, manifests itself and require that we retain the terms $\dot{v}_r$ and $\dot{v}_\varphi$ in Eqs. (4.60)–(4.61), see the explanation on the page 173. Indeed, Eq. (4.92) would yield an infinite $\hat{\mathbf{B}}$ for any non-zero $\hat{\mathbf{W}}$ in absence of $\dot{\mathbf{A}}$ and $\dot{\mathbf{B}}$, provided that $\alpha = \Omega_p = 0$. On the other hand, if viscous damping and/or relativistic precession is high enough, so that $\omega \ll \min\{\alpha\Omega, \Omega_p\}$, we can neglect the influence of $\dot{\mathbf{A}}$ and $\dot{\mathbf{B}}$, and accordingly, set $\omega = 0$ inside the parenthesis in question. Below we formulate a condition for when it is possible to do, in the form of a restriction on the wavelength of the twist perturbation.

Similarly, the square brackets on the RHS of (4.93) contain terms that are small compared to Ω. For this reason, we keep the terms inside the square brackets as they are, and only consider the relation $\delta^2 r U^\varphi = \Omega h^2$. The two remaining square brackets, however, both include leading terms much larger than the corrections due to viscosity and/or deviation from Keplerian dynamics. This implies that here we may neglect these corrections.

Finally, we return to the terms in square brackets before $\hat{\mathbf{B}}$ on the LHS of (4.92). Regardless of the relation between ω and $\alpha\Omega$, or between ω and Ω_p, we can always drop the terms ω^2, $2\alpha\omega\Omega$ and $\alpha^2\Omega^2$ as lower order terms compared to ω, $\alpha\Omega$ and Ω_p.[6]

Taking all of the above into account, we arrive at the following dispersion relation

$$(\omega + \mathrm{i}\alpha\Omega + \Omega_p)[-\mathrm{i}(\omega + \Omega_{LT}) + \alpha\Omega(kh)^2/2] = -\mathrm{i}\Omega^2(kh)^2/4,$$

which immediately allows us to conclude that in order for the condition $\omega \ll \Omega$ to be satisfied, it is necessary to make the additional assumption that $kh \ll 1$, i.e. that in our model, the twist perturbations must have a wavelength much larger than the disc thickness. This, in turn, allows us to omit the term $\sim \alpha$ on the LHS in the square brackets. The dispersion relation takes the final form

$$(\omega + \mathrm{i}\alpha\Omega + \Omega_p)(\omega + \Omega_{LT}) = \Omega^2(kh)^2/4, \tag{4.94}$$

[6]Retaining the term ω^2 in (4.92) and considering the inviscid Newtonian limit for the set of Eqs. (4.92)–(4.93), we may obtain a cubic equation with respect to ω, and check that it always has three real roots, one always being of the order of $\sim \Omega$, even for $kh \ll 1$, which violates the restriction of slow evolution of the twist imposed in our model.

and can be found in the Appendix to Zhuravlev et al. (2014), where a general solution is presented. Here we restrict ourselves to consider certain limiting cases.

4.4.1.1 A Newtonian Viscous Disc

In order to study the strictly Newtonian dynamics of twist perturbation, we set $\Omega_{LT} = 0$ and $\Omega_p = 0$ in Eq. (4.94). We obtain the following solution

$$2\frac{\omega_{1,2}}{\Omega} = -\mathrm{i}\alpha \pm (k^2h^2 - \alpha^2)^{1/2}. \quad (4.95)$$

In the short wavelength limit, or equivalently, in the limit of sufficiently small viscosity, when $\alpha \ll kh$, we arrive at the following dispersion equation

$$2\frac{\omega_{1,2}}{\Omega} = -\mathrm{i}\alpha \pm kh, \quad (4.96)$$

which is typical for waves propagating with phase speed equal to half the sound speed in a disc and dissipating on a characteristic timescale $\sim \alpha^{-1}$ times larger than the Keplerian dynamical time. Since the phase speed $\sim h\Omega$ does not depend on k, there is no dispersion and an arbitrary twist perturbation propagates in the disc conserving its shape. This kind of twist perturbation is also called a 'bending wave'.

In the opposite case, when $\alpha \gg kh$, i.e. when long wavelength perturbations are considered or, equivalently, the disc is highly viscous, we obtain the couple of imaginary solutions

$$\frac{\omega_{1,2}}{\Omega} = -\mathrm{i}\alpha, \quad -\mathrm{i}(kh)^2/(4\alpha). \quad (4.97)$$

The meaning of the first root can be understood noting that it remains non-zero in the limit $kh \to 0$, which describes the motion of fluid elements in absence of radial projection of the vertical pressure gradient, but with account for viscosity. In the inviscid limit fluid elements move freely, which corresponds to nothing but epicyclic oscillations. As discussed on p. 176, see Eq. (4.69), epicyclic motion is described by the equations for velocity perturbations (4.60) and (4.61), together with the assumption that $\mathbf{W} = \alpha = 0$. The non-zero viscosity causes damping of the epicyclic oscillations. It can be verified that in the Newtonian limit, Eqs. (4.89)–(4.90), with their RHSs set equal to zero yield the same decrement. Thus, ω_1^{-1} is a damping timescale for epicyclic oscillations. Only the second root is relevant to twist dynamics despite the fact that its absolute value is much smaller than that of the first root.

As long as the viscosity is sufficiently high, $\omega \ll \alpha\Omega$, we may set $\omega = 0$ in the square brackets on the LHS of Eq. (4.92), or, equivalently, in the first parentheses on the LHS of (4.94), see the comments to Eqs. (4.92)–(4.93). It can be verified that in the latter case we immediately arrive at the solution ω_2 in Eq. (4.97). Thus, in the limit $\alpha \gg kh$, the terms $\dot{\mathbf{A}}$ and $\dot{\mathbf{B}}$ can be omitted, and twist dynamics is controlled by $\dot{\mathbf{W}}$ in the left part of the (twist) equation (4.91). The dispersion relation for ω_2 has a form, commonly produced by an equation of diffusion type. Hence, as far as $\alpha \gg kh$ any twist perturbation arising in a disc at some instant diffuses in the disc. The corresponding diffusion coefficient is

$$\mathscr{D} = \Omega h^2/(4\alpha) \propto \alpha^{-1} . \tag{4.98}$$

For $\alpha \ll 1$ the twist perturbation propagates in the disc substantially faster, than the disc matter spirals inwards. Importantly, the diffusion coefficient depends on viscosity in an unusual way, $\mathscr{D} \propto \alpha^{-1}$, due to Keplerian resonance, which is the reason why the first parentheses in (4.94) contain only the term $i\alpha\Omega$ in the limit considered here. Physically, such a dependence is caused by the fact that the magnitudes of the velocity perturbations induced by disc twist increase inversely proportional to the value of the viscosity. In turn, the increase of the amplitudes of v_r and v_φ enhances the asymmetry of the mass density distribution along the vertical direction in the disc and, consequently, the torque associated with the central gravitational force acting on the disc, see the introduction to this chapter.

An analysis of the limiting cases of twist dynamics in a Newtonian viscous disc as described here, as well as a confirmation of the conclusions with the help of hydrodynamic simulations, was provided by Nelson and Papaloizou (1999). The dispersion relation (4.95) can be found also in Lodato and Pringle (2007).

4.4.1.2 A Formally Inviscid Weakly Relativistic Disc

Let us now consider the opposite case when viscosity is small enough for relativistic effects to become important, i.e. when $\Omega_p \gg \alpha\Omega$.

Setting $\alpha = 0$ in Eq. (4.94) (assuming for simplicity that $\Omega_{LT} = 0$), we arrive at the following solution

$$2\omega_{1,2} = -\Omega_p \pm ((\Omega kh)^2 + \Omega_p^2)^{1/2}. \tag{4.99}$$

Again, in the short wavelength limit or, equivalently, in the limit of slow Einstein precession when $\Omega_p \ll \Omega kh$, we obtain

$$2\omega_{1,2} = -\Omega_p \pm \Omega kh, \tag{4.100}$$

which recovers (4.96) with the correction that viscous damping is replaced by uniform precession.

In the case of small k or, equivalently, fast Einstein precession when $\Omega_p \gg \Omega kh$, we have two real solutions (cf. Eq. (4.97))

$$\omega_{1,2} = -\Omega_p,\ (\Omega kh)^2/(4\Omega_p). \tag{4.101}$$

Note that in order to change from (4.96) and (4.97) to (4.100) and (4.101), it is sufficient to make the replacement $i\alpha\Omega \to \Omega_p$.

Similar to what was found in the previous section, the first root of (4.101) is not relevant to the dynamics of twist perturbations. Instead, it describes relativistic precession of elliptical orbits of free fluid elements, and can be obtained from Eqs. (4.89)–(4.90) for $\alpha = 0$ and with the RHS set equal to zero. However, in this case the root ω_2 does not correspond to the diffusion dispersion relation, but to a wave with dispersion, since its phase speed $\propto k$. This means that a wave packet spreads as it propagates in the disc. Additionally, in this case the speed of a wave packet is much less than the typical sound speed in a disc. It was noted in the previous section that the terms $\dot{\mathbf{A}}$ and $\dot{\mathbf{B}}$ in Eqs. (4.89)–(4.90) are negligible in the diffusion limit $\alpha \gg kh$, and only the term $\dot{\mathbf{W}}$ on the LHS of the twist equation (4.91) remains to be responsible for the dynamics of the twist perturbation. Note that the same situation occurs in the regime $\Omega_p \gg \Omega kh$ considered here.

Finally, we note that the phase speed of the twist perturbation, as well as its dispersion rate, increases inversely proportional to Ω_p. Similarly to this, the rate of diffusion of the twist perturbation increases inversely proportional to α in the case of a Newtonian viscous disc.

4.4.2 *Reduction of the Set of Equations for Velocity Perturbations*

The review of the non-stationary dynamics of twist perturbation carried out in the previous section makes it clear that the condition of slow evolution of the twisted disc shape, i.e. the condition $u \ll 1$, inherent in the theory of twisted discs, is equivalent to discarding the term ω^2 in square brackets on the LHS of Eq. (4.92), see the footnote on p. 183. Note that this remains valid without the additional simplifying assumptions $\alpha \ll 1$ and $r^{-1} \ll 1$, used only to avoid cumbersome calculations. At the same time, the discarding of ω^2 reduces the set of twist equations with respect to τ down to the second order. The question arises whether this could have been done from the beginning when writing down the set of Eqs. (4.86)–(4.88). The discussion below suggests such a possibility.

Let us use the following combination of Eqs. (4.86) and (4.87), which does not include terms containing the variable $\mathbf{A}$,

$$\dot{\mathbf{B}} - \frac{2\tilde{\Omega}}{(i-\alpha)\Omega}\dot{\mathbf{A}} =$$

$$\left[1 + \frac{\kappa^2}{(i-\alpha)^2\Omega^2}\right](i-\alpha)\Omega\mathbf{B} - \left[(i+\alpha)U^\varphi\Omega - \frac{3i\alpha}{i-\alpha}K_1(U^\tau)^2U^\varphi\tilde{\Omega}\right]\mathbf{W}'. \tag{4.102}$$

As expected (see the discussion after Eqs. (4.60)–(4.61) on p. 173)), the coefficient in front of **B** on the RHS of Eq. (4.102),

$$\mathscr{C} \equiv \left[1 + \frac{\kappa^2}{(i-\alpha)^2\Omega^2}\right](i-\alpha)\Omega,$$

vanishes in the limit $\alpha \to 0$, $r^{-1} \to 0$. At the same time, the LHS of Eq. (4.102) is always small with respect to $\sim \Omega B$, since $u \ll 1$. For this reason, it effects the dynamics of twist perturbation only as far as $|\mathscr{C}| \ll \Omega$. This means that the LHS of Eq. (4.102) can be used only when $|\mathscr{C}| \ll \Omega$, and (4.86) has the form

$$-i\Omega\mathbf{A} + \frac{\Omega}{2}\mathbf{B} = \Omega\, O(\max\{u\mathbf{A}, r^{-1}\mathbf{B}, \alpha U^\varphi \mathbf{W}'\}). \tag{4.103}$$

Equation (4.103) gives that in the leading order with respect to all small parameters appearing in it, the following equality holds

$$\dot{\mathbf{B}} = 2i\dot{\mathbf{A}}. \tag{4.104}$$

By the same reasoning, it is sufficient to take the coefficient in front of $\dot{\mathbf{A}}$ on the LHS of Eq. (4.102) in the inviscid Newtonian limit. Hence, we find that Eq. (4.102) can be written in the form

$$2\dot{\mathbf{B}} = \left[1 + \frac{\kappa^2}{(i-\alpha)^2\Omega^2}\right](i-\alpha)\Omega\mathbf{B} - \left[(i+\alpha)U^\varphi\Omega - \frac{3i\alpha}{i-\alpha}K_1(U^\tau)^2U^\varphi\tilde{\Omega}\right]\mathbf{W}'. \tag{4.105}$$

The analysis performed above allows us to conclude that equation (4.105) is equivalent to the more general set of the two equations (4.86)–(4.87), provided that $u \ll 1$ ($\omega \ll \Omega$). It can be verified that equation (4.105) together with the twist equation (4.88) reproduces the local dispersion relation (4.94). Additionally, in the low frequency case, when $\omega \ll \min\{\alpha\Omega, \Omega_p\}$, the dynamics of twist perturbation is provided by the balance of the first and the second terms on the RHS of Eq. (4.105), which implies that $\dot{\mathbf{B}} = 0$.

So, in fact, the reduced equation for perturbations of the velocity (4.105) and the twist Eq. (4.88) together fully describe all regimes of non-stationary twist dynamics in a geometrically thin disc with α-prescription of viscosity. Note that a reduction of the more general Eqs. (4.60) and (4.61) can be performed along the same lines.

4.5 Stationary Twisted Disc

4.5.1 Main Equation and Boundary Conditions

We now consider stationary solutions to the set of Eqs. (4.86)–(4.88). The main goal of this section is to calculate the shape of a stationary twisted disc.

Expressing **B** using **W**′ from expression (4.105) for $\dot{\mathbf{B}} = 0$, and substituting the result into (4.88), assuming that $\dot{\mathbf{W}} = 0$, we obtain the following equation

$$\frac{K_1}{r_S^{1/2} L} \frac{d}{dr_S} \left(\frac{r_S^{3/2} L}{K_1 U^\tau} f^*(\alpha, r_S) \frac{d\mathbf{W}}{dr_S} \right) - 3\alpha U^\tau (1 - L^{-1}) \frac{d\mathbf{W}}{dr_S} + \frac{4ia}{\delta^2 K_1^3 r_S^3 U^\varphi} \mathbf{W} = 0, \tag{4.106}$$

where the asterisk denotes complex conjugation and

$$f(\alpha, r_S) = (1 + \alpha^2 - 3i\alpha K_1^2) \frac{r_S(i - \alpha)}{\alpha r_S(\alpha + 2i) - 6} + \alpha. \tag{4.107}$$

We note that Eq. (4.106) was written after switching to the Schwarzschild radial coordinate r_S. In what follows, we wish to consider only the case $a > 0$, i.e. a prograde disc. Apparently, the problem has two free parameters. The first of these is the combination $\tilde{\delta} \equiv \delta_*/\sqrt{|a|}$. Obviously, $\tilde{\delta}$ ranges from 0 to ∞ and characterizes the relative roles of the hydrodynamic and gravitomagnetic forces acting on the disc rings. Secondly, (4.106) contains the disc viscosity parameter $0 < \alpha < 1$. Equation (4.106) in the strict Newtonian limit with non-zero viscosity reproduces the corresponding equation (2.10) from Kumar and Pringle (1985) and, additionally, with post-Newtonian corrections, reproduces equation (33) from Ivanov and Illarionov (1997), which was confirmed in Zhuravlev and Ivanov (2011) (see paragraph 4.1 therein).

The coefficients of Eq. (4.106) have a singular point at the inner edge of the disc at $r_S = \bar{r}_S \equiv 6$, where L vanishes. The regularity of the solution at $\bar{r}_S$ must yield a condition for the function **W**. Using this condition as the initial one, we can integrate (4.106) from $\bar{r}_S$ to infinity and to obtain the form of the stationary twisted disc. We expand Eq. (4.106) in a power-series of small $x_0 = r_S - \bar{r}_S \ll 1$. In practice, to do this, all quantities that take non-zero values at $\bar{r}_S$ should be set exactly equal to these values and L should be expanded to the main order in x_0. From (4.74), we find

$$L \approx \frac{x_0^2}{72}, \tag{4.108}$$

whence we see that another quantity in (4.106) that vanishes at the inner disc edge, δ, can be written as

$$\delta = \delta_{ms} x_0^{2\epsilon},$$

where ϵ is the power-law exponent L in Eq. (4.82). Accordingly, δ_{ms} is also given by Eq. (4.82), which is evaluated at $\bar{r}_S$ and into which we now substitute the coefficient 72^{-1} from (4.108) instead of L.

After that, it is easy to obtain the equation valid for $x_0 \ll 1$,

$$\frac{d}{dx_0}\left(x_0^2 \frac{d\mathbf{W}}{dx_0}\right) + C_1 x_0^{2-4\epsilon}\mathbf{W} + C_2 \frac{d\mathbf{W}}{dx_0} = 0, \tag{4.109}$$

where

$$C_1 = -\frac{2\mathrm{i}}{f(\alpha, r_S)} \frac{U^\tau}{U^\varphi} \frac{\Omega_{LT}}{K_1^3 r_S \delta_{ms}^2}$$

and

$$C_2 = -\frac{216\alpha}{f(\alpha, r_S)} \frac{(U^\tau)^2}{r_S}$$

are taken at $\bar{r}_S$. We see that for any finite viscosity the last term in (4.109) becomes dominant sufficiently close to the disc edge. Therefore, the boundary condition can be straightforwardly written as

$$\left.\frac{d\mathbf{W}}{dx_0}\right|_{\bar{r}_S} = 0. \tag{4.110}$$

On the other hand, from (4.109) with $\alpha = 0$, we obtain a simpler equation, the solution to which is a Bessel function:

$$\mathbf{W} = C x_0^{-1/2} J_{\frac{1}{2-4\epsilon}}(z), \tag{4.111}$$

where

$$z = \sqrt{C_1}\frac{x_0^{1-2\epsilon}}{1-2\epsilon}. \tag{4.112}$$

As $x_0 \to 0$ (4.111) tends to a non-zero constant but with a zero derivative with respect to x_0. Therefore, in this case we return to condition (4.110).

Due to the linearity of the problem, it is sufficient to take an arbitrary non-zero value of $\mathbf{W}$ in $\bar{r}_S$, to set the first derivative of $\mathbf{W}$ at $\bar{r}_S$ equal to zero, and, with these boundary conditions, integrate (4.106) to infinity. The modulus and phase of $\mathbf{W}$ give the profiles $\beta(r_S)$ and $\gamma(r_S)$ for a stationary twisted disc. In what follows, we normalize the profile β to unity at infinity.

4.5.2 A Disc with Marginally Small Viscosity

We consider a disc with very low viscosity separately. Clearly, it is possible to treat the accretion disc analytically, formally setting $\alpha \to 0$, if simultaneously $\dot{M} \to 0$. In such a disc, $U^r \to 0$. It will, however, still have definite profiles of Σ and h.

In addition, to obtain an analytical solution, we consider the case $\tilde{\delta} \ll 1$; in other words, we assume a sufficiently thin disc around a rapidly rotating black hole.

Setting $\alpha = 0$ in (4.106) yields

$$\frac{d}{dr_S}\left(b\frac{d}{dr_S}\mathbf{W}\right) + \lambda\mathbf{W} = 0, \tag{4.113}$$

where

$$b = \frac{r_S^{5/2}L}{K_1 U^\tau}, \quad \lambda = \frac{24aL}{\delta^2 K_1^4 U^\varphi r_S^{5/2}}. \tag{4.114}$$

The coefficients in (4.113) take real values, and therefore, there exist real solutions to this equation. This means that in the absence of viscosity in a stationary twisted disc $\gamma = const$, and may be set equal to zero by the corresponding choice of reference frame. Therefore, the variable $\mathbf{W}$ is identical to the angle β in this section.

4.5.2.1 The Shape of the Disc Near Its Inner Edge

Earlier in this chapter, we have already presented the solution near the inner edge of an inviscid disc (see Eq. (4.111)). The constant C_1 has in this case the explicit form:

$$C_1 = \frac{24aU^\tau}{r_S^5 K_1^3 U^\varphi \delta_{ms}^2}, \tag{4.115}$$

for $r_S = \bar{r}_S$.

Using the well-known approximation to the Bessel function for a small argument, we obtain a relation between the constant C in (4.111) and the value of $\mathbf{W}$ at $\bar{r}_S$, $\mathbf{W}(r_S) \equiv \mathbf{W}_0$:

$$C = \Gamma\left(\frac{3-4\epsilon}{2(1-2\epsilon)}\right)\left(\frac{\sqrt{\chi}}{2(1-2\epsilon)}\right)^{-1/2(1-2\epsilon)}\mathbf{W}_0, \tag{4.116}$$

where $\Gamma(x)$ is the gamma-function.

In addition, we need the asymptotic to (4.111) for $z \gg 1$. Clearly, z can be large, even for $x \ll 1$, since $\sqrt{C_1} \sim \tilde{\delta}^{-1} \gg 1$. Hence, for $z \gg 1$ we obtain

$$\mathbf{W} \approx C\sqrt{\frac{2}{\pi x z}} \cos\left(z - \frac{\pi}{2}\frac{1-\epsilon}{1-2\epsilon}\right). \tag{4.117}$$

4.5.2.2 The Shape of the Disc at Large Distances

We now consider Eq. (4.106) for $r_S \gg 1$ and $\alpha \to 0$. Importantly, we cannot set all variables in (4.106) to their Newtonian values and make the viscosity simultaneously vanish. This already follows from the fact that then $f(\alpha, R) \to 1/(2\alpha) \to \infty$. Physically, this reflects the fact that, as we mentioned above, in the absence of viscosity in a strictly Newtonian potential, a Keplerian resonance occurs when the circular and epicyclic frequencies coincide, and perturbations in the twisted disc grow infinitely due to the action of the radial projection of the vertical pressure gradient. Therefore, a stationary twist is impossible in this case. Taking the next-order term in the expansion of $f(\alpha, r_S)$ in small r_S^{-1} into account, we obtain

$$f(\alpha, r_S) \approx \frac{1}{2\alpha\left(1 + \frac{3\mathrm{i}}{\alpha r_S}\right)}. \tag{4.118}$$

As $\alpha \to 0$, $f(\alpha, r_S)$ now remains finite at any finite r_S. Nevertheless, it makes the leading contribution due to relativistic effects, and all other variables in (4.106) can now be set equal to their Newtonian values $U^\tau = 1$, $U^\varphi = r_S^{-1/2}$, $L = 1$ and $K_1 = 1$. Moreover, we neglect the weak dependence of δ on r_S far from the black hole and set $\delta = \delta_*$.

After that, by introducing the new independent variable $x_1 \equiv r_S^{-1/2} \ll 1$, we obtain the equation

$$x_1 \frac{d^2}{dx_1^2}\mathbf{W} - 2\frac{d}{dx_1}\mathbf{W} + 96\tilde{\delta}^{-2} x_1^4 \mathbf{W} = 0. \tag{4.119}$$

The solution to (4.119) can again be expressed in terms of a Bessel function:

$$\mathbf{W} = x_1^{3/2}(A_1 J_{-3/5}(z_1) + A_2 J_{3/5}(z_1)), \tag{4.120}$$

where

$$z_1 = \frac{8}{5}\sqrt{6}\tilde{\delta}^{-1} x_1^{5/2}, \tag{4.121}$$

A_1 and A_2 are constants.

When r_S is large enough so that $z_1 \ll 1$, the first and second terms respectively in (4.120), multiplied by $x_1^{3/2}$, tend to a non-zero constant and to zero. This allows us to express the constant A_1 in terms of the value of $\mathbf{W}$ at infinity, $\mathbf{W}_\infty$:

$$\mathbf{W}_\infty = \left(\frac{5}{4\sqrt{6}}\right)^{3/5} \frac{\tilde{\delta}^{3/5}}{\Gamma(2/5)} A_1. \tag{4.122}$$

In the opposite case, at $z_1 \gg 1$, i.e. closer to the black hole, we obtain another asymptotic form:

$$\mathbf{W} \approx \sqrt{\frac{5\tilde{\delta}}{2\pi\sqrt{24}}} r_S^{-1/8} \left[A_1 \cos\left(z_1 + \frac{\pi}{20}\right) + A_2 \sin\left(z_1 - \frac{\pi}{20}\right)\right]. \tag{4.123}$$

4.5.2.3 A WKBJ-Solution for the Disc Shape

Throughout the disc the asymptotic solutions (4.117) and (4.123) can be matched by a WKBJ-solution to Eq. (4.113). Indeed, since we are considering the case $\tilde{\delta} \ll 1$, the ratio of λ and b in (4.113), $\tilde{\lambda} = \lambda/b$, is large at all r_S such that z and z_1 are large.

The WKBJ-solution has the form

$$\mathbf{W} \approx \frac{C_3}{(\lambda b)^{1/4}} \cos\left(\int_{\bar{r}_S}^{r_S} \sqrt{\tilde{\lambda}} dr_S + \phi_{WKBJ}\right), \tag{4.124}$$

where the constants C_3 and ϕ_{WKBJ} should be chosen so that (4.124) is smoothly matched with formula (4.117) in the corresponding region. It can be verified that this yields

$$\phi_{WKBJ} = -\frac{\pi}{2}\frac{1-\epsilon}{1-2\epsilon} \tag{4.125}$$

and

$$C_3 = 6^{1/4} \sqrt{\frac{1-2\epsilon}{\pi K_1 U^\tau}} C, \tag{4.126}$$

where we assume that K_1 and U^τ are evaluated at $r_S = \bar{r}_S = 6$ and $L \approx x^2/72$ near $\bar{r}_S$.

Further, in the limit $r_S \to \infty$ we can set λ and b before the cosine in (4.124) equal to their Newtonian values. In addition, the integral in (4.124) can be represented as $I(r_S) \equiv \int_{\bar{r}_S}^{r_S} \sqrt{\tilde{\lambda}} dr_S = I - \int_{r_S}^{\infty} \sqrt{\tilde{\lambda}} dr_S$, where $I = \int_{\bar{r}_S}^{\infty} \sqrt{\tilde{\lambda}} dr_S$. Taking into account that the Newtonian value $\tilde{\lambda} = 24\tilde{\delta}^{-2} R^{-9/4}$, we have that $\int_{r_S}^{\infty} \sqrt{\tilde{\lambda}} dr_S \approx$

$\frac{8\sqrt{6}}{5}\tilde{\delta}^{-1}r_S^{-5/4}$, and therefore

$$\mathbf{W} \approx C_3 \frac{\tilde{\delta}^{1/2}}{24^{1/4}} \cos\left(\frac{8\sqrt{6}}{5}\tilde{\delta}^{-1}r_S^{-5/4} - I - \phi_{WKBJ}\right). \tag{4.127}$$

Solution (4.127) must be smoothly matched with expression (4.123) in the corresponding region, which yields the values of constants A_1 and A_2. It can be verified that these are

$$A_1 = \sqrt{\frac{2\pi}{5}} C_3 \cos\left(I + \phi_{WKBJ} - \frac{\pi}{20}\right) \Big/ \cos\frac{\pi}{10},$$

$$A_2 = \sqrt{\frac{2\pi}{5}} C_3 \sin\left(I + \phi_{WKBJ} + \frac{\pi}{20}\right) \Big/ \cos\frac{\pi}{10}. \tag{4.128}$$

Thus, Eqs. (4.111), (4.124) and (4.120) together with coefficients (4.126), (4.128) and phase (4.125), determine the shape of an inviscid stationary relativistic twisted disc at all distances in the range from $r_S = \bar{r}_S$ to $r_S = \infty$.

4.5.2.4 Resonance Solutions in a Low-Viscosity Disc

We note that Eqs. (4.116), (4.126), (4.128) and (4.122) provide a relation between $\mathbf{W}_0$ and $\mathbf{W}_\infty$:

$$\mathbf{W}_\infty = C_{tot}(\tilde{\delta})\mathbf{W}_0, \tag{4.129}$$

where the explicit form of $C_{tot}(\tilde{\delta})$ follows from these formulas. In particular, as follows from (4.122) and (4.128), $C_{tot}(\tilde{\delta}) \propto \cos(I + \phi_{WKBJ} - \frac{\pi}{20})$.

We hence conclude that for some discrete set of $\tilde{\delta}$ for which $\cos(I + \phi_{WKBJ} - \frac{\pi}{20}) = 0$ so $\mathbf{W}_\infty = 0$ despite that $\mathbf{W}_0 \neq 0$.

From Eq. (4.114), it is possible to represent the integral I in the form $I = \tilde{\delta}^{-1}\tilde{I}$, where $\tilde{I}$ does not depend on $\tilde{\delta}$. This allows us to write the singular values of $\tilde{\delta}$ explicitly:

$$\tilde{\delta}_k = \frac{\tilde{I}}{\frac{\pi}{2}\left(\frac{11}{10} + \frac{1-\epsilon}{1-2\epsilon} + 2k\right)}, \tag{4.130}$$

where k is an integer number.

The values of $\tilde{\delta}_k$ correspond to a balance between the external gravitomagnetic force and the internal pressure gradient in the disc that leads to disc twist even if the matter flowing into the disc at infinity moves in the equatorial plane of the black hole. Note that, naturally, for these $\tilde{\delta}_k$ there also exists a solution in the

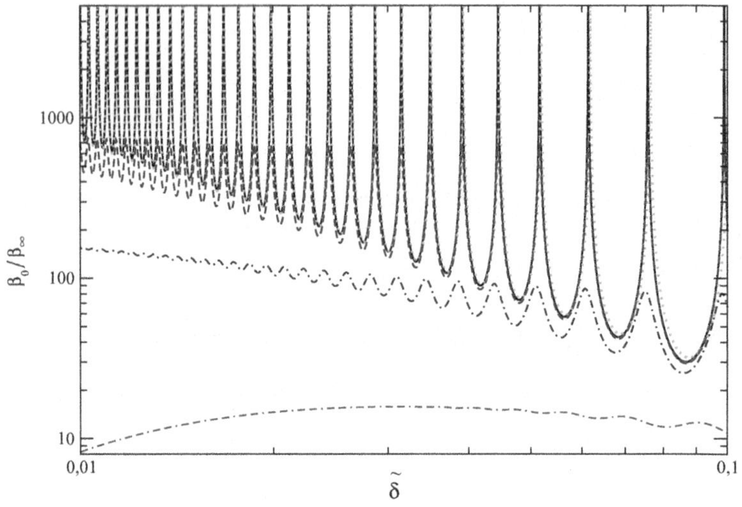

Fig. 4.6 Ratio of the tilt angle of the inner disc edge to the tilt at infinity, β_0/β_∞, as a function of the parameter $\tilde{\delta}$. The solid curve shows the numerical solution to equation (4.106) with $\alpha = 0$. The dotted curve represents the analytical dependence $C_{tot}^{-1}(\tilde{\delta})$, where C_{tot} is given by equation (4.129). The dashed, dash-dotted and dash-dot-dashed curves are obtained by numerical integration of Eq. (4.106) with $\alpha = 10^{-4}$, 10^{-3} and 10^{-2}, respectively

form of a flat disc lying entirely in the black hole equatorial plane. This non-uniqueness of the solution disappears with any appearance of low viscosity in the disc, for which $\mathbf{W}_\infty = 0$ always implies $\mathbf{W}_0 = 0$. For small $\alpha \ll 1$, the disc 'feels' these 'resonance' solutions, and its inner parts deviate significantly from the equatorial plane of the black hole, even when the outer parts of the disc lie almost in the equatorial plane. Figure 4.6 shows the curve corresponding to the analytical solution (4.129), as well as several curves for a viscous twisted disc obtained by integrating the original equation (4.106). We see that already for $\alpha = 10^{-3}$ the discussed resonances are almost entirely suppressed.

4.5.3 Disc Behavior in the Plane of the Parameters α and $\tilde{\delta}$

In conclusion, we present a full study of regimes of behavior of a stationary twisted relativistic disc near a rotating black hole. It is convenient to show the results of a numerical integration of Eq. (4.106) in the plane of the free parameters of the problem, $\tilde{\delta}$ and α. The first parameter varies in the range $10^{-3} < \tilde{\delta} < 10$ and the second parameter in the range $0 < \alpha < 1$. As follows from Fig. 4.7, at small $\tilde{\delta}$, i.e. when the gravitomagnetic force exceeds the internal forces in a twisted disc, it either lies in the equatorial plane of the black hole, i.e. $\beta_0/\beta_\infty \to 0$, or, conversely, the tilt of its rings strongly increases in the inner parts of the disc, with oscillations of $\beta(r_S)$

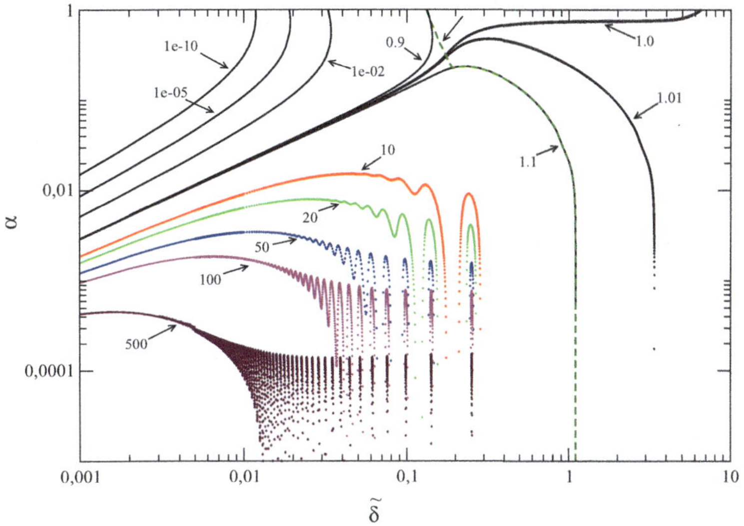

Fig. 4.7 Contours of constant ratio β_0/β_∞ in the parameter plane $(\tilde{\delta}, \alpha)$. The numbers show the value of β_0/β_∞ for each curve. The dashed curve in the right part of the figure separates the region where the change in β with r_S is more than 10% of β_∞ (to the left) from the region where the disc twist is insignificant, and β deviates by less than 10% from β_∞ (to the right)

along the radial coordinate. Note that for low viscosity, these oscillations become so strong that the corresponding gradient of the tilt angle, β', leads to supersonic perturbations of the velocity components, v_r and v^φ, at heights of the order of the disc thickness, $\xi \sim h$. This, in turn, must lead to the generation of various hydrodynamic instabilities and sound waves, which cause additional disc heating (and hence also an increase in $\tilde{\delta}$), as well as growth of α. These processes should partially suppress the oscillations of β discussed above.

Disc alignment into the equatorial plane of the black hole occurs at sufficiently high viscosity, when the condition $\alpha > \tilde{\delta}$ is satisfied with a large margin, and is referred to as the Bardeen-Petterson effect (Bardeen and Petterson 1975). It is seen from Fig. 4.7 that this effect occurs only in sufficiently viscous and thin discs. But already for $\tilde{\delta} \sim \alpha$ the ratio β_0/β_∞ becomes of the order of unity, which means absence of disc alignment. At the same time, oscillations of β disappear. Figure 4.8 shows the profiles of $\beta(r_S)$ when $\beta_0/\beta_\infty = 1$ for several $\tilde{\delta}$. We see that for not very small $\tilde{\delta}$, the twisted disc has a sufficiently smooth shape, which suggests the possibility of the existence of such configurations in nature. We note that β behaves non-monotonically: it first decreases and then increases with the decrease of r_S. The latter can have important implications both for the disc structure itself and for its observational manifestations. For example, the hot inner regions of such a disc should illuminate its outer parts much stronger compared to the flat disc case. Clearly, this is due to the disc inner parts being tilted with respect to the outer parts.

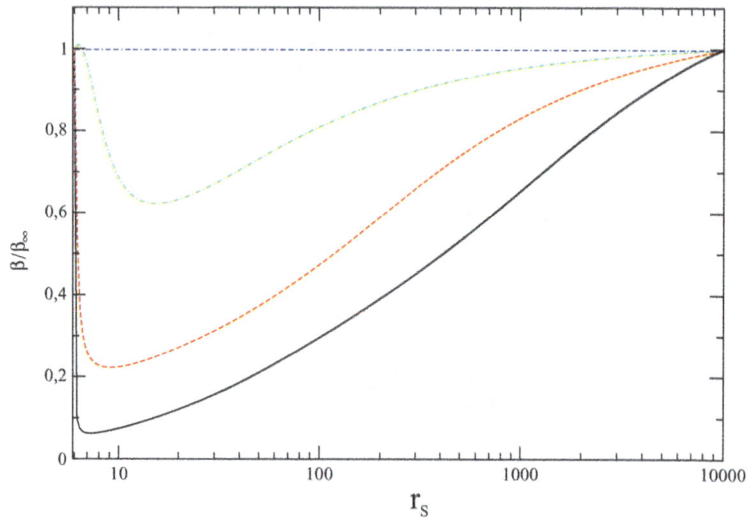

Fig. 4.8 The dependence of β on r_S along the curve in Fig. 4.7 for which $\beta_0/\beta_\infty = 1$. The parameter $\tilde{\delta}$ takes the values 10^{-3}, 10^{-2}, 10^{-1} and 1 for the solid, dashed, dash-dotted and dash-dot-dashed curves, respectively

In the region where $\tilde{\delta}$ is of the order of or greater than unity, the action of the gravitomagnetic force becomes insignificant, and the disc is weakly twisted. In Fig. 4.7, the area to the right of the dashed line is where $\beta(r_S)$ deviates from β_∞ by less than 10%. It is also worth noting that for $\tilde{\delta} > 0.1$ the Bardeen-Petterson effect is completely absent for any α.

4.6 Conclusions

We have presented a detailed technical derivation of the governing equations for the evolution of the shape of a relativistic twisted disc, as well as for perturbations of the velocity and density inside it. Only three simplifying assumptions have been used: the smallness of the disc aspect ratio, $\delta \ll 1$, the slowness of the black hole rotation, $a \ll 1$, and the smallness of the tilt of the disc rings with respect to the equatorial plane of the black hole, $\beta \ll 1$. This allowed us to formulate Eqs. (4.60), (4.61) and (4.64) for three variables describing Eulerian perturbations of the azimuthal velocity, v^r and v^φ and the geometrical form of the disc, Z. In general, the dependence of v^r and v^φ on the twisted coordinates r, ξ and τ, and the dependence of Z on r and τ should be found. In accordance with Eq. (4.30), all these variables depend harmonically on the azimuthal coordinate. The governing equations contain the profiles of the background solution, representing an accretion disc with similar radial and vertical structure but lying in the equatorial plane of the black hole. We note once again that not only the twisted disc but also the background itself can be non-stationary, since when deriving the set of Eqs. (4.60), (4.61), (4.64), only one assumption about the background, the smallness of $\delta \ll 1$, was used. Therefore, in addition, the twist equations enable us to study the evolution of tilted/twisted gaseous tori/rings near a rotating black hole as they are spreading in the radial direction, in other words, as non-stationary accretion proceeds due to turbulent viscosity.

In the particular case of a stationary, vertically isothermal background with α-parametrization of the viscosity, the twist equations have been reduced to the simpler equations (4.86), (4.87) and (4.88) for the complex amplitudes **A** and **B** describing the velocity perturbations, and **W** describing the disc geometry, which depend only on r and τ. Here, the solution for a flat relativistic disc, which was presented in detail in Chap. 3, was utilized. The corresponding stationary problem can be described by a second-order linear differential equation for **W** (see Eq. (4.106)). An analytic integration of this equation for a formally inviscid disc with $\tilde{\delta} \ll 1$ enabled us to find the singular resonance solutions for a discrete set $\tilde{\delta}_k$, which in fact corresponds to an instability in a flat non-tilted disc, where the latter can acquire a twisted shape near the black hole, even with its outer part lying in the black hole equatorial plane. This instability, however, rapidly disappears already for $\alpha \sim 10^{-3}$ and for $\alpha > \tilde{\delta}$, provided that $\tilde{\delta} < 0.1$, numerical calculations show the Bardeen-Petterson effect. At the same time, already for $\alpha \sim \tilde{\delta}$, alignment of the inner parts of the disc into the equatorial plane of the black hole does no longer occur, and for $\tilde{\delta} \geq 0.1$ smooth but non-monotonic profiles $\beta(r)$ appear (see Fig. 4.8), which suggests their stability under perturbations and the possibility of their realization in nature. The last effect is confirmed by the first numerical simulations of tilted thin relativistic accretion discs with $\delta \sim \alpha \sim a \sim 0.1$ carried out in the recent papers by Teixeira et al. (2014) and Zhuravlev et al. (2014). In these papers, a comparison with the semi-analytic model based on the solution of the set of Eqs. (4.60), (4.61), (4.64) was also done for a slightly tilted vertically barotropic torus.

Observational confirmation of the existence of twisted accretion discs around rotating black holes has just started to emerge. Apparently, one of the most direct pieces of evidence of their existence is the observation of maser sources at subparsec scales in the disc around a supermassive black hole in the nucleus of NGC 4258 (Neufeld and Maloney 1995; Herrnstein et al. 1996). The subsequent modeling in Martin (2008) and Caproni et al. (2007) showed that the disc twist in this case can be due to the Bardeen-Petterson effect. In the recent paper by Wu et al. (2013), observations of jets in the nucleus of NGC 4248 were used to independently estimate the black hole Kerr parameter $a \sim 0.7$ and, in a similar model, to calculate the radius of the disc alignment into the equatorial plane of the black hole in agreement with observations. Additional but more indirect arguments favoring the presence of twisted discs in galactic nuclei were obtained, for example, in Cadez et al. (2003) and Cadez and Calvani (2005), where the observed profiles of the X-ray iron line K_α were calculated for different accretion disc models. It was concluded that in many cases, the observed line profile can be more easily explained in the model of a twisted disc than in the model of a flat disc, with some specific radial intensity distribution. In Wu et al. (2010), a similar modeling of hydrogen Balmer lines was performed. These should arise due to the heating of the outer parts of a twisted disc by hard emission from its inner parts, which have much larger tilt angles than in the case of a flat disc. The presence of twisted discs is also suspected in binary stellar systems with black holes. For example, this could be the case in the two microquasars, GROJ1655-40 and V4641 Sgr, in which a tilt of the jets relative to the orbital plane was discovered (see Martin et al. (2008b,a)).

As mentioned above, Eqs. (4.60), (4.61) and (4.64) also describe the non-stationary dynamics of a torus tilted with respect to the equatorial plane of a black hole. If $\delta > \alpha$, the action of the gravitomagnetic force must lead to solid-body precession of the torus, since in this case the twist (also called bending) waves, propagating at almost the speed of sound, smear out the dependence of γ on r due to the Lense-Thirring effect. Similar non-stationary models are invoked to explain the variability of Balmer line profiles, as well as the precession of jets in active galactic nuclei (see, e.g., Caproni et al. (2004)). In many papers, precessing tori are used to explain low-frequency quasi-periodic oscillations in X-ray binary systems (see, e.g., Veledina et al. (2013)). Of special interest is the modeling of observational manifestations of a tilted accretion disc around the black hole in the centre of our Galaxy (Dexter and C. 2013).

The theory of relativistic twisted discs presented here can also be successfully applied both to constructing self-consistent models of individual objects, and to making further theoretical predictions regarding the dynamics of accretion flows around rotating black holes.

References

Bardeen JM, Petterson JA (1975) Astrophys J 195:65
Bardeen JM, Press WH, Teukolsky SA (1972) Astrophys J 178:347
Cadez A, Calvani M (2005) Mon Not R Astron Soc 363:177
Cadez A, Brajnik M, Gomboc A, Calvani M, Fanton C (2003) Astron Astrophys 403:29
Caproni A, Mosquera Cuesta HJ, Abraham Z (2004) Astrophys J 616:L99
Caproni A, Abraham Z, Livio M, Mosquera Cuesta HJ (2007) Mon Not R Astron Soc 379:135
Demianski M, Ivanov PB (1997) Astron Astrophys 324:829
Dexter J, C FP (2013) Mon Not R Astron Soc 432:2252
Hatchett SP, Begelman MC, Sarazin CL (1981) Astrophys J 247:677
Herrnstein JR, Greenhill LJ, Moran JM (1996) Astrophys J Lett 468:17
Ipser GR (1996) Astrophys J 458:508
Ivanov PB, Illarionov AF (1997) Mon Not R Astron Soc 285:394
Kato S (1990) Publ Astron Soc Jpn 42:99
Kumar S, Pringle JE (1985) Mon Not R Astron Soc 213:435
Landau LD, Lifshitz EM (1971) The classical theory of fields. Pergamon Press, Oxford (Paragraph 6, Problem 1)
Lodato G, Pringle JE (2007) Mon Not R Astron Soc 381:1287
Martin RG (2008) Mon Not R Astron Soc 387:830
Martin RG, Reis RC, Pringle JE (2008a) Mon Not R Astron Soc Lett 391:L15
Martin RG, Tout CA, Pringle JE (2008b) Mon Not R Astron Soc 387:188

Nelson RP, Papaloizou JCB (1999) Mon Not R Astron Soc 309:929
Neufeld DA, Maloney PR (1995) Astrophys J Lett 447:L17
Okazaki AT, Kato S, Fukue J (1987) Publ Astron Soc Jpn 39:457
Papaloizou JCB, Lin DNC (1995) Astrophys J 438:841
Papaloizou JCB, Pringle JE (1983) Mon Not R Astron Soc 202:1181
Petterson JA (1977) Astrophys J 214:550
Petterson JA (1978) Astrophys J 226:253
Teixeira D, Fragile PC, Zhuravlev VV, Ivanov PB (2014) Astrophys J 796:103
Thorne KS, Price RH (1986) Macdonald DA, Black holes: the membrane paradigm. Yale University Press, Yale (Chapter 3, Paragraph A)
Veledina A, Poutanen J, Ingram A (2013) Astrophys J 778:165
Wu S-M, Chen L, Yuan F (2010) Mon Not R Astron Soc 402:537
Wu Q, Yan H, Yi Z (2013) Mon Not R Astron Soc 436:1278
Zhuravlev VV, Ivanov PB (2011) Mon Not R Astron Soc 415:2122
Zhuravlev VV, Ivanov PB, Fragile PC, Teixeira D (2014) Astrophys J 796:104

Chapter 5
Structure of Accretion Discs in Lensed QSOs

Pavel Abolmasov, Nikolay Shakura, and Anna Chashkina

Abstract As early as in 1937, Zwicky wrote about gravitational lenses acting as 'space telescopes', allowing the observation of faint and distant objects, the fluxes from which may be considerably enhanced due to the lensing. It is clear today that gravitational lensing may be helpful in performing another important task, one of the main purposes of telescopic observations, namely, increasing spatial resolution. The images of strongly lensed QSOs are affected by microlensing effects in the halo of the lensing galaxy. In contrast to the classical strong lensing, these effects are sensitive to the size and form of an object. The purpose of this chapter is to give a general introduction to **quasar microlensing** and to illustrate the capabilities of the method, with a review of the latest results in this field, concentrating especially on the results obtained in our three recent papers.

5.1 Introduction: Gravitational Lensing and Microlensing

5.1.1 Light Bending by a Thin Gravitational Lens

Gravitational lensing is deviation of light from a straight path when travelling in a gravitational field. A light ray can trace very complex curves in strong gravitational fields, turning by many hundreds of degrees (this takes place, for example, in the vicinities of black holes, see Sect. 5.2.4). As a rule, however, gravitational lenses distort the direction of light propagation and, consequently, the apparent source position, by no more than a few arcseconds or tens of arcseconds. In doing so, one source may yield several images displaced with respect to each other. The largest

P. Abolmasov (✉) · A. Chashkina
Sternberg Astronomical Institute, Lomonosov Moscow State University, Moscow, Russia

Tuorla Observatory, University of Turku, Piikkio, Finland

N. Shakura
Sternberg Astronomical Institute, Lomonosov Moscow State University, Moscow, Russia

Kazan Federal University, Kazan, Russia

N. Shakura (ed.), *Accretion Flows in Astrophysics*, Astrophysics and Space Science Library 454, https://doi.org/10.1007/978-3-319-93009-1_5

separations between images are observed when clusters and groups of galaxies act as lenses, as in SDSS J1004+4112 and, particularly, SDSS J1029+2623 (Inada et al. 2006), where the separation between the images is currently the largest observed at approximately $22\overset{''}{.}5$. These angular distances are quite reachable for large telescopes, and the largest separations may be visible even using amateur instruments, ($20''$ is the size of the apparent disc of Mars at its closest approaches. Lensed QSOs are, however, too faint to observe through amateur instruments), but sometimes (for example, in the case of QSO J2237+0305) we have to work nearly at the limit of the angular resolution for ground-based instruments ($\sim 1''$).

We may talk about the following kinds of gravitational lensing:

- *weak lensing*, which produces a single, slightly amplified and deformed, image of the source
- *strong lensing*, which forms a few resolvable images of a single object, and
- *microlensing*, in which images of the object cannot be resolved, providing nothing else but total fluxes as the only observable characteristics.

This is an observational classification affected by the sensitivity and resolution of the devices in use. A fairly detailed and up-to-date introduction to these three kinds of lensing may be found in the proceedings of the 33rd Saas-Fee summer school: Schneider (2006, 2005), Kochanek (2006), and Wambsganss (2006). Clusters of galaxies often act as weak lenses, where the weak lensing can be used to reconstruct the mass distribution in the cluster. There are also many examples of strong lensing; at least a few dozens of QSOs are luckily situated behind foreground galaxies and are strongly lensed by them. As we will see below, characteristic separations between the images may amount to several arcseconds. The stars of our Galaxy separate images approximately by milliarcseconds, thus potentially acting as microlenses.

The smallness of the deflection angles simplifies the general task of describing strong lensing, and offers a possibility to write down the lens equation in an algebraic form. *Thin lens approximation* is an important simplification: the extent of the lens along the line of sight is considered to be much smaller than the distances between the lens, the source, and the observer. This approximation is analogous to that used in geometric optics. The thin lens approximation is also known as the thin screen or 'single-screen approximation', referring to a flat 'screen' perpendicular to the line of sight, on which all the masses affecting the propagation of light are meant to be concentrated. Each elementary mass ΔM on the 'screen' changes the direction of a light ray by some small angle towards the lens,

$$\Delta\alpha = \frac{4G\Delta M}{pc^2}, \tag{5.1}$$

where p is the minimal distance between a passing photon and an elementary mass, $p \gg \dfrac{2GM}{c^2}$. The combined effect of multiple masses is a vector sum, or an integral over all the elementary masses (see the formula (5.4) below).

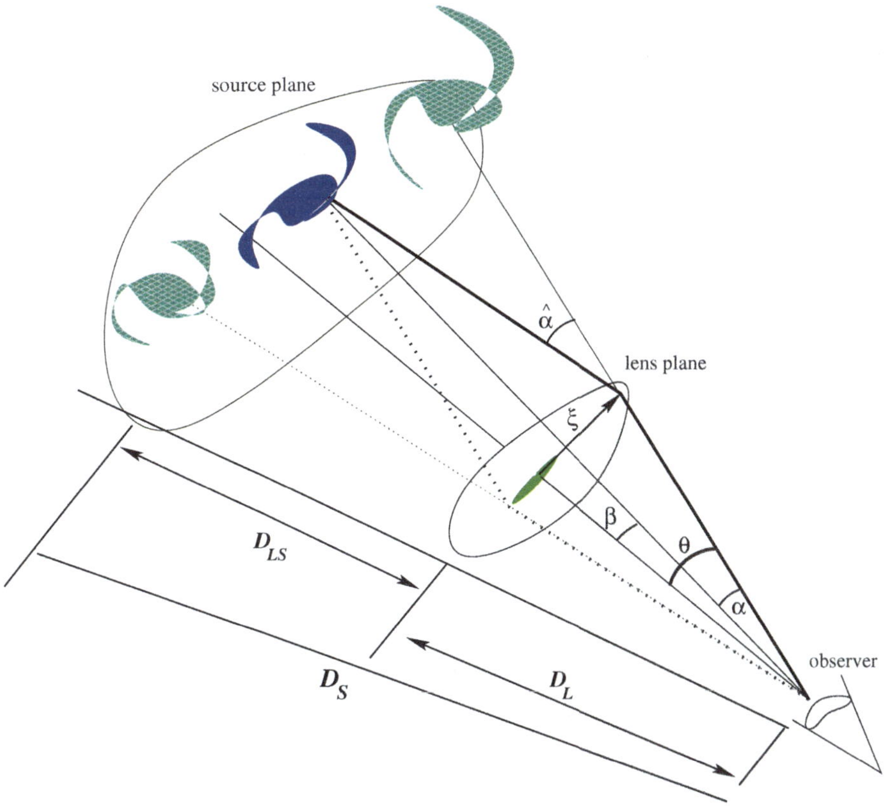

Fig. 5.1 The formation of images in the thin lens approximation. The bold lines (solid and dotted) depict schematically the paths of photons and the thin lines connect the observer with the false images. The dotted and solid lines correspond to two different images at the opposite sides of the lens in the picture frame. The source position (that would be seen in the absence of the lens) is shown in blue and the observed images are represented in teal. The lower image is upside down

The smallness of the lens size allows us to consider the path of light as a polyline composed of two segments (see the diagram in Fig. 5.1). Let us designate the distance between the source and the observer as D_S, the distance between the lens and the observer as D_L, and the distance between the lens and the source as D_{LS}. Note that on cosmological length scales, we have to use 'angular-size distances' (see, for example, Zeldovich and Novikov 1975), which differ from the comoving distance by a factor of $1+z$ (unchangeable for two static objects in the expanding universe), where z is the redshift (the redshifts of the lens and the source will be designated as z_L and z_S, respectively). Generally speaking, the distance D_S to the source is not equal to the sum of the distances D_{LS} and D_L. However, the

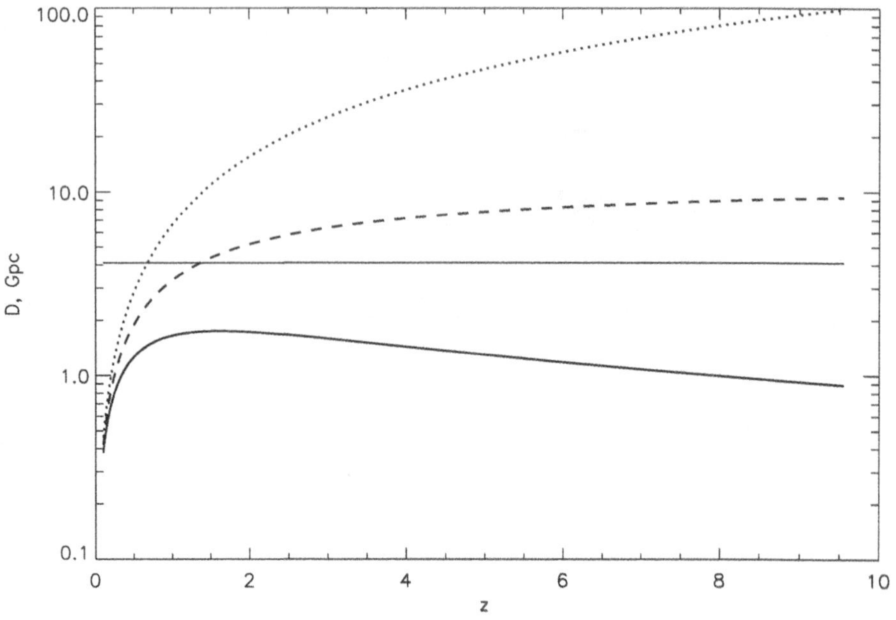

Fig. 5.2 Three different distances used in the standard cosmological Λ-CDM model ($\Omega_\Lambda = 0.7$, $\Omega_k = 0$) as functions of redshift: the photometric distance (dotted line), the comoving distance (dashed line) and the angular-size distance (solid line). The horizontal line corresponds to a Hubble length of $D_H = c/H_0 \simeq 4.1$ Gpc

relationship among the three distances in a flat universe has a fairly simple form:

$$D_S = D_{LS} + \frac{1+z_L}{1+z_S} D_L \tag{5.2}$$

More general aspects concerning distances in cosmology are considered, for example, in Hogg (1999). Note that the angular-size distance is a non-monotonous function of redshift, having a maximum at $z_{\max} \sim 1.6$ (corresponding to a distance of $D_{\max} \simeq 1.75$ Gpc, see Fig. 5.2) in the standard ΛCDM cosmological model with $\Omega_\Lambda \sim 0.7$ and a modern Hubble constant of $H_0 = 70\,\mathrm{km\,s^{-1}/Mpc}$. As a result, more distant objects may have larger apparent sizes in a certain range of redshifts, being easier to resolve than closer objects. Image scales of distant QSOs and galaxies with redshifts of $z \sim 0.5$–8 vary from $\sim$5 kpc/$''$ at the ends of the interval to $\sim$8 kpc/$''$ near $z_{\max}$. The two other characteristic distances, namely, the comoving (physical) distance and the photometric distance, monotonously grow with redshift.

For most problems of gravitational lensing, the thin screen approximation is valid with good accuracy. If we take into account that the lens is extended along the line of sight, this yields a next-order correction in respect to deviations; hence, systematic uncertainties connected with the limitations of thin lens approximation are of the next order in l/D, where l is the size of the source along the line of sight and D is the distance to it. There is only one case of astrophysical importance when deviations from the thin lens approximation may have a noticeable value, namely, the lensing by clusters of galaxies. All characteristic angles of the problem are small, allowing us to sum up linearly the effects of all the gravitational forces bending the light ray. However, the lens equation is non-linear in the general case, which leads to multiple images and specific peculiarities in microlensing curves, giving ultimately a possibility to examine the spatial structure of sources with the use of microlensing (for more details, see Sect. 5.2.1 below).

5.1.2 *Transformation of the Plane of the Sky*

The mathematical apparatus for gravitational lensing is considered in a more strict and detailed manner in the monograph by Zakharov (see Zakharov 1997 and references therein, and also the overview Zakharov and Sazhin 1998). There are two aspects of the problem, which we touch upon here only briefly, not going into details: (1) it is convenient to use the method of potentials when using the thin lens approximation, namely, to examine the shift in position of a point-like source as a gradient of a scalar function (potential) that is given in the plane of the sky; (2) the lens equation can be obtained using the minimization of a functional—as a rule, optical length of a travel time along a null geodesic, in accordance with Fermat's principle (see Blandford and Narayan 1986).

Lensing can be treated as single-valued mapping from the plane of the sky, in which the lens is situated (or, simply, the *lens plane*; the other name is "the image plane") and where the coordinates are given by the two-dimensional vector $\boldsymbol{\theta}$, onto the source plane (vector $\boldsymbol{\beta}$). Generally speaking, the deflection angle $\hat{\boldsymbol{\alpha}}$ is also a vector, which is calculated, as we mentioned before, as a sum (integration) of the deviations created by the mass elements:

$$\hat{\boldsymbol{\alpha}}(\boldsymbol{\xi}) = \sum_i \Delta\alpha_i = \sum_i \frac{4Gm_i}{c^2} \frac{\boldsymbol{\xi} - \boldsymbol{\xi}'}{|\boldsymbol{\xi} - \boldsymbol{\xi}'|^2}. \tag{5.3}$$

The right-hand side contains a sum of terms similar to (5.1), which are directed away from the mass m_i in the lens plane. Let us introduce the surface density of the lens $\Sigma(\boldsymbol{\xi})$ at the point $\boldsymbol{\xi}$ in the form $\mathrm{d}m = \Sigma(\boldsymbol{\xi})\mathrm{d}\boldsymbol{\xi}$, where $\mathrm{d}m$ is the mass per unit area in the lens plane. When changing from summation to integration, we obtain:

$$\hat{\boldsymbol{\alpha}}(\boldsymbol{\xi}) = \frac{4G}{c^2} \int \frac{\boldsymbol{\xi} - \boldsymbol{\xi}'}{|\boldsymbol{\xi}' - \boldsymbol{\xi}|^2} \Sigma(\boldsymbol{\xi}') d\boldsymbol{\xi}'. \tag{5.4}$$

In case of circular symmetry,

$$\hat{\boldsymbol{\alpha}}(\boldsymbol{\xi}) = \frac{4GM(\xi)}{c^2} \frac{\boldsymbol{\xi}}{\xi^2} \tag{5.5}$$

where

$$M(\xi) = 2\pi \int_0^{\xi} \Sigma(\xi')\xi' \mathrm{d}\xi'. \tag{5.6}$$

All the angles in the problem are small, which allows us to significantly simplify the calculations. As can be seen from Fig. 5.1, $D_{LS}\hat{\boldsymbol{\alpha}} = D_S\boldsymbol{\alpha}$ implying that the relationship between the position of the source (given by vector $\boldsymbol{\beta}$) and the position of its image $\boldsymbol{\theta} = \boldsymbol{\xi}/D_L$ has the form

$$\boldsymbol{\theta} = \boldsymbol{\beta} + \boldsymbol{\alpha} = \boldsymbol{\beta} + \frac{D_{LS}}{D_S}\hat{\boldsymbol{\alpha}}(D_L\boldsymbol{\theta}). \tag{5.7}$$

It can be seen that the position of the source $\boldsymbol{\beta}$ is a single-valued function of the image position $\boldsymbol{\theta}$. The inverse mapping $\boldsymbol{\beta} \to \boldsymbol{\theta}$ is not single-valued in the general case as several images may correspond to one source. This property is valid even for the simplest models, such as a point-like lens, and is due to the fact that the equation is not linear with respect to $\boldsymbol{\theta}$. The dotted line in Fig. 5.1 illustrates the path of a photon that takes part in the formation of the second image. The relationship (5.7) is known as *the lens equation* if considered as an equation (or, more precisely, a system of two equations) relative to $\boldsymbol{\theta}$.

5.1.3 Symmetric Lenses

5.1.3.1 Point-Like Lenses

A lens that has a circular symmetry ($\Sigma = \Sigma(|\theta|) = \Sigma(\theta)$) shifts the apparent position of a source along the straight line passing through the centre of lens symmetry. From this point onwards, we will refer to it simply as *a symmetric lens*. From Eq. (5.5), the relationship between the apparent shift of the image relative to the source and the deflection angle may be derived:

$$\boldsymbol{\alpha}(\boldsymbol{\theta}) = \frac{D_{LS}}{D_S}\hat{\boldsymbol{\alpha}}(\boldsymbol{\theta}) = \frac{4GM}{c^2}\frac{D_{LS}}{D_S D_L}\frac{\boldsymbol{\theta}}{|\boldsymbol{\theta}|^2}, \tag{5.8}$$

which makes it possible to move from (5.7) to the equation for a point-like lens:

$$\boldsymbol{\beta} = \boldsymbol{\theta} - \theta_{\mathrm{Ein}}^2 \frac{\boldsymbol{\theta}}{|\boldsymbol{\theta}|^2}, \tag{5.9}$$

where θ_{Ein} is the characteristic angular radius known as the Einstein-Chwolson radius (Einstein 1936; Chwolson 1924):

$$\theta_{\text{Ein}} = \sqrt{\frac{4GM}{c^2}\frac{D_{LS}}{D_S D_L}}. \tag{5.10}$$

For a point-like lens, the physical meaning of θ_{Ein} corresponds to the angular radius of *the Einstein ring*, a ring image of the point-like source that arises when the observer, the lens and the source are lying on the same straight line.

Equation (5.9) has two solutions,

$$\theta_{1,2} = \frac{\beta}{2} \pm \frac{1}{2}\sqrt{\beta^2 + 4\theta_{\text{Ein}}^2}. \tag{5.11}$$

The different signs of the solutions mean that the images arise on different sides of the lens. As the source moves away from the lens, one image is approaching the source, while the other is moving towards the lens. The violation of coaxial alignment leads to the splitting of the ring into two distorted images, one of them strictly within and the other strictly outside of θ_{Ein}. For a point-like lens, Eq. (5.9) has two solutions,

$$\theta_{1,2} = \frac{\beta}{2} \pm \frac{1}{2}\sqrt{\beta^2 + 4\theta_{\text{Ein}}^2}. \tag{5.12}$$

These two solutions with the different signs correspond to two images arising on different sides of the lens. As the source moves away from the lens, the position of one image is approaching the source, meanwhile the other is approaching the lens.

Einstein rings are observed as a particular case of strong lensing by galaxies (Kochanek et al. 2001). In a more general case, the Einstein-Chwolson radius still remains a characteristic angular scale for splitting images in lensing.

It is useful to estimate characteristic angular scales of the effect for different cases. In particular, for stars in the halo of our galaxy ($D_L \sim D_S \sim D_{LS} \sim 10\,\text{kpc}$, $M \sim \text{M}_\odot$), $\theta_E \sim 10^{-3\prime\prime}$, for massive galaxies at cosmological distances ($D_L \sim D_S \sim D_{LS} \sim 1\,\text{Gpc}$, $M \sim 10^{12}\text{M}_\odot$), $\theta_E \sim 1''$. Individual stars in remote galaxies produce deviations on the order of microarcseconds.

5.1.3.2 Time Scales

In some cases, it is important to verify the time scale defined by the relative tangential velocities. As a rule, the motion of the source relative to the lens is important in case of microlensing by stellar mass objects. The characteristic time scale, namely, the time of passing θ_E, amounts to several years for stellar mass lenses at cosmological distances. This time scale for stars of our galaxy is considerably less, about a few days. In order to estimate it correctly, we need to know the proper motions of the source and the lens, $v_\mu^{S,L}$:

$$v_\mu^{S,L} = \frac{v_t^{S,L}}{(1 + z_{S,L})D_{S,L}}, \tag{5.13}$$

where $v_t^{S,L}$ are the tangential velocities of the source and the lens measured in a comoving reference frame. Remember that the proper motion is an angular displacement per unit time. When studying the structure of the source, it is convenient to use the effective velocity $v_{\rm eff}$, which is defined as a transverse physical distance by which the source moves with respect to the direction towards the lens (see also Sect. 5.2.1). The factor $1 + z$ accounts for the effects of cosmological time dilation. Knowing the relative proper motion $v_\mu = v_{\rm eff}/D_S$, it is easy to estimate the time scale as $t \sim \theta_{\rm Ein}/v_\mu$:

$$\begin{aligned} t &\sim \sqrt{\frac{4GM}{c^2} \times \frac{D_{LS} D_S}{D_L}} \frac{1}{v_{eff}} \simeq \\ &\simeq 13.5 \left(\frac{M}{\rm M_\odot}\right)^{1/2} \left(\frac{D_{LS} D_S}{D_L {\rm Gpc}}\right)^{1/2} \frac{1000\,{\rm km\,s^{-1}}}{v_{eff}} .{\rm yr} \end{aligned} \tag{5.14}$$

5.1.3.3 The Isothermal Sphere

Another important particular case is the lensing by an isothermal sphere (see the review by Kochanek (2006), as well as Kormann et al. (1994), where the more general case of an isothermal ellipsoid is considered). In this case, the surface density depends on R inversely, as $\Sigma(R) \propto R^{-1}$, while the mass grows linearly with the radius (and thus is unlimited indicating that this solution, generally speaking, has little physical meaning). The simplicity of the isothermal sphere model lies in the fact that the dispersion of velocities σ^2 is constant for virial motions. Herewith, the expression for the mass within a sphere of radius ξ may be written as

$$M(\xi) = \frac{\sigma^2}{G}\xi. \tag{5.15}$$

The above formula follows from the virial theorem as the potential energy for a test particle with the mass m is $-GMm/\xi$, and the mean kinetic energy is $m\sigma^2/2$.

Substituting (5.15) into the lens equation (5.9) yields one or two solutions obeying the equation:

$$\theta = \beta + \frac{4\sigma^2}{c^2}\frac{D_{LS}}{D_S} \times \frac{\boldsymbol{\theta}}{|\theta|}. \tag{5.16}$$

The primary image is shifted relative to the source by a fixed angle in the outward direction from the lens centre, while the secondary image appears only if $\beta < \frac{4\sigma^2}{c^2}\frac{D_{LS}}{D_S}$. The isothermal sphere is the simplest model used to describe the gravitational fields of elliptical galaxies. By order of magnitude, $\Delta\theta \sim 1''$, becoming greater with the growing dispersion of velocities in the lensing galaxy. Formula (5.16) also illustrates the principal role of massive galaxies among gravitational lenses (Kochanek and Keeton 1997). The image is significantly amplified in the area (solid angle) proportional to σ^4, but only if the isothermal law is valid at distances larger than the Einstein-Chwolson radius.

5.1.4 Asymmetric Lenses

The use of centrally symmetric models for extended lenses is almost never justified (Kochanek 2006) since small deviations from symmetry in the mass distribution result in a qualitative change of the pattern, giving rise to various singularities and multiplication of images, the number of which increases from two or three to five or even more. At the same time, non-symmetric models are more complex from a mathematical point of view, having a greater number of parameters, which hinders their application to observational data.

In some cases, there are analytic solutions to the lens equation. A fairly complete catalogue of models used is given by Keeton (2001). For most models, the number of solutions amounts to three or five (depending on the source position), but it also may be higher. As a rule, one of the images is attenuated and cannot be actually observed, implying that the majority of strong lenses produce two or four images of the point-like background object. The existence and brightness of the central attenuated image are sensitive to the form of potential near the galactic centre. This can be demonstrated, for example, by Eq. (5.9), which has a singularity at zero for a point-like lens that leads to the finite limit $\lim_{\theta\to 0}\alpha(\theta)$ and, consequently, to the chance of 'losing' one solution to the equation.

Each of the observed images is distorted and amplified, which may be described in terms of a linear transformation of the plane $\boldsymbol{\beta} \to \boldsymbol{\theta}$,

$$A_{ij} = \frac{\partial\beta_i}{\partial\theta_j} = \delta_{ij} - \frac{\partial\alpha_i}{\partial\theta_j}. \tag{5.17}$$

Here, $i, j = 1..2$, and δ_{ij} is the Kronecker delta symbol (1 for $i = j$, 0 for $i \neq j$). The matrix A_{ij} contains all the information regarding the deformation of a source with an infinitely small angular size. Traditionally, the contributions due to convergence $\varkappa$ and 'shear' γ are separated:

$$A = (1 - \varkappa)\begin{pmatrix} 1 & 0 \\ 0 & 1 \end{pmatrix} - \gamma\begin{pmatrix} \cos 2\psi & \sin 2\psi \\ \sin 2\psi & -\cos 2\psi \end{pmatrix}. \tag{5.18}$$

This decomposition exists and is unique for any symmetric matrix A; it may be perceived as a definition of the values $\varkappa$, γ, and ψ. For convergence, the following relationship is also valid:

$$\varkappa = \frac{4\pi G}{c^2}\frac{D_{LS}D_L}{D_S}\Sigma \tag{5.19}$$

The values $\varkappa$, γ, and ψ have a clear physical meaning that is well illustrated by expression (5.18): a source having circular form is mapped to an ellipse with the semi-major and semi-minor axes of $1-\varkappa+\gamma$ and $1-\varkappa-\gamma$, respectively, turned by the angle ψ on the celestial sphere. The physical meaning of the matrix determinant is the ratio of the solid angles of the source and image. As the effects we are considering are expected to preserve the radiation intensity, $1/\det A$ indicates also to what extent the observed flux from a point-like source changes (the amplification factor). A negative Jacobian would mean that the image is 'turned inside out', that is, no combination of motions and stretches can identify such an image with the source.

$$\det A = (1-\varkappa)^2 - \gamma^2 \tag{5.20}$$

At the points where $\det A = 0$, generally speaking, the area of an infinitely small object increases by an infinite factor. Since lensing does not change the intensity, the ratio of the solid angles is equal to the ratio of the observed fluxes, which is particularly important in case of microlensing. Hence, the amplification factor is

$$\mu = \frac{1}{(1-\varkappa)^2 - \gamma^2}. \tag{5.21}$$

In particular, if we know the relationship $\alpha(\theta)$ for a symmetric lens, then $\varkappa = \frac{(\theta\alpha(\theta))'}{2\theta}$, $\gamma = \frac{\theta}{2}\left(\frac{\alpha(\theta)}{\theta}\right)'$, and $\psi = \varphi$. The amplification factor for an image in this case can be easily calculated as $\mu = \frac{\theta d\theta}{\beta d\beta}$. This is the stretching factor in the source plane for a narrow ring with radius β.

Figure 5.3 shows the amplification curve for a point-like lens and the positions of the source and the two images. The maximum amplification corresponds to the coalescence of the two images into an Einstein ring; as the source moves away from the lens, one image is approaching the source, while the other (weakened) one is approaching the lens.

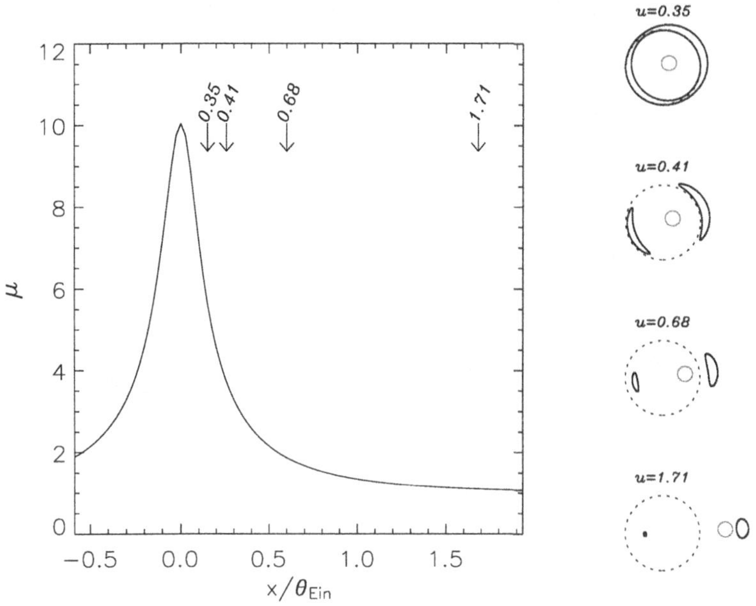

Fig. 5.3 Lensing by a point-like mass. Left: the amplification curve for an object passing at a distance of $0.1\theta_{\mathrm{Ein}}$ from the lens. Right: the positions and shapes of two images of the object are schematically shown for four selected values: $u = \beta/\theta_{Ein}$. The diameter of the source is $0.2\theta_{Ein}$, which is greater than the minimal angular distance to the lens; that is why the images coalesce into the "Einstein ring" near the amplification maximum

5.1.4.1 Caustics

For a point-like source, amplification diverges when the determinant A becomes zero. It is obvious that an extended object will be amplified by some finite value determined by its size and the behaviour of $\det A$ close to zero.[1]

If the lens parameters are smooth functions of the coordinates, the properties of the zeros of the determinant $\det A$ are described by the bifurcation theory (Arnold et al. 2003) for the time delay Δt (the time for the photon to pass along the trajectory; see also Sect. 5.1.4.2), which is a smooth function of the source position on the celestial sphere. In this case, the coordinates of the source can be treated as two external control parameters.

[1]For a fold caustic considered below, the amplification decreases inversely proportional to the distance to the caustic in the image plane, $\mu \propto \Delta\theta^{-1}$, or $\mu \propto \Delta\beta^{-1/2}$ (see later this section). The integral of this expression over a restricted solid angle converges. For a more complex singularity (for example, a cusp), the finiteness of amplification follows from the asymptotic behaviour of amplification which decreases inversely proportional to the distance to this singularity (a point), or slower (Gaudi and Petters 2002).

A fold caustic is the simplest kind of singularity (A_2) considered in catastrophe theory, which corresponds to the occurrence or disappearance of a pair of solutions to the lens equation for small variations in the source position. In the plane $\boldsymbol{\beta}$, it is generally a smooth curve limited by other singularities (two caustics almost always meet in an A_3-singularity called a cusp). In what follows, we will imply a fold-type singularity when speaking about caustics. A vanishing Jacobian means that an image of the source is infinitely stretched in some direction. This direction is perpendicular to the caustic. Deformation of the image in the direction of caustic can be neglected. Therefore, in the lens equation, instead of the vectors $\boldsymbol{\beta}$ and $\boldsymbol{\theta}$, we may use the scalars $\Delta\beta$ and $\Delta\theta$ which become zero at the caustic. In these terms, $\det A = d\Delta\beta/d\Delta\theta$. An asymptotic expression for amplification near the caustic may be obtained using the expansion of the lens equation in powers of $\Delta\theta$, the distance from the image to singularity. When $\det A$ approaches zero, the lens equation in the direction perpendicular to the caustic may be written as $\Delta\beta = C \times (\Delta\theta)^2 + O(\Delta\theta)^4$, where C is some constant and $\Delta\beta$ is the angular distance from the object to the caustic. This gives, in a quadratic approximation, two solutions with the amplification coefficients $\mu \simeq 1/\det A \simeq d\Delta\theta/d\Delta\beta \propto \Delta\beta^{-1/2}$. The specific form of the coefficients in this dependence is not very important; what is essential is the asymptotic form of the amplification for an object that crosses the straight caustic:

$$\mu(\Delta\beta) = \mu_0 + \frac{\mu_1}{\sqrt{\Delta\beta}}. \tag{5.22}$$

Here, μ_0 is the amplification on the side of the singularity where the number of images is fewer by two and μ_1 is the coefficient characterizing its amplitude.

5.1.4.2 Delays

Catastrophe theory allows us to treat any bifurcation as the occurrence of a pair of extrema for some function defined in the space of all task parameters (internal ones, such as the coordinates of the image, and control ones, namely, the source position). In case of lensing, the time needed to pass the trajectory characterized by a given set of control parameters may be chosen as such a function. Herewith, the images may correspond not only to minima and maxima of this function but also to saddle points. This behavior may be described in terms of Fermat's principle, as was done, for example, in the paper by Blandford and Narayan (1986).

In connection to this, it is interesting to mention QSOs, which are variable objects with 'red' noise power spectra that resemble Bernoulli random walk (Webb and Malkan 2000; Collier and Peterson 2001). Variability has been detected on almost all scales, from a few days to time intervals comparable to the time spans of the observations (decades). As a rule, variability becomes uncorrelated on time scales of tens or hundreds of days (depending on the BH mass). On shorter time scales, the variability has an 'infrared' power spectrum with a slope greater than 2.

Delays are characterized by time scales of the order of $\frac{D_L}{c}\theta^2_{Ein} \sim$ $\sim 100 \frac{M}{10^{12}M_\odot}\frac{D_{LS}}{D_S}$ days. For this reason, if an observing run is long enough, it is convenient to use the variability of individual images to measure these delays with the help of cross-correlation of individual light curves. The resulting values can be used to verify the model of the lens and to determine the Hubble constant independent of distance scale (Kochanek and Schechter 2004).

5.1.4.3 Optical Depth

Microlensing optical depth is an important factor that characterizes the probability of lensing by foreground objects and specifies the general pattern of microlensing. The optical depth may be defined as the convergence obtained as a result of integration over all the screens (all the distances at which the lensing masses are localized):

$$\tau = \frac{4\pi G}{c^2}\frac{1}{D_S}\int D_{\mathrm{LS}} D_{\mathrm{L}} d\Sigma. \tag{5.23}$$

The distances $D_{\mathrm{L,LS}}$ generally change along the line of sight. It is easy to verify that τ, when defined in this manner, is calculated as ordinary optical depth with the area within the 'Einstein ring' as a cross-section: $\sigma = \pi(D_L\theta_{\mathrm{Ein}})^2$. We may say that for $\tau \ll 1$, the optical depth determines the probability for a background source to appear within the Einstein ring of some microlens. Since the cross-section is linear in mass, the superposition of cross-sections and optical depths will in general be approximately linear, too. A linear approximation is sufficient to estimate the number of microlensing events in our galaxy, where $\tau \sim 10^{-6}$ (see, for example, Hamadache et al. 2006). If a galaxy with fixed parameters is moving away, its optical depth will grow with D_L, reaching a noticeable value at cosmological distances:

$$\tau_{\mathrm{gal}} \simeq \frac{4\pi G\Sigma}{c^2}\frac{D_{LS}D_L}{D_S} \simeq 0.06\frac{\Sigma}{100\mathrm{M}_\odot\ pc^{-2}}\frac{D_{LS}}{D_S}\frac{D_L}{1\,\mathrm{Gpc}}. \tag{5.24}$$

This determines, among other things, the importance of taking microlensing effects into account in case of strong lensing (see Sect. 5.2.1 below) and the nonlinear character of this microlensing. If $\tau \gtrsim 1$ for some reason, chances are high that there are several lenses that affect the position, form and amplification of the object image. In this case, the definition of optical depth given above has no physical meaning.

There is another case when we have to be cautious in using the value of (5.23), namely, when applying it to the optical depth of the universe as a whole when considering the lensing of sources with high z_S. A direct estimate using the flat

ΛCDM-model yields:

$$\begin{aligned}\tau(z) &= \frac{4\pi G}{c^2}\int_0^z \frac{D_{LS}D_L}{D_S}\frac{d\Sigma}{dz}dz = \\ &= \frac{4\pi G}{c^2}\int_0^z \frac{D_{LS}D_L}{D_S}\rho(z)\frac{dD_L^C}{dz}dz = \\ &= \frac{4\pi G\rho_{cr}\Omega_M}{c^2}\int_0^z \frac{D_L D_{LS}}{D_S}\frac{dD_L^C}{dz}(1+z)^3 dz,\end{aligned} \tag{5.25}$$

where $\rho(z) = \Omega_M\rho_{cr} \times (1+z)^3$ is the local density of inhomogeneous matter, $\rho_{cr} = \dfrac{3c^2}{8\pi G}D_H^{-2}$ is the critical density, and $D_H = c/H_0 \simeq 4.1$ Gpc is the Hubble length. The distances designated by C superscripts are physical distances. It can be shown that for the flat ΛCDM model, $dD_L^C/dz = D_H\left(\Omega_\Lambda + (1+z)^3\Omega_M\right)^{-1/2}$, where $\Omega_{\Lambda,M}$ are the densities of dark energy and inhomogeneous matter in terms of ρ_{cr}.

$$\begin{aligned}\tau(z) = \frac{3}{2}\Omega_M\int_0^z (1+z)^3\frac{D_L}{D_H} \times \left(1 - \frac{1+z_L}{1+z_S}\frac{D_L}{D_S}\right) \times \\ \times \frac{dz}{\sqrt{\Omega_\Lambda + (1+z)^3\Omega_M}}\end{aligned} \tag{5.26}$$

An estimate for a source at $z_S \sim 10$ proves to be approximately 0.1, which is two or three orders of magnitude higher compared with calculations including the inhomogeneous distribution of matter in galaxies and clusters of galaxies, $\tau \sim 10^{-3}$–10^{-4} (Hilbert et al. 2007).

5.1.5 Microlensing

Remember that we talk about microlensing in those cases when individual images cannot be resolved and photometry is the only source of information. Total amplification μ becomes the most essential value. The amplification for an extended source is calculated as

$$\mu_{\rm tot} = \frac{\int I\mu d\Omega}{\int I d\Omega}, \tag{5.27}$$

where the integration is performed over the solid angle in the source plane, $\mu = 1/\det A$ and I is the intensity. The integral of intensity over the source plane yields the total flux.

5.1.5.1 Amplification by a Point-Like Lens: Classical Microlensing

As noted above, a point-like lens produces two images. The amplification factors for these two images can be found as the ratios of the area elements of the image and the source

$$\mu_{1,2} = \frac{\theta_{1,2} d\theta_{1,2}}{\beta d\beta}. \tag{5.28}$$

Substitution of (5.11) into this expression yields

$$\mu_{1,2} = \frac{1}{4}\left(1 \pm \frac{\beta}{\sqrt{\beta^2 + \theta_E^2}},\right) \times \left(\beta \pm \sqrt{\beta^2 + \theta_E^2}\right) \tag{5.29}$$

$$\mu_{1,2} = \frac{1}{2} \pm \frac{1}{4}\left(\sqrt{1 + \frac{4}{u^2}} + \frac{1}{\sqrt{1 + \frac{4}{u^2}}}\right), \tag{5.30}$$

where $u = \frac{\beta}{\theta_{Ein}}$. It can easily be seen that the algebraic sum of two amplifications is equal to 1, whereas the sum of moduli gives the total amplification factor for a point-like source placed at an angular distance β from the lens (Einstein 1936):

$$\mu = \frac{u^2 + 2}{u\sqrt{u^2 + 4}}. \tag{5.31}$$

This amplification may be expressed as a function of time for the case when the source and the lens have a constant relative velocity. In the general case, the object passes at a finite angular distance β_0 from the lens, which corresponds to the maximal amplification possible in microlensing. A time scale is specified by the time needed to pass the Einstein-Chwolson radius: $t_{Ein} = \theta_{Ein} D_S / v_{eff}$.

5.1.5.2 Multiple Microlensing and Caustic Network

The point-like lens case is degenerate since the determinant det A is equal to zero only if $\beta = 0$ (this is by the way valid for any axially symmetric lens). The character of the transition to multiple lensing may be clearly illustrated using the following example. Let us take a point-like lens and add a constant shear γ into the expression for $\boldsymbol{\alpha}(\boldsymbol{\theta})$ in a manner such that

$$\beta_x = \theta_x \times \left(1 - \frac{\theta_E^2}{\theta^2} - \gamma\right), \tag{5.32}$$

$$\beta_y = \theta_y \times \left(1 - \frac{\theta_E^2}{\theta^2} + \gamma\right). \tag{5.33}$$

The celestial sphere image proves to be additionally stretched in the direction of the y-axis. The combination of the lensing by the point-like mass and the artificially introduced deformation leads to the occurrence of areas in the plane $\boldsymbol{\beta}$, where there are three or five images. The boundary between them represents on the plane a curve that encloses an area with the size $\sim \gamma\theta_E$. On this curve, the Jacobian of the transformation of the plane also becomes zero: $\det A = 0$. For a small perturbation $\gamma \ll 1$, this curve is an astroid composed of four caustics and four cusps (see also Zakharov and Sazhin 1998; Chang and Refsdal 1984). A source located within the astroid gives rise to two extra images, the intensities of which decrease when moving away from the caustics, reaching a minimum at the centre of the pattern.

If a fairly wide binary acts as a lens, each of the two stars in the binary perturbs the lensing pattern of the other. Then, $\gamma \simeq \theta_{E,2}^2/\theta_2^2$, where the subscript '2' indicates that this value is defined with respect to the second lens. If the distance is estimated as $\theta_2 \lesssim \sqrt{\theta_{Ein}\theta_{E,2}}$, the perturbations are too large for the point-like lens model or the perturbed 'astroid' model described above to be used. If there are numerous lenses at distances of the order of θ_E from each other, the astroids merge to form a single caustic network and the number of images of the source increases manyfold.

An example amplification map for multiple microlensing is shown in Fig. 5.4. The lensing was calculated for an area measuring approximately 15 microarcseconds, and the Einstein-Chwolson radius for one solar mass was taken to be 2.85 ms, which corresponds to a characteristic distance to the lenses of the order

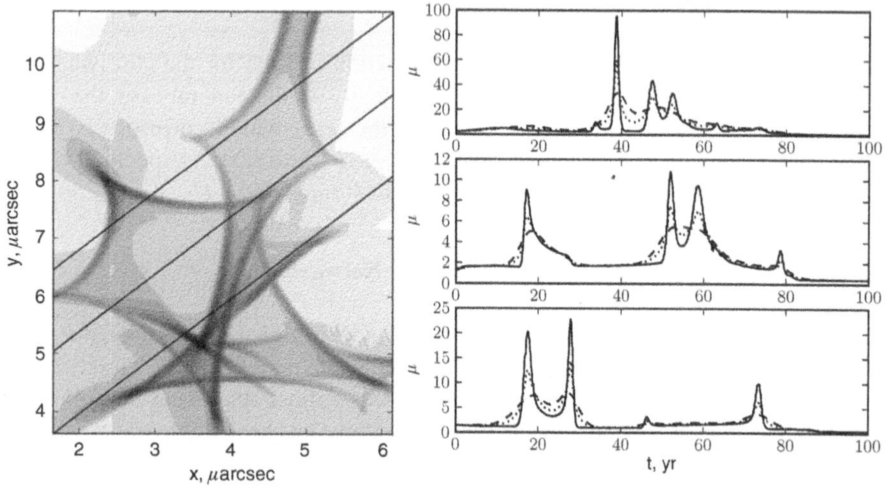

Fig. 5.4 Left: an example of an amplification map calculated using a reverse ray tracing technique. Right: the model light curves for an object passing along the straight lines shown on the map (left). The amplification curves are shown as solid lines for a very small source (comparable in size with the map resolution), dashed lines for a source with a Gaussian two-dimensional brightness distribution, dotted lines for a source with an intensity distribution corresponding to the standard accretion disc model. The half-light radii for both non-trivial intensity distributions are the same

of 1 Gpc. Masses of 13 lenses are randomly distributed in accordance with a power-law slope of -1.3 (which is somewhat lower than the Salpeter function slope; see, for example, Bastian et al. 2010). The intensity distribution of the object has a characteristic angular size of approximately 0.13 ms, which corresponds to the accretion disc around a black hole with the mass of $10^8 M_\odot$ at a redshift of approximately 2 (more details, as well as the accretion disc intensity distribution, can be found below in Sect. 5.2.3).

5.2 Microlensing of Accretion Discs

5.2.1 Microlensing of Strongly Lensed Quasars

Microlensing of strongly lensed QSOs by stars of the lensing galaxy is an important and nearly unique case of microlensing at high optical depth. The optical depth of a chosen galaxy with surface density Σ does not depend on the characteristic masses of its stars or other microlenses and can be expressed as

$$\tau_{gal} \simeq 0.06 \frac{\Sigma_*}{100 M_\odot \; pc^{-2}} \frac{D_{LS}}{D_S} \frac{D_L}{1\,\mathrm{Gpc}}. \tag{5.34}$$

Here, Σ_* is the surface density of inhomogeneous matter, such as stars and compact objects. Dark matter is treated here as a substance that, although contributing to the mass of the galaxy, is distributed smoothly, thus having no effect on microlensing. Nevertheless, some models (involving, for example, mirror matter that is able to form compact configurations; see Berezhiani et al. 2005) predict that particles of dark matter can accumulate to form 'dark stars' or substellar mass objects. Microlensing by hypothetical objects of the dark halo was considered in Paczynski (1986).

The characteristic size of the caustic network and, hence, the time scales of microlensing variability depend on the masses of the microlenses. The fact that the lensed QSOs do not demonstrate flux variations on time scales shorter than months indicates that the lenses are stellar-mass objects, not hypothetical 'dark stars' with masses three orders of magnitude lower.

The radioquasar QSO 0957+561 (Walsh et al. 1979) was the first discovered strongly lensed QSO. Today a hundred or so such objects are known and reasonably well studied (Muñoz et al. 1998), among them 10–15 radioloud, which is in agreement with the proportion between radio-bright and radio-quiet quasars ($\sim$7–8%, see Ivezić et al. 2002).

Knowing the positions of the images, we may try to restore the mass distribution in the lensing galaxy. If this proves to be too ambitious a task, we may at least impose limits on it. The fluxes observed from individual images can differ noticeably from those predicted in the strong lens model. There are two main factors leading to occurrence of *flux anomalies* detected from lensed QSO images:

absorption in the lensing galaxy and lensing on smaller scales (by satellites or the spiral arms of the lensing galaxy, or even by individual stars within this galaxy). Therefore, the main method for studying microlensing effects is the analysis of anomalous fluxes from individual images, especially their variability, since interstellar extinction changes appreciably only on the larger, parsec-scale, distances. Examination of the variability at different wavelengths may help eliminate fluctuations in interstellar extinction. An analysis of variability of the images of SDSS 0924+0219 provides evidence for an insignificant role of variable extinction in the formation of flux anomalies, at least for this particular object (Floyd et al. 2009).

However, QSOs are variable objects; the only chance for us to confidently distinguish the variability due to microlensing from the intrinsic brightness variations of the object would be to significantly resolve uncorrelated components in the variability of individual images. Therefore, microlensing is a disturbance when studying intrinsic QSO variability and time delays between individual images. Vice versa, intrinsic variability becomes a disturbance if the aim is to find and analyze the variations of flux anomalies.

QSO J2237+0305 (aka the Einstein cross, or Huchra's lens) is an extremely favourable object for studying microlensing effects. The object is unique, firstly, in the sense that the lens is a nearby galaxy at $z = 0.039$. This redshift corresponds to a distance of approximately 160 Mpc, whereas most strong lenses known so far are distant objects at $z \sim 0.5$ and $D_{\rm L} \sim 1$ Gpc. Secondly, the lens is a low-mass spiral galaxy ($M \sim 2 \times 10^{10} {\rm M}_{\odot}$, according to Ferreras et al. 2005). Thirdly, the QSO with a good accuracy (of the order of a tenth of an arcsecond) is directly behind the galactic nucleus. Finally, the intrinsic variability of the QSO is fairly weak (of the order of 0.1 mag) and the delays between images are negligible, a few hours up to maybe days rather than dozens or hundreds of days as characteristic for most of the lensed QSOs (Eigenbrod et al. 2008).

The proximity of the lens implies, among other things, a high proper motion (approximately inversely proportional to the distance $D_{\rm L}$), while the small mass of the lens indicates that the Einstein radius is comparatively small and the images are rather close to the nucleus. As a consequence, the optical depth for microlensing is fairly high, $\tau \sim 1$. The combination of these factors provides a basis for the expectation that we might detect several microlensing events per year (crossings of caustics or other singularities). This was predicted as early as in Chang and Refsdal (1979). Since then, the magnitude of the effect has been repeatedly estimated under different assumptions.

Uncorrelated variability of the images of the Einstein cross was found already in Irwin et al. (1989) on the basis of a high spatial resolution photometric observational run about a month long. Since then, the object has been observed with different telescopes on a regular basis (this will be discussed in more detail in Sect. 5.3.1).

Evidently, to the familiar forms of observational information including energy distribution in the spectrum, intrinsic variability, and polarization, we may now add one more method, microlensing, which is sensitive to the intensity distribution on the celestial sphere on a length scale of the Einstein-Chwolson radius (of the order of microarcseconds and smaller). The sensitivity of the method can be clearly illustrated using a straight caustic model, which has been applied on repeated occasions to describe peaks on amplification curves of lensed QSOs. The structure of the fold caustic naturally appears when examining the vicinity of a line in the source plane, where the Jacobian of the mapping vanishes (see Sect. 5.1.2 above). The amplification factor for a point-like source is the following function of the distance between the source and the caustic:

$$\mu(x, y) = \mu_0 + \mu_1 \sqrt{\frac{\zeta_0}{y}}. \tag{5.35}$$

Here, μ_0 is the amplification factor without the caustic; $\mu_1 \sim 1$ is the so-called caustic strength, or amplitude; y is the distance between the source and the caustic; ζ_0 is some characteristic angular scale, which is close to the typical angles θ_E for lensing stars. A source having a finite size R will be amplified approximately by a factor of $\Delta y \times \mu \sim \sqrt{R}$. Already this estimate alone leads us to conclude that the pattern of a microlensing event, especially the amplitude with which the amplification μ changes, is sensitive to the size and form of the source, hence being capable of providing additional information on the object. Information about accretion disc sizes in QSOs is not only highly desired as a test for the standard accretion disc model (see Sect. 5.2.2) and as a probe for the parameters of the central black hole (see Sect. 5.2.4), but also is unlikely to be obtained in a more direct way. A distant QSO ($D_S \sim 1$ Gpc) observed in the optical range has an accretion disc with an angular size of the order of one microarcsecond. As has been proved earlier in Sect. 5.1.3.1, this value has the same order of magnitude as the Einstein-Chwolson radius for a stellar mass lens at a distance of $D_{\mathrm{L}} \sim 1$ Gpc. Thus, microlensing effects allow us to reach angular resolutions unprecedented in the history of observational astrophysics.

Since our main goal is to study the structure of emitting objects, we will use distances in the source plane (obtained when multiplying the angular distances by D_{S}). The distance normalization $\zeta_0 = \sqrt{1-\varkappa} \times D_{\mathrm{S}}\theta_{Ein}$ is the stellar Einstein-Chwolson radius projected onto the source plane with allowance for strong lensing, which may be treated as a transformation of the plane independent of microlensing. This normalization was used, for example, in Witt et al. (1993).

When considering an accretion disc, it is convenient to use the system of coordinates α, β oriented along the semi-major and semi-minor axes of the disc projection onto the plane of the sky (see Abolmasov and Shakura 2012b). Along with that, we use the coordinate system x, y aligned with the caustic and rotated relative to the coordinate system α, β by some angle ψ (see diagram in Fig. 5.5).

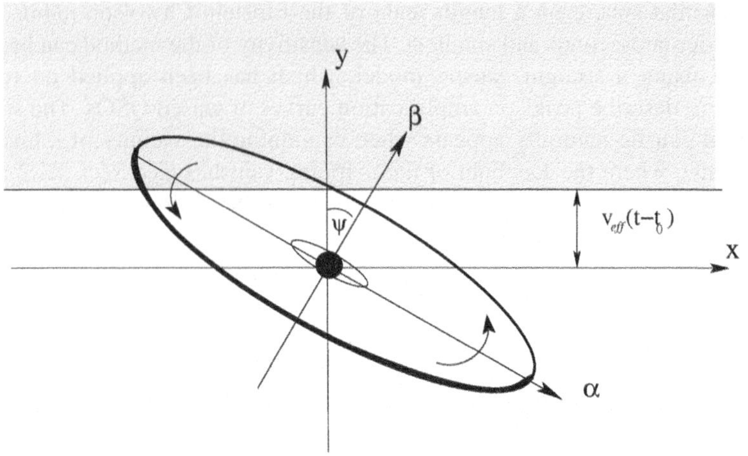

Fig. 5.5 A qualitative diagram of an accretion disc lensed by a straight caustic. The grey sector shows the amplified area behind the caustic. The side of the disc closest to the observer is outlined in bold. The two coordinate systems used in the calculation are shown. Illustration taken from Abolmasov and Shakura (2012b)

If there is a relative motion of an extended object and the caustic, the full amplification at time t can be calculated as (see formula (18) in Abolmasov and Shakura 2012b)

$$\mu(t) = \mu_1\sqrt{\zeta_0} \times \frac{\int I_1(y) \times \Delta y^{-1/2} \times \Theta(\Delta y)dy}{\int I_1(y)dy} + \mu_0, \tag{5.36}$$

where $\Theta(x)$ is the Heaviside function, $\Theta(x > 0) = 1$ and $\Theta(x < 0) = 0$, and

$$I_1(y) = \int_{-\infty}^{+\infty} I(x, y)dx. \tag{5.37}$$

Here, $\Delta y = y - v_{\rm eff}(t - t_0)$. The effective velocity of transverse motion (see also Sect. 5.1.3.2) can be understood as the distance in the source plane per unit time of the observer, and is composed of the peculiar velocities of the source $\boldsymbol{v}_S$, the lens $\boldsymbol{v}_L$, and the observer $\boldsymbol{v}_o$ projected on the normal to the caustic specified by the unit vector $\boldsymbol{n}$:

$$v_{eff} = \frac{\boldsymbol{v}_S \cdot \boldsymbol{n}}{1 + z_S} - \frac{\boldsymbol{v}_L \cdot \boldsymbol{n}}{1 + z_L}\frac{D_S}{D_L} + \frac{\boldsymbol{v}_o \cdot \boldsymbol{n}}{1 + z_L}\frac{D_{LS}}{D_L}. \tag{5.38}$$

The motion of the lens is generally dominant since the apparent peculiar velocities decrease as $\propto (1 + z)^{-1}$ due to time dilation. The amplitudes of the peculiar velocities (in the co-moving restframes) change only slightly with the redshift, remaining of the order of $\sim$500 km s^{-1} (Raychaudhury and Saslaw 1996). The overwhelming majority of galaxies have peculiar velocities less than $v_{\rm max} \simeq 2000\,{\rm km\,s^{-1}}$, which allows us to estimate the maximal physically justified velocity $v_{\rm eff}$ as

$$v_{\rm eff} \lesssim v_{\rm max}\sqrt{(1+z_{\rm S})^{-2} + (1+z_{\rm L})^{-2} \times (D_{\rm S}/D_{\rm L})^2}. \tag{5.39}$$

This estimate corresponds to the situation when $\boldsymbol{v}_{\rm S}$ is perpendicular to $\boldsymbol{v}_{\rm L}$, or to an average over all the possible orientations of $\boldsymbol{v}_{\rm S}$ and $\boldsymbol{v}_{\rm L}$. If the proper motions of the lens and the source are directed in precisely opposite directions, the maximal effective velocity value may grow up to $v_{\rm eff,\,max} \sim v_{max}\left((1+z_{\rm S})^{-1} + (1+z_{\rm L})^{-1} \times (D_{\rm S}/D_{\rm L})\right)$. As a rule, $v_{\rm eff} \sim v_{\rm max}$, the time of passage of one Einstein-Chwolson radius being $t_{Ein} \sim D_S\theta_{\rm Ein}/v_{\rm eff} \sim 30$ year. Due to the factor D_S/D_L, this characteristic time appears significantly shorter (and the passages through caustics more frequent) for lenses fairly close to observer. The Einstein cross is unique since the lens is so close to us, its redshift (~0.04) corresponding to a distance of $D_L \simeq 160\,{\rm Mpc}$. When the distance to the source is of the order of 1 Gpc, we might expect only one event during a period of several years.

5.2.2 *The Standard Accretion Disc Model*

The standard thin disc model was introduced in the works by Shakura (1972), Shakura and Sunyaev (1973) and generalised to the relativistic case in Novikov and Thorne (1973), Page and Thorne (1974) and Riffert and Herold (1995). It is the discussion about the nature of quasars that gave rise to the relativistic model. In contrast to binaries, where generally a wide spectral range from optics to ultraviolet is available where the nonrelativistic model is applicable within the accuracy of $\sim GM_{BH}/Rc^2 \sim 10^{-3}$, the optical discs of active galactic nuclei (AGNs) are relatively small in comparison with the inner edges of their accretion discs (no larger than some hundreds of GM_{BH}/c^2). Therefore, QSOs and other galactic nuclei may demonstrate relativistic effects even at optical wavelengths. However, this depends on the mass and the accretion rate, both values for AGNs being able to change by some orders of magnitude, within a range of millions to billions of solar masses and from $\sim 10^{-2}$ to $\gtrsim 10{\rm M}_\odot\,{\rm year}^{-1}$, respectively.

The main assumptions of the standard accretion disc theory are:

- axially symmetric and stationary disc
- small geometric thickness of the disc (which allows us to separate variables in the dynamic equations, calculating independently the vertical and the radial structure of the disc)
- all the other velocity components are small in comparison with v_φ; radial velocity is used only in the equation for angular momentum transfer
- the optical depth of the disc is large

- there is no substantial heat transfer in the radial direction
- the component $T_{r\varphi}$ of the viscous stress tensor, when integrated over the vertical coordinate z, is proportional to the pressure of the disc P integrated over z: $T^{r\varphi} = \alpha\Pi$, where α is a dimensionless constant normally smaller than unity
- the boundary condition is: $T_{r\varphi} = 0$ at the radius of the innermost stable orbit of the black hole.

Viscous stresses may be due to magnetic fields or turbulence. In both cases, the constancy of α is ensured by *equipartition* understood as proportionality between the energy density of the magnetic field $B^2/8\pi$, or turbulent motions $\rho\langle v_t^2\rangle$, and the thermal energy density. Thus, the turbulent component of the viscous parameter can be estimated as the mean turbulent Mach number squared, $\alpha_t \sim \langle M_t^2\rangle$.

The standard thin disc implies a zero boundary condition at its inner edge which corresponds to the innermost stable orbit in case of a BH: $T_{r\varphi}(R_{in}) = 0$. A nontrivial boundary condition at the inner edge of the disc may arise in case of accretion onto a rapidly rotating compact object with a strong magnetic field (see Siuniaev and Shakura 1977; it is important here that the inner radius of the disc should be greater than the corotation radius). The applicability of this solution for accretion onto supermassive BHs was discussed in Agol and Krolik (2000). Large-scale magnetic fields confined by the pressure in the disc in the so-called 'magnetically-arrested disc' accretion regime, may create a torque acting on the inner edge of the disc. The angular momentum flux produced by magnetic stresses at the inner edge of the disc cannot significantly exceed the angular momentum flux produced by viscous forces since the pressure of the magnetic field is restricted by the pressure in the disc.

Large optical thickness of the disc makes it possible to calculate its spectrum and brightness distribution using the local black-body approximation. The bolometric radiation flux is determined by the local dissipation of energy in the deep layers of the disc

$$Q = \frac{3}{8\pi}\frac{GM\dot{M}}{R^3} f\,(r, a)\,. \tag{5.40}$$

Here, $-1 < a < 1$ is the rotation parameter of the BH, which is considered to be negative when the angular momentum of the disc and of the BH are oppositely directed. The small letter r, here and elsewhere, designates distances normalized to the BH mass, $r = Rc^2/GM$. For convenience, the accretion rate $\dot{M}$ may also be expressed in the dimensionless form, $\dot{M} = \frac{L_{Edd}}{c^2}\dot{m}$, where $L_{Edd} = \frac{4\pi GM}{\varkappa_T c}$ is the Eddington luminosity limit and $\varkappa_T$ is the opacity due to electron scattering.

The correction factor $f(r, a)$ includes the effect of the presence of an inner edge as well as the relativistic effects in the thin disc. In the simplest case of the standard Shakura-Sunyaev thin disc,

$$f = 1 - \sqrt{\frac{r_{in}}{r}}. \tag{5.41}$$

In reality, the situation is more complex even for thin disc due to the transonic nature of the flow and to relativistic effects. More details can be found in Chap. 1 of this book and in Sect. 5.2.4 below.

Here, $r_{in} = r_{ISCO}(a)$ is the radius of the 'innermost stable circular orbit', within which stable circular orbits are not possible and the assumption of a Keplerian velocity profile in the disc cannot apply; accretion of matter without angular momentum transfer begins and viscous stresses vanish. The dependence $r_{ISCO}(a)$ and the correction factor for the relativistic case will be considered in Sect. 5.2.4.

Since the standard disc is optically thick, the radiation from its surface may be considered thermal with the temperature determined by local energy release. Equating the temperature of the local black-body radiation to the effective temperature ($\sigma T^4 = Q$, where σ is the Stefan-Boltzmann constant), the locally observed monochromatic intensity at frequency ν may be written as follows:

$$I_\nu = \frac{2h\nu^3}{c^2} \frac{1}{\exp\left(\frac{h\nu}{kT(r)}\right) - 1}. \tag{5.42}$$

Here, h, c and k are the Planck constant, the speed of light and the Boltzmann constant, respectively. The integrated flux from the accretion disc, which is observed at the angle i to the normal and situated at redshift z (which corresponds to some distance $D(z)$), may be determined by integration over solid angle. Let us denote the frequency in the reference frame of the QSO as $\nu_{\rm em}$ and the frequency received by the observer as $\nu_{\rm obs} = \nu_{\rm em}/(1+z)$ (for the moment, we take into account only the cosmological redshift; the motion of the radiating matter in the disc and strong gravity effects will be considered in Sect. 5.2.4 below). The total observed monochromatic flux is

$$\begin{aligned} F_\nu &= \int I_\nu d\Omega = \\ &= 2\pi \cos i \frac{1}{(1+z)^3 D^2} \int_{R_{in}}^{+\infty} I_\nu^{em} R dR. \end{aligned} \tag{5.43}$$

Here, we pass from integration over solid angle Ω to integration over disc radius R. Provided that the relativistic effects in the vicinity of the BH may be neglected, an element of the solid angle is $d\Omega = RdRd\varphi/D^2$, where D is the angular-size distance (see Sect. 5.2.4). The observed intensity differs from the intensity in the comoving system by a factor of $(1+z)^{-3}$ related to the cosmological expansion of the universe. When passing from one reference frame to another, the value I_ν/ν^3 is preserved (see, for example, Rybicki and Lightman 1986).

$$F_\nu \propto \nu_{obs}^{1/3} (GM)^{4/3} \dot{m}^{2/3} \cos i \times \frac{1}{D^2 \times (1+z)^{8/3}}. \tag{5.44}$$

The proportionality factor is provided, for example, in our paper (Abolmasov and Shakura 2012a). This estimate is correct far away from the limits of integration ($kT_{min} \ll h\nu \ll kT_{\max}$). From the point of view of spectral energy distribution, this is an intermediate asymptotic scaling $F_\nu \propto \nu^{1/3}$. It is important to mention that formula (5.44) may also be used to estimate BH masses. The mass is involved in combination with the dimensionless accretion rate, $F_\nu \propto \left(M^2\dot{m}\right)^{2/3}$.

5.2.3 *Intensity Distribution*

The data of QSO microlensing make it possible to examine the spatial structure of the disc, regardless of the integrated spectrum, even on angular scales that are currently too small for us to resolve directly. The statistics of anomalous fluxes and the analysis of amplification curves (see Sect. 5.3.1) provide information concerning mainly *the half-light radii* $R_{1/2}$ defined as

$$\frac{\int_{R_{in}}^{R_{1/2}} I(R)RdR}{\int_{R_{in}}^{+\infty} I(R)RdR} = \frac{1}{2}. \tag{5.45}$$

The physical meaning of this value is fairly obvious: half of the observed flux comes from within a circle with radius $R_{1/2}$. This definition is applicable to any source if there is a clear identification of the centre. The monochromatic intensity (5.42) may be written, with a precision up to an inessential common multiplier, as

$$I_\nu \propto \frac{1}{\exp\left(\left(\frac{R}{R_d}\right)^{3/4} f^{-1/4}\right) - 1}, \tag{5.46}$$

where R_d is the radial scale of the disc (see, for example, Morgan et al. 2010):

$$R_d \simeq 10.2 \left(\frac{\lambda_{em}}{0.25\text{mcm}}\right)^{4/3} \left(\frac{M}{10^9 \text{M}_\odot}\right)^{-1/3} \dot{m}^{1/3} \times \frac{GM}{c^2}, \tag{5.47}$$

and $\lambda_{\rm em}$ is the radiation wavelength in the QSO reference frame. If the effect of the outer and inner edges of the disc is considered negligible, then $R_{1/2} \simeq 2.44 R_d$. This relation is fulfilled with good accuracy in the spectral range in which the intermediate asymptotics is valid. For a disc with finite outer size and nonzero inner radius, the relationship $R_{1/2}(R_d)$ is weaker and saturates at long and short wavelengths. Therefore, the half-light radius on long and short wavelengths is usually shorter and longer, respectively, than the radius calculated using the power-law asymptotics (see Fig. 5.6).

Not calling into question the conclusions of Mortonson et al. (2005), let us note that the accretion disc has several spatial scales, the smaller of them having an advantage in microlensing. When crossing the straight caustic, an item the size of ΔR will be amplified as $\mu \propto (\Delta R)^{-1/2}$. The intensity distribution in the inner parts of the disc, where $h\nu \ll kT$, is approximately described by a power

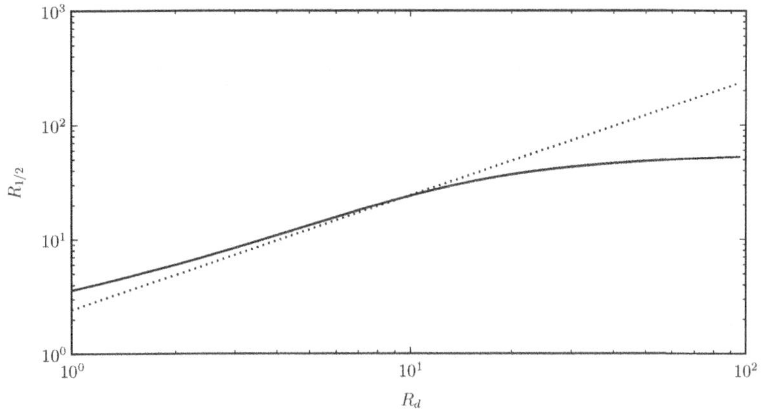

Fig. 5.6 Solid curve: the half-light radius as a function of the characteristic radius R_d of a standard disc that radiates locally as a black body (the outer radius is $R_{out} = 100$). Dotted line: the approximation $R_{1/2} = 2.44 R_d$. The radii are given in arbitrary units

law, $I_\nu \simeq 2\nu^2 kT/c^2 \propto R^{-3/4}$. The integration over solid angle near the caustic demonstrates that the amplification factor variations related to the inner edge of the disc are larger by approximately a factor of $(R_d/R_{in})^{1/4}$ than the amplification variations related to the rest of the disc. However, these become apparent only in close vicinity to the caustic crossing events. Numerical calculations by Jaroszynski et al. (1992) show that the caustic crossings may be sensitive to the structure of the inner edge of the disc.

The radius of the inner edge of the disc is

$$R_{\rm in} = r_{\rm in}(a)\frac{GM}{c^2}, \tag{5.48}$$

where the normalized value $r_{\rm in}$ is determined using formula (5.56) from the next section. The shape of the lightcurve maximum is determined mainly by the parameter X:

$$\begin{aligned} X = \frac{R_{\rm d}}{R_{\rm in}} = \frac{1}{r_{in}(a)}\left(\frac{k}{h}\frac{\lambda}{1+z}\right)^{4/3}\left(\frac{3}{2}\dot{m}\frac{c}{GM\sigma\varkappa_{\rm T}}\right)^{1/3} \simeq \\ \simeq 92\left(\frac{\lambda/1\mu}{1+z}\right)^{4/3}\left(\frac{\dot{m}}{2.5}\right)^{1/3}\left(\frac{M}{10^9{\rm M}_\odot}\right)^{-1/3} r_{\rm in}^{-1}(a). \end{aligned} \tag{5.49}$$

We provide below convenient to use formulas that do not take into account relativistic effects. However, the effects of general relativity are of the same order of magnitude as the effects caused by the presence of the inner edge of the disc, the former becoming more important for a disc tilted to the line of sight. Thus, the formulas given below (see Abolmasov and Shakura 2012b) should be used with caution.

To calculate the amplification curve, it is convenient first to integrate the brightness distribution with respect to the x coordinate:

$$I_1(y) = \int I(x,y)dx \propto \\ \propto \sqrt{\frac{J}{K}} y \int_{t_{in}}^{+\infty} \frac{dt}{\exp\left(\left(\frac{y^2}{r_d^2} J \times (1+t^2)\right)^{3/8} f^{-1/4}\right) - 1}, \tag{5.50}$$

where t is a dimensionless variable of integration.

$$K = K(i, \psi) = \cos^2\psi + \frac{\sin^2\psi}{\cos^2 i}, \tag{5.51}$$

which can be viewed as a factor of disc area decrease due to inclination,

$$J = J(i, \psi) = \sin^2\psi + \frac{\cos^2\psi}{\cos^2 i} - \frac{\sin^2\psi\cos^2\psi}{K(i,\psi)} \tan^4 i, \tag{5.52}$$

$$f = 1 - \left(\frac{1}{J}\frac{r_{in}^2}{y^2}\frac{1}{1+t^2}\right)^{1/4} \tag{5.53}$$

is the temperature correction factor we already know, and

$$t_{in} = \begin{cases} 0 & \text{if } y \geq r_{in}/\sqrt{J} \\ \sqrt{1 - \frac{1}{J}\left(\frac{r_{in}}{y}\right)^2} & \text{if } y < r_{in}/\sqrt{J}. \end{cases} \tag{5.54}$$

Generally speaking, the inclination and the radial scale of the disc are degenerate: the y coordinate is included in combinations with the radial scales only as $y\sqrt{J}/r_d$ and $y\sqrt{J}/r_{in}$. Examples of calculated amplification curves for discs with different positions of the inner edge are shown in Fig. 5.7.

5.2.4 Account for General Relativity Effects and Location of the Inner Edge of the Disc

To find the value $r_{\rm ISCO}$ as a function of the rotation parameter a for a rotating black hole, we should solve the equation (see Bardeen et al. 1972):

$$r^2 - 6r + 8ar^{1/2} - 3a^2 = 0. \tag{5.55}$$

The solution to this equation may be written as follows:

$$r_{\rm ISCO} = 3 + Z_2 - \text{sign}\,(a)\sqrt{(3 - Z_1)(3 + Z_1 + 2Z_2)}, \tag{5.56}$$

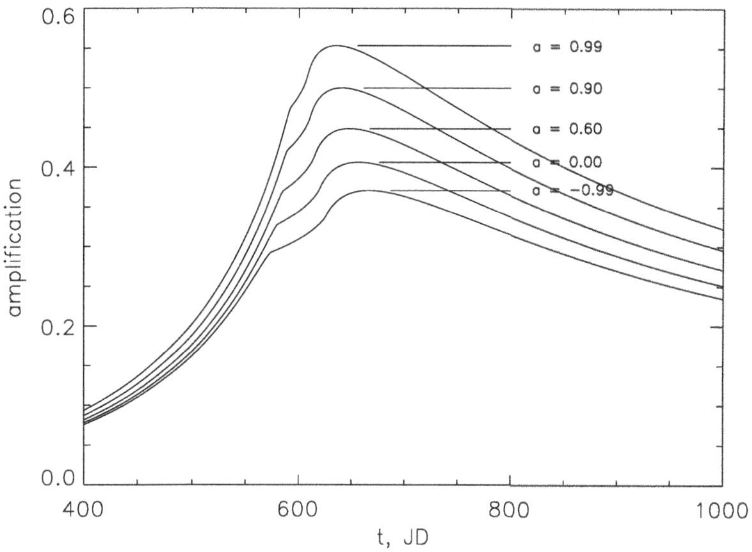

Fig. 5.7 Amplification curves for the simplified disc model (without relativistic effects). The radial scale of the disc is $R_{\rm in} = 13.46GM/c^2$, the Kerr spin parameter ranges from −0.99 to 0.99, where a negative sign means counterrotation. Illustration from Abolmasov and Shakura (2012b)

where

$$Z_1 = 1 + (1-a^2)^{1/3}\left((1+a)^{1/3} + (1-a)^{1/3}\right), \tag{5.57}$$

and

$$Z_2 = \sqrt{3a^2 + Z_1^2}. \tag{5.58}$$

The correction factor f was analytically obtained for the relativistic thin accretion disc in Page and Thorne (1974):

$$\begin{aligned} f = \tfrac{3}{2}\tfrac{1}{x^2(x^3-3x+2a)} \times \\ \times \left(x - x_0 - \tfrac{3}{2} a \ln \tfrac{x}{x_0} - A_1 - A_2 - A_3\right). \end{aligned} \tag{5.59}$$

Here:

$$A_1 = \frac{3(x_1-a)^2}{x_1(x_1-x_2)(x_1-x_3)} \ln\left(\frac{x-x_1}{x-x_1}\right), \tag{5.60}$$

$$A_2 = \frac{3(x_2-a)^2}{x_2(x_2-x_3)(x_2-x_1)} \ln\left(\frac{x-x_2}{x-x_2}\right), \tag{5.61}$$

$$A_3 = \frac{3(x_3 - a)^2}{x_3(x_3 - x_1)(x_3 - x_2)} \ln\left(\frac{x - x_3}{x - x_3}\right), \tag{5.62}$$

$$x = \sqrt{r} \tag{5.63}$$

$$x_0 = \sqrt{r_{ISCO}} \tag{5.64}$$

$$x_1 = 2\cos\left(\frac{1}{3}\mathrm{arccos\ a} - \frac{\pi}{3}\right) \tag{5.65}$$

$$x_2 = 2\cos\left(\frac{1}{3}\mathrm{arccos\ a} + \frac{\pi}{3}\right) \tag{5.66}$$

$$x_3 = -2\cos\left(\frac{1}{3}\mathrm{arccos\ a}\right) \tag{5.67}$$

These $x_{1,2,3}$ values are the three solutions to the cubic equation $x^3 - 3x + 2a = 0$ (Page and Thorne 1974) expressed in trigonometric form. An analogous notation will be used in Sect. 5.3.3.2 for the position of the spherization radius.

It should be noted here that the domain of applicability of the thin disc model is limited by several potentially important effects:

- disc inclination with respect to the BH rotation axis; it was shown that the Bardeen-Petterson effect (Bardeen and Petterson 1975), which tends to align the accretion disc with the equatorial plane of the BH, operates only for high viscosity and small thickness of the disc (see Ivanov and Illarionov 1997; Zhuravlev and Ivanov 2011). In the general case, however, the disc may remain inclined while approaching the innermost stable circular orbit; or it may pass to an essentially nonlinear mode of alignment, with formation of shockwaves having velocities of the order of the Keplerian velocity $\sim v_K \sin i$
- deviations from the approximation of the geometrically thin, optically thick disc; different models predict formation of a corona or advection-dominated flows for low accretion rates (Narayan and Yi 1995; Meyer et al. 2000) and thick advective discs for high accretion rates (Sadowski 2011); formation of essentially supercritical flows with winds is also possible (Shakura and Sunyaev 1973; Poutanen et al. 2007)
- deviations from the model of local blackbody radiation; the use of stellar atmospheres (Kolykhalov and Sunyaev 1984) cannot explain many of the observational properties of QSOs such as the almost total absence of the Lyman jump; electron scattering and some other effects that may influence the spatial properties of QSO discs will be considered in Sect. 5.3.1.1.

However, even a thin disc in perfect alignment with the BH does not prevent certain effects to dramatically alter its apparent brightness distribution. These are bending of photon trajectories and relativistic aberration. The photon trajectories were analytically calculated using elliptic integrals in Dexter and Agol (2009) as a part of the `geokerr` software, which is available at http://www.astro.washington.edu/users/agol/geokerr. We calculated the forms of geodesics using `geokerr`, after which the Doppler factor (the ratio of the frequency in the observer reference frame to the frequency in the reference frame comoving with the disc) was determined from the law of conservation of energy and angular momentum of the photon along the trajectory as follows (Abolmasov and Shakura 2012b):

$$\delta = \frac{k_t}{k_i u^i} = \frac{1}{u^t\,(1+\Omega l)}. \tag{5.68}$$

Here, Ω is the orbital frequency in the disc, $l = a\sin i$ is the specific angular momentum that is determined through the impact parameter of a given point on the celestial sphere (see Fig. 5.5). The time component of the four-velocity is determined from the normalization condition ($g_{ik}u^i u^k = -1$) as follows:

$$\begin{aligned} u^t &= \left(-(g_{tt} + 2g_{t\varphi}\Omega + g_{\varphi\varphi}\Omega^2)\right)^{-1/2} = \\ &= \left(\rho^2\frac{\Delta}{\Sigma^2} - \frac{\Sigma^2}{\rho^2}\,(\Omega - \omega_{LT})^2\right)^{-1/2}. \end{aligned} \tag{5.69}$$

The Keplerian frequency as measured by a distant observer is

$$\Omega = \frac{1}{r^{3/2} + a}. \tag{5.70}$$

The invariance of the value I_ν/ν^3 results in the fact that the relativistic Doppler effect appears only in the form of radiation frequency shift:

$$\begin{aligned} I_\nu &\simeq \Delta\nu_{obs}\delta^3 I^0_\nu(\nu_{em}) \propto \\ &\propto \left(\exp\left(\tfrac{1}{\delta}\,(r/r_d)^{3/4}\, f(r,a)^{1/4}\right) - 1\right)^{-1}. \end{aligned} \tag{5.71}$$

At large distances ($r \gg r_{\rm in}$ and $r \gg r_{\rm d}$), the Doppler effect remains essential, producing a noticeable intensity asymmetry, due to the strong (exponential) behaviour of the Planck curve for $h\nu \gg kT$. If we expand the expression for monochromatic intensity in powers of δ^{-1} for small r^{-1}, we obtain $\delta^{-1} \simeq 1 + \frac{\sin i}{\sqrt{r}}$, which implies that the intensity contrast between the approaching and the receding sides of the disc increases with distance as follows

$$\frac{I^+ - I^-}{I^+ + I^-} \simeq (r/r_d)^{3/4}\,\frac{\sin(i)}{\sqrt{r}} \propto r^{1/4}. \tag{5.72}$$

5.2.5 *Applications to Actual Lightcurves: QSO J2237+0305 and SBS J1520+530*

We have chosen the lightcurves of two objects, QSO J2237+0305, also known as 'the Einstein cross', and SBS J1520+530 (Abolmasov and Shakura 2012b). The main properties of these two objects are given in Table 5.1. Detailed studies of QSO J2237+0305 were carried out under the *OGLE* project (Woźniak et al. 2000) targeted to search for classical microlensing events in our galaxy. Lightcurves were obtained for all the four images observed by *OGLE-II* and *OGLE-III*; all the data are available at http://ogle.astrouw.edu.pl. Figure 5.8 shows the lightcurves of the QSO J2237+0305 images based on *OGLE* photometric data. So far, the search for correlated variability of individual images has not been successful, although there are upper limits for the delays between individual images (Vakulik et al. 2006). Certain parts of the lightcurves visually resemble caustic crossings in shape; for this reason, the straight caustic model is sometimes used to describe the amplification

Table 5.1 The observational data for QSO J2237+0305 (A and C images) and SBS J1520+530 used in Abolmasov and Shakura (2012b)

	SBS J1520+530	QSO J2237+0305 (A)	QSO J2237+0305 (C)
z_S	1.855	1.695	
z_L	0.72	0.039	
Range of dates (V), JD-2450000		1400–1650	1200–1650
Number of observations (V)	–	53/52	83
Range of dates (R), JD-2450000	1200–3000	1450–1510	–
Number of observations (R)	253	51/49	–

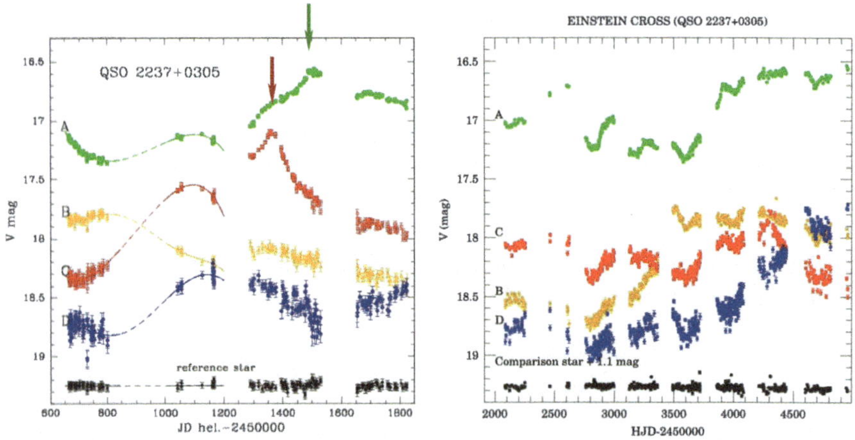

Fig. 5.8 Magnitudes of the QSO J2237+0305 images according to *OGLE-II,III* data. The arrows indicate two events that seem to correspond to caustic crossings. Illustrations were taken from Woźniak et al. (2000) and Udalski et al. (2006)

curves (Bogdanov and Cherepashchuk 2004; Shalyapin et al. 2002; Gil-Merino et al. 2006).

For the other examined object, SBS J1520+530 with only two detected images, the optical depth due to lensing is assumed to be less than that for the "Einstein cross", the lens being situated much further away. This object is known to be variable; the time delay between the two images is measured with a good accuracy as $\Delta t \simeq 130$ days. Knowing the lag between the variability of the two components, we may separate the variability due to microlensing by shifting one time series by Δt relative to the other. After that, we may interpolate the shifted time series for A onto a time grid of the series for B and vice versa, thus obtaining a series of B/A flux ratios (keeping both interpolated curves to prevent information loss, although slightly violating statistics as the data set becomes redundant; see Fig. 5.9).

The fitting was performed using the dynamical method in the space of the following parameters: the caustic crossing moment t_0, the caustic parameters $\mu_{0,1}$, the radial scale of the disc r_d, the shape parameter X (see Sect. 5.2.3), and the effective transverse velocity v_{eff}. Since the use of the relativistic model implied the

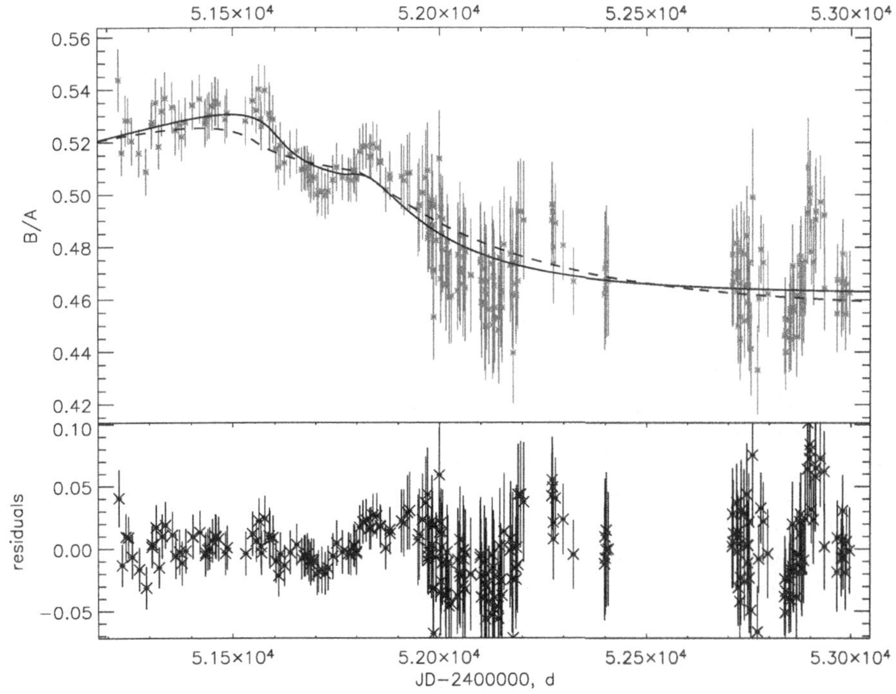

Fig. 5.9 Top panel: The observed flux ratios of the QSO SBS J1520+530 B and A images and the model fitting: the simplified model without relativistic effects (dashed curve) and the Kerr disc model (solid curve). The optimal parameters of the Kerr model: $a = -0.5$, $i = 80°$, $\psi = 330°$. Lower panel: the residuals with respect to the Kerr model. Illustration from Abolmasov and Shakura (2012b)

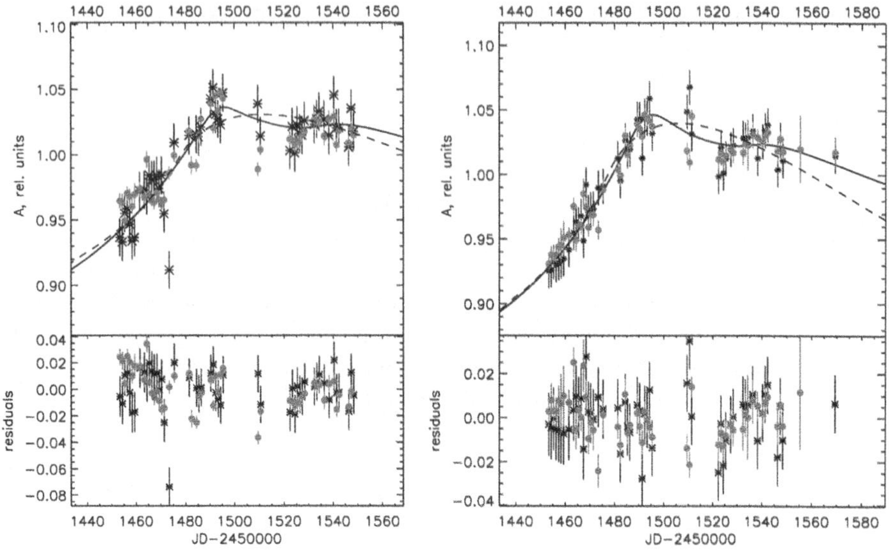

Fig. 5.10 One of the amplification events detected for QSO J2237+0305 (image A). Left panel: R-band data; right panel: V-band data. The dashed curve represents the simplified model, the solid curve takes into account relativistic effects. The optimal model parameters are: $a = 0.2$, $i = 70°$, $\psi = 96°$. Asterisks and grey circles correspond to different photometric methods. Illustration from Abolmasov and Shakura (2012b)

calculation of intensity maps for a set of values of a, r_d, and inclination i, a grid of intensity maps was produced first and then a search for an optimal solution for every map was performed. For a more detailed description of the results see Abolmasov and Shakura (2012b), where Tables 2 and 3 provide the optimal parameter values. The example of a fitted lightcurve near one of the 'events' is given in Fig. 5.10.

As a whole, the obtained solutions demonstrate that some interesting things may be happening near the inner edge of the disc: either the inner parts are strongly inclined with respect to the observer or they are simply brighter and more distinct against the background of the disc, in a greater extent than we would expect. Either alternative may be connected with the misalignment between the BH and accretion disc planes. Almost certainly, there is an appreciable angle between these two planes. Alignment processes are still poorly understood and may involve formation of shockwaves within inner parts of accretion discs or weakly interacting precessing rings, which may produce effects similar to the observed. At the same time, a satisfactory result is obtained for SBS J1520+530 when using a model involving two caustics or two accretion discs (see below the final paragraph of Sect. 5.3.1).

5.3 The Problem of Large Radii and Observational Evidence for Supercritical Accretion

5.3.1 Observed Estimates of Disc Sizes

As early as 1979, it was proposed to use the microlensing effects produced by the stars of the lensing galaxy to study the spatial structure of QSO discs (Chang and Refsdal 1979). The effect itself was discovered 10 years later (Irwin et al. 1989) but proved to be difficult to use: consecutive fitting of the microlensing lightcurves beyond the approximation of a single straight caustic (see above) requires huge computer power and nontrivial techniques for comparison between observational data and simulated results. Today, the standard technique for modelling microlensing processes is the construction of two-dimensional *amplification maps* by means of inverse ray-tracing using random distribution of lensing stars over the celestial sphere (Wambsganss 2006). The masses and positions of the stars are chosen randomly. A light ray trajectory is constructed for every point of some field of view in the image space (as if the ray was emitted from the eye of the observer. Hence the name 'inverse ray-tracing'). The Jacobian of the transformation of the plane of the sky, $\det A$, is then determined (see Sect. 5.1.4), allowing us to obtain the amplification factor μ. The amplification map for an extended source is obtained from the amplification map for a point-like source using two-dimensional convolution with source intensity distribution. Subsequently, individual model lightcurves can be obtained by varying the initial coordinates of the source and the components of relative proper motion.

The distribution of amplification μ is sensitive to the size of the source (which determines mainly the maximal possible value of μ) and to the microlensing optical depth (Kofman et al. 1997). Some other parameters of stellar population, such as the mass distribution of stars, are also of importance. The mean stellar mass itself does not in general affect the distribution in μ for a point-like source, but affects the characteristic angular scale of the caustic network.

The opportunity to confidently measure QSO sizes did not appear until the beginning of the new millennium. *Anomalous fluxes*, or more precisely, anomalies of fluxes (the ratios of observed amplifications to those obtained in the strong lens model) were considered in the optical and X-ray range for a sample of ten objects in Pooley et al. (2007). The statistics of amplifications due to microlensing is directly linked to the object size (in terms of the mean Einstein-Chwolson radius). The goal of Pooley et al. (2007) was to solve the inverse problem of reconstructing the possible mean disc sizes to reproduce the observed distribution over flux anomalies.

The results obtained by Pooley et al. (2007) made it possible to derive several nontrivial conclusions about the structure of QSO discs: (1) the size of an average QSO in the near UV range is considerably larger than predicted by the standard multicolour thin disc theory (see Sect. 5.2.2 above); (2) the sizes of X-ray-emitting regions are much smaller and comparable to the radius of the innermost stable circular orbit.

The former conclusion received further qualitative support in Morgan et al. (2010), although the magnitude of the effect turned out to be somewhat lower: the QSO disc sizes appeared on average approximately a factor of three larger than predicted by theory. It is worthwhile to indicate more precisely which values were compared with each other. Along with disc sizes estimated through microlensing effects, I-band photometric fluxes measured by the Hubble Space Telescope were used. The monochromatic flux at the given frequency $\nu_{\rm obs}$ was calculated using formula (5.44) for the standard disc.

For the broadband spectrum of an accretion disc, we may recalculate the flux F_ν to I-band magnitude with good accuracy. From the above expression (5.44) for monochromatic flux, we may derive the value $M^2\dot{m}$, on which the characteristic radius depends according to formula (5.47). In this manner, a working formula for determining the disc size was obtained (Morgan et al. 2010)

$$R_I \simeq 2.83 \times 10^{15} \frac{1}{\sqrt{\cos i}} \frac{D}{D_H} \times \left(\frac{\lambda_I}{1\rm mcm}\right)^{3/2} \times 10^{-0.2(I-19)}\,\rm cm. \qquad (5.73)$$

We will use these photometric data and disc sizes in Sect. 5.3.3.2 below. Morgan et al. (2010) interpret the observed inconsistency as a result of strong deviations from the standard model, potentially related to a very low actual efficiency of accretion: the observed intensity of the disc is much smaller than expected for its estimated angular size.

As quasar microlensing effects depend on wavelength (since the monochromatic disc sizes depend on wavelength), it is often referred to as *chromatic lensing* in contrast to strong lensing and microlensing produced by an isolated lens, where chromatic effects are not important. Flux anomalies from lensed QSOs depend appreciably on wavelength, although not always in the manner predicted by the standard model (Floyd et al. 2009).

Formula (5.47) implies that the approximate relationship $R \propto \lambda^{4/3}$ should be valid for the standard model in a wide range of wavelengths (mainly, from optics to UV). As a matter of fact, the maximal slope, which is 4/3, should be observed on a limited interval between the inner and outer edges of the disc, decreasing at both longer and shorter wavelengths (see Fig. 5.6). Many alternative models of disc accretion also predict power-law dependencies $R(\lambda)$, although with different exponents; for this reason, it is convenient to introduce *the structural parameter* $\zeta = d\ln R/d\ln\lambda$. For a power-law function $R(\lambda)$, the structural parameter does not depend on wavelength and $R \propto \lambda^\zeta$.

Blackburne et al. (2011) present an analysis of multi-wavelength data for 12 objects. The dependencies $R(\lambda)$ were compared with theoretical predictions (flux-based radii were calculated using formula (5.73) for the standard disc). This work confirms in general the abnormally large 'disc' sizes for most of the sources. However, a stunning diversity of $R(\lambda)$ curves becomes apparent. As a rule, the size of the disc usually increases with wavelength, although this does not seem to be a general property (for example, the disc size of WFI J2033–4723 decreases slightly with wavelength). For more than a half of all objects, $\zeta \simeq 0$; some, however, are

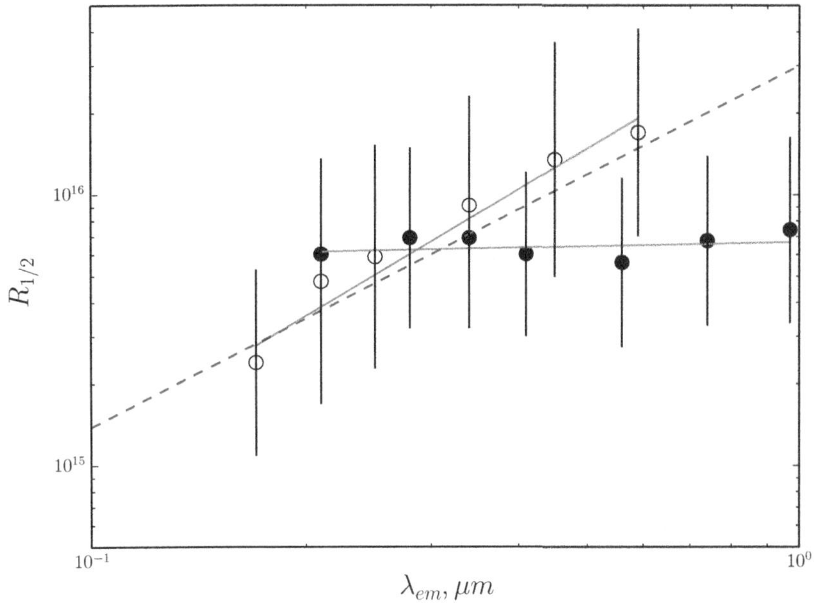

Fig. 5.11 The half-light radius as a function of wavelength for MGJ 0414+0534 (open circles) and HE J1113–1641 (solid circles). The data were taken from Blackburne et al. (2011). The grey solid lines show power-law fits to the data; for comparison, the dashed red line shows a power law with slope 4/3

well fitted by the standard disc model (this concerns both the dependence slope and the normalization). Among the latter, the most prominent is MGJ 0414+0534, the only radioloud object in the sample, with the greatest BH mass of approximately $2 \times 10^9 M_\odot$ (Bate et al. 2008). Figure 5.11 shows how the size changes with observed wavelength for two QSOs with very different ζ. One of them is MGJ 0414+0534 mentioned above, for which power-law fitting yields $\zeta = 1.55 \pm 0.4$. The dependence $R_{1/2}(\lambda)$ for the second object, HE J1113–1641, is almost flat with $\zeta = 0.05 \pm 0.20$.

The chromatic lensing data for many objects may be described in terms of the approximate values of ζ estimated by fitting the observed $R(\lambda)$ dependence. There is a clear correlation between ζ and BH mass (see Fig. 5.12): the dependencies $R(\lambda)$ are almost flat for low masses, $M \lesssim 3 \times 10^8 M_\odot$ (on average, however, $\zeta \gtrsim 0$); for massive objects with $M \gtrsim 10^9 M_\odot$, at least in some cases $\zeta \gtrsim 1$. The latter class of objects includes MG J0414+0534 and the Einstein cross.

Along with the accretion disc model with scattering in an extended envelope which will be described below in more detail, different models were proposed to explain the abnormally large and 'grey' QSO sizes. Most of these models suggest that the disc emits locally as a blackbody. It is easiest to assume that the slope $\zeta < 4/3$ arises simply as a consequence of another (steeper) dependence of the temperature on the radius in the disc. This may be, for example, an accretion disc

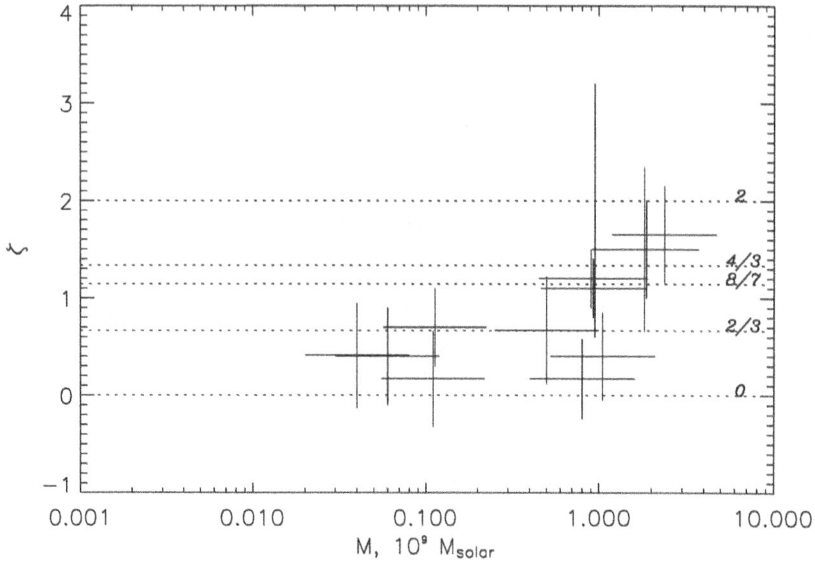

Fig. 5.12 The structural parameter ζ as a function of BH mass. This parameter is calculated using the QSO sizes published in Blackburne et al. (2011). The illustration was taken from Abolmasov and Shakura (2012a)

with a nontrivial boundary condition at the inner edge (Agol and Krolik 2000), for which $\zeta \simeq 8/7$. Such a model was suggested in Floyd et al. (2009) to explain the multiwavelength properties of SDSS 0924+0219. It seems that $\zeta < 1$ for many objects, implying that there should be some other solution. It is worthwhile to note also the work by Yan et al. (2014), in which the objects with small ζ are supposedly considered to be related to binary black holes. If it is a fairly close binary, two accretion discs having similar sizes and luminosities might be observed. With that, the size of the emitting area measured through microlensing will be close to the size of the binary and will no longer depend on wavelength. If the separation between the BHs in this binary is of the order of a few tenths of a parsec (which is comparable with the sizes of accretion discs themselves), many properties in brightness distribution of lensed QSOs may be qualitatively explained quite well. Particularly, the unusual shape of the SBS J1520+530 amplification curve, which is difficult to explain in the model with one caustic and one accretion disc, is easily explained in terms of a binary.

5.3.1.1 Scattering in a Disc Atmosphere

Atmospheres of accretion discs are fairly hot and rarefied. For this reason, their spectra should clearly show the effects of electron scattering. These effects are especially interesting in those cases when an appreciable part of energy dissipating in the disc is eventually transferred into heat above the effective photosphere. If the local viscous stresses scale with gas pressure (see Blaes et al. (2011); this may be considered as a generalization of the α law in the Shakura-Sunyaev disc model; see Sect. 5.2.2 above), a translucent atmosphere becomes approximately isothermal. To prove this, let us consider the transfer equation for the radiation energy density u

$$\frac{d}{dz}\left(D\frac{d}{dz}u\right) = \alpha\Omega p. \tag{5.74}$$

Here, $D = \frac{c}{3\varkappa\rho}$ is the radiation diffusion coefficient, Ω is the angular frequency of rotation (dependent on radius but constant along the vertical coordinate), p is the gas pressure, $\varkappa$ is the Rosseland mean opacity, and the optical depth is $d\tau = -\varkappa\rho dz$. Assume that $\varkappa = const$ (which is the case, in particular, if the electron scattering is dominant). Dividing both parts of the equation by $\varkappa\rho$, we obtain:

$$\frac{d^2}{d\tau^2}u = \frac{\alpha}{\varkappa}\frac{p}{\rho}\Omega. \tag{5.75}$$

For gas pressure, $p/\rho \propto T$. At the same time, $u = aT^4$, where a is a constant related to the Stefan-Boltzmann constant. Therefore, we can write

$$\frac{d^2u}{d\tau^2} = Ku^{1/4}, \tag{5.76}$$

where K is a constant. It is easy to see that the solution of this equation is $u \propto \tau^{2/3}$, or $T \propto \tau^{1/6}$, which is a very weak dependence. Numerical computations of accretion disc atmospheres (see, for example, Hubeny and Hubeny 1998) also predict that the temperature reaches a plateau in the outer layers of the atmosphere, for $\tau \lesssim 5$.

Scattering, even coherent, leads to deviations in the radiation field from the thermal law. An example demonstrating the emergence of these deviations is given in Mihalas (1978, Chapter 6.1). Coherent scattering may be treated as an additional source of absorption plus an additional contribution to the source function proportional to the mean intensity. Such a description corresponds to the case of isotropic coherent scattering, reproducing some essential features of actual hot atmospheres. For example, the radiation that escapes from a semi-infinite isothermal atmosphere would have lower intensity for the same colour temperature. With that, the monochromatic intensity would change by some factor depending on the contributions of different opacity sources (see Mihalas 1978, Section 6.1):

$$I_\nu(\tau = 0) \simeq \lambda_\nu^{1/2} \times B_\nu, \tag{5.77}$$

where $\lambda_\nu = \alpha_{abs}/(\alpha_{abs} + \alpha_{sc})$ is the relative contribution of true absorption to opacity, $\alpha_{sc} = \rho\varkappa_{sc}$ is the linear coefficient of absorption due to scattering, and α_{abs} is the absorption coefficient due to true absorption. The observed intensity I_ν is lower than the blackbody intensity B_ν of the same temperature by the *the dilution factor* d_ν defined as $I_\nu = d_\nu B_\nu$. Further on, we will consider dilution factor averaged over frequencies ($I = d \cdot B$). Since the integrated flux that escapes from the disc should remain unchanged, the dilution factor $d < 1$ implies that the actual temperature, which determines the shape of the spectrum and may be identified with the colour temperature for the dilution factor independent of frequency, is higher than the effective temperature by a factor of $d^{-1/4}$.

We may follow qualitatively the effects arising due to electron scattering, including coherent scattering, using the following simplified model. Assume that the disc radiates locally as a blackbody but its radiation is diluted; this means that the effective temperature is not coincident with the colour temperature, and

$$\sigma T_{eff}^4 = d(r)\sigma T_{colour}^4, \tag{5.78}$$

where $d(r)$ is the dilution factor. The effective temperature depends only on the energy released in the disc, changing with radius approximately as $T_{eff} \propto r^{-3/4}$. If we assume that the dilution factor changes with radius also according to a power law, $d = d_0\,(r/r_0)^\gamma$, then both the slope of the accretion disc spectrum (defined as $F_\nu \propto \nu^\beta$) and the disc brightness distribution law will change. We obtain $T_{colour} \propto r^{-(3+\gamma)/4}$, and for the structural parameter introduced in Sect. 5.3.1

$$\zeta = \frac{4}{3+\gamma}. \tag{5.79}$$

Integration of the diluted spectrum over frequency allows us to estimate the slope of the integrated spectrum far away from the maximal and minimal frequencies:

$$\beta = \frac{1-\gamma}{3+\gamma}. \tag{5.80}$$

For any accretion disc that radiates locally as a black body with a power-law temperature dependence on radius, the relationship $\beta = 3-2\zeta$ is valid. Meanwhile, for any disc with $T_{eff} \propto r^{-3/4}$ and a power-law dependence of the dilution factor on radius, the above formulas yield $\beta = \zeta - 1$. For another interesting case of a 'passive disc', in which most of the angular momentum is carried from the inner edge, whereas the angular momentum transfer by accreting matter may be considered negligible, $T_{eff} \propto r^{-7/8}$ and $\beta = \frac{3}{2}\zeta - 1$ (Fig. 5.13).

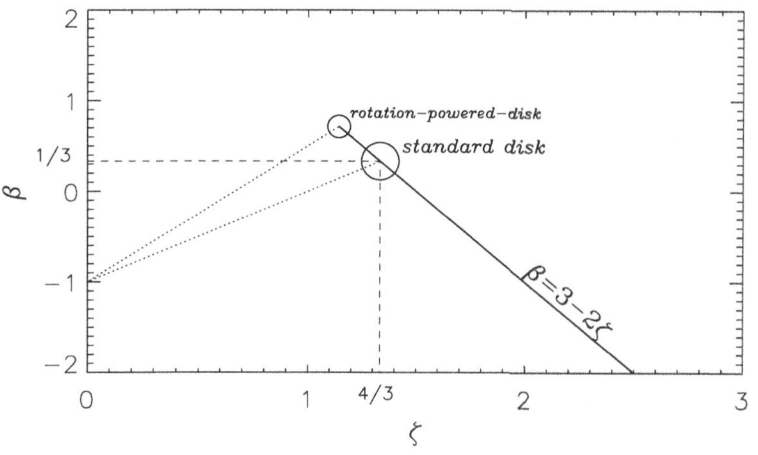

Fig. 5.13 Structural parameter ζ and spectral slope β for the local blackbody models (solid line) and for the models with diluted blackbody radiation (dotted lines) with a fixed effective temperature distribution in the disc $T_{\text{eff}} \propto r^{-3/4}$

5.3.2 Nonlocal Scattering

We will consider the envelope as extended and the scattering as nonlocal in those cases when a photon is emitted and undergoes the latest scattering at essentially different distances from the centre of the disc (the BH). This is possible, for example, if the disc gives rise to an optically thick wind.

An extended scattering atmosphere around an accretion disc does not affect significantly its spectrum as long as the atmosphere remains optically thin with regard to true absorption and comptonization effects are not too strong. Due to geometrical reasons, only a small part of all the photons emitted by the disc and scattered in an extended atmosphere returns back. For this reason, the above effects leading to an increase in the colour temperature are not important. On the other hand, multiple scattering in an extended envelope totally changes the spatial properties of the object, whose apparent size is now determined by the size of the photosphere rather than the disc. The object will have similar intensity distributions at all the wavelengths if the opacity does not depend on wavelength. The radiation will always be appreciably diluted, with the apparent size of the object larger than the 'blackbody' radius of the disc.

Thus, the use of a scattering envelope allows us to solve two problems, namely, to explain large apparent QSO radii and weak dependencies of these radii on wavelength. Three essentially different models of such an envelope can be proposed: (1) a static corona consisting of gas with virial temperature, (2) gravitationally unbound outflow from the accretion disc, and (3) a gravitationally bound inflow feeding the accretion disc, which we will consider later in Sect. 5.3.4. Since we deal with bright objects characterized by high radiation density near the disc, there is a problem of Compton cooling in the first model (Pietrini and Krolik 1995). Moreover, there are serious reasons to suggest that intense accretion is accompanied by winds and ejections, the optical depth of which due to scattering may be greater than unity. For simplicity, let us consider a model of a spherically symmetric stationary wind (see also Abolmasov and Shakura 2012a).

For constant wind velocity v_w and a total mass loss rate through the wind $\dot{M}_\mathrm{w}$, the wind density changes as follows:

$$\rho = \frac{\dot{M}_\mathrm{w}}{4\pi R^2 v_\mathrm{w}}. \tag{5.81}$$

The radial optical depth is obtained as an integral of opacity times density $\varkappa\rho$ over the line of sight from the observer to a given distance R:

$$\tau(R) = \int_R^{+\infty} \varkappa\rho(R) dR = \frac{\varkappa \dot{M}_w}{4\pi v_w R}. \tag{5.82}$$

The size of the photosphere can be found as the radius where $\tau = 1$

$$R_1 = R(\tau = 1) = \frac{f_\mathrm{w}\varkappa \dot{M}}{4\pi v_\mathrm{w}}, \tag{5.83}$$

where $f_\mathrm{w} \lesssim 1$ is the fraction of accreting matter ejected from the disc. More detailed calculations indicate that the brightness distribution in the atmosphere with this density profile has an optically thick core and power-law optically thin wings, the resulting half-light radius being approximately $R_{1/2} \simeq 1.06 R_1$.

5.3.2.1 Radiation Transfer in an Extended Atmosphere

An optically thick wind forms an envelope with density decreasing according to the power law $\rho \propto R^{-2}$. In the general case, the radiation transfer in the atmosphere of this disc is described by the following equation (see Mihalas (1978, Chapter 7.6), the case of a 'grey extended atmosphere'):

$$\chi \frac{\partial}{\partial R}(I) + \frac{1-\chi^2}{R}\frac{\partial}{\partial \chi}(I) = -\varkappa\rho(R)(S - I), \tag{5.84}$$

Here, $I = I(R, \chi)$ is the monochromatic intensity, $\chi = \cos\theta$ is the cosine of the angle θ between the radiation propagation direction and the radius vector, and $S = S(R)$ is the source function. Since we consider isotropic scattering, the source function can be equated to the intensity J averaged over angles. As a simple approximation to describe the intensity distribution over angles, the method of moments is often used, in which the dependence of intensity on the angle θ is parametrized by three (or more) moments:

$$J = \frac{1}{2}\int_{-1}^{+1} I d\chi \tag{5.85}$$

is the zeroth moment having the physical meaning of intensity averaged over all directions. Note that $d\chi = d\cos\theta = \sin\theta d\theta$.

$$H = \frac{1}{2}\int_{-1}^{+1} I\chi d\chi \tag{5.86}$$

is the first moment having a physical meaning related to the radiative flux of energy (up to a factor of 4π).

$$K = \frac{1}{2}\int_{-1}^{+1} I\chi^2 d\chi \tag{5.87}$$

is the second moment, related to radiation pressure. The moments of orders higher than two are considered equal to zero. Moreover, it is convenient to make use of the so called Eddington approximation, in which $K = fJ$. This proportionality is present, for example, in an isotropic radiation field where $f = 1/3$, and also in radiation field formed by a point-like source, where $f = 1$. By fixing f we can, first, integrate Eq. (5.84) over χ. The second of the two independent equations of radiative transfer is obtained by multiplying (5.84) by χ and then integrating over χ. This yields a closed system of equations in the Eddington approximation:

$$\begin{cases} \frac{1}{R^2}\frac{d}{dR}\left(R^2 H\right) = 0 \\ \frac{d}{dR}(fJ) + \frac{3f-1}{R} J = -\varkappa\rho H \end{cases} \tag{5.88}$$

It can easily be seen that the system of equations can be essentially simplified both for $f = 1/3$ and for $f = 1$. The first approximation works well at large optical depths (deep in the envelope), the second applies at small optical depths (where radiation propagates essentially radially). In the first case, the mean intensity J changes as $J \propto C_1 R^{-3}$. In the second case, $J \propto C_2 R^{-2} \times (1 + \tau)$, where $C_{1,2}$ are some constants. The assumption $C_1 = C_2 = H(\tau = 1)$ might satisfy both asymptotics. However, the total flux cannot be kept constant in this case (the first equation (5.88) is not fulfilled any more); for this reason, we suggest the following modified law for the source function:

$$S(r) = H_0 r^{-2} \times \left(1 + d \times r^{-1/2} + r^{-1}\right), \tag{5.89}$$

where $r = R/R_1 = 1/\tau$ and d is a free parameter. If we know how the source function depends on the (radial) optical depth τ, we can calculate the intensity as a function of the impact parameter P:

$$I = \varkappa_T \int_{-\infty}^{+\infty} S\left(\sqrt{P^2+l^2}\right) e^{-\tau_l(P,l)} \rho \left(\sqrt{P^2+l^2}\right) dl, \tag{5.90}$$

where $\varkappa_T$ is the opacity due to scattering and τ_l is the optical depth along the line of sight,

$$\begin{aligned} \tau_l &= \varkappa_T \int_{-\infty}^{l} \rho \left(\sqrt{P^2+l^2}\right) dl = \\ &= \tfrac{R_1}{P}\left(\operatorname{arctg}\tfrac{l}{P} + \pi/2\right). \end{aligned} \tag{5.91}$$

Substituting the expression for the source function, we can express the observed intensity distribution as

$$I(p) = H_0 \times \left(u_2(p) + u_3(p) + d \times u_{5/2}(p)\right), \tag{5.92}$$

where $x = l/R_1$ and $p = P/R_1$ and

$$u_k(p) = p^{-(k+1)} \int_0^{\pi} e^{-\theta/p} \sin^k \theta d\theta. \tag{5.93}$$

The parameter d is chosen so that the solution should reproduce the total luminosity $2\pi \int_0^{+\infty} I(p) p dp = 4\pi H_0$. Numerical estimates yield $d \simeq -0.097$ (Fig. 5.14).

Finally, the half-light radius $R_{1/2}$ can be calculated for the obtained intensity distribution using formula (5.45). Numerical calculations result in $R_{1/2} \simeq 1.063 R_1$, where R_1 is the radius of the photosphere at which the radial optical depth is $\tau = 1$.

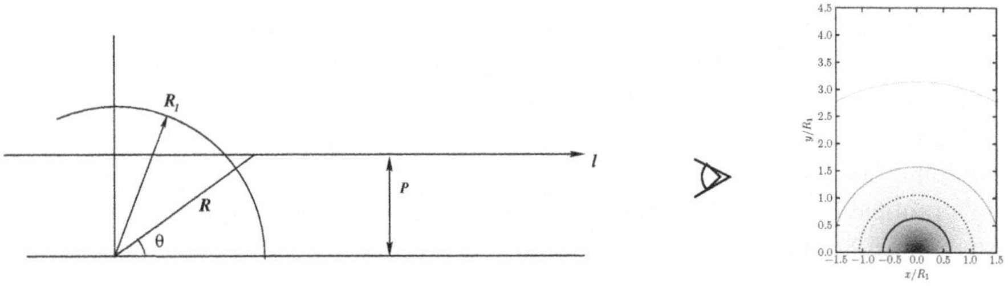

Fig. 5.14 Left: diagram illustrating the calculation of the intensity distribution in Sect. 5.3.2.1. Here, P is the impact parameter, l is the variable of integration. Right: the intensity distribution for the model in use. The levels of equal optical depth τ_l (0.1, 1, 2 and 5) along the line of sight are shown by solid lines, and the half-light radius $R_{1/2}$ by a dotted line

5.3.3 The Supercritical Accretion Regime

As repeatedly mentioned above, the standard accretion disc model has its applicability limits, the first of which, the Eddington limit, was pointed out as early as 1973 by Shakura and Sunyaev (1973).

The Eddington limit (Eddington 1925) arises due to radiation pressure. An isotropic source coincident with the gravitating centre produces a dynamical effect on the surrounding matter through the pressure of its radiation. For an opacity $\varkappa$, this radiation pressure force acting upon a unit mass is

$$\varkappa F = \frac{\varkappa L}{4\pi R^2}. \quad (5.94)$$

If this force is balanced by the gravitational force (which is also inversely proportional to the distance squared, thus eliminating R in the classical approximation), matter becomes gravitationally unbound, which leads to wind launching and may halt the accretion process itself. In the spherically symmetric case, this luminosity limit is

$$L_{\mathrm{Edd}} = \frac{4\pi GMc}{\varkappa}. \quad (5.95)$$

The luminosity released during accretion is $L = \eta \dot{M} c^2$, where η is the *accretion efficiency*, which is independent of $\dot{M}$ and M for the standard disc but is sensitive to the position of the inner edge of the disc (and, respectively, to the rotation parameter a). The accretion is considered supercritical if the luminosity $L > L_{\mathrm{Edd}}$, although this is a fairly uncertain boundary, depending on both the geometry of the accretion flow (the Eddington limit is obtained in the assumption of spherical symmetry) and its dynamics. In the case of an accretion disc, it is more correct to locally compare the vertical component of gravity with the radiation pressure. A version of such a local approach to the Eddington limit was used in our paper: Abolmasov and Shakura (2012a). In such a local approach, it is convenient to normalize the accretion rate to an 'Eddington' value that does not contain the global accretion efficiency η:

$$\dot{M}^* = \frac{L_{\mathrm{Edd}}}{c^2} = \frac{4\pi GM}{c\varkappa}, \quad (5.96)$$

Radiation pressure in the inner parts of the accretion disc dominates over gas pressure. This is the so called 'zone A' of the standard disc, in which the transition to supercritical accretion occurs for all reasonable values of BH masses. Here, the disc thickness H is determined from the balance between the radiation pressure force and the vertical component of the BH gravity

$$\frac{\varkappa}{c} Q = \frac{\varkappa}{c} \frac{3}{8\pi} \frac{GM\dot{M}}{R^3} f = \frac{GM}{R^3} H, \quad (5.97)$$

which yields in the nonrelativistic approximation

$$H = \frac{3}{8\pi}\frac{\varkappa \dot{M}}{c}\left(1 - \sqrt{\frac{R_{in}}{R}}\right) = \frac{3}{2}\frac{GM}{c^2}\dot{m}\left(1 - \sqrt{\frac{R_{in}}{R}}\right). \qquad (5.98)$$

It can easily be seen that an increase in accretion rate gives rise to the occurrence of a thick disc. When the total luminosity exceeds the Eddington limit, part of the released radiation energy is converted to kinetic energy of the flow, leading to formation of massive noncollimated outflows. If we assume that the transition to supercritical accretion occurs under the condition $H(R) > R$, we may identify the maximal radius at which the disc still remains locally supercritical. Traditionally, this radius $R_{\rm sph}$ is named the spherization radius (Shakura and Sunyaev 1973) as the thicker the disc the closer it approaches spherical symmetry. The simplest nonrelativistic case, which takes into account the presence of the inner edge of the disc, allows us to estimate $r_{\rm sph}$ as the largest solution to the equation

$$\frac{r}{1 - 1/\sqrt{r}} = \frac{3}{2}\frac{\dot{m}}{r_{in}}. \qquad (5.99)$$

This equation has a positive solution starting from $\dot{m} = 4.5 r_{\rm in}$, which may be considered as the critical value of the accretion rate. Solving the above equation allows us to express the spherization radius as

$$R_{\rm sph} = \frac{3}{2}\frac{GM}{c^2}\dot{m} \times \psi^2(\dot{m}/x_{\rm in}). \qquad (5.100)$$

Here, the correction factor ψ, calculated in the assumption of zero stress, accounts for the presence of the inner edge of the disc. When Eq. (5.99) has three real roots, the relevant is the largest of them

$$\psi(x) = \frac{2}{\sqrt{3}}\cos\left(\frac{1}{3}\arccos\left(-\frac{3}{\sqrt{2x}}\right)\right). \qquad (5.101)$$

For $\dot{m} \gg 1$, the function $\psi(\dot{m}/x_{in})$ quickly approaches unity, and

$$R_{\rm sph} \simeq \frac{3}{2}\dot{m} \times \frac{GM}{c^2}, \qquad (5.102)$$

which is consistent with the classical definition of the spherization radius (Shakura and Sunyaev 1973).

5.3.3.1 Size of the Photosphere

Assume that a QSO accretion disc is surrounded by a wind which is spherically symmetric and carries away an appreciable part of the accreting matter (the outflow rate is $f_{\mathrm{w}}\dot{M}$, where $f_{\mathrm{w}} \lesssim 1$ is the mass fraction ejected in the wind). Since most of the mass is lost by the disc at radii close to the spherization radius, the velocity of the outflow is close to virial at R_{sph}

$$v_{\mathrm{w}} = \beta_{\mathrm{w}}\sqrt{2GM/R_{\mathrm{sph}}} = \frac{2}{\sqrt{3}}\frac{\beta_{\mathrm{w}}}{\psi}\dot{m}^{-1/2}c, \tag{5.103}$$

where $\beta_{\mathrm{w}} \simeq 1$ is a scaling dimensionless constant of the order of unity, and ψ is the correction factor determined using expression (5.101). It can be shown that the use of the nonrelativistic formula is quite reasonable since post-Newtonian corrections to the expression (5.103) are generally within a couple per cent. The radial optical depth due to scattering is

$$\tau = \int_R^{+\infty} \varkappa_T \rho dR. \tag{5.104}$$

The continuity equation allows us to estimate the density in the wind as $\rho = \frac{\dot{M}}{4\pi v_{\mathrm{w}} R^2}$. Substituting the expression for density and setting $\tau = 1$, we obtain the following estimate for the photospheric radius:

$$R_1 = \frac{\sqrt{3}}{2}\frac{f_{\mathrm{w}}}{\beta}\dot{m}^{3/2}\psi \times \frac{GM}{c^2}. \tag{5.105}$$

This estimate is valid if the local Eddington limit is violated somewhere in the disc.

As we have already mentioned, a transition to the supercritical accretion regime may be treated as a local process (energy flux becomes sufficient to throw a test particle away to infinity), even if a large geometrical thickness of the disc restricts the applicability of this approach. Considered locally, the Eddington limit depends on GR corrections which make the effective vertical gravity near the equatorial plane two or three times larger. Moreover, the effects related to disc thickness should be accounted for: the deviations from the Kepler rotation law, dependence of angular velocity on height, and radial heat advection. Taken together, these effects seem to raise the value of the limit by a factor of a few, slightly decreasing the size of the spherization radius (Abolmasov and Chashkina 2015) (Fig. 5.15).

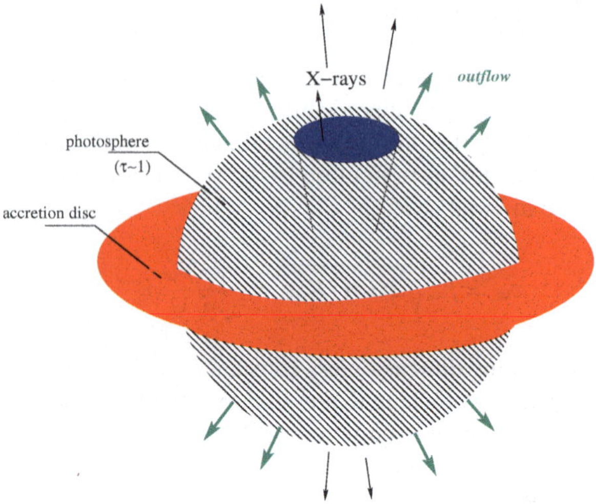

Fig. 5.15 The accretion disc, the wind and the scattering envelope produced by the wind. The area in red is the disc with size depending on the wavelength. The envelope (hatched) is unable to hide the disc at long wavelengths. It seems that there is a channel in the envelope through which (directly or after reflection/scattering) the X-rays from the inner parts of the disc may be observed

5.3.3.2 Evidence for Supercritical Accretion

Let us assume that a QSO is the source of a wind producing a scattering envelope, the optical depth and the size of which, in general, depend on the wind strength, and increases with the rate of matter ejection and thus with the accretion rate. Some QSOs (those hidden below scattering envelopes) will have significantly larger effective disc sizes depending more weakly on wavelength, since the size of the scattering photosphere does not depend on wavelength. Most probably, the envelopes are actually nonspherical and translucent. However, for simplicity we will use the approximation of a spherical envelope with a clearly outlined photosphere. Despite the fact that in reality such a photosphere should be fairly fuzzy, with strong limb darkening, all the intensity distribution and limb darkening effects are almost independent of wavelength. The size of the photosphere does not depend on wavelength as long as scattering by free electrons remains the main source of opacity. We estimated the accretion rates and the masses of the objects from the sample of Morgan et al. (2010), for which disc sizes through microlensing are known and fluxes in the photometric I band are available. The accretion rates and BH masses were determined using simultaneous solution of Eqs. (5.44) and (5.105) for M and $\dot{m}$. Unfortunately, the original tables and plots contained a calculation error; the correct results are given in Abolmasov and Shakura (2013) and in this book in Table 5.2 and Fig. 5.16. Figure 5.17 shows the corresponding accretion rates in solar masses per year. The masses obtained are essentially lower than the 'virial' masses $M_{\rm vir}$ calculated using the reverberation method or through emission line widths

Table 5.2 The accretion rates of microlensed QSOs as estimated from their sizes in the framework of the scattering envelope model

Object	z	M_{vir}, $10^9 M_\odot$	$R_{1/2}$, 10^{15} cm		$\dot{m}$	M_S, $10^9 M_\odot$	$\dot{M}$, $M_\odot$ year^{-1}	ζ	I_{corr}
			I	II					
Q J0158-4325	1.29	0.16	1–4	–	<65	>0.065	<10	–	19.09±0.12
Q J0158-4325*	1.29	0.16	5–30	–	5–9	80–300	0.1–0.15	–	19.09±0.12
HE J0435-1223	1.689	0.5	2–40	4–21	90–1300	0.007–0.024	5–20	0.7±0.6	20.76±0.25
SDSS 0924+0219	1.52297	0.11	1–6	2.2–10	170–800	0.005–0.011	2–4	0.46±0.3	21.24±0.25
FB J0951+2635	1.24603	0.89	12–80	–	60–300	0.11–0.25	40–100	–	17.16±0.11
SDSS J1004+4112	1.73995	0.39	1–4	–	40–150	0.017–0.031	3–6	–	20.97±0.22
HE J1104-1805	2.3192	2.37	10–30	–	40–130	0.18–0.30	30–50	1.65±0.5	18.17±0.31
PG 1115+080	1.73549	1.23	40–190	25–110	500–2600	0.011–0.025	30–70	0.4±0.5	19.52±0.27
RXJ 1131-1231	0.654	0.06	3–8	1.0–6	360–900	0.0022–0.0035	3–5	0.4±0.5	20.73±0.4
SDSS J1138+0314	2.44375	0.04	0.5–8	3–26	30–400	0.008–0.028	2–8	0.4±0.5	21.97±0.19
SBS J1520+530	1.855	0.88	8–20	–	70–170	0.07–0.11	20–30	–	18.92±0.13
QSO J2237+0305	1.695	0.9	5–20	–	<100	>0.17	<40	1.15±0.2	17.90±0.44
MGJ 0414+0534	2.639	1.82	–	–	–	–	–	1.5±0.4	–
RXJ 0911+0551	2.799	0.8	–	–	–	–	–	0.17±0.4	–

The virial masses and the half-light radii were taken from Morgan et al. (2010) and Blackburne et al. (2011) and given in separate columns. This table also contains the relative accretion rates, the masses M_S and the corresponding accretion rates in absolute units ($M_\odot$ year^{-1}) calculated using photometric and microlensing data for $a = 0.9$. All ζ values are taken from Abolmasov and Shakura (2012a) (and references therein). If several estimates of a structural parameter are available for an object, these are averaged according to the weights inversely proportional to the uncertainties

*—Morgan et al. (2012), I—Morgan et al. (2010), II—Blackburne et al. (2011)

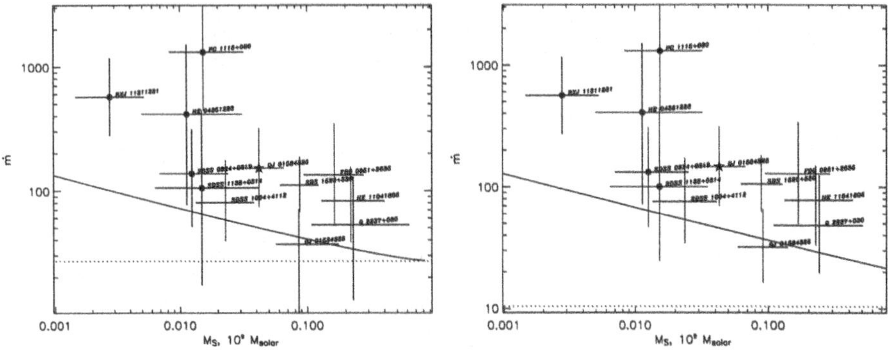

Fig. 5.16 The masses and the dimensionless accretion rates obtained in a self-consistent way in the framework of the scattering envelope model for $a = 0.9$ (left) and $a = 0$ (right). The dotted horizontal lines indicate the critical accretion rates, the solid lines show the $\dot{m}$ values for which the apparent sizes of accretion discs and envelopes become comparable. The uncertainties correspond to a significance level of 1σ, considering the uncertainties in determining the source sizes and the magnitudes. The asterisk indicates that the data were taken from Morgan et al. (2012); all the other data were taken from Morgan et al. (2010). This illustration together with the next one were taken from Abolmasov and Shakura (2013)

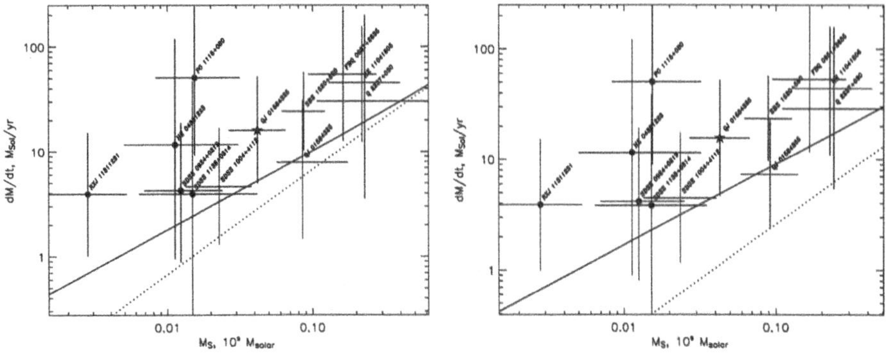

Fig. 5.17 The masses and the dimensional accretion rates obtained in a self-consistent way in the framework of the scattering envelope model for $a = 0.9$ (left) and $a = 0$ (right). The notation is the same as in the previous figure

(Vestergaard and Peterson 2006), while the accretion rates are suspiciously high. If we abandon the use of self-consistent mass estimates in favour of virial masses, the accretion rates become $\dot{m} \sim 10$–100, consistent with some independent estimates of maximal accretion rates in distant QSOs (Collin et al. 2002). Using (5.44) to estimate BH masses seems to be incorrect since this formula is valid for the standard disc, not taking into account neither the decrease of accretion rate within spherization radius nor any of the true absorption processes affecting the observed QSO spectrum. Two circumstances should be noted: first, the optical I_{corr} magnitudes are determined using a strong lens model with an amplification factor of tens or hundreds, potentially vulnerable to systematic errors. Second, the QSO radiation which is detected in the optical to near infrared range, has been emitted in UV, thus being very sensitive to absorption by interstellar dust (one magnitude of absorption in the V band corresponds to absorption of two to four magnitudes at 0.1–0.2 μm). The presence of dust in galactic centres is rather expected, and taking this into consideration enables a substantial decrease of the accretion rates and increase of the BH masses, making them consistent with virial estimates. For this, dust extinction of the order of 1–2 mag is sufficient, which corresponds to absorption lower than ~ 0.5 mag at optical wavelengths. Taking into account absorption A_ν yields a correction for the observed flux $F_\nu^{\rm obs} = 10^{-0.4A_\nu} F_\nu^{em}$. Estimates for the accretion rate and the BH mass will depend on the absorption as

$$\dot{m} \propto 10^{-0.3A_\nu}, \tag{5.106}$$

$$M \propto 10^{0.45A_\nu}, \tag{5.107}$$

and

$$\dot{M} \propto 10^{0.15A_\nu}. \tag{5.108}$$

On the other hand, interstellar extinction does not affect the shape of the amplification curves and therefore cannot affect the observed values of the ζ parameter. Possible contribution of systematic uncertainties due to interstellar extinction was considered for SDSS 0924+0219 in Floyd et al. (2009).

Systematic uncertainties in our calculations may be introduced by other reasons as well. First, the formulas in use contain the dimensionless coefficients $f_{\rm w}$ (the fraction of accreting matter ejected in the form of a wind) and $\beta_{\rm w}$ (the ratio of the terminal wind velocity to the escape velocity at the spherization radius). The first coefficient is less than unity by definition. The second coefficient may be more or less than unity, depending on how the momentum and energy are redistributed in the outflowing matter. If the wind from the spherization radius is accelerated approximately radially, its velocity is:

$$v(R) \simeq \sqrt{2\left(\frac{L}{L_{Edd}} - 1\right) \times \left(\frac{GM}{R_{sph}} - \frac{GM}{R}\right)}. \tag{5.109}$$

Close to the Eddington limit, $L/L_{\rm Edd} \sim \dot{m}_{\rm cr}\eta \sim 4.5$. Therefore, $\beta_{\rm w} \simeq \sqrt{L/L_{\rm Edd} - 1} \sim 2$. Lower values of $\beta_{\rm w}$ may arise in cases when the main contribution to the expansion of the envelope comes from matter launched at relatively large radii in the disc (this result is supported by numerical calculations of Sadowski et al. 2014). In this case, a decrease in velocity approximately by a factor of 10 ($\beta_{\rm W} \sim 0.1$) makes it possible to reconcile, on average, the mass estimates with the virial ones as the resulting quantities scale with the parameter as

$$\dot{m} \propto \beta_{\rm W}, \tag{5.110}$$

$$M \propto \beta_{\rm W}^{-1/2}, \tag{5.111}$$

and

$$\dot{M} \propto \beta_{\rm W}^{1/2}. \tag{5.112}$$

Another very important effect, which is able to appreciably change the expected envelope size, is the presence of additional opacity mechanisms. The accretion rates and masses depend on the opacity $\varkappa$ as follows[2]:

$$\dot{m} \propto \varkappa^{-2} \tag{5.113}$$

$$M \propto \varkappa \tag{5.114}$$

$$\dot{M} \propto \varkappa^{-1}. \tag{5.115}$$

The presence of additional absorption mechanisms allows us to explain why the mean value ζ is not zero for QSOs with envelopes. For instance, free-free scattering in a comparatively hot wind ($h\nu \ll kT$) results in the following dependence of the size of the photosphere on wavelength (see Cassinelli and Hartmann 1977):

$$R_1 \propto \lambda^{\frac{2}{2n-3m/2-1}}, \tag{5.116}$$

if $\rho \propto r^{-n}$ and $T \propto r^{-m}$. In particular, $\zeta = 8/9$ for $n = 2$ and $m = 1/2$.

A lot remains unclear in the schematic scenario outlined above. In particular, the X-rays we observe come from a source that is much more compact than the envelope. We may assume that the shape of the envelope differs from spherical and has a channel to release X-rays (as shown in Fig. 5.15). It is also unclear how the low masses obtained in Abolmasov and Shakura (2012a) can be reconciled with the virial mass estimates. Maybe, further numerical calculations will help finding the answer. Note for example that an optically thick wind subjected to different instabilities may become highly inhomogeneous near a photosphere arising at a distance of the order of hundred gravitational radii (Takeuchi et al. 2013). This differs considerably from radii calculated according to formula (5.105).

---

[2]The photospheric size is proportional to $R_1 \propto \varkappa \dot{M}$, while the product $M\dot{M} \propto F_\nu^{3/2}$ does not depend on opacity; the Thomson opacity $\varkappa_T$, which is included in the Eddington luminosity normalization, is considered fixed.

5.3.4 *Scattering by Inflowing Matter*

As we could see above, the observed properties of lensed QSOs may be explained using a scattering envelope model if the wind velocity is assumed to be very small, less than one tenth of the virial velocity at the spherization radius. Such a small dimensionless factor, far less than unity, is unlikely to arise during wind acceleration: this requires the total mechanical energy of the wind to be equal to zero with a precision of the order of $\beta_{\rm w} \ll 1$, too much of a fine tuning to represent reality. Only the smallness of the total mechanical energy can ensure a low wind velocity (compared to the virial velocity in the acceleration area) at large distances. On the other hand, during accretion of matter with high angular momentum, it is quite natural to expect very small radial velocities determined by loss or viscous transfer of the angular momentum. If both inflows and outflows exist, it is more reasonable to expect the disc radiation to be scattered by the inflowing gas rather than by the wind. The angular momentum of the accreting matter definitely needs to be high to form an accretion disc.

Many active galactic nuclei (AGNs), including QSOs, are surrounded by gas-dust tori (Elitzur 2008) and areas radiating in broad emission lines, BLR (Peterson 2006). The gas-dust torus seems to be a reservoir from which the active galactic nucleus receives matter to accrete. The data available for well-studied galactic nuclei, such as NGC5548 (Kollatschny and Zetzl 2013), indicate that the geometries of the BLR resemble tori or geometrically thick discs with axis ratios different for different emission lines. On the other hand, there is evidence (Doroshenko et al. 2012; Grier et al. 2013) that the gas in BLR has, on average, radial velocities directed inwards. This contradicts the widespread conception that broad lines form in the disc wind (see, for example, Korista 1999).

A possible contribution of this inflowing scattering substance to the observed spatial properties of the QSO 'disc' is estimated in Abolmasov (2017). A reasonable fit to the observational data can be achieved by assuming a radial inflow velocity of the order of one per cent of the virial, the size of the scattering area being of the order of 10^4 gravitational radii. Assuming that the accretion rates of the objects of the sample are close in absolute values and constitute a few solar masses per year (consistent with the observed values within an order of magnitude), we can reproduce the separation of objects into two groups with different values of ζ (see Fig. 5.12) and absolute values of the observed radii. However, for this to work, the contribution of the scattered radiation should be significant, which requires large geometrical thickness and high scattering efficiency.

This approach is able to qualitatively explain the fact that the ζ parameter is noticeably larger than zero for most of the objects, although appreciably less than that for the standard disc. This rather expected effect may be illustrated using the following simplified model. Let us assume there is a disc for which the half-light radius is $R_{\rm d} = R_{\rm d}(\lambda)$. If, instead of the disc, we see a halo with intensity distribution $I \propto R^{-2}$ between $R_{\rm d}$ and some fixed outer radius $R_{\rm out}$, the half-light radius of the halo will be $R_{1/2} \simeq \sqrt{R_\mathrm{d} R_\mathrm{out}} \propto R_\mathrm{d}^{1/2}$. In terms of the slope of the dependence $R_{1/2}(\lambda)$, this means ζ becomes equal to half of the structure parameter for R_d.

The simplified assumption of coherent scattering by a geometrically thick accretion flow is not able to conclusively resolve the problem. The solution to the problem seems to be directly related to other problems of accretion onto supermassive black holes. First, there are inconsistencies in the spectra of QSOs: the predictions of the standard theory of multi-temperature accretion disc do not fit well with the shapes of QSO spectra in the range 0.1–1 μm, the area of the 'big blue bump', where it is supposed to work. It seems quite reasonable (this possibility was considered by Lawrence 2012) that the observed radiation of the 'big blue bump' is not emitted by the surface of the accretion disc itself. Instead, it is emitted further away from the BH, in the gas ionized and heated by harder disc radiation. In this case, the shape of the observed spectrum is determined by local microscopic processes, namely, ionization, radiation transfer, and cooling, rather than by energy release per unit area of the emitting surface. This model allows us to explain why the QSO spectra in UV and optics are approximately of constant shape and why there is no strong Lyman jump at all. If emitting regions are ionized by comparatively hard radiation with energies of a few tens of eV, recombination will not result in the emergence of any appreciable Lyman emission jump since Lyman continuum quanta will be emitted mainly at high optical depths, being effectively reprocessed into Lyα quanta (this effect of a 'Lyman greenhouse' was considered in Abolmasov and Poutanen 2017). It seems that to accurately explain the spatial properties of the emitting areas in lensed QSOs it is necessary to develop a detailed physical model for scattering and reprocessing of the radiation of the disc, capable of reproducing both the continuum and the emission-line components of QSO emission together with their spatial properties. Such a unified spectral model might be applied not only to the photometric estimates of accretion discs and their spectral properties but also to the data on microlensing-induced spectral variability of QSOs, that are currently being gradually accumulated (O'Dowd et al. 2015; Braibant et al. 2016; Guerras et al. 2013a). In principle, a much larger amount of information on the emitting area could be derived from the evolution of emission line profiles since amplification of areas with different radial velocities, velocity dispersions and physical conditions do not occur simultaneously.

5.4 Conclusions

It appears as if the opportunities provided by QSO microlensing are far from being exhausted and further advance in this area will require further improvement in the methods involved in the analysis of the observational data and elimination of the numerous sources of systematical errors (see, for example, Vernardos and Fluke (2014) where such biases are discussed for strong lensing models). It is likely that not only the size but also the disc brightness distribution may soon be reproduced on the basis of the amplification curves, allowing us to directly verify models of accretion flows and scattering envelopes.

The effects of broad emission-line microlensing could become an important source of information. It has now become clear that microlensing often distorts line profiles, affecting one wing more than the other. The regions producing such lines should be spherically asymmetric, resemble thick discs and, likely, rotate fast enough (Braibant et al. 2014). The FeII and FeIII emission line blends seem to form approximately in the same area as the UV continuum radiation (Guerras et al. 2013b). Unfortunately, there have been no effects detected so far which might be related to chromatic lensing of IR radiation from QSOs since the available data on far IR variability are very scarce.

The question of what may be the nature of the X-ray emission in QSOs requires special attention. The sizes of the X-ray sources estimated using microlensing data often turn out comparable to the assumed size of the innermost stable orbit (Morgan et al. 2012). Possibly, the X-rays could be caused by shocks arising between the innermost stable orbit and the event horizon, in the area that emits the energy stored in an accretion disc of finite thickness (see Introduction in Abolmasov 2014 and the discussion therein). The source of the X-rays may also be some sort of hot corona, although very compact, with a size of the order of the radius of the innermost stable circular orbit (Cao 2009).

References

Abolmasov P (2014) The thickness of a weakly magnetized accretion flow inside the last stable orbit of a Kerr black hole. Mon Not R Astron Soc 445:1269–1287. https://doi.org/10.1093/mnras/stu1753, 1408.6449

Abolmasov P (2017) Apparent quasar disc sizes in the "bird's nest" paradigm. Astron Astrophys 600:A79. https://doi.org/10.1051/0004-6361/201628842, 1701.08957

Abolmasov P, Chashkina A (2015) On the Eddington limit for relativistic accretion discs. Mon Not R Astron Soc 454:3432–3444. https://doi.org/10.1093/mnras/stv2229, 1509.07261

Abolmasov P, Poutanen J (2017) Gamma-ray opacity of the anisotropic stratified broad-line regions in blazars. Mon Not R Astron Soc 464:152–169. https://doi.org/10.1093/mnras/stw2326, 1609.03350

Abolmasov P, Shakura NI (2012a) Microlensing evidence for super-Eddington disc accretion in quasars. Mon Not R Astron Soc 427:1867–1876. https://doi.org/10.1111/j.1365-2966.2012.21881.x, 1208.1678

Abolmasov P, Shakura NI (2012b) Resolving the inner structure of QSO discs through fold-caustic-crossing events. Mon Not R Astron Soc 423:676–693. https://doi.org/10.1111/j.1365-2966.2012.20904.x, 1203.2656

Abolmasov P, Shakura NI (2013) Erratum: microlensing evidence for super-Eddington disc accretion in quasars. Mon Not R Astron Soc 434:906–908. https://doi.org/10.1093/mnras/stt976

Agol E, Krolik JH (2000) Magnetic stress at the marginally stable orbit: altered disk structure, radiation, and black hole spin evolution. Astrophys J 528:161–170. https://doi.org/10.1086/308177, arXiv:astro-ph/9908049

Arnold VI, Wassermann GS, Thomas RK (2003) Catastrophe theory. Springer, Berlin/Heidelberg

Bardeen JM, Petterson JA (1975) The lense-thirring effect and accretion disks around Kerr black holes. Astrophys J 195:L65+. https://doi.org/10.1086/181711
Bardeen JM, Press WH, Teukolsky SA (1972) Rotating black holes: locally nonrotating frames, energy extraction, and scalar synchrotron radiation. Astrophys J 178:347–370. https://doi.org/10.1086/151796
Bastian N, Covey KR, Meyer MR (2010) A Universal stellar initial mass function? A critical look at variations. Annu Rev Astron Astrophys 48:339–389. https://doi.org/10.1146/annurev-astro-082708-101642, 1001.2965
Bate NF, Floyd DJE, Webster RL, Wyithe JSB (2008) A microlensing study of the accretion disc in the quasar MG 0414+0534. Mon Not R Astron Soc 391:1955–1960. https://doi.org/10.1111/j.1365-2966.2008.14020.x, 0810.1092
Berezhiani Z, Ciarcelluti P, Comelli D, Villante FL (2005) Structure formation with mirror dark matter. Int J Mod Phys D 14:107–119. https://doi.org/10.1142/S0218271805005165, arXiv:astro-ph/0312605
Blackburne JA, Pooley D, Rappaport S, Schechter PL (2011) Sizes and temperature profiles of quasar accretion disks from chromatic microlensing. Astrophys J 729:34. https://doi.org/10.1088/0004-637X/729/1/34, 1007.1665
Blaes O, Krolik JH, Hirose S, Shabaltas N (2011) Dissipation and vertical energy transport in radiation-dominated accretion disks. Astrophys J 733:110. https://doi.org/10.1088/0004-637X/733/2/110, 1103.5052
Blandford R, Narayan R (1986) Fermat's principle, caustics, and the classification of gravitational lens images. Astrophys J 310:568–582. https://doi.org/10.1086/164709
Bogdanov MB, Cherepashchuk AM (2004) Analysis of a high-amplitude event in component A of the gravitational lens QSO 2237 + 0305. Astron Rep 48:261–266. https://doi.org/10.1134/1.1704671
Braibant L, Hutsemékers D, Sluse D, Anguita T, García-Vergara CJ (2014) Microlensing of the broad-line region in the quadruply imaged quasar HE0435–1223. Astron Astrophys 565:L11. https://doi.org/10.1051/0004-6361/201423633, 1405.5014
Braibant L, Hutsemékers D, Sluse D, Anguita T (2016) The different origins of high- and low-ionization broad emission lines revealed by gravitational microlensing in the Einstein cross. Astron Astrophys 592:A23. https://doi.org/10.1051/0004-6361/201628594, 1606.01734
Cao X (2009) An accretion disc-corona model for X-ray spectra of active galactic nuclei. Mon Not R Astron Soc 394:207–213. https://doi.org/10.1111/j.1365-2966.2008.14347.x, 0812.1828
Cassinelli JP, Hartmann L (1977) The effect of winds and coronae of hot stars on the infrared and radio continua. Astrophys J 212:488–493. https://doi.org/10.1086/155068
Chang K, Refsdal S (1979) Flux variations of QSO 0957+561 A, B and image splitting by stars near the light path. Nature 282:561–564. https://doi.org/10.1038/282561a0
Chang K, Refsdal S (1984) Star disturbances in gravitational lens galaxies. Astron Astrophys 132:168–178
Chwolson O (1924) Über eine mögliche Form fiktiver Doppelsterne. Astron Nachr 221:329
Collier S, Peterson BM (2001) Characteristic ultraviolet/optical timescales in active galactic nuclei. Astrophys J 555:775–785. https://doi.org/10.1086/321517
Collin S, Boisson C, Mouchet M, Dumont A, Coupé S, Porquet D, Rokaki E (2002) Are quasars accreting at super-Eddington rates? Astron Astrophys 388:771–786. https://doi.org/10.1051/0004-6361:20020550, arXiv:astro-ph/0203439
Dexter J, Agol E (2009) A fast new public code for computing photon orbits in a Kerr spacetime. Astrophys J 696:1616–1629. https://doi.org/10.1088/0004-637X/696/2/1616, 0903.0620
Doroshenko VT, Sergeev SG, Klimanov SA, Pronik VI, Efimov YS (2012) Broad-line region kinematics and black hole mass in Markarian 6. Mon Not R Astron Soc 426:416–426. https://doi.org/10.1111/j.1365-2966.2012.20843.x, 1203.2084
Eddington AS (1925) A limiting case in the theory of radiative equilibrium. Mon Not R Astron Soc 85:408
Eigenbrod A, Courbin F, Sluse D, Meylan G, Agol E (2008) Microlensing variability in the gravitationally lensed quasar QSO 2237+0305 = the Einstein cross. I. Spectrophotometric

monitoring with the VLT. Astron Astrophys 480:647–661. https://doi.org/10.1051/0004-6361:20078703, 0709.2828

Einstein A (1936) Lens-like action of a star by the deviation of light in the gravitational field. Science 84:506–507. https://doi.org/10.1126/science.84.2188.506

Elitzur M (2008) The toroidal obscuration of active galactic nuclei. New A Rev 52:274–288. https://doi.org/10.1016/j.newar.2008.06.010, 0805.3699

Ferreras I, Saha P, Williams LLR (2005) Stellar and total mass in early-type lensing galaxies. Astrophys J 623:L5–L8. https://doi.org/10.1086/429995, arXiv:astro-ph/0503168

Floyd DJE, Bate NF, Webster RL (2009) The accretion disc in the quasar SDSS J0924+0219. Mon Not R Astron Soc 398:233–239. https://doi.org/10.1111/j.1365-2966.2009.15045.x, 0905.2651

Gaudi BS, Petters AO (2002) Gravitational microlensing near caustics. I. Folds. Astrophys J 574:970–984. https://doi.org/10.1086/341063, arXiv:astro-ph/0112531

Gil-Merino R, González-Cadelo J, Goicoechea LJ, Shalyapin VN, Lewis GF (2006) Is there a caustic crossing in the lensed quasar Q2237+0305 observational data record? Mon Not R Astron Soc 371:1478–1482. https://doi.org/10.1111/j.1365-2966.2006.10782.x, arXiv:astro-ph/0607162

Grier CJ, Peterson BM, Horne K, Bentz MC, Pogge RW, Denney KD, De Rosa G, Martini P, Kochanek CS, Zu Y, Shappee B, Siverd R, Beatty TG, Sergeev SG, Kaspi S, Araya Salvo C, Bird JC, Bord DJ, Borman GA, Che X, Chen C, Cohen SA, Dietrich M, Doroshenko VT, Efimov YS, Free N, Ginsburg I, Henderson CB, King AL, Mogren K, Molina M, Mosquera AM, Nazarov SV, Okhmat DN, Pejcha O, Rafter S, Shields JC, Skowron J, Szczygiel DM, Valluri M, van Saders JL (2013) The structure of the broad-line region in active galactic nuclei. I. Reconstructed velocity-delay maps. Astrophys J 764:47. https://doi.org/10.1088/0004-637X/764/1/47, 1210.2397

Guerras E, Mediavilla E, Jimenez-Vicente J, Kochanek CS, Muñoz JA, Falco E, Motta V (2013a) Microlensing of quasar broad emission lines: constraints on broad line region size. Astrophys J 764:160. https://doi.org/10.1088/0004-637X/764/2/160, 1207.2042

Guerras E, Mediavilla E, Jimenez-Vicente J, Kochanek CS, Muñoz JA, Falco E, Motta V, Rojas K (2013b) Microlensing of quasar ultraviolet iron emission. Astrophys J 778:123. https://doi.org/10.1088/0004-637X/778/2/123, 1309.2603

Hamadache C, Le Guillou L, Tisserand P, Afonso C, Albert JN, Andersen J, Ansari R, Aubourg É, Bareyre P, Beaulieu JP, Charlot X, Coutures C, Ferlet R, Fouqué P, Glicenstein JF, Goldman B, Gould A, Graff D, Gros M, Haissinski J, de Kat J, Lesquoy É, Loup C, Magneville C, Marquette JB, Maurice É, Maury A, Milsztajn A, Moniez M, Palanque-Delabrouille N, Perdereau O, Rahal YR, Rich J, Spiro M, Vidal-Madjar A, Vigroux L, Zylberajch S (2006) Galactic bulge microlensing optical depth from EROS-2. Astron Astrophys 454:185–199. https://doi.org/10.1051/0004-6361:20064893, arXiv:astro-ph/0601510

Hilbert S, White SDM, Hartlap J, Schneider P (2007) Strong lensing optical depths in a ΛCDM universe. Mon Not R Astron Soc 382:121–132. https://doi.org/10.1111/j.1365-2966.2007.12391.x, arXiv:astro-ph/0703803

Hogg DW (1999) Distance measures in cosmology. ArXiv Astrophysics. e-prints 9905116. arXiv:astro-ph/9905116

Hubeny I, Hubeny V (1998) Non-LTE models and theoretical spectra of accretion disks in active galactic nuclei. II. Vertical structure of the disk. Astrophys J 505:558–576. https://doi.org/10.1086/306207, arXiv:astro-ph/9804288

Inada N, Oguri M, Morokuma T, Doi M, Yasuda N, Becker RH, Richards GT, Kochanek CS, Kayo I, Konishi K, Utsunomiya H, Shin MS, Strauss MA, Sheldon ES, York DG, Hennawi JF, Schneider DP, Dai X, Fukugita M (2006) SDSS J1029+2623: a gravitationally lensed quasar with an image separation of 22.5". Astrophys J 653:L97–L100. https://doi.org/10.1086/510671, arXiv:astro-ph/0611275

Irwin MJ, Webster RL, Hewett PC, Corrigan RT, Jedrzejewski RI (1989) Photometric variations in the Q2237 + 0305 system - first detection of a microlensing event. Astron J 98:1989–1994. https://doi.org/10.1086/115272

Ivanov PB, Illarionov AF (1997) The oscillatory shape of the stationary twisted disc around a Kerr black hole. Mon Not R Astron Soc 285:394–402

Ivezić Ž, Menou K, Knapp GR, Strauss MA, Lupton RH, Vanden Berk DE, Richards GT, Tremonti C, Weinstein MA, Anderson S, Bahcall NA, Becker RH, Bernardi M, Blanton M, Eisenstein D, Fan X, Finkbeiner D, Finlator K, Frieman J, Gunn JE, Hall PB, Kim RSJ, Kinkhabwala A, Narayanan VK, Rockosi CM, Schlegel D, Schneider DP, Strateva I, SubbaRao M, Thakar AR, Voges W, White RL, Yanny B, Brinkmann J, Doi M, Fukugita M, Hennessy GS, Munn JA, Nichol RC, York DG (2002) Optical and radio properties of extragalactic sources observed by the FIRST survey and the Sloan digital sky survey. Astron J 124:2364–2400. https://doi.org/10.1086/344069, astro-ph/0202408

Jaroszynski M, Wambsganss J, Paczynski B (1992) Microlensed light curves for thin accretion disks around Schwarzschild and Kerr black holes. Astrophys J 396:L65–L68. https://doi.org/10.1086/186518

Keeton CR (2001) A catalog of mass models for gravitational lensing. ArXiv Astrophysics. e-prints arXiv:astro-ph/0102341

Kochanek CS (2006) Part 2: Strong gravitational lensing. In: Meylan G, Jetzer P, North P, Schneider P, Kochanek CS, Wambsganss J (eds) Saas-Fee advanced course 33: gravitational lensing: strong, weak and micro. Springer, Berlin, pp 91–268

Kochanek CS, Keeton CR (1997) Gravitational lensing limits on early-type galaxies. In: Arnaboldi M, Da Costa GS, Saha P (eds) The nature of elliptical galaxies, 2nd Stromlo symposium. Astronomical society of the pacific conference series, vol 116, p 21. arXiv:astro-ph/9611217

Kochanek CS, Schechter PL (2004) The hubble constant from gravitational lens time delays. Measuring and modeling the universe. Cambridge University Press, Cambridge, p 117. astro-ph/0306040

Kochanek CS, Keeton CR, McLeod BA (2001) The importance of Einstein rings. Astrophys J 547:50–59. https://doi.org/10.1086/318350, arXiv:astro-ph/0006116

Kofman L, Kaiser N, Lee MH, Babul A (1997) Statistics of gravitational microlensing magnification. I. Two-dimensional lens distribution. Astrophys J 489:508–+. https://doi.org/10.1086/304791, arXiv:astro-ph/9608138

Kollatschny W, Zetzl M (2013) Vertical broad-line region structure in nearby active galactic nuclei. Astron Astrophys 558:A26. https://doi.org/10.1051/0004-6361/201321685, 1308.1902

Kolykhalov PI, Sunyaev RA (1984) Radiation of accretion disks in quasars and galactic nuclei. Adv Space Res 3:249–254 https://doi.org/10.1016/0273-1177(84)90100-5

Korista K (1999) What's emitting the broad emission lines? In: Ferland G, Baldwin J (eds) Quasars and cosmology. Astronomical society of the pacific conference series, vol 162. Astronomical Society of the Pacific, San Francisco, p 165, astro-ph/9812043

Kormann R, Schneider P, Bartelmann M (1994) Isothermal elliptical gravitational lens models. Astron Astrophys 284:285–299

Lawrence A (2012) The UV peak in active galactic nuclei: a false continuum from blurred reflection? Mon Not R Astron Soc 423:451–463. https://doi.org/10.1111/j.1365-2966.2012.20889.x, 1110.0854

Meyer F, Liu BF, Meyer-Hofmeister E (2000) Evaporation: the change from accretion via a thin disk to a coronal flow. Astron Astrophys 361:175–188. astro-ph/0007091

Mihalas D (1978) Stellar atmospheres, 2nd edn. W.H. Freeman and Co., San Francisco

Morgan CW, Kochanek CS, Morgan ND, Falco EE (2010) The Quasar accretion disk size-black hole mass relation. Astrophys J 712:1129–1136. https://doi.org/10.1088/0004-637X/712/2/1129, 1002.4160

Morgan CW, Hainline LJ, Chen B, Tewes M, Kochanek CS, Dai X, Kozlowski S, Blackburne JA, Mosquera AM, Chartas G, Courbin F, Meylan G (2012) Further evidence that quasar X-ray emitting regions are compact: X-ray and optical microlensing in the lensed quasar Q J0158–4325. Astrophys J 756:52. https://doi.org/10.1088/0004-637X/756/1/52, 1205.4727

Mortonson MJ, Schechter PL, Wambsganss J (2005) Size is everything: universal features of quasar microlensing with extended sources. Astrophys J 628:594–603. https://doi.org/10.1086/431195, arXiv:astro-ph/0408195

Muñoz JA, Falco EE, Kochanek CS, Lehár J, McLeod BA, Impey CD, Rix H, Peng CY (1998) The castles project. Astrophys Space Sci 263:51–54. https://doi.org/10.1023/A:1002120921330, arXiv:astro-ph/9902131

Narayan R, Yi I (1995) Advection-dominated accretion: underfed black holes and neutron stars. Astrophys J 452:710–+. https://doi.org/10.1086/176343, arXiv:astro-ph/9411059

Novikov ID, Thorne KS (1973) Astrophysics of black holes. In: Black holes (Les Astres Occlus). Gordon&Breach, Paris, pp 343–450

O'Dowd MJ, Bate NF, Webster RL, Labrie K, Rogers J (2015) Microlensing constraints on broad absorption and emission line flows in the quasar H1413+117. Astrophys J 813:62. https://doi.org/10.1088/0004-637X/813/1/62, 1504.07160

Paczynski B (1986) Gravitational microlensing at large optical depth. Astrophys J 301:503–516. https://doi.org/10.1086/163919

Page DN, Thorne KS (1974) Disk-Accretion onto a black hole. Time-averaged structure of accretion disk. Astrophys J 191:499–506. https://doi.org/10.1086/152990

Peterson BM (2006) The broad-line region in active galactic nuclei. In: Alloin D (ed) Physics of active galactic nuclei at all scales. Lecture notes in physics, vol 693. Springer, Berlin, p 77. https://doi.org/10.1007/3-540-34621-X_3

Pietrini P, Krolik JH (1995) The inverse Compton thermostat in hot plasmas near accreting black holes. Astrophys J 447:526. https://doi.org/10.1086/175897, astro-ph/9501093

Pooley D, Blackburne JA, Rappaport S, Schechter PL (2007) X-ray and optical flux ratio anomalies in quadruply lensed quasars. I. Zooming in on quasar emission regions. Astrophys J 661:19–29. https://doi.org/10.1086/512115, arXiv:astro-ph/0607655

Poutanen J, Lipunova G, Fabrika S, Butkevich AG, Abolmasov P (2007) Supercritically accreting stellar mass black holes as ultraluminous X-ray sources. Mon Not R Astron Soc 377:1187–1194. https://doi.org/10.1111/j.1365-2966.2007.11668.x, astro-ph/0609274

Raychaudhury S, Saslaw WC (1996) The observed distribution function of peculiar velocities of galaxies. Astrophys J 461:514–+. https://doi.org/10.1086/177078, arXiv:astro-ph/9602001

Riffert H, Herold H (1995) Relativistic accretion disk structure revisited. Astrophys J 450:508–+. https://doi.org/10.1086/176161

Rybicki GB, Lightman AP (1986) Radiative processes in astrophysics. Wiley, Weinheim

Sadowski A (2011) Slim accretion disks around black holes. ArXiv e-prints 1108.0396

Sadowski A, Narayan R, Tchekhovskoy A, Abarca D, Zhu Y, McKinney JC (2014) Global simulations of axisymmetric radiative black hole accretion disks in general relativity with a sub-grid magnetic dynamo. ArXiv e-prints 1407.4421

Schneider P (2005) Weak gravitational lensing. astro-ph/0509252 astro-ph/0509252

Schneider P (2006) Part 1: Introduction to gravitational lensing and cosmology. In: Meylan G, Jetzer P, North P, Schneider P, Kochanek CS, Wambsganss J (eds) Saas-Fee advanced course 33: gravitational lensing: strong, weak and micro. Springer, Berlin, pp 1–89

Shakura NI (1972) Disk model of gas accretion on a relativistic star in a close binary system. Astron Rep 49:921

Shakura NI, Sunyaev RA (1973) Black holes in binary systems. Observational appearance. Astron Astrophys 24:337–355

Shalyapin VN, Goicoechea LJ, Alcalde D, Mediavilla E, Muñoz JA, Gil-Merino R (2002) The nature and size of the optical continuum source in QSO 2237+0305. Astrophys J 579:127–135. https://doi.org/10.1086/342753, arXiv:astro-ph/0207236

Siuniaev RA, Shakura NI (1977) Disk reservoirs in binary systems and prospects for observing them. Pis'ma Astron Zh 3:262–266

Takeuchi S, Ohsuga K, Mineshige S (2013) Clumpy outflows from supercritical accretion flow. Publ Astron Soc Jpn 65:88. https://doi.org/10.1093/pasj/65.4.88, 1305.1023

Udalski A, Szymanski MK, Kubiak M, Pietrzynski G, Soszynski I, Zebrun K, Szewczyk O, Wyrzykowski L, Ulaczyk K, Wiêckowski T (2006) The optical gravitational lensing experiment. OGLE-III long term monitoring of the gravitational lens QSO 2237+0305. Acta Astron 56:293–305, arXiv:astro-ph/0701300

Vakulik V, Schild R, Dudinov V, Nuritdinov S, Tsvetkova V, Burkhonov O, Akhunov T (2006) Observational determination of the time delays in gravitational lens system <ASTROBJ>Q2237+0305</ASTROBJ>. Astron Astrophys 447:905–913. https://doi.org/10.1051/0004-6361:20053574, astro-ph/0509545

Vernardos G, Fluke CJ (2014) The effect of macromodel uncertainties on microlensing modelling of lensed quasars. Mon Not R Astron Soc 445:1223–1234. https://doi.org/10.1093/mnras/stu1833, 1409.1640

Vestergaard M, Peterson BM (2006) Determining central black hole masses in distant active galaxies and quasars. II. Improved optical and UV scaling relationships. Astrophys J 641:689–709. https://doi.org/10.1086/500572, arXiv:astro-ph/0601303

Walsh D, Carswell RF, Weymann RJ (1979) 0957 + 561 A, B - Twin quasistellar objects or gravitational lens. Nature 279:381–384. https://doi.org/10.1038/279381a0

Wambsganss J (2006) Part 4: Gravitational microlensing. In: G Meylan, P Jetzer, P North, P Schneider, C S Kochanek, & J Wambsganss (ed) Saas-Fee advanced course 33: gravitational lensing: strong, weak and micro. Springer, Heidelberg, pp 453–540

Webb W, Malkan M (2000) Rapid optical variability in active galactic nuclei and quasars. Astrophys J 540:652–677. https://doi.org/10.1086/309341

Witt HJ, Kayser R, Refsdal S (1993) Microlensing predictions for the Einstein cross 2237+0305. Astron Astrophys 268:501–510

Woźniak PR, Alard C, Udalski A, Szymański M, Kubiak M, Pietrzyński G, Zebruń K (2000) The optical gravitational lensing experiment monitoring of QSO 2237+0305. Astrophys J 529:88–92. https://doi.org/10.1086/308258, arXiv:astro-ph/9904329

Yan CS, Lu Y, Yu Q, Mao S, Wambsganss J (2014) Microlensing of sub-parsec massive binary black holes in lensed QSOs: light curves and size-wavelength relation. Astrophys J 784:100. https://doi.org/10.1088/0004-637X/784/2/100, 1402.2504

Zakharov AF (1997) Gravitational lenses and microlenses (in Russian). Yanus-K, Moscow

Zakharov AF, Sazhin MV (1998) Reviews of topical problems: gravitational microlensing. Phys Usp 41:945–982. https://doi.org/10.1070/PU1998v041n10ABEH000460

Zeldovich IB, Novikov ID (1975) Structure and evolution of the universe. Izdatel'stvo Nauka, Moscow (in Russian), 736 p

Zhuravlev VV, Ivanov PB (2011) A fully relativistic twisted disc around a slowly rotating Kerr black hole: derivation of dynamical equations and the shape of stationary configurations. Mon Not R Astron Soc 415:2122–2144. https://doi.org/10.1111/j.1365-2966.2011.18830.x, 1103.5739

Chapter 6
Transient Dynamics of Perturbations in Astrophysical Discs

Dmitry Razdoburdin and Viacheslav Zhuravlev

Abstract This chapter reviews some aspects of one of the major unsolved problems in understanding astrophysical (in particular, accretion) discs: whether the disc interiors may be effectively viscous in spite of the absence of magnetorotational instability. In this case, a rotational homogeneous inviscid flow with a Keplerian angular velocity profile is spectrally stable, making the transient growth of perturbations a candidate mechanism for energy transfer from regular motion to perturbations. Transient perturbations differ qualitatively from perturbation modes and can grow substantially in shear flows due to the non-normality of their dynamical evolution operator. Since the eigenvectors of this operator, alias perturbation modes, are mutually nonorthogonal, they can mutually interfere, resulting in transient growth of their linear combinations. Physically, a growing transient perturbation is a leading spiral whose branches are shrunk as a result of the differential rotation of the flow. This chapter discusses in detail the transient growth of vortex shear harmonics in the spatially local limit as well as methods for identifying the optimal (fastest growth) perturbations. Special attention is given to obtaining such solutions variationally, by integrating the direct and adjoint equations forwards and backwards in time, respectively. The material is presented in a newcomer-friendly style.

6.1 Introduction: Modal and Non-modal Analysis of Perturbations

A salient feature of disc accretion is that it is impossible without a dissipation mechanism for the differential rotation energy of matter. It is the internal friction in the disc, i.e. irreversible interaction of its adjacent rings, that leads to the transformation of gravitational energy of the accreting matter into heat and electromagnetic

D. Razdoburdin · V. Zhuravlev (✉)
Sternberg Astronomical Institute, Lomonosov Moscow State University, Moscow, Russia
e-mail: zhuravlev@sai.msu.ru

N. Shakura (ed.), *Accretion Flows in Astrophysics*, Astrophysics and Space Science Library 454, https://doi.org/10.1007/978-3-319-93009-1_6

radiation, which simultaneously allows the matter to flow towards the centre and the angular momentum to flow outwards to the disc periphery.

Direct dissipation is already possible due to the microscopic viscosity of the gas (plasma). However, in astrophysical conditions it turns out to be absolutely insufficient to explain the observed properties of discs. Essentially, discs are too large for the characteristic accretion time, t_ν, to be explained by microscopic viscosity. For example, in protoplanetary discs with typical size $L \sim 10$ a.u., where the kinematic viscosity is estimated to be $\nu_m \sim 10^7\,\mathrm{cm^2/s}$, the accretion time is $t_\nu = L^2/\nu \sim 10^{13}$ years, (see Section 3.3.2 in Armitage 2009). Apparently, t_ν is several orders of magnitude larger than the age of the Universe. At the same time, observations of gas-dust discs around young stars suggest that their lifetime is as short as only a few million years (see, for example, the review by Youdin and Kenyon (2013)). A similar conclusion is obtained for hot accretion discs around, in particular, black holes in close binary systems. In this case, for much smaller scales, $L \sim 10^{10}\,\mathrm{cm}$, and somewhat smaller viscosity of the hydrogen plasma $\nu_m \sim 10^5\,\mathrm{cm^2/s}$, we get $t_\nu \sim 3 \times 10^7$ years, which exceeds by many orders of magnitude, for example, the duration of X-ray Nova outbursts caused by non-stationary disc accretion (see the review by Remillard and McClintock (2006)).

At the same time, it is known from statistical hydromechanics (see the discussion of the Reynolds equations in Monin and Yaglom (1971), v. 1, Ch. 3) that the presence of significant correlating fluctuations of the velocity components in a flow is equivalent to the presence of a high effective viscosity, that exceeds the microscopic viscosity because the mixing scale of matter in the flow is much larger than the free path length of individual particles. In turn, the high effective viscosity enhances the angular momentum transfer towards the disc periphery, thus decreasing t_ν to the observed values. The perturbations under discussion can generally be regular: for example, accretion can be due to tidal waves generated in the disc by the secondary companion of a binary system (see Menou 2000). However, it is more natural to assume that these perturbations are generated by *turbulence* in the fluid. The turbulence takes its energy on the one hand from the rotational motion of matter on large scales, and on the other hand, via interaction of perturbation components with different wave numbers and cascades this energy to small scales where its direct dissipation into heat occurs due to microscopic viscosity.

It is important to recognize that energy transfer from a regular flow to perturbations should be mediated by some *linear* mechanism that follows from the dynamics of small perturbations described by linearized hydrodynamic equations. This can be rigorously proved for vortex fluid motion using the Navier-Stokes equations (see Schmid and Henningson (2001), Section 1.4, as well as Henningson and Reddy (1994)). Therefore, a first natural step in a theoretical study of turbulence generation in a given (stationary) flow is to search for exponentially growing linear perturbations against a steady-state background. Such perturbations are usually referred to as *modes*, and the corresponding analysis is dubbed *modal* or spectral analysis of perturbations, since it is used to determine eigenvalues of the corresponding dynamical operator of the problem: (complex) mode frequencies.

Turbulence arising from growing modes is called supercritical . In astrophysical flows with Keplerian angular frequency, spectral (magneto-rotational) instability with corresponding supercritical (MHD) turbulence has been found in analytical and numerical calculations Balbus and Hawley (1991), Hawley et al. (1995) and Stone et al. (1996) (see also reviews Balbus and Hawley (1998) and Balbus (2003)) for discs with a frozen seed magnetic field. Nevertheless, the magneto-rotational instability does not operate in cold low-ionized discs. Protoplanetary discs, accretion discs in quiescent states of cataclysmic variables and the outer parts of accretion discs in active galactic nuclei provide examples. Thus, it would be very important to show that differential rotation alone is capable of exciting turbulence in Keplerian discs. This property of Keplerian flows is universal, unlike the presence of a seed magnetic field together with sufficiently high degree of ionization of matter, or the existence of flow inhomogeneities due to the vorticity jump (see, for example, the review by Fridman and Bisikalo (2008)), or the appearance of radial velocity gradients (see the review by Kurbatov et al. (2014)), of vertical and/or horizontal gradients of some thermodynamic values (see, for example, Lovelace et al. 1999; Klahr and Hubbard 2014). However, generation of turbulence in a *homogeneous* Keplerian flow without magnetic field remains questionable so far.

The main difficulty here is that such a flow is spectrally stable: the specific angular momentum for the Keplerian rotation increases with radial distance from the centre, therefore according to the Rayleigh criterion (Rayleigh (1916) and Landau and Lifshitz (1987), v. 6, paragraph 27) the growth of axially symmetric modes is impossible; in turn, non-axisymmetric modes cannot grow since the necessary Rayleigh condition on the existence of extremum of vorticity in the background flow (Rayleigh 1880; Charney et al. 1950) is not fulfilled. In spite of that (and as follows from laboratory experiments and numerical simulations), turbulence arises in spectrally stable flows as well. In this case it is called subcritical . The plane-parallel Couette flow provides the simplest and the most prominent example (see the classical monographs by Drazin and Reid (1981) and Joseph (1976)).

In the theory of hydrodynamic stability, the transition of some flow (with non-zero microscopic viscosity) to a turbulent state is usually characterized by a set of critical Reynolds numbers Re (see Section 1.3.2 in the book Schmid and Henningson (2001)). The smallest of them is the number $\mathrm{Re_E}$ such that at $\mathrm{Re} < \mathrm{Re_E}$ there are no initial perturbations, irrespective of their amplitudes, whose energy would grow at the initial time $t = 0$. $\mathrm{Re_E}$ can be derived from the Reynolds-Orr energy equation (see Section 1.4 in Schmid and Henningson 2001). For a Couette flow $\mathrm{Re_E} \sim 20$. For $\mathrm{Re} > \mathrm{Re_E}$ initially growing perturbations at $t = 0$ arise, but as long as $\mathrm{Re} < \mathrm{Re_G}$, again there are no initial perturbations with an amplitude that would not decay at $t \to \infty$. This is the definition of the second critical number $\mathrm{Re_G} > \mathrm{Re_E}$. Finally, at higher values $\mathrm{Re} > \mathrm{Re_G}$ perturbations appear that can sustain their amplitude at all times, and starting from some $\mathrm{Re_T} > \mathrm{Re_G}$ the transition to a turbulent state is experimentally observed. For a Couette flow $\mathrm{Re_T} \sim 360$. The largest of the critical Reynolds numbers is $\mathrm{Re_L} > \mathrm{Re_T}$, starting from which growing modes arise, i.e. the flow becomes spectrally unstable. For a Couette flow, as well as for a Keplerian flow of interest here, $\mathrm{Re_L} = \infty$. However, the case of Keplerian flow is different in that up to the present time, the value of $\mathrm{Re_G}$ remains unknown, and $\mathrm{Re_T}$ has not been measured neither theoretically nor experimentally.

On the one hand, a general opinion has emerged that for Keplerian flows $\mathrm{Re_G} = \mathrm{Re_T} \to \infty$. This is based on the indirect argument that (locally) the action of the tidal and Coriolis forces on the perturbation, which are absent in a Couette flow, strongly stabilizes the shear flow (see Fig. 9 in the review by Balbus and Hawley (1998), in which the results from Balbus et al. (1996) are shown). This conclusion is supported by local numerical simulations Hawley et al. (1999), Shen et al. (2006) and series of laboratory experiments Ji et al. (2006), Schartman et al. (2009) and Schartman et al. (2012), in which stability of a quasi-Keplerian flow was observed up to $\mathrm{Re} = 2 \times 10^6$. Here we assume the quasi-Keplerian flow to be a so-called anti-cyclonic flow (see, for example, the definition in Lesur and Longaretti (2005)), where the specific angular momentum increases while the angular velocity itself, in contrast, decreases towards the periphery.[1]

On the other hand, in a cyclonic flow subcritical turbulence is observed at finite, although large values $\mathrm{Re_T}$, see Taylor (1936), Wendt (1933) on experiments with spectrally stable Taylor-Couette flows, as well as their analysis in astrophysical context in Zeldovich (1981) and later in Richard and Zahn (1999). In addition, negative results obtained in numerical experiments mentioned above can be explained by insufficient numerical resolution, as discussed in Longaretti (2002). In a subsequent paper by Lesur and Longaretti (2005), the dynamics of perturbations in cyclonic and anti-cyclonic flows was compared numerically. It was concluded that the required numerical resolution in the second case is much higher than in the first case, and the current computational power is insufficient to discover turbulence in a Keplerian flow; also it is impossible to argue that the stabilizing action of the Coriolis force in this case excludes the existence of a finite value of $\mathrm{Re_T} < \infty$. At last, another laboratory experiment presented in Paoletti and Lathrop (2011) and Paoletti et al. (2012) shows the appearance of subcritical turbulence and angular momentum transfer outwards in a quasi-Keplerian flow. The contradictory results claimed by different experimental groups show the complexity of the experiment due to inevitable arising of secondary flows induced by experimental tools. Presently, the influence of axial boundaries on the laboratory flow is discussed (see Avila 2012; Edlund and Ji 2014).

Anyway, it can be stated that of all types of homogeneous rotating flows, quasi-Keplerian (anti-cyclonic) flows turns out to be the most stable relative to finite-amplitude perturbations. Nevertheless, the smallness of microscopic viscosity in astrophysical conditions mentioned above simultaneously means that huge Reynolds numbers should exist in the discs. For example, if in the protoplanetary disc discussed above we take its thickness $H \sim 0.05L = 0.5\,\mathrm{a.u.}$ as the natural limiting scale of the problem, which corresponds to the sound velocity in the disc at this radius $c_s \sim 0.5\,\mathrm{km/s}$, we get $\mathrm{Re} \approx 10^{10}$. In other astrophysical discs Re can be even higher. Apparently, considering all negative results, the possibility of

[1]In a cyclonic flow both these quantities increase with distance from the centre.

turbulence in astrophysical Keplerian flows still spans several orders of magnitude: $10^6 < \mathrm{Re_T} < 10^{10}$.

Thus, a search for the critical value $\mathrm{Re_T}$ for Keplerian flows continues, and in the present chapter we will discuss in detail the necessary condition for turbulence and/or enhanced angular momentum transfer to the disc periphery—the transition of energy from a regular flow to perturbations in such a flow. As mentioned above, this transition should be mediated by a linear mechanism. Here, since a Keplerian flow is spectrally stable, only (small) perturbations different from modes can provide such a mechanism. The existence of such transiently growing non-modal perturbations in a shear flow was suggested already in papers by Kelvin (1887) and Orr (1907a,b). In astrophysics, this problem was studied in stellar dynamics (see Goldreich and Lynden-Bell 1965; Julian and Toomre 1966). However, in the context of hydrodynamic stability, rigorous treatment of such perturbations and methods to determine them were elaborated only in the 1990s and were dubbed *non-modal* perturbation analysis. To stress the inapplicability here of traditional modal analysis, the corresponding concept of the transition to subcritical turbulence due to transient growth of perturbations was called the *bypass* transition . The non-modal analysis of perturbations was formulated in Farrell (1988), Butler and Farrell (1992), Reddy et al. (1993), Reddy and Henningson (1993), (see also the reviews by Trefethen et al. (1993), Schmid (2007) and the book by Schmid and Henningson (2001)). These papers showed that mathematically the non-modal growth is due to non-orthogonality of the perturbation modes. If modes with a physically motivated norm are non-orthogonal to each other, their linear combinations can grow in this norm, in spite of each separate mode being decaying, as in a spectrally stable flow (see Fig. 6.5 in Sect. 6.3.1). In turn, the modes are non-orthogonal due to *non-normality* of the linear dynamical operator governing the perturbation evolution (see the introductory information about the operators in the same section). A non-normal operator does not commute with its adjoint operator, which is due to a non-zero velocity shear in the regular flow (see the concluding part of Sect. 6.3.4 below for more detail). Here, the higher Re, the higher the degree of non-orthogonality of the modes to each other and, correspondingly, the higher *transient growth* is possible. The papers mentioned above argue that the maximum possible transient growth of perturbations during a fixed time, called the optimal growth, is determined by the norm of a dynamical operator that can be obtained by calculating singular vectors of the operator (see Sect. 6.3.1 for more detail). Finally, the operator norm is tightly related to the notion of the operator's pseudospectrum (see Trefethen et al. 1993; Schmid and Henningson 2001).

Later this method was applied to astrophysical flows in Ioannou and Kakouris (2001), Yecko (2004), Mukhopadhyay et al. (2005), where different models were used to search for optimal perturbations demonstrating the optimal growth. In particular, it was shown that for a Keplerian velocity profile, the growth can be substantial only starting from $\mathrm{Re} \sim 10^6$, while in a similar setup for an iso-momentum profile and a Couette flow the growth starts already at $\mathrm{Re} \sim 10^3$ (see the discussion in Mukhopadhyay et al. 2005). Here, the papers by Meseguer (2002) and Maretzke et al. (2014) should also be mentioned, where the transient dynamics in a spectrally stable Taylor-Couette flow including both cyclonic and anti-cyclonic regimes is discussed. A correlation was found in Meseguer (2002) between the experimentally obtained stability boundary in a laminar flow (see Coles 1965) and the optimal growth value; Maretzke et al. (2014) found that for one and the same Re number, in a quasi-Keplerian regime the transient growth is minimal. Using the correlation from Meseguer (2002), the authors Maretzke et al. (2014) estimated $\mathrm{Re}_T \sim 10^5$ for the quasi-Keplerian regime. As in the numerical experiments Balbus et al. (1996), Hawley et al. (1999) and Shen et al. (2006) mentioned above, the effective Re, caused by the numerical viscosity, were hardly above $\sim 10^4 \div 10^5$, it is not surprising that in these studies the Keplerian profile was stable against perturbations.

Presently, there are in addition a lot of astrophysical studies of the transient growth of local perturbations using the Lagrangian method, where a transformation to the reference frame co-moving with the shear is performed and separate shear harmonics are considered (see Sect. 6.2.2). It was found that in the local space limit, transiently growing vortex shear harmonics emit wave shear harmonics of various type (depending on the compressibility or certain inhomogeneities in the flow) at the moment of swing (see Sect. 6.2.2), which themselves demonstrate non-modal growth (Lominadze et al. 1988; Fridman 1989; Chagelishvili et al. 1997, 2003; Tevzadze et al. 2003; Afshordi et al. 2005; Bodo et al. 2005; Tevzadze et al. 2008; Heinemann and Papaloizou 2009a,b; Tevzadze et al. 2010; Volponi 2010; Salhi and Pieri 2014).

Finally, Umurhan et al. (2006) and Rebusco et al. (2009) investigated the non-linear transient dynamics of three-dimensional perturbations taking into account the global structure of the flow in the model of a geometrically thin disc with α-viscosity. As in Ioannou and Kakouris (2001), these papers discussed the possibility of exciting non-modal perturbations by weak turbulence, already present in the disc, and giving rise to low effective viscosity parametrized by the α-parameter. In Sect. 6.2.3 we will also consider the influence of the effective viscosity on the transient growth of vortices on different scales in comparison to the disc thickness. Thus, the transient growth of perturbations can be discussed not only in the context of the bypass transition of a laminar flow to turbulence, but as a mechanism to enhance the angular momentum transfer in a disc with pre-existing weak turbulence producing low viscosity. In the final case this turbulence can be mathematically treated as an external stochastic perturbation in a shear flow, which transits into a quasi-stationary state with significant amplitude increase of perturbations due to non-normality of the linear operator governing their dynamics (see Ioannou and Kakouris 2001).

The purpose of this chapter is to consider in detail the transient growth phenomenon using the simplest example of two-dimensional adiabatic perturbations in a homogeneous rotating shear flow with a quasi-Keplerian angular velocity profile. In Sect. 6.2 we present an analysis of the shear vortex harmonics that are responsible for the transient growth in the spatially local treatment of the problem, and discuss the mechanism of perturbation growth using them as an example. Sections 6.3 and 6.4 are mainly devoted to methods of studying the non-modal perturbation growth as well as to finding the optimal perturbations with maximum growth. Two methods of obtaining the optimal growth curve are presented: a matrix and a variational one. The variational method is less applied, especially in astrophysical studies (see Zhuravlev and Razdoburdin 2014). However, it is essentially more universal than the matrix method. For example, using this method, we calculated in this chapter one of the optimal transient perturbations in a geometrically thin quasi-Keplerian flow with free boundaries (Fig. 6.2) as well as the most unstable perturbation mode (Fig. 6.1), which we discuss in detail in the concluding part of Sect. 6.4.2. A comparison of Figs. 6.2 and 6.1 shows that these two types of perturbations are indeed qualitatively different: the transient spiral is wound up by the flow and its amplitude increases, while the modal spiral rotates as a solid body and demonstrates a monotonic but very weak growth due to the low instability increment. Here, the phase velocity of the modal spiral is such that its corotation radius, at which the energy is transferred from the regular flow, lies inside the flow.

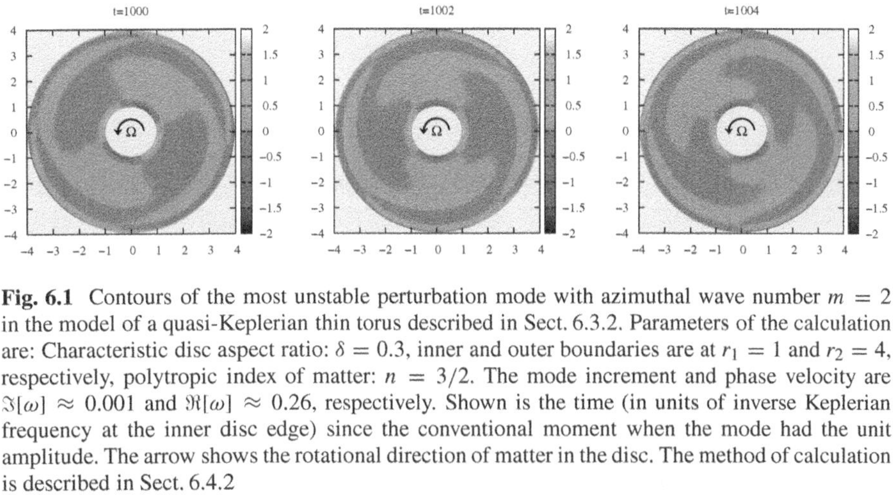

Fig. 6.1 Contours of the most unstable perturbation mode with azimuthal wave number $m = 2$ in the model of a quasi-Keplerian thin torus described in Sect. 6.3.2. Parameters of the calculation are: Characteristic disc aspect ratio: $\delta = 0.3$, inner and outer boundaries are at $r_1 = 1$ and $r_2 = 4$, respectively, polytropic index of matter: $n = 3/2$. The mode increment and phase velocity are $\Im[\omega] \approx 0.001$ and $\Re[\omega] \approx 0.26$, respectively. Shown is the time (in units of inverse Keplerian frequency at the inner disc edge) since the conventional moment when the mode had the unit amplitude. The arrow shows the rotational direction of matter in the disc. The method of calculation is described in Sect. 6.4.2

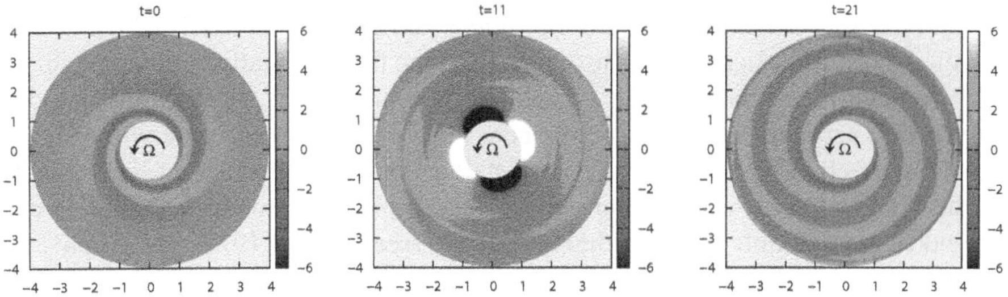

Fig. 6.2 Contours of the perturbation $m = 2$, demonstrating a maximum possible transient growth of acoustic energy at time $t_{opt} = 10$ counted from the beginning of the perturbation evolution in units of the inverse Keplerian frequency at the inner disc edge. The initial perturbation has conventionally unity amplitude and the model of the flow is the same as that in Fig. 6.1. The method of calculation is described in Sect. 6.4.2

6.2 Analytical Treatment of Two-Dimensional Vortices

6.2.1 Adiabatic Perturbations in a Rotational Shear Flow

Consider first the dynamics of small adiabatic perturbations in a perfect fluid with an isentropic equation of state. Perturbations will be described using the Euler approach, i.e. as variations of physical quantities such as density ρ, velocity $\mathbf{v}$ and pressure p at a given point of space at a given time in the perturbed flow relative to the unperturbed background.[2] For simplicity we assume no entropy gradients in the fluid. Then, on the right-hand side of the Euler equations it is convenient to switch from the pressure gradient to the enthalpy gradient. Indeed, under constant entropy the enthalpy differential per unit mass is $dh = dp/\rho$ (see Landau and Lifshitz 1980), and this is valid in both the background and perturbed flows. Therefore, for the Euler perturbations we get $\delta(\nabla p/\rho) = \nabla\delta h$. Making use of this relation, we write down the equations for $\delta\rho$, δh and $\delta\mathbf{v}$ (see also Landau and Lifshitz (1987), paragraph 26) in the form:

$$\frac{\partial \delta\mathbf{v}}{\partial t} + (\mathbf{v}\cdot\nabla)\delta\mathbf{v} + (\delta\mathbf{v}\cdot\nabla)\mathbf{v} = -\nabla\delta h, \tag{6.1}$$

$$\frac{\partial \delta\rho}{\partial t} + \nabla\cdot(\rho\delta\mathbf{v}) + \nabla\cdot(\delta\rho\mathbf{v}) = 0, \tag{6.2}$$

where we have assumed that $\mathbf{v}$ and ρ are the velocity and density of the unperturbed (background) flow, which itself can evolve in time. Equations (6.1) and (6.2) are linear since perturbations are small, and all quadratic terms are omitted.

6.2.1.1 The Model and Basic Equations

To write down the projections of the corresponding equations, let us specify the model we wish to consider to illustrate the transient dynamics. First of all, we assume that the background flow is *stationary* and purely rotational, which is well satisfied in astrophysical discs. This means that the flow is axially symmetric, and it is convenient to use the cylindric coordinate system (r, φ, z) in which the velocity has only an azimuthal non-zero component $\mathbf{v} = (0, v_\varphi, 0)$. Below we will also use the angular velocity of the flow, $\Omega = v_\varphi/r$. It is important to note that isentropicity of the fluid (which is a particular case of barotropicity) immediately implies that v_φ and Ω depend only on the radial coordinate (see Tassoul 1978, paragraph 4.3). At the same time, the density in Eqs. (6.1) and (6.2) is a function of both r and z:

[2]See the monograph by Pringle and King (2007) concerning applications of hydrodynamics to astrophysical problems, in particular, on the application of the theory of hydrodynamic perturbations.

$\rho = \rho(r, z)$. Most interesting, from a phenomenological point of view, is the case of a geometrically thin disc, where $H(r)/r \ll 1$ and H is the disc semi-thickness. The thin-disc approximation will be useful here to find how the density ρ changes with height above the equatorial disc plane. Let us use the hydrostatic equilibrium condition in the background flow:

$$\frac{\partial h}{\partial z} = -\Omega^2(r)z, \tag{6.3}$$

where the vertical gravity acceleration due to the central gravitating body around which the disc rotates stands on the right-hand side. This acceleration is written here ignoring quadratic corrections in the small parameter z/r. Integrating (6.3) with the condition $h(z = H) = 0$ yields the vertical enthalpy distribution:

$$h = \frac{1}{2}(\Omega H)^2 \left(1 - \frac{z^2}{H^2}\right). \tag{6.4}$$

Next, due to the constant entropy assumption $p \propto \rho^\gamma$, where $\gamma = 1 + 1/n$ is the adiabatic index of matter written via the polytropic index n. This means that the square of the sound velocity in the background flow is $a^2 = \gamma p/\rho$, and the density will be mainly dependent on z as follows:

$$a^2 \propto \left(1 - \frac{z^2}{H^2}\right), \quad \rho \propto \left(1 - \frac{z^2}{H^2}\right)^n. \tag{6.5}$$

Finally, for simplicity we will consider only perturbations in which $\delta\mathbf{v}$ *is independent* of z. Generally, this very strong assumption needs justification. In particular, it is relevant to ask: if we choose initial perturbations with such a property, will this be conserved in the further evolution, and if not, how rapidly will this assumption be violated? The answer depends on the vertical disc structure. For example, in Okazaki et al. (1987) it was shown that in the particular case of isothermal vertical density distribution ($n \to \infty$), small perturbations with a homogeneous velocity field in z are exact solutions to Eqs. (6.1) and (6.2). In the more general case with finite n this is no longer the case. However, for example three-dimensional simulations of barotropic toroidal flows indicate that the most unstable perturbations there depend only weakly on z (see Frank and Robertson 1988). This can be related to the fact that when the angular velocity is independent of z, the Reynolds stresses, responsible for the energy transfer from the main flow to the perturbations, do not depend on the vertical component of the velocity perturbation (Kojima 1989; Kojima et al. 1989). At last, the three-dynamical study of transient dynamics of vortices in a Keplerian flow by Yecko (2004) also shows that the most rapidly growing perturbations in a vertically non-stratified medium are almost independent of z (see also Maretzke et al. 2014). Now, looking at the vertical radial and azimuthal projections of (6.1), we see that our assumption implies independence of δh on z, and therefore the right-hand side of the vertical projection of (6.1) vanishes. Then, if we additionally assume that vertical velocity perturbations are absent initially, $\delta v_z = 0$, these will remain absent. Therefore, the perturbed flow, as well as the background flow, will remain in vertical hydrostatic equilibrium. It can be shown that the assumption of vertical hydrostatic equilibrium in the perturbed flow is equivalent to the assumption of a homogeneous velocity perturbation field in the z direction, i.e. one assumption always follows from the other. At the same time, if the fluid is not isentropic and there is a radial entropy gradient in the disc, the simplifying assumptions made above are insufficient to set δv_z to zero.

We have thus come to the conclusion that we will deal with a flat velocity perturbation field, i.e. $\delta \mathbf{v} = \{\delta v_r, \delta v_\varphi, 0\}$, with δv_r and δv_φ, like δh, being dependent on the radial and azimuthal coordinates only. However, it is important to emphasize that this is not the case for $\delta\rho$ that enters the continuity equation (6.2). Here it is convenient to use the relation between the pressure and density variations in an isentropic fluid, $dp = a^2 d\rho$, which is a consequence of the barotropic equation of state. Due to the universal character of this relation, small Eulerian perturbations will be related in the same way, i.e. $\delta p = a^2 \delta\rho$, where a^2 is the speed of sound in the background flow. Consequently,

$$\delta\rho = (\rho/a^2)\delta h, \tag{6.6}$$

and this expression will be plugged into (6.2), after which only background quantities in Eq. (6.2) will depend on the radial coordinate. When integrating Eq. (6.2) in its new form over z, we should keep in mind that

$$\int_{-H}^{H} \frac{\rho}{a^2} dz = \sqrt{\pi}\frac{\Gamma(n)}{\Gamma(n+1/2)}\frac{\rho}{a^2}\bigg|_{z=0}, \quad \int_{-H}^{H} \rho dz \equiv \Sigma = \sqrt{\pi}\frac{\Gamma(n+1)}{\Gamma(n+3/2)}\rho|_{z=0}, \tag{6.7}$$

where we have used relation (6.5) and introduced the surface density Σ.

Using the fundamental property of the gamma-function, $\Gamma(\mathbf{z}+1) = \mathbf{z}\Gamma(\mathbf{z})$, in (6.2), we can explicitly write down the set of Eqs. (6.1), (6.2) for azimuthal complex Fourier harmonics $\delta v_r, \delta v_\varphi, \delta h \propto \exp(im\varphi)$

$$\frac{\partial \delta v_r}{\partial t} = -im\Omega\, \delta v_r + 2\Omega \delta v_\varphi - \frac{\partial \delta h}{\partial r}, \tag{6.8}$$

$$\frac{\partial \delta v_\varphi}{\partial t} = -\frac{\kappa^2}{2\Omega}\delta v_r - im\Omega\, \delta v_\varphi - \frac{im}{r}\delta h, \tag{6.9}$$

$$\frac{\partial \delta h}{\partial t} = -\frac{a_*^2}{r\Sigma}\frac{\partial}{\partial r}(r\Sigma\delta v_r) - \frac{ima_*^2}{r}\delta v_\varphi - im\Omega\, \delta h, \tag{6.10}$$

where $a_*^2 \equiv na_{eq}^2/(n+1/2)$, and a_{eq} is the background speed of sound in the equatorial disc plane. In addition, $\kappa^2 = (2\Omega/r)d/dr(\Omega r^2)$ is the square of the epicyclic frequency, i.e. the frequency of free oscillation of the fluid in the (r, φ) plane, which can be easily checked by writing (6.8), (6.9) for $\delta h = 0$ and substituting there the solution $\delta v_r, \delta v_\varphi \propto \exp(-i\omega t)$. We would like to point out that reducing the three-dimensional problem to an effectively two-dimensional one in a thin disc clearly can be performed by simply changing from volume density to surface density, and replacing the polytropic index with $n + 1/2$ as in the original, not integrated over z equations, as was first shown in Churilov and Shuhman (1981).

6.2.1.2 Types of Perturbations

The set of Eqs. (6.8)–(6.10) describes the dynamics of two types of perturbations inside the disc which are possible in the two-dimensional formulation of the problem: vortices and density waves.[3] The separation between them for transient perturbations will be described below in the local framework that allows the simplest physical interpretation of the behavior of perturbations in a differentially rotating flow. In addition, when there are free radial boundaries in the background flow (for example, in a disc with finite radial extension when at some inner and outer radii Σ vanishes and the shear acquires a super-Keplerian angular velocity gradient), surface gravity waves arise near the boundaries (see papers Blaes and Glatzel (1986), Glatzel (1987a,b)). This occurs because of the presence of a somewhat significant radial pressure gradient in the flow, equivalent to a non-zero gravitational acceleration, which gives rise to waves similar to ocean waves running over the free surfaces (or radial density jumps).

6.2.1.3 On Perturbation Modes

These types of perturbations were studied in detail in the 1980s using the spectral method, when the set of Eqs. (6.8)–(6.10) was solved for particular temporal Fourier harmonics $\propto \exp(-i\omega t)$ called *modes* (see the reviews by Narayan and Goodman (1989) and Narayan (1991)). In this analysis, the local dispersion relation gives only real values of ω in all astrophysically important cases where $\Omega(r)$ is such that the specific angular momentum Ωr^2 increases with radius outwards. This means local stability of the discs and prohibits exponential growth of small-scale perturbations, which is also in accordance with the well-known Rayleigh criterion for the particular case of axially symmetric perturbations (see paragraph 27 in Landau and Lifshitz (1987)). Unlike this case, the global setup of the problem for axially non-symmetric modes, when the set of differential equations with respect to the radial coordinate with the corresponding boundary conditions at the inner disc radius and at infinity (or at the outer disc boundary) is solved, yields a discrete set of ω, where there can be complex frequencies as well (see, for example, Papaloizou

[3]Density waves are also frequently referred to as inertial-acoustic waves.

and Pringle 1984, 1985, 1987; Glatzel 1987a,b; Goldreich et al. 1986; Kojima 1986; Kato 1987; Sekiya and Miyama 1988, etc.) The non-zero real part of the frequency corresponds to the angular velocity of *solid-body* rotation of a given mode in the flow. Generally, solid-body azimuthal motion of perturbations of constant phase with the same azimuthal velocity $\Re[\omega]/m$ at all r is the main distinctive feature of modes distinguishing them from other types of perturbations. Here $\Re$ means the real part of the frequency ω. A non-zero imaginary part of the frequency, $\Im[\omega]$, means that the (canonical, see Friedman and Schutz (1978)) energy and angular momentum are exchanged between this mode and either the background flow (Goldreich and Narayan 1985; Drury 1985; Narayan et al. 1987; Papaloizou and Pringle 1987) or the mode with (canonical) energy of the opposite sign (Glatzel 1987b, 1988; Savonije and Heemskerk 1990). In the literature, the first mechanism is also referred to as the Landau mechanism, and the second one—as mode coupling. The energy exchange in both cases is resonant, i.e. always occurs in the so-called critical layer at the radius where $\omega = m\Re[\Omega]$, which is called the corotation radius. For a detailed discussion of the physics of these resonant mechanisms of mode growth (decay), see the monograph by Stepanyants and Fabrikant (1989). Nevertheless, in flows with almost Keplerian rotation both the mode coupling and their interaction with the background are extremely slow, and the corresponding increments even for a substantial disc aspect ratio $H/r \sim 0.1$ is only one hundred thousandth of the characteristic Keplerian frequency (see Zhuravlev and Shakura 2007a,b). This result led to the general conclusion that at least in the simplest barotropic discs the modes cannot underly any hydrodynamic activity and, in particular, cannot induce turbulence or any other variant of enhanced angular momentum transfer to the flow periphery.

6.2.1.4 On Perturbation Measurements

To conclude this section, let us discuss the problem of perturbation measurements. Indeed, in the present chapter we are interested in how strongly can some perturbations grow in a given time interval. To describe this quantitatively, it is necessary to introduce a norm of perturbations which would characterize the amplitudes of $\delta v_r, \delta v_\varphi, \delta h$ at a given time. This should be a real and positive definite quantity.

The most natural choice is the total acoustic energy of the perturbation in the disc. A derivation of the expression for the acoustic energy density can be found for example in a 65 of Landau and Lifshitz (1987). Here we rewrite this expression in a more convienient way:

$$E = \frac{1}{2}\int r dr d\varphi dz \left(\rho_0|\delta\mathbf{v}|^2 + \frac{a^2}{\rho_0}|\delta\rho|^2\right) = \frac{1}{2}\int r dr d\varphi dz \left(\rho_0|\delta\mathbf{v}|^2 + \frac{\rho_0}{a^2}|\delta h|^2\right). \tag{6.11}$$

In the second term we use perturbation of enthalpy δh instead of perturbation of density $\delta\rho$. These two quantities are connected via the expression: $\delta\rho = \delta p/a^2 =$

$\rho_0 \delta h / a^2$. The integrated expression does not depend on the azimuthal coordinate, and the integral over the vertical coordinate can easily be calculated with the help of the expressions (6.7). Thus we get the final form of the expression for the total acoustic energy of the perturbation:

$$E = \pi \int \Sigma \left(|\delta v_r|^2 + |\delta v_\varphi|^2 + \frac{|\delta h|^2}{a_*^2} \right) r \, dr \tag{6.12}$$

After taking the derivative of (6.12) with respect to time and making use of Eqs. (6.8)–(6.10), we obtain (see also expression (8) from Savonije and Heemskerk (1990)):

$$\frac{dE}{dt} = -2\pi \int \frac{d\Omega}{dr} r \Sigma \Re[\delta v_r \delta v_\varphi^*] \, r \, dr - 2\pi r \Sigma \Re[\delta v_r \delta h^*] \, |_{r_1, r_2}, \tag{6.13}$$

where the symbol $*$ means complex conjugation and r_1 and r_2 are the inner and outer boundaries of the flow, respectively. Here r_2 can be at infinity. As $\Sigma \to 0$ at the flow boundaries, the second term on the right-hand side of (6.13) disappears, and we see that E can be variable precisely in a differentially rotating flow. Without rotation or for solid-state rotation E remains constant in time. It is important to note that the increase/decrease of E will imply that the average flow amplitudes δv_r, δv_φ and δh, also increase/decrease, since (6.12) contains squares of modules of these values taken with the same signs. Note that for modes, Eq. (6.13) implies

$$\frac{dE}{dt} \propto \exp(2\Im[\omega] t), \tag{6.14}$$

i.e. the small increments obtained for quasi-Keplerian flows allow us to conclude that the total acoustic energy of modes there $E \simeq const$ on dynamic $\sim \Omega^{-1}$ and acoustic $\sim (\Omega H / r)^{-1}$ time scales.

Our task now is to understand how E can change over such time intervals for arbitrary perturbations. By introducing the perturbation vector $\mathbf{q}(t)$ as a set of functions $\{\delta v_r(r),\ \delta v_\varphi(r), \delta h(r)\}$ taken at some time t, the norm of the perturbation can be chosen as

$$||\mathbf{q}(t)||^2 = E(t). \tag{6.15}$$

6.2.2 Local Approximation: Transition to Shear Harmonics

The easiest solution to the problem formulated above can be obtained in the spatially local approximation. In this approximation it is assumed that the characteristic scale of perturbations, λ, is a small fraction of some fiducial radial coordinate r_0 around which the dynamics of the perturbation is studied, $\lambda \ll r_0$. We introduce the new radial variable $x \equiv r - r_0 \ll r_0$ and also the new azimuthal variable $y \equiv r_0(\varphi - \Omega_0 t) \ll r_0$, where $\Omega_0 \equiv \Omega(r_0)$ is the angular velocity of rotation of the new coordinate system. Here in Eqs. (6.8)–(6.10) only the leading terms in small x are retained. In practice, this means that only the linear dependence on x has to be taken into account in the angular velocity profile:

$$\Omega = \left.\frac{d\Omega}{dx}\right|_{r_0} x = -q\Omega_0 \frac{x}{r_0} \ll \Omega_0, \tag{6.16}$$

where $q \equiv -(r/\Omega)(d\Omega/dr)|_{r=r_0}$ and $\Omega(x = 0) = 0$, since we are working in the frame rotating with angular velocity Ω_0. The corresponding linear background velocity is $v_y^{loc} = r_0\Omega = -q\Omega_0 x$.

Next, on the right-hand side of Eqs. (6.8)–(6.10) we only keep terms of the order up to $\sim x/\lambda$ and drop the terms $\sim x/r_0$ and lower. For clarity, we also write down the coefficient before δv_r in the term from (6.9) that includes κ^2:

$$-\frac{\kappa^2}{2\Omega} = -2\Omega - r\frac{d\Omega}{dr} = 2q\Omega_0\frac{x}{r_0} + (r_0 + x)\frac{q\Omega_0}{r_0} = 3q\Omega_0\frac{x}{r_0} + q\Omega_0$$

and we find that it is sufficient to take into account only the term $q\Omega_0$. Next, bearing in mind that the new reference frame is not inertial, it is necessary to add the perturbed Coriolis force components $2\Omega_0\delta v_\varphi$ to the right-hand side of (6.8) and $-2\Omega_0\delta v_r$ to the right-hand side of (6.9).

After substituting $im \to \partial/\partial\varphi$ in the set (6.8)–(6.10), i.e. after returning back to the arbitrary dependence of the Eulerian perturbations on φ and by denoting the local analogues of perturbations of the velocity components as u_x, u_y and W, respectively, we arrive at the following equations:

$$\left(\frac{\partial}{\partial t} - q\Omega_0 x\frac{\partial}{\partial y}\right) u_x - 2\Omega_0 u_y = -\frac{\partial W}{\partial x}, \tag{6.17}$$

$$\left(\frac{\partial}{\partial t} - q\Omega_0 x\frac{\partial}{\partial y}\right) u_y + (2-q)\Omega_0 u_x = -\frac{\partial W}{\partial y}, \tag{6.18}$$

$$\left(\frac{\partial}{\partial t} - q\Omega_0 x\frac{\partial}{\partial y}\right) W + a_*^2\left(\frac{\partial u_x}{\partial x} + \frac{\partial u_y}{\partial y}\right) = 0. \tag{6.19}$$

The set of Eqs. (6.17)–(6.19) was first derived in Goldreich and Lynden-Bell (1965)[4] (see also Regev and Umurhan (2008), where it is described for different background flow models).

[4]Even earlier, in the context of lunar dynamics, the local approach to study the motion of matter was utilized by Hill (1878).

6.2.2.1 Transition to Shear Harmonics

A convenient property of the set of Eqs. (6.17)–(6.19) is that by changing to variables corresponding to the co-moving shear reference frame, it is possible to make it homogeneous in both x and y, which, in turn, enables us to split any arbitrary perturbation into individual spatial Fourier harmonics (SFHs) with certain wave numbers k_x and k_y. Indeed, let us introduce the new *dimensionless* variables $x' = \Omega_0 x/a_*$, $y' = \Omega_0(y + q\Omega_0 xt)/a_*$, $t' = \Omega_0 t$.[5] Such a substitution corresponds to a change of partial derivatives according to the rule

$$\frac{a_*}{\Omega_0}\frac{\partial}{\partial x} = \frac{\partial}{\partial x'} + qt'\frac{\partial}{\partial y'}, \quad \frac{a_*}{\Omega_0}\frac{\partial}{\partial y} = \frac{\partial}{\partial y'}, \quad \Omega_0^{-1}\frac{\partial}{\partial t} = \frac{\partial}{\partial t'} + qx'\frac{\partial}{\partial y'} \tag{6.20}$$

Making use of (6.20), we arrive at a set of equations in which all coefficients depend only on t'. We now substitute into this system SFH written in the form

$$f = \hat{f}(k_x, k_y, t')\exp(\mathrm{i}k_x x' + \mathrm{i}k_y y'), \tag{6.21}$$

where f is any unknown variable, $\hat{f}$ is its Fourier amplitude, k_x and k_y are the dimensionless wave numbers along axes x' and y', respectively, expressed in units Ω_0/a_*. Changing back to variables x, y in particular solutions (6.21) reveals that they represent perturbations periodic in space whose phase forms a plane front with orientation depending on time for $k_y \neq 0$. The dimensionless wave number along x has the form

$$\tilde{k}_x(t) \equiv k_x + qk_y t \tag{6.22}$$

and changes with time: the wave vector turns around during advection by the shear flow, which was first noted by Kelvin (1887) and Orr (1907a,b) so the SFH are often called *shear harmonics*. We directly note that for $\tilde{k}_x < 0$ the wave vector is directed inwards, and on the global scale for Fourier harmonics with wave number m this corresponds to so-called *leading* spirals with arms pointing turned in the disc rotation direction. Inversely, the case $\tilde{k}_x > 0$ corresponds to *trailing* spirals with arms pointing oppositely to the disc rotation. If at the initial time $k_x < 0$, the arms of the initially leading spiral are deformed and shortened by the flow, and the so-called *swing* moment occurs, t_s, when the wave vector of SFH is strictly azimuthal and $\tilde{k}_x(t_s) = 0$, after which the spiral becomes trailing, and its arms are stretched by the flow (see Fig. 6.2). This process is well-known in the dynamics of stellar galactic discs (see paragraph 6.3.2 in Binney and Tremaine (2008)).

[5]Due to the vertical hydrostatic equilibrium in the disc, this means that we express the length in units of its semi-thickness, $H = a_*/\Omega_0$.

Thus, for SFH we arrive at the following set of ordinary differential equations:

$$\frac{d\hat{u}_x}{dt} = 2\hat{u}_y - \mathrm{i}\,\tilde{k}_x(t)\hat{W}, \tag{6.23}$$

$$\frac{d\hat{u}_y}{dt} = -(2-q)\hat{u}_x - \mathrm{i}\,k_y\hat{W}, \tag{6.24}$$

$$\frac{d\hat{W}}{dt} = -\mathrm{i}\,(\,\tilde{k}_x(t)\hat{u}_x + k_y\hat{u}_y\,), \tag{6.25}$$

where $\hat{u}_x$ and $\hat{u}_y$ are expressed in units a_* and $\hat{W}$ in units a_*^2. Here and below we will omit the prime for the time variable notation.

6.2.2.2 Potential Vorticity

Equations (6.23)–(6.25) have an important property: the quantity

$$I = \tilde{k}_x(t)\hat{u}_y - k_y\hat{u}_x + \mathrm{i}(2-q)\hat{W} \tag{6.26}$$

is the invariant of motion, which can be easily verified by direct calculation of dI/dt.

It turns out that I (to the multiplication factor i) is SFH of the Eulerian perturbation of the potential vorticity. The potential vorticity $\boldsymbol{\zeta}$, which is by definition the vorticity itself divided by density, $\boldsymbol{\zeta} \equiv \boldsymbol{\omega}/\rho$ (see Johnson and Gammie 2005), is conserved in all fluid elements in plane-parallel barotropic flows. Therefore, for its Eulerian perturbation we have

$$\delta\left(\frac{d\boldsymbol{\zeta}}{dt}\right) = \frac{d\delta\boldsymbol{\zeta}}{dt} + (\delta\mathbf{v}\nabla)\boldsymbol{\zeta}_0 = 0, \tag{6.27}$$

where $\boldsymbol{\zeta}_0$ is the potential vorticity of the background flow. As in both background and perturbed flows the velocity fields are plane-parallel, the vorticity has only one non-zero z-component, which we will consider scalar below.

Next, by definition (in a non-rotating cylindrical coordinate system), the potential vorticity in the background flow is

$$\zeta_0 = (r\Sigma)^{-1} d/dr(\Omega r^2) = \kappa^2/(2\Omega\Sigma) = (2-q)\Omega/\Sigma, \tag{6.28}$$

and should be constant in the local space approximation in use, since the velocity shear is then constant, cf. (6.16). Therefore, the second term in the last equality in (6.27) vanishes, and we see that $\delta\zeta$ is indeed conserved. Apparently, the first two terms in (6.26) arise due to perturbation of the vorticity itself, which is equal to the curl of the velocity perturbation, and the third term emerges due to the non-zero density perturbation represented by the dimensionless quantity $\hat{W}$ (the coefficient $2 - q$ here arises due to multiplication by the constant background vorticity, cf. (6.28)).

6.2.2.3 Inhomogeneous Wave Equations: Density Waves and Vortices

We now differentiate Eq. (6.24) with respect to t and take into account the relations following from Eqs. (6.23), (6.25), as well as the definition (6.26), to obtain a new equation:

$$\frac{d^2\hat{u}_y}{dt^2} + K(t)\hat{u}_y = \tilde{k}_x(t)I, \tag{6.29}$$

where $K(t) \equiv \tilde{k}_x^2(t) + k_y^2 + 2(2-q)$. Apparently, (6.29) represents a detached *wave* equation for azimuthal velocity component perturbation, $\hat{u}_y$, with an *inhomogeneous* part $\sim I$ (see Bodo et al. 2005).

In a similar way, from (6.23) and (6.25) we derive two equations of the same type:

$$\frac{d^2\hat{u}_x}{dt^2} + K(t)\hat{u}_x + 2\mathrm{i}qk_y\hat{W} = -k_yI, \tag{6.30}$$

$$\frac{d^2\hat{W}}{dt^2} + K(t)\hat{W} + 2\mathrm{i}qk_y\hat{u}_x = -2\mathrm{i}I, \tag{6.31}$$

which can be separated by changing variables $\hat{u}_\pm = (\hat{u}_x \pm \hat{W})/2$ (see Heinemann and Papaloizou 2009a).

Let us now consider in more detail, for example, Eq. (6.29). Its general solution is the sum of the general solution of the corresponding homogeneous equation and a partial solution of the inhomogeneous equation. First, we consider both these solutions in the solid-body rotation limit, i.e. without shear, $q = 0$. Then all coefficients in (6.29) turn constant and

- the homogeneous equation has partial fundamental solutions $\hat{u}_y^{(dw)} \propto \exp(\pm\mathrm{i}\omega t)$ with frequency $\omega = \sqrt{K}$, corresponding to the density waves propagating in opposite directions,
- the partial solution with non-zero right-hand part can be taken as the constant $\hat{u}_y^v = (k_x/K)I$. In other words, $u_y^{(v)}$ corresponds to the zero frequency $\omega = 0$ and represents a static perturbation. This perturbation, apparently, has a non-zero vorticity and corresponds to a vortex (it is possible to show that divergence of the velocity perturbation for this solution vanishes, by taking the similar solution for $\hat{u}_x$ from Eq. (6.30), $\hat{u}_x^v$, and checking that the combination $k_x\hat{u}_x^v + k_y\hat{u}_y^v = 0$).

6.2.2.4 Amplification of Density Waves

Accounting for the non-zero shear, the density wave frequency becomes a function of time. For example, for leading/trailing spirals this frequency gradually decreases/increases with a simultaneous wavelength increase/decrease, which, in turn, in the absence of viscosity, leads to a monotonic decrease/increase in the energy and amplitude of the density waves. Such growth of density-wave amplitudes was studied in Chagelishvili et al. (1994, 1997). The reason for this growth can be understood from the fact that due to the axial symmetry of the background flow, the canonical angular momentum of the wave, J_c, should be conserved (see Friedman and Schutz 1978). From here we obtain that, following equation (52) from Friedman and Schutz (1978), the canonical energy, $E_c \sim \omega J_c$, *linearly* increases starting from some sufficiently long time, since $\omega = \sqrt{K}$ (see above). The conservation of J_c for the local perturbation considered here is discussed in Section 3.2 of Heinemann and Papaloizou (2009a). Unlike J_c, the canonical energy itself in this case is not conserved any more, since the time-variable frequency makes the problem inhomogeneous in time. This growth (or decrease) of the energy, despite that the wave frequency ω is present here, is already essentially non-modal, since ω is a function of time, which, in turn, is connected precisely to the deformation of SFH by the shear flow.

In the present chapter, however, we will be more interested in the 'classical' variant of non-modal growth, which is called 'transient' in the literature. In the simplest model considered here it is represented by the vortex solution which for $q \neq 0$ becomes dynamical and, oppositely to the waves, is aperiodic.

6.2.2.5 The Vortex Existence Criterion

Before discussing in detail the behavior of the vortex solution, let us analyze the justification for the decoupling of perturbations in waves and vortices made above in the presence of a shear. Indeed, immediately after $\tilde{k}_x$ becoming variable, the solution $\hat{u}_y^v$ does not exactly satisfy Eq. (6.29) anymore, since a non-zero second derivative of $\hat{u}_y^v$ appears. Moreover, in the limit $\tilde{k}_x \to 0$ Eq. (6.29) becomes homogeneous, and its solution describes density waves only. The region, in which $\tilde{k}_x \to 0$, corresponds to the swing of SFH, and thus we see that the vortex solution becomes poorly defined there: the vortex must share wave properties. This means that we cannot neglect the second time derivative in Eq. (6.29) anymore for slowly evolving solutions. In other words, $\hat{u}_y^v$ cannot be considered, even approximately, as a solution of Eq. (6.29). Let us discuss in more detail the criterion of decoupling of density waves and vortices in a shear flow.

In order to do this, we use the fact that vortex dynamics is possible only in subsonic flows (see Landau and Lifshitz (1987), end of Sect. 10). In the considered case of an infinite flow this means that the difference in the fluid velocity on the characteristic scale of the problem must be smaller than the sound velocity. The characteristic spatial scale is determined by the instant spatial period of SFH in the radial direction, $\lambda_x \sim H|\tilde{k}_x|^{-1}$. For the infinitesimal perturbations considered here, its is sufficient to apply the condition of vortex dynamics for the background flow, and then the velocity difference is given simply by the change in the flow azimuthal velocity, i.e. for a flow with constant shear we get

$$\lambda_x q \Omega_0 / a_* = \frac{q}{|\tilde{k}_x|} \ll 1, \tag{6.32}$$

Thus, the spatial radial period of the vortex harmonics must be smaller than the disc thickness. It is important to note that the condition (6.32) does not directly contain the azimuthal wave number k_y, and hence perturbations can be vortex even if their azimuthal spatial scale exceeds the disc thickness. In connection with this, it is most important to consider the case of initially leading spirals, i.e. SFH with $k_x < 0$. For such spirals, the swing occurs at

$$t_s = -k_x/(q k_y) > 0, \tag{6.33}$$

i.e. when $\tilde{k}_x = 0$. Clearly, if the initial spiral was vortex-like, and therefore $|k_x| \gg 1$, and its evolution was initially described by the approximate solution $\hat{u}_y^v$, then in some time interval around t_s the vortex approximation is not valid, and the complete Eq. (6.29) should be integrated. Let us call this time interval *the swing interval* and obtain the condition under which its duration will be much shorter than the characteristic time for evolution of SFH, determined by the time of spiral unwinding, t_s (see Zhuravlev and Razdoburdin 2014).

The moments in time at which the vortex approximation breaks down can be estimated from the limiting case of equality in the condition (6.32):

$$t_{s_1,s_2} = t_s \left(1 \pm \frac{q}{k_x}\right), \tag{6.34}$$

from where we see that the swing interval is much shorter than the evolution time of the entire vortex spiral, $t_{s_2} - t_{s_1} \ll t_s$, once

$$|k_x| \gg 2q, \tag{6.35}$$

which does not contain k_y. The condition (6.35) implies that to study vortex dynamics, we can use the solution $\hat{u}_y^v$ each time when at the initial moment the spiral is sufficiently strongly wound irrespective of the value of k_y, i.e. in both the true short-wave limit $k_y \gg 1$ and the long-wave limit $k_y \ll 1$. In the last case, the vortices will be referred to as 'large-scale'. Here we exclude the case $k_y \sim 1$, since as was shown numerically in Chagelishvili et al. (1997), Bodo et al. (2005) and analytically studied in the WKB approximation in Heinemann and Papaloizou (2009a), in this case during the swing the vortex additionally generate a pair of density waves corresponding to trailing spirals and propagating inside and outside the disc. This process is asymmetric, since only density wave generation is possible by vortices, and not vice versa. In Heinemann and Papaloizou (2009a) analytical expressions for the amplitude and phase of the generated wave were obtained. It was shown that its amplitude is proportional, at first, to the vortex vorticity I, and at second, to the combination $\epsilon^{-1/2}\exp(-4\pi/\epsilon)$ (see formula (53) in Heinemann and Papaloizou (2009a)). Here ϵ is the small WKB parameter

$$\epsilon = \frac{q k_y}{k_y^2 + \kappa^2/\Omega_0^2}, \tag{6.36}$$

were, again, $\kappa^2/\Omega_0^2 = 2(2-q)$. Expression (6.36) implies that the excitation of density waves is exponentially suppressed in both the short-wave and long-wave limits and is significant only for $k_y \sim 1$ (here we specify that we will not consider the extreme cases where $q \ll 1$, and therefore $\epsilon \ll 1$ even for $k_y \sim 1$, as well as when $q \to 2$, and hence $\epsilon \gtrsim 1$ even for $k_y \ll 1$).

Thus, the vortex solution of Eq. (6.29) exists when the condition (6.35) holds together with the requirement $k_y \ll 1$ or $k_y \gg 1$, which excludes density wave generation with non-zero vorticity during the swing of a vortex SFH. At the same time, these restrictions provide a criterion to separate waves and vortices in the perturbed flow. Indeed, under such constraints density waves with zero vorticity propagate in the flow independently of vortices and represent the high-frequency branch of solutions to Eq. (6.29) with zero right-hand side. Similarly, for example, sound and wind exist independently in the Earth atmosphere.

6.2.2.6 Vortex Solution

Below we will only consider the evolution of vortex SFH in a shear flow. To conclude Sect. 6.2.2, we also obtain vortex solutions for $\hat{u}_x$ and $\hat{W}$. This can be done most easily by neglecting the second time derivatives of $\hat{u}_x$ and $\hat{W}$ in Eqs. (6.30) and (6.31), as has been done with Eq. (6.29) to obtain u_y^v. Thus, we will have for all three quantities:

$$\hat{u}_x^v = -\frac{K+4q}{K^2+4q^2k_y^2} k_y I, \tag{6.37}$$

$$\hat{u}_y^v = \frac{\tilde{k}_x}{K} I, \tag{6.38}$$

$$\hat{W}^v = 2\mathrm{i}\,\frac{q k_y^2 - K}{K^2+4q^2k_y^2} I, \tag{6.39}$$

It is important to note that the existence of an aperiodic vortex solution in the form (6.37)–(6.39) is possible because of the main simplifying assumption on the local constant velocity shear which provides the existence of time invariant I. This enables us to reduce the set of three homogeneous first-order equations (6.23)–(6.25) to one inhomogeneous second-order equation (6.29). (Other dynamical variables can be obtained from the known solution $\hat{u}_y(t)$, which gives two independent wave solutions (the general solution to the corresponding homogeneous equation) and one aperiodic vortex solution (the partial solution (6.29)). However, taking into account the gradient of velocity shear in the flow, the invariant I disappears, and a reduction of the set of Eqs. (6.23)–(6.25) becomes impossible. From this set we will thus need to obtain directly three independent solutions, two of which, as before, will correspond to the density waves, and the third solution will describe the vortex wave called the Rossby wave (see the discussion in paragraph 4 in Bodo et al. (2005)).[6]

6.2.3 Vortex Amplification Factor

To measure the growth of local perturbations, the average density of their acoustic energy can be taken as the local analogue of norm (6.12):

$$E = \frac{1}{2\bar{S}} \int\limits_S \left((\Re[u_x])^2 + (\Re[u_y])^2 + \frac{(\Re[W])^2}{a_*^2} \right) \mathrm{d}x\mathrm{d}y. \tag{6.40}$$

where $\bar{S}$ is the area of the integration region S.

After substituting the dimensionless SFH (6.21) into (6.40) and integrating over their spatial period we obtain the local variant of norm (6.15):

$$||\mathbf{q}||^2 = \frac{1}{2}\left(|\hat{u}_x|^2 + |\hat{u}_y|^2 + |\hat{W}|^2 \right). \tag{6.41}$$

Making use of the vortex solution for SFH (6.37)–(6.39), we get the norm in the following form:

$$||\mathbf{q}||^2 = \left[\frac{\tilde{k}_x^2}{K^2} + \frac{4 + k_y^2}{K^2 + 4q^2 k_y^2} \right] I^2. \tag{6.42}$$

Below we shall utilize the growth factor as the main quantity characterizing perturbation dynamics:

$$g(t) \equiv \frac{||\mathbf{q(t)}||^2}{||\mathbf{q(0)}||^2}, \tag{6.43}$$

which is, in other words, the norm of a perturbation with respect to its initial value.

[6]See Brekhovskikh and Goncharov (1985), paragraph 43, for a discussion of Rossby waves arising due to the gradient of the velocity shear (the gradient of vorticity) in an incompressible rotating flow.

- **Short-wave perturbations.** For $k_y \gg 1$ we can in any case omit the factor 4 in (6.42) in the numerator of the second term, the term $4q^2k_y^2$ in the denominator of the second term, as well as the term $2(2-q) = \kappa^2/\Omega_0^2$ in the quantity K. Then

$$g \approx \frac{k_x^2 + k_y^2}{\tilde{k}_x^2 + k_y^2}, \tag{6.44}$$

which is the result obtained in Lominadze et al. (1988) (see also formula 4 in Afshordi et al. (2005)). Expression (6.44) shows that SFH initially taken as a leading spiral with $k_x < 0$ increases in amplitude until time (6.33), and at the swing moment, when $\tilde{k}_x = 0$, reaches maximum in the norm and then decays. The energy transfer from the background flow to perturbations is described in detail in terms of fluid particles in Chagelishvili et al. (1996) (see Fig. 2 therein). Similar to the well-known lift-up effect (see the book by Schmid and Henningson (2001), paragraph 2.3.3 for more detail), it is based on 'pickup' of fluid particles by the main flow as they move into a region with different shear velocity. However, it also has an important additional ingredient being interaction of particles with each other at the planes of pressure extrema, resulting in growth of their velocity respective to the background flow, even in situations where the lift-up effect does not work.

6.2.3.1 On the Transient Growth Mechanism

Here we present an additional discussion clarifying the transient growth mechanism. As mentioned in the Introduction and discussed in Sect. 6.2.2, a differentially rotating flow shortens the length of the leading spiral arms of a transiently growing vortex until the swing moment (see Fig. 6.2). Due to the barotropicity of the perturbed flow, the velocity circulation along a fluid contour coinciding with the spiral arm boundary must be constant. Consequently, the contour shortening must lead to a compensating increase in gas velocity along the spiral's boundary. Consider this suggestion more rigorously in the local space limit (see the scheme in Fig. 6.3). Let us calculate the velocity circulation for the most simple fluid contour. Without perturbations, this is naturally a parallelogram with one pair of sides (call them the base of the parallelogram) along the background stream lines, i.e. parallel to the y axis and symmetrical on both sides from the level $x = 0$. The condition that these sides move synchronously with the fluid automatically implies that the entire contour is co-moving with the background flow, since the velocity in the flow is linear in x. Now let us switch to the reference frame co-moving with the shear, in which Eqs. (6.23)–(6.25) were written. In this frame, the background velocity together with the velocity circulation along the given contour are zero. Next, taking into account small perturbations, the velocity circulation must change, strictly speaking, for two reasons. Firstly, a velocity perturbation arises, $\mathbf{u}$ (as determined in the shear reference frame), and secondly, even the contour taken at

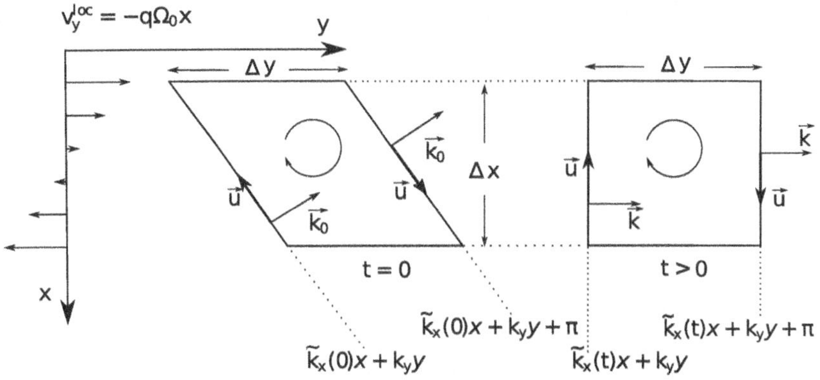

Fig. 6.3 Illustration of the physical reasons for transient growth of two-dimensional vortices in the local space limit (see Sect. 6.2.2). We consider here the case of a short-wave ($k_y \gg 1$) vortex SFH with $k_x < 0$. A liquid contour co-moving with the background flow at two instants is shown: at the initial time $t = 0$ and at the time of the SFH swing when $\tilde{k}_x = 0$. See text (Sect. 6.2.3.1) for an explanation of why it is possible to ignore deformation of the contour by perturbations. At $t = 0$ the contour has the form of a parallelogram with one pair of sides along the y-axis symmetrically relative to $x = 0$ and another pair along two SFH fronts, with the phase difference between them equal to π. Here, $\mathbf{u}$ is the velocity perturbation vector, $\mathbf{k_0}$ and $\mathbf{k}$ show the SFH wave vector at different time moments. Δx and Δy are the parallelogram's height and base, respectively

the time $t = 0$ as a parallelogram starts being deformed due to additional shifts caused by perturbations. In the second case, however, for the small perturbations considered here, only the contribution due to the corresponding change in the background velocity circulation will be important. But this addition is absent, since in the shear reference frame the background velocity is zero at all points. Thus, all we need to do is to calculate the circulation $\mathbf{u}$ along a contour co-moving with the background flow. At time $t = 0$ we take it such that the parallelogram sides coincide with the SFH front lines separated by the phase π (see Fig. 6.3, where the initial front direction is denoted by the wave vector $\mathbf{k_0}$). As in the shear frame SFH, by definition, has constant space phase front lines, it is clear that at times $t > 0$ they remain coinciding with the contour's sides. Now, note that we consider the case $k_y \gg 1$. Therefore $\hat{W}^v \to 0$, and from (6.25) we derive the orthogonality condition $\mathbf{u} \perp \mathbf{k}$. Consequently, the velocity perturbation is directed along the parallelogram's sides and always points to their going around. As for the parallelogram's bases, their contribution to the circulation will be mutually cancelled, since along them the projection of the velocity $\mathbf{u}$ does not change, while the going around direction becomes opposite. With account for the above considerations, the perturbed flow circulation in the co-moving shear frame for the left contour in Fig. 6.3 reads:

$$\mathscr{C}|_{t=0} = 2\Delta y \left(1 + \frac{k_y^2}{k_x^2}\right)^{1/2} |\mathbf{u}|_{t=0}.$$

For the right contour in Fig. 6.3 taken at the spiral swing moment, we similarly find:

$$\mathscr{C}|_{t=t_s} = 2\Delta x\, |\mathbf{u}|_{t=t_s}.$$

By equating these two expressions, we see that the circulation conservation law yields for the vortex SFH with $k_y \gg 1$:

$$g(t_s) = \frac{|\mathbf{u}(t_s)|^2}{|\mathbf{u}(0)|^2} = \frac{k_x^2 + k_y^2}{k_y^2}, \tag{6.45}$$

This coincides with the result following from (6.44) for the spiral swing time.

Thus, we have been convinced that the transient growth of a vortex is in fact due to its perimeter (its 'size') shortening by the background shear flow with constant velocity circulation, $\mathscr{C} = const$, along this perimeter. It is important to note that $\mathscr{C}$, as well as the corresponding vorticity flux, is the measure of the vortex rotation. Therefore, it is appropriate to compare it with a body compressing with angular momentum conservation, since in that case the body's angular velocity increases inversely with the moment of inertia, $\omega_{rot} \propto I_{rot}^{-1}$, and the rotation energy $E_{rot} = 1/2 I_{rot}\omega_{rot}^2 \propto I_{rot}^{-1}$ increases with time. In our case, the background flow does work on shortening the vortex size and thus transfers it the kinetic energy.

Finally, note also that as the differential rotation is purely shear, i.e. occurs with zero divergence of the background flow, the area subtended by the contour considered above must keep constant. Indeed, the area of the parallelogram is the product of its base (which is constant since the flow in homogeneous in y) times its height (which is constant since there is no radial background velocity). Therefore, due to the constant $\mathscr{C}$ and hence the vorticity perturbation flux through the contour, the vorticity perturbation itself is constant. The same conclusion was obtained in Sect. 6.2.2 from the discussion of the invariant (6.26).

6.2.3.2 Estimation of Optimal Growth

Knowing the physical mechanism of the transient vortex growth, let us return to expression (6.44) for their growth factor in the case of short azimuthal wavelength. Clearly, the growth factor of an individual SFH is a function of three arguments, $g = g(k_x, k_y, t)$. However, it is possible to consider a more general characteristic of the transient dynamics which is called the optimal growth of perturbations G. By definition,

$$G \equiv \max_{\forall k_x}\{g\}. \tag{6.46}$$

Formula (6.46) gives the maximum possible amplification among all vortices with given k_y which can occur during a time interval t. Note that below we will also employ an analogue to (6.46) used for the global space problem described by the set of Eqs. (6.8)–(6.10) (see formula (6.94)), where the value G will be determined for all perturbations with fixed azimuthal wave number m.

There are rigorous mathematical algorithms to search for the optimal growth, which we will discuss in the next Section. Here, for analytical estimates in the local space limit, it will be sufficient to recognize that since the growth factor $g(k_x, k_y, t)$ of a certain SFH has maximum at $\tilde{k}_x = 0$, it is reasonable to suppose that G can be estimated as

$$G \approx g(k_x = -k_y q t), \tag{6.47}$$

in other words, to adopt that of all SFH with given k_y, the harmonics that swings at time t reaches maximum possible growth by this time.

Making use of definition (6.47), from (6.44) we obtain the simple expression:

$$G_1 \approx (qt)^2, \tag{6.48}$$

which can be also found in Afshordi et al. (2005) (see formula (5) therein). Note that in that paper corrections to G_1 due to non-zero vertical projection of the wave vector and a finite value of k_y were also obtained. As we see, in a sufficiently long time it is possible to reach arbitrarily large amplitude growth of small-scale vortices $k_y \gg 1$. This growth, however, is power-law and not exponential, as would be expected from a modal instability of the flow.

- **Long-wave perturbations.** We now turn to another limiting case where $k_y \ll 1$ and the azimuthal space period of SFH is much larger than the disc thickness (see Zhuravlev and Razdoburdin 2014). In this case, in the second term in (6.42) we omit k_y^2 in the numerator and $4q^2k_y^2$ in the denominator, and also assume that $K = \tilde{k}_x^2 + \kappa^2/\Omega_0^2$. Here, by the condition (6.35), we see that $||\mathbf{q}(\mathbf{0})||^2 \approx k_x^{-2}$.

  Then, for the SFH growth factor we obtain

$$g \approx k_x^2 \frac{\tilde{k}_x^2 + 4}{(\tilde{k}_x^2 + \kappa^2/\Omega_0^2)^2}, \tag{6.49}$$

  This quantity increases for $\tilde{k}_x$ decreasing with time, i.e., similar to the short-wave vortices, transient growth occurs for $k_x < 0$. Note that now the maximum g, attained during the spiral swing, is proportional to the square of the value k_x itself, but not to the square of the ratio k_x/k_y, as in the case of the short wavelength vortices (cf. (6.44)). In addition, another important difference is that now g depends on the epicyclic frequency as κ^{-4}. Such a strong dependence can be important in discs with super-Keplerian angular velocity gradient. In thin discs this can occur in the inner regions of relativistic discs, where $\kappa \to 0$ when approaching their inner boundary.

Following the definition (6.47), we obtain from (6.49) the corresponding optimal growth factor:

$$G_2 \approx \frac{4\Omega_0^4}{\kappa^4} k_y^2 (qt)^2. \tag{6.50}$$

Note that both (6.48) and (6.50) are valid only for sufficiently large timespans since in order to obtain this expression we used the condition $k_x = -qk_y t$, but at the same time the condition $k_x \gg 1$ must hold, as required by (6.35). Formula (6.50) shows that for rotation profiles weakly different from Keplerians, when $\kappa \sim \Omega_0$, for equal time intervals $G_2 \ll G_1$, since the azimuthal wave number now explicitly entering the optimal growth factor is small, $k_y \ll 1$.[7] Therefore, in the local space limit considered here, small-scale vortices extract energy from the flow more efficiently than large-scale ones. However, it is interesting to learn which of them can display the highest growth over the entire time interval. In an inviscid flow $G_{1,2} \to \infty$ mostly due to small-scale SFH, as we just noted. Nevertheless, a shear flow can have noticeable effective viscosity due to, for example, some weak turbulence. Then the dependence $G(t)$ turns out to have a global maximum G_{max} corresponding to the maximum possible non-modal growth of perturbations irrespective of the time intervals we have considered so far. Physically, the decrease of $G(t)$ after some long time is related to the fact that more tightly wound spirals have larger swing times t_s. This in turn means smaller radial scale of the perturbations and hence smaller dissipation time of perturbations due to viscosity. Ultimately, the leading transient spirals faster start decaying than growing due to unwinding by the flow. It is the value G_{max} for cases $k_y \gg 1$ and $k_y \ll 1$ that we would like to compare below.

6.2.3.3 Role of Viscosity

The effect of viscosity on the maximum possible transient growth of vortices can be estimated as follows (an accurate viscosity calculation is a much more complicated problem, which was solved in Razdoburdin and Zhuravlev (2017)). For sufficiently long time intervals $qt \gg 1$ we have $k_x \gg k_y$ for any of the two limits of k_y we consider. Therefore, in a shearless flow the spiral would decay in the characteristic viscous time $\Delta t_\nu \sim \lambda_x^2/\nu$, where ν is the kinematic viscosity coefficient. Performing a standard viscosity parametrization using the Shakura-Sunyaev α-parameter, $\nu = \alpha a_* H$, we get that $\Delta t_\nu \sim (\Omega_0^{-1} \alpha k_x^2)^{-1}$ rapidly decreases with increasing $|k_x|$. At the same time, the larger $|k_x|$, the longer the transient growth time of the spiral, $\Delta t_{tg} \sim |k_x/(qk_y)|$. Simultaneously with arising of a shear in the flow, the spiral starts unwinding, and therefore the viscous dissipation is delayed. Thus, the equality

[7] In Sect. 6.4.2 below we calculate G in the global problem (see Fig. 6.14), which implies that as $m \to 1$, the difference in the transient growth rate between vortices with azimuthal wavelength shorter and longer than the disc thickness is significantly smaller.

of these characteristic times, $\Delta t_{tg} = \Delta t_{\nu}$, gives the lower limit on the duration of the transient growth of vortices in a viscous flow. Using it we obtain:

$$\max(\Delta t_{tg}) \gtrsim \alpha^{-1/3}(qk_y)^{-2/3} \tag{6.51}$$

It can be verified that expression (6.51) reproduces the estimate made in Afshordi et al. (2005) (see formula (81) therein).

The upper limit on the optimal growth time (6.51), $G_{max} \equiv G(\max(\Delta t_{tg}))$, is given by its inviscid value taken for G_1 or G_2. We then obtain that for $k_y \gg 1$

$$(G_{max})_1 \approx \alpha^{-2/3} q^{2/3} k_y^{-4/3}, \tag{6.52}$$

(see also formula (83) in Afshordi et al. (2005)). At the same time, for $k_y \ll 1$ we have

$$(G_{max})_2 \approx \frac{4\Omega_0^4}{\kappa^4} \alpha^{-2/3} q^{2/3} k_y^{2/3}. \tag{6.53}$$

This result is shown in Fig. 6.4 for some small α and several different shears q: Keplerian and super-Keplerian. We see that even for the Keplerian shear, when $\kappa = \Omega_0$, for k_y different from 1, $(G_{max})_2 \gtrsim (G_{max})_1$. This occurs because the large-scale vortices are significantly less dissipative, which more than compensate

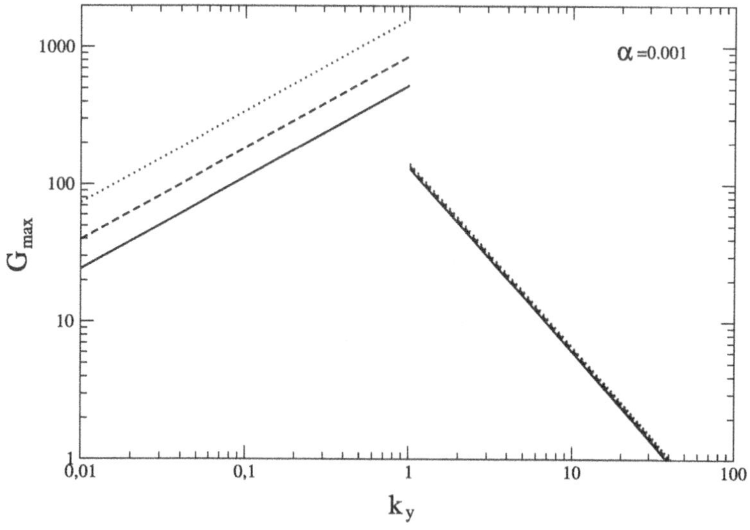

Fig. 6.4 Estimate of the maximum possible transient growth of acoustic energy in a disc with efficient viscosity $\alpha = 0.001$. The solid, dashed and dotted lines correspond to $q = 1.5, 1.6$ and 1.7, respectively. The three uppermost right and left curves are obtained using formula (6.52) and (6.53), respectively

of their low growth rate compared to the small-scale vortices. Note also that despite $(G_{max})_2$ decreasing with decreasing k_y, this occurs at lower rate compared to the case of $(G_{max})_1$ decreasing with increasing k_y. As a result, the integral transient growth of large-scale vortices at all k_y increases in comparison with small-scale ones. An even more significant advantage of large-scale vortices appears for super-Keplerian shears, when $q > 3/2$, due to $(G_{max})_2 \propto \kappa^{-4}$ (see the comment after formula (6.49)). Clearly, the deviation from $q = 3/2$ by several per cents would increase the transient growth rate of perturbations by a factor of a few.

As discussed in Zhuravlev and Razdoburdin (2014), the estimate (6.53) is in reasonable agreement with exact calculations of the optimal growth rate in thin discs in the global space limit for low azimuthal wave numbers m. Thus, large-scale vortices are also able to provide additional transportation of angular momentum to the periphery of a disc with pre-existing weak turbulence.

In Sect. 6.3 we provide a rigorous mathematical justification of algorithms to search for the most rapidly growing perturbations in shear flows. Such perturbations will be called optimal, and the corresponding amplification, as we already mentioned, will be referred to as the optimal growth G. The solutions presented in the Introduction and shown in Figs. 6.1 and 6.2 were obtained using one of these algorithms. We will also provide another example of calculation of G by solving the general set of Eqs. (6.8)–(6.10) in a geometrically thin disc (see Fig. 6.14 below). When discussing mathematical aspects of the non-modal dynamics of perturbations in shear flows, already in the introductory part to the next section we will see that the transient growth phenomenon can be treated as a consequence of *non-orthogonality* of perturbation modes, which will be evident, in particular, from consideration of simple analogues presented in Figs. 6.5 and 6.6.

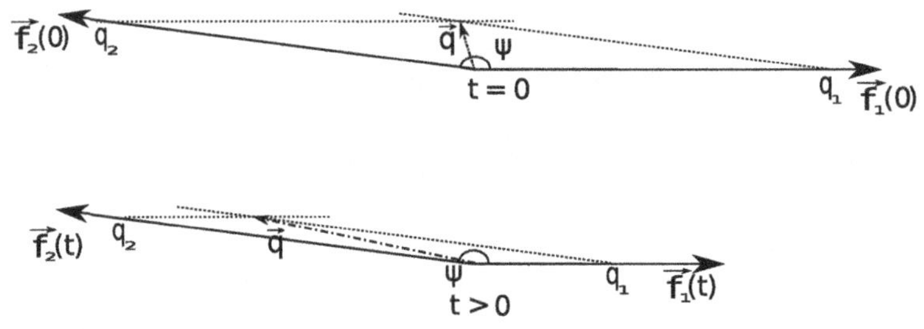

Fig. 6.5 The increase in the sum of two non-orthogonal vectors, $\mathbf{q} = \mathbf{f}_1 + \mathbf{f}_2$, with decreasing lengths but conserved angle between each other. It is assumed that $q_1 = q_2 = 1$

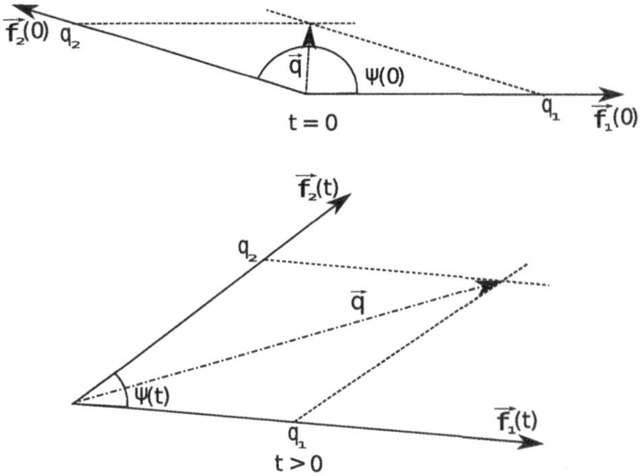

Fig. 6.6 The increase in the sum of two non-orthogonal vectors, $\mathbf{q} = \mathbf{f}_1 + \mathbf{f}_2$, with conserved lengths but changing angle between each other. It is assumed that $q_1 = q_2 = 1$

6.3 Calculation of Optimal Perturbations

6.3.1 Definition and Properties of Singular Vectors

General solutions to the initial value problem of evolution of small perturbations, as described by the general Eqs. (6.1)–(6.2), supplemented by appropriate boundary conditions, can be conveniently studied using abstract concepts of the functional space of the so-called state vectors of the system, as well as the notion of linear operators acting on these vectors. In Sect. 6.2.1, in addition to the set of Eqs. (6.8)–(6.10), we have already introduced the particular case of the state vector as a set of azimuthal Fourier harmonics of Eulerian perturbations $\mathbf{q}(t) \equiv \{\delta v_r(r), \delta v_\varphi(r), \delta h(r)\}$, taken at some fixed instant t. In this section we will take on the original general case when $\mathbf{q}(t) \equiv \{\delta \mathbf{v}(\mathbf{r}), \delta h(\mathbf{r}), \delta \rho(\mathbf{r})\}$.

Let us consider some properties of a dynamical operator $\mathbf{Z}$ acting in the Hylbert space of vectors $\mathbf{q}$ and corresponding to the set of Eqs. (6.1)–(6.2). This operator transforms the initial perturbation vector $\mathbf{q}(0)$ to the consecutive vector $\mathbf{q}(t)$, i.e. in the operator form the set of equations can be written as

$$\mathbf{q}(t) = \mathbf{Z}\mathbf{q}(0). \tag{6.54}$$

Since the operator $\mathbf{Z}$ controls dynamics of linear perturbations in a flow it is often called *dynamical operator*. First, let us recall some terms of linear operator theory.

6.3.1.1 Linear Operators: From the Particular to the General

There are a lot of linear operator types. Let us list those of them that we will need below, from the more particular to the more general case. We start with *positive definite* operators, for which the inner product $(\mathbf{Z}\mathbf{q}, \mathbf{q}) > 0$ for any vector $\mathbf{q}$. By definition, the eigenvalues of a positive definite operator are positive. Indeed, by multiplying the equation $\mathbf{Z}\mathbf{q} = \lambda\mathbf{q}$ with $\mathbf{q}$, we see that its left-hand side is positive, and the right-hand side is the product of the eigenvalue and a positive value, hence the positive eigenvalue.

Self-adjoint (Hermitian) operators, which are identical to their adjoint operators, $\mathbf{Z} = \mathbf{Z}^\dagger$ (Korn and Korn 1968, paragraph 14.4), are most frequently used in different physical problems. Eigenvalues of a self-adjoined operator are real values (Korn and Korn 1968, paragraph 14.8).

In turn, self-adjoined operators are a particular case of *normal operators* . An operator $\mathbf{Z}$ is called normal if it commutes with its adjoint operator: $\mathbf{Z}\mathbf{Z}^\dagger = \mathbf{Z}^\dagger\mathbf{Z}$ (Korn and Korn 1968, paragraph 14.4). All eigenvalues of a normal operator are complex conjugates of its adjoint operator's eigenvalues. Eigenfunctions of the operators $\mathbf{Z}$ and $\mathbf{Z}^\dagger$ coincide. Additionally, eigenvectors of a normal operator corresponding to different eigenvalues are orthogonal (Korn and Korn 1968, paragraph 14.8). Therefore, to calculate the *operator norm* of these operators, it is sufficient to find their eigenvalues. We recall that the norm of an operator $\mathbf{Z}$ mapping a Hylbert space H into itself is the number (Vilenkin et al. 1972, Ch. 1)

$$||\mathbf{Z}|| = \sup_{\mathbf{x}\in H} \frac{||\mathbf{Z}\mathbf{x}||}{||\mathbf{x}||} \tag{6.55}$$

The norm of the dynamical operator is very useful, because it allows us to calculate the limit of the vector's norm growth under the action of this operator.

For a normal operator this problem is solved quite easily. To illustrate this, we (following Schmid 2007) consider an important particular case in which the operator $\mathbf{Z}$ can be represented as an operator exponent: $\mathbf{Z} = \mathrm{e}^{\mathbf{A}t}$ (see Sect. 6.3.3.1 for more detail). The operator $\mathbf{A}$ is time-independent, and its eigenvalues are traditionally denoted as $\{-\mathrm{i}\omega_1, -\mathrm{i}\omega_2, \ldots - \mathrm{i}\omega_N\}$, where ω can take both real and complex values. In this case, eigenvalues of the operator $\mathbf{Z}$ are $\{\mathrm{e}^{-\mathrm{i}\omega_1 t}, \mathrm{e}^{-\mathrm{i}\omega_2 t}, \ldots \mathrm{e}^{-\mathrm{i}\omega_N t}\}$.

Let us denote the set of eigenvectors of the operator $\mathbf{Z}$ as $\mathbf{y}_j$. All vectors in the set $\mathbf{y}_j$ are orthogonal to each other due to the normality of the operator $\mathbf{Z}$. Moreover, all these vectors can be considered to be orthonormal. This means that the scalar product of two different vectors from the set is equal to zero, while the square of each vector is equal to unity: $(\mathbf{y}_j, \mathbf{y}_k) = \delta_{jk}$. We now consider some vector $\mathbf{x}$ with the following decomposition over the eigenvectors of $\mathbf{Z}$: $\mathbf{x} = \sum \xi_j \mathbf{y}_j$. Thus, we get that

$$\mathbf{Z}\mathbf{x} = \sum \xi_j \mathbf{y}_j \mathrm{e}^{-\mathrm{i}\omega_j t} \tag{6.56}$$

This is the so called spectral representation of a linear operator (see Korn and Korn 1968). With the help of this representation, the norm of a normal operator can easily be found. Since eigenvectors of normal operators are orthonormal, the expression (6.55) can be rewritten as:

$$||\mathbf{Z}||^2 = \sup_{\xi_j} \frac{\sqrt{\sum \xi_j^2 e^{2\Im[\omega_j]t}}}{\sqrt{\sum \xi_j^2}} \tag{6.57}$$

It is easy to see that the maximum value of the operator norm is reached when the weights of all eigenvectors, except the one corresponding to the maximal imaginary part of the eigenvalue, vanish.

$$||\mathbf{Z}|| = e^{2\omega_{max}t}, \tag{6.58}$$

where $\omega_{max} = \max_j \left(\Im\,[\omega_j]\right)$. Thus we have shown that to calculate the operator norm for a normal operator, it is enough to find the eigenvalue with the maximal imaginary part. However, as will be shown below, the operator for linear perturbations in shear flows is not normal, so calculating the eigenvalue with the maximal imaginary part is not sufficient to find the maximal possible growth of perturbations.

Finally, the most general are the *non-normal* operators , i.e. those that do not commute with their adjoint operator: $\mathbf{Z}\mathbf{Z}^\dagger \neq \mathbf{Z}^\dagger\mathbf{Z}$. Eigenvalues of these operators can be both purely real and complex, and their eigenvectors are non-orthogonal to each other. The non-orthogonality of the eigenvectors complicates calculation of the operator's norm. For this reason, the energy of a combination of modes is not equal to the sum of the energy of each mode, i.e. the Parceval rule is not valid and non-zero cross terms appear. In other words, due to interference in time between non-orthogonal modes, perturbations described by such an operator can increase even if there are no growing modes. This energy growth of perturbations, which is mathematically related to the non-normality of the dynamical operator, was dubbed the transient growth of perturbations. In the context of stability of hydrodynamical flows, non-normal operators and examples were discussed in Farrell and Ioannou (1996a), as well as in Sects. 6.3 and 6.4 of Schmid and Henningson (2001).

The dynamical operator can relate to the different variants described in this section for different parameters of the problem. For example in Sect. 6.3.4.2, we will show that for solid-state rotation the operator is normal, and for all other rotation laws—non normal.

6.3.1.2 Simple Geometrical Example of the Non-orthogonality of Eigenvectors

A simple geometrical example can illustrate the transient growth mechanism. Let us introduce two vectors on the plane (x, y) symbolizing two perturbation modes. We write them in the form of two complex numbers, $\mathbf{f}_1 = f_0 e^{-i\omega_1 t}$, $\mathbf{f}_2 = f_0 e^{-i\omega_2 t + i\psi}$, where the numbers $\omega_{1,2}$ can be complex as well. In this form the analogy between $\mathbf{f}_{1,2}$ and perturbation modes will be the most clear. The real and imaginary part of each of the vectors $\mathbf{f}_{1,2}$ yields the x- and y- vector components, respectively. Clearly, $\Re[\omega_{1,2}]$ corresponds to the angular velocity with which both vectors rotate on the plane, and $\Im[\omega]_{1,2}$ corresponds to the rate of change of their respective lengths. Below we will assume that imaginary parts of $\omega_{1,2}$ are negative, which corresponds to the shortening of $\mathbf{f}_{1,2}$. We recall that in the case of modes, real parts give angular velocities of the solid-body rotation of the spiral pattern in the flow (see Fig. 6.1), and imaginary parts give their decay rate, in analogy with a spectrally stable flow. In addition, we will assume that at time $t = 0$ the vectors have the same length f_0 and the angle between them is ψ.

Now take the vector $\mathbf{q} = \mathbf{f_1} + \mathbf{f_2}$ and calculate the quantity similar to (6.43), which gives the rate of change of the square of the length $\mathbf{q}$ with time:

$$g = \frac{e^{2\Im[\omega_1]t} + e^{2\Im[\omega_2]t} + 2e^{\Im[\omega_1+\omega_2]t}\cos(\Re[\omega_1 - \omega_2]t + \psi)}{2(1+\cos\psi)}. \tag{6.59}$$

This shows that for angles close to π the denominator in (6.59) is small, and any insignificant increase in the numerator will lead to a large increase in g. Consider two particular examples. In the first case assume that $\Re[\omega_{1,2}] = 0$, and in the second case that $\Im[\omega_{1,2}] = 0$. For simplicity, assume $\cos\psi \approx -1 + \epsilon$, where $\epsilon \ll 1$.

Then for the case $\Re[\omega_{1,2}] = 0$ we see that if we additionally admit a large difference in decrements, $|\Im[\omega_1]| \gg |\Im[\omega_2]|$, after some long time g will be

$$g \approx \frac{e^{2\Im[\omega_2]t}}{2\epsilon}, \tag{6.60}$$

which corresponds to $g \gg 1$ on time intervals such that $|\Im[\omega_1]t| \gg 1$ but simultaneously $|\Im[\omega_2]t| \ll 1$. This means that despite the decrease in length of each particular vector, in the case of strong non-orthogonality (which is characterized by strong difference of ϵ from 1), their sum exhibits a transient growth up to values $\sim \epsilon^{-1}$ (Fig. 6.5). Only at later times will g decrease again at a rate determined by the most slowly decreasing vector. A similar effect takes place for transient perturbations, which can be represented as a sum of decaying modes with zero phase velocity.

In the opposite case $\Im[\omega_{1,2}] = 0$, from (6.59) the following approximate formula can be derived:

$$g \approx \frac{1 - \cos(\Re[\omega_1 - \omega_2]t)}{\epsilon}, \tag{6.61}$$

which is valid when the value of cosine in the numerator is not too close to unity. Apparently, unlike the example with the sum of non-orthogonal vectors with decreasing length (when the length $\mathbf{q}$ first increases to a maximum and then monotonically decreases down to zero at $t \to \infty$), the length of the sum of the rotating vectors exhibits an oscillating growth, by returning many times to ever increasing values $\sim \epsilon^{-1}$ in equal time intervals $\sim |\Re\omega_1 - \Re\omega_2|^{-1}$, as is evident from the illustration in Fig. 6.6. Unlike the first case, it would be inappropriate to refer to this second possible variant of the mode superposition growth as 'transient growth', as we did, for example when analyzing local SFH in Sect. 6.2.3. Therefore, it is more appropriate to call it 'non-modal growth'. One example of such a non-modal growth of a superposition of neutral modes with non-zero phase velocities is considered in Sect. 6.3.2 and was studied in Razdoburdin and Zhuravlev (2012).

6.3.1.3 Singular Vectors

We have thus just demonstrated how non-orthogonality of the modes leads to transient growth of perturbations. In many physical and astrophysical problems, the evolution of linear perturbations is determined precisely by non-normal operators with non-orthogonal eigenvectors. Here the non-normality of $\mathbf{Z}$ is provided by a shear in the background flow. We can verify this by deriving the set of *adjoint* dynamical equations corresponding to the action of the adjoint operator $\mathbf{Z}^\dagger$ (see Sect. 6.3.4.1).

It follows that knowledge of a non-normal operators eigenvalues only is insufficient to fully describe possible (transient) growth of perturbations in the system. In addition, the pair inner products ('angles') between the eigenvectors on the chosen norm of perturbations must be known. One more potential complication to the problem with a non-normal dynamical operator is that it becomes impossible to guarantee the completeness of the set of its eigenvectors, and hence, to guarantee the adequacy of the solution of the problem when using the eigenvectors as a basis for decomposition of an arbitrary perturbation.

For all these reasons, in order to compute the maximal transient growth rate of perturbations, below we will use the technique of *singular* values and vectors. As will be shown, the singular vectors form a complete orthonormal set, which allows us to employ them as a basis to describe the evolution of perturbations. Moreover, the singular values, unlike eigenvalues, enable us to calculate the perturbation energy growth at any given time even for non-normal operators.

The non-negative real number σ is called the singular number of a linear operator $\mathbf{Z}$ if there are such vectors $\mathbf{u}$ and $\mathbf{v}$ of unit length that

$$\begin{aligned} \mathbf{Z}\mathbf{v} &= \sigma\mathbf{u} \\ \mathbf{Z}^\dagger\mathbf{u} &= \sigma\mathbf{v} \end{aligned} \tag{6.62}$$

The vectors $\mathbf{u}$ and $\mathbf{v}$ are called the left and right singular vectors, respectively, corresponding to the singular value σ.

Note that the singular values and vectors are related to the eigenvalues and eigenvectors of the composed self-adjoint operators $\mathbf{Z}\mathbf{Z}^\dagger$ and $\mathbf{Z}^\dagger\mathbf{Z}$. To see this, let the operator $\mathbf{Z}^{\dagger}$ act on vector $\mathbf{Zv}$ and the operator $\mathbf{Z}$ act on vector $\mathbf{Z}^{\dagger}\mathbf{u}$ and then use the definition (6.62):

$$\mathbf{Z}^{\dagger}(\mathbf{Zv}) = \mathbf{Z}^{\dagger}(\sigma\mathbf{u}) = \sigma\mathbf{Z}^{\dagger}\mathbf{u} = \sigma^2\mathbf{v} \tag{6.63}$$

$$\mathbf{Z}\left(\mathbf{Z}^{\dagger}\mathbf{u}\right) = \mathbf{Z}(\sigma\mathbf{v}) = \sigma\mathbf{Zv} = \sigma^2\mathbf{u} \tag{6.64}$$

Thus, vectors $\mathbf{v}$ and $\mathbf{u}$ are eigenvectors of the operators $\mathbf{Z}^{\dagger}\mathbf{Z}$ and $\mathbf{ZZ}^{\dagger}$, respectively. And since they are eigenvectors of self-adjoint operators, they form a complete orthonormal set of functions. The squares of the singular values are eigenvalues of the composite operators.

The operators $\mathbf{ZZ}^{\dagger}$ and $\mathbf{Z}^{\dagger}\mathbf{Z}$ are positive definite, since for any vector $\mathbf{f}$ the inequalities $\left(\mathbf{f}, \mathbf{ZZ}^{\dagger}\mathbf{f}\right) = \left(\mathbf{Z}^{\dagger}\mathbf{f}, \mathbf{Z}^{\dagger}\mathbf{f}\right) > 0$ and $\left(\mathbf{f}, \mathbf{Z}^{\dagger}\mathbf{Zf}\right) = (\mathbf{Zf}, \mathbf{Zf}) > 0$ hold. And since all eigenvalues of a positive definite operator are positive, the singular values are real.

Now, if we rewrite the expression for the operator norm (6.55) with the help of definition (6.62) for some unit norm vector $\mathbf{x}$ decomposed over an orthogonal set of singular vectors: $\mathbf{x} = \sum\limits_j \xi_j \mathbf{v}_j$

$$||\mathbf{Z}||^2 = \sup_{\xi_j} ||\sum_j \xi_j \sigma_j \mathbf{u}_j||^2 = \sup_{\xi_j} \sum_j \left(\xi_j \sigma_j\right)^2 = \max_j \sigma_j^2 \tag{6.65}$$

Thus, the norm of the operator $\mathbf{Z}$ is limited by the maximum singular value of this operator. Going back to the dynamics of perturbations, we conclude that the maximal growth of a perturbation is limited by the maximal singular value of the dynamical operator. In most physical problems, it is natural to assume that the maximal singular value is finite (i.e. the perturbation can not demonstrate infinite growth during a finite time interval), and the set of singular values is discrete.

The singular values and corresponding singular vectors are usually numbered in order of decrease (see Golub and Reinsch 1970). So, the growth of a perturbation is limited by the first singular value of the corresponding dynamic operator, and the first singular vector is the perturbation that exhibit such a growth.

The above considerations imply that to calculate the maximum possible perturbation growth rate it is sufficient to calculate the first singular value, called the optimal growth in the literature, and the right singular vector corresponding to this value will be the sought for (optimal) perturbation demonstrating the maximum possible growth rate. Below we present two methods of calculation of singular values and corresponding singular vectors.

6.3.2 The Matrix Method for Optimal Solutions

The first method to calculate singular vectors will be referred to here as the *matrix* method. It is based on singular value decomposition of the matrix of a dynamical operator. As a rule, the set of eigenvectors is used as the basis for the matrix calculation.

Note that there is another possible variant, which was used, for example, in Ioannou and Kakouris (2001), when the space is covered by a grid of points, and each perturbation is given by a column of numbers corresponding to the values of the perturbation at these points. A dynamical operator corresponds to a matrix obtained by a difference approximation to the derivatives in the dynamical equations. A singular-value decomposition of this matrix enables us to calculate the singular vectors at the grid points. The large size of the operator matrix is a shortcoming of this approach, which requires a lot of time to calculate the singular value decomposition. An advantage is that it is not necessary to calculate the operator's eigenvectors. In this section, we describe the matrix method in the eigenvector basis.

The problem is to find the linear combination of the dynamical operator modes whose norm exhibits the largest growth by the given time. Assume that the sequence of eigenvectors $\{\mathbf{f}_1, \mathbf{f}_2, \mathbf{f}_3 \ldots \mathbf{f}_N\}$ and the corresponding eigenvalues $\{e^{-i\omega_1 t}, e^{-i\omega_2 t}, e^{-i\omega_3 t} \ldots e^{-i\omega_N t)}\}$ of the operator $\mathbf{Z}$ are known. In the space of linear combinations of eigenvectors, the representation of an arbitrary perturbation vector has the form (see paragraph 4.3.2 and Section 4.4 in Schmid and Henningson (2001) for more detail)

$$\mathbf{q} = \sum_{j=1}^{N} \kappa^j \hat{f}_j, \tag{6.66}$$

where the numbers $\{\kappa^1, \kappa^2, \kappa^3 \ldots \kappa^N\}$ are coordinates of the vector $\mathbf{q}$ in the eigenvector basis. Note that the time dependence of $\mathbf{q}$ is essentially in its coordinates.

The inner product of two vectors $\mathbf{q}$ and $\mathbf{g}$ in this representation can be calculated from the known coordinates using the metric matrix $\mathbf{M}$:

$$(\mathbf{q}, \mathbf{g}) = \sum_{i,j} \left(\mathbf{q}^\dagger\right)^i M_{ij} \mathbf{g}^j, \tag{6.67}$$

where the elements of the metric matrix are equal to the inner product of the eigenvectors:

$$M_{ij} = \left(\mathbf{f}_i, \mathbf{f}_j\right) \tag{6.68}$$

Note that the matrix $\mathbf{M}$ is positive definite.

Now the problem of calculation of the maximum possible perturbation growth is reduced to finding the values κ^j for which the growth of the perturbation norm, determined using these values according to formula (6.66), is maximal at the given moment in time.

The representation of an operator **Z** in the eigenvector basis can be easily calculated by letting this operator act on the basis element:

$$\mathbf{Z}\mathbf{f}_j = \mathbf{f}_j(\tau) = \mathrm{e}^{-\mathrm{i}\omega_j\tau}\mathbf{f}_j, \tag{6.69}$$

Therefore, in the set of basis eigenvectors, an operator can be represented by a diagonal matrix **P** with complex exponents on the main diagonal: $\mathbf{P} = diag\{\mathrm{e}^{-\mathrm{i}\omega_1\tau}, \mathrm{e}^{-\mathrm{i}\omega_2\tau}, \mathrm{e}^{-\mathrm{i}\omega_3\tau} \ldots \mathrm{e}^{-\mathrm{i}\omega_N\tau}\}$

Next, let us use the first equality from definition (6.62), $\mathbf{Z}\mathbf{v} = \sigma\mathbf{u}$, and rewrite it in the matrix form:

$$\mathbf{P} = \mathbf{U}\boldsymbol{\Sigma}\mathbf{V}^{-1} \tag{6.70}$$

The matrix $\boldsymbol{\Sigma}$ is diagonal with the singular values on the diagonal, $\boldsymbol{\Sigma} = diag\{\sigma_1, \sigma_2, \sigma_3 \ldots \sigma_N\}$. Columns of matrices **U** and **V** represent right and left singular vectors, respectively.

Now let us write the inner product for two arbitrary singular vectors **q** and **g** as

$$(\mathbf{q}, \mathbf{g}) = \sum_{i,j} \left(\mathbf{q}^\dagger\right)^i M_{ij}\mathbf{g}^j = \sum_i \left((\mathbf{F}\mathbf{q})^\dagger\right)^i (\mathbf{F}\mathbf{g})^i, \tag{6.71}$$

where the matrix **F** is the *Cholesky decomposition* of the metric matrix $\mathbf{M} = \mathbf{F}^T\mathbf{F}$ (for a more detailed description of this decomposition see for example Sect. 4.2 in Golub and Van Loan (1996)). As the matrix **M** is positive definite, its Cholesky decomposition always exists and is unique.

Sets of singular vectors are orthonormal. Therefore the following relations for matrices **V** and **U** hold:

$$\mathbf{V}^\dagger\mathbf{F}^T\mathbf{F}\mathbf{V} = \mathbf{I}, \tag{6.72}$$

$$\mathbf{U}^\dagger\mathbf{F}^T\mathbf{F}\mathbf{U} = \mathbf{I}, \tag{6.73}$$

where **I** is the identity matrix.

Thus, matrices inverse to **V** and **U** are expressed through Hermitian-conjugate as follows:

$$\mathbf{V}^{-1} = \mathbf{V}^\dagger\mathbf{F}^T\mathbf{F}, \tag{6.74}$$

$$\mathbf{U}^{-1} = \mathbf{U}^\dagger\mathbf{F}^T\mathbf{F}. \tag{6.75}$$

Making use of these relations in (6.70) yields

$$\mathbf{P} = \mathbf{U}\Sigma\mathbf{V}^{\dagger}\mathbf{F}^{T}\mathbf{F} = \mathbf{F}^{-1}\mathbf{F}\mathbf{U}\Sigma\mathbf{V}^{\dagger}\mathbf{F}^{T}\mathbf{F}. \tag{6.76}$$

Rewrite this in the form:

$$\mathbf{FPF}^{-1} = (\mathbf{FU})\,\boldsymbol{\Sigma}\,(\mathbf{FV})^{\dagger} \equiv \tilde{\mathbf{U}}\boldsymbol{\Sigma}\tilde{\mathbf{V}}^{\dagger}. \tag{6.77}$$

Now it is clear that the right-hand side of this equality becomes the same as the so-called *singular value decomposition* (SVD) of the matrix $\mathbf{FPF}^{-1}$. Recall that the singular value decomposition is a factorization of a matrix in the form $\tilde{\mathbf{U}}\boldsymbol{\Sigma}\tilde{\mathbf{V}}^{\dagger}$ where $\tilde{\mathbf{U}}$ and $\tilde{\mathbf{V}}$ are orthogonal matrices and $\boldsymbol{\Sigma}$ is a diagonal matrix with positive numbers on the main diagonal (see Golub and Van Loan (1996) for more details). This factorization exists for any real matrix and is unique. It is easy to be convinced that the matrices $\tilde{\mathbf{U}}$, $\tilde{\mathbf{V}}$ and $\boldsymbol{\Sigma}$ satisfy the singular value decomposition conditions, and therefore to calculate singular values and vectors it is sufficient to perform this decomposition for the matrix $\mathbf{FPF}^{-1}$. The singular value decomposition procedure is a standard tool in many linear algebra software packages (for example in the GNU Scientific Library).

The original matrices $\mathbf{U}$ and $\mathbf{V}$ are calculated using $\mathbf{F}^{-1}$: $\mathbf{U} = \mathbf{F}^{-1}\tilde{\mathbf{U}}$, $\mathbf{V} = \mathbf{F}^{-1}\tilde{\mathbf{V}}$. The maximum number on the diagonal of the matrix $\boldsymbol{\Sigma}$ is the first singular value at time t, and the corresponding column of the matrix $\mathbf{V}$ is the first singular vector in the eigenvector basis.

6.3.2.1 Illustration of the Matrix Method

The matrix method has been used in many studies on stability of laboratory flows (see, for example, Butler and Farrell 1992; Reddy and Henningson 1993; Hanifi et al. 1996; Meseguer 2002; Malik et al. 2006; Maretzke et al. 2014) and in astrophysical papers Yecko (2004), Mukhopadhyay et al. (2005), Zhuravlev and Shakura (2009). Here, we elucidate it by a simple semi-analytical study (Razdoburdin and Zhuravlev 2012), where the eigenvector basis[8] is calculated in the WKB approximation in a geometrically thin and barotropic quasi-Keplerian torus with free boundaries. For simplicity, only the modes whose corotation radius is outside the outer boundary of the torus are considered. (See Sect. 6.2.1 for a discussion of the mechanism of energy exchange between the modes and the background flow at the corotation radius in the context of the spectral problem corresponding to Eqs. (6.8)–(6.10)). When the corotation radius is outside the flow, the energy of the modes is conserved. This means that they do not show exponential growth or decay, i.e. their frequencies ω are real values (see expression (6.14)). These perturbations are

[8]Henceforth, the eigenvectors of an operator $\mathbf{Z}$ multiplied by the eigenvalues, i.e. by the time dependence $e^{-i\omega t}$, will be referred to as perturbation modes.

referred to as *neutral* modes. Nevertheless, due to their mutual non-orthogonality, in other words, due to the non-orthogonality of the eigenvectors of the dynamical operator acting on the perturbations, we expect a non-modal growth of their linear combinations (see the analogy in Fig. 6.6 and comments to it in the text).

The modes we wish to obtain below physically correspond to inertial-acoustic waves, which form a solid-body rotating pattern in the disc, i.e. whose azimuthal projection of the wave vector that is constant in time and space. Here, as will be seen from the WKB analysis, their characteristic radial wavelength is close to the disc thickness H. As for their characteristic azimuthal scale, λ_φ, it can be both larger and smaller than H, determined by the azimuthal wave number m entering the set of Eqs. (6.8)–(6.10). Results concerning the optimal perturbation growth will be presented for the case $\lambda_\varphi \gg H$ (see Fig. 6.8).

We will see that in that case the optimal perturbation does not have the form of a spiral unwound by the flow, which we discussed in the context of the transient growth of vortices (see Fig. 6.2), but is a wave packet initially located at the outer boundary of the torus and further propagating towards its inner boundary. At the moment of reflection from the inner boundary, its total acoustic energy reaches maximum and then decreases while the packet goes back to the flow periphery. After reflection from the outer boundary the process repeats. Thus, the non-modal growth in this case is oscillating rather than transient, as must be the case according to the analogy shown in Fig. 6.6.

6.3.2.2 Background Flow

Consider a toroidal flow of finite radial size as a model background flow. The azimuthal velocity component will correspond to the power-law angular velocity radial profile:

$$\Omega = \Omega_0 \left(\frac{r}{r_0}\right)^{-q}, \tag{6.78}$$

where r_0 is the distance to the gravitating centre in the equatorial plane of the torus at which rotation has the Keplerian frequency Ω_0, $2 > q > 3/2$. Assume that matter moves in the external Newtonian gravitational potential produced by a central point-like mass:

$$\Phi = -\frac{\Omega_0^2 r_0^3}{(r^2 + z^2)^{1/2}}.$$

As will be clear below, in this case the parameter q characterizes the torus thickness which approaches zero as the angular velocity profile becomes close to Keplerian. As in Sect. 6.2.1, we use here the polytropic equation of state and write the force balance using the enthalpy h:

$$\begin{aligned} \frac{\partial h}{\partial r} &= \Omega^2 r - \frac{\partial \Phi}{\partial r}, \\ \frac{\partial h}{\partial z} &= -\frac{\partial \Phi}{\partial z}, \end{aligned} \tag{6.79}$$

where the first and the second equations correspond to the projection of the Euler equation on the radial and vertical direction, respectively. The joint integration of (6.79) yields

$$h(r,z) = \frac{\Omega_0^2 r_0^3}{(r^2+z^2)^{1/2}} + \frac{\Omega_0^2 r_0^{2q}}{2(1-q)} r^{2(1-q)} + C,$$

where the integration constant C is determined from the condition that $h(r_1, 0) = 0$ at the inner boundary of the torus $r_1 < r_0$.

Then, in dimensionless coordinates $\hat{x} \equiv r/r_0$, $\hat{y} \equiv z/r_0$ we obtain

$$h = (\Omega_0 r_0)^2 \left[(\hat{x}^2 + \hat{y}^2)^{-1/2} - \hat{x}_1^{-1} + \frac{1}{2(q-1)} \left(\hat{x}_1^{-2(q-1)} - \hat{x}^{-2(q-1)} \right) \right]. \tag{6.80}$$

Here $\hat{x}_1 \equiv r_1/r_0$. The enthalpy distribution (6.80) also gives the outer radial boundary of the torus $\hat{x}_2 > 1$, where $h(\hat{x}_2, 0) = 0$. The quantity $\hat{x}_d = \hat{x}_2 - \hat{x}_1$ will be called the radial extension of the flow.

Now it is not difficult to move to the case of the quasi-Keplerian, geometrically thin torus of interest here: $q = \frac{3}{2} + \frac{\epsilon^2}{2}$, $\epsilon \ll 1$. Using this assumption, the enthalpy profile can be simplified to

$$\frac{h}{\Omega_0^2 r_0^2} = \frac{\hat{H}^2}{2\hat{x}^3} \left[1 - \left(\frac{\hat{y}}{\hat{H}} \right)^2 \right], \tag{6.81}$$

where $\hat{H}(x)$ is the dimensionless thickness of the torus in units of r_0:

$$\hat{H} = \delta\, \hat{x} \left[\frac{\hat{x}_1(1+\ln\hat{x}) - \hat{x}(1+\ln\hat{x}_1)}{\hat{x}_1 - 1 - \ln\hat{x}_1} \right]^{1/2} \tag{6.82}$$

Here, we have introduced a descriptive small parameter

$$\delta \equiv \hat{H}(\hat{x}=1) = 2^{1/2}\epsilon \left(1 - \frac{1+\ln\hat{x}_1}{\hat{x}_1} \right)^{1/2} \ll 1,$$

that defines the characteristic aspect ratio of the disc-like torus with $\delta \ll \hat{x}_d$. It is not difficult to make sure that expression (6.81) is equivalent to (6.4).

Equations (6.81), (6.82) fully determine the quasi-Keplerian background flow which will be used to illustrate the matrix method of determination of the non-modal growth of the superposition of modes. In the next section, we will solve the spectral problem for such a flow, i.e. we will find the perturbation mode profiles.

6.3.2.3 Modes

Modes are non-stationary perturbations with exponential time dependence $\propto \exp(-i\omega t)$. They are also solutions to the operator equation (6.54) determining evolution of a linear perturbation in the flow. This means that modes are state vectors which we obtain by the operator $\mathbf{Z}$ acting on its eigenvectors $\mathbf{f}_i$:

$$\mathbf{f}_i(t) = \mathbf{Z}\mathbf{f}_i = \mathrm{e}^{-i\omega t}\mathbf{f}_i.$$

Again, the numbers $\exp(-i\omega t)$ are eigenvalues of $\mathbf{Z}$ that we will need to find along with its eigenvectors.

In practice, we will not use the equation exactly in the form (6.54), but instead derive an equivalent ordinary differential equation of the second order in the radial coordinate for an Eulerian enthalpy perturbation. As everywhere in this chapter, we will assume that hydrostatic equilibrium always holds, i.e. that $\delta v_z = 0$. As we deal with a thin torus, $\delta \ll 1$, our perturbations taken originally in the form of azimuthal Fourier harmonics $\propto \exp(im\varphi)$ satisfy the set of Eqs. (6.8)–(6.10), which contains the background variables integrated over z (see Sect. 6.2.1). The modal analysis implies the substitution $\partial/\partial t \to i\omega$, after which from (6.8) and (6.9) we find that complex Fourier harmonics of the Eulerian velocity perturbations, which are denoted here as $\mathbf{v}_r$ and $\mathbf{v}_\varphi$, are expressed through the Fourier harmonics of the enthalpy perturbation, which is denoted here as $\mathbf{W}$, as follows:

$$\mathbf{v}_r = \frac{i}{D}\left[\bar{\omega}\frac{d\mathbf{W}}{d\hat{x}} - \frac{2m\Omega\mathbf{W}}{\hat{x}}\right], \tag{6.83}$$

$$\mathbf{v}_\varphi = \frac{1}{D}\left[\frac{\kappa^2}{2\Omega}\frac{d\mathbf{W}}{d\hat{x}} - \frac{m\bar{\omega}\mathbf{W}}{\hat{x}}\right], \tag{6.84}$$

where $D \equiv \kappa^2 - \bar{\omega}^2$, $\kappa^2 = \frac{2\Omega}{x}\frac{d}{d\hat{x}}\left(\Omega\hat{x}^2\right)$ is, as usually, the epicyclic frequency squared, and $\bar{\omega} \equiv \omega - m\Omega$ is the shifted frequency. Below in this section we assume that all frequencies are in units of the frequency Ω_0 and time is in units of Ω_0^{-1}.

Plugging (6.83) and (6.84) into the continuity equation (6.10), we obtain the following equation for $\mathbf{W}$:

$$\frac{D}{\hat{x}\Sigma}\frac{d}{d\hat{x}}\left(\frac{\hat{x}\Sigma}{D}\frac{d\mathbf{W}}{d\hat{x}}\right) - \left[\frac{2m}{\bar{\omega}}\frac{D}{\hat{x}\Sigma}\frac{d}{d\hat{x}}\left(\frac{\Omega\Sigma}{D}\right) + (n+1/2)\frac{D}{h_*} + \frac{m^2}{\hat{x}^2}\right]\mathbf{W} = 0, \tag{6.85}$$

where

$$\Sigma(r) = \int_{-H}^{H} \rho dz \propto \hat{H} \left(\frac{\hat{H}^2}{\hat{x}^3}\right)^n \quad \text{and} \quad h_* = \frac{\hat{H}^2}{2\hat{x}^3}. \tag{6.86}$$

Here h_* is the dimensionless background enthalpy in the equatorial disc plane (cf. (6.81)). To reproduce the surface density dependence on r given above, $\Sigma(r)$, it is enough to recall that $\Sigma \sim H\rho|_{z=0}$, and $\rho \sim h^n$ for a polytropic equation of state. Equation (6.85), as well as its more general analogue for three-dimensional perturbation modes, is often used in the literature. Their derivation and analysis can be found, for example, in papers Goldreich et al. (1986), Kato (1987), Sekiya and Miyama (1988), Kojima (1989), Kato (2001).

As we have already mentioned, the integration of Eq. (6.85) is complicated by resonances: corotational, where $\bar{\omega} = 0$, and Lindblad resonances, where $D = 0$. These points are singular for (6.85). However, in order to illustrate the matrix method of optimization, we will restrict ourselves to calculation of only part of the modes with resonances lying outside the outer boundary of the flow, $\hat{x}_2$. The condition that the inner Lindblad resonance lies at $\hat{x} > \hat{x}_2$ implies

$$\omega < (m-1)\Omega(\hat{x}_2), \tag{6.87}$$

where in the condition $D = 0$ we have set $\kappa \approx \Omega$ due to the nearly Keplerian angular velocity profile in a thin torus. Recall also that ω is a real value. Note that for $m = 1$ the inner Lindblad resonance is at $\hat{x} = 0$, and hence there are no modes with $m = 1$ satisfying the condition (6.87). For this reason, we will consider only modes with $m > 1$. Thus, under the restrictions made, the term $\propto D/h_* \sim \delta^{-2}$ will be large everywhere in the flow, and therefore the solution to the equation can be found in the WKB approximation.

A WKB solution to Eq. (6.85) can be written as

$$\mathbf{W} = \mathbf{C}_0 S_1 \cos(S_0 + \varphi_0), \tag{6.88}$$

in which $S_0 \sim \delta^{-1}$, and $S_1 \sim \delta^0$.

Plugging (6.88) into (6.85) yields its decomposition in δ. By collecting terms with similar powers of δ, namely, δ^{-2} and δ^{-1}, we find the explicit form of functions S_0 and S_1:

$$S_0 = \int_{\hat{x}_1}^{\hat{x}} \left((n+1/2)\frac{-D}{h_*} - \frac{m^2}{\hat{x}^2}\right)^{1/2} d\hat{x},$$

$$S_1 = \left(\frac{-D}{\hat{x}\Sigma}\right)^{1/2} \left((n+1/2)\frac{-D}{h_*} - \frac{m^2}{\hat{x}^2}\right)^{-1/4}.$$

The phase φ_0 is fixed by the boundary conditions.

The WKB solution (6.88) is irregular at the boundary points $\hat{x}_1$ and $\hat{x}_2$ at which $h_* \to 0$. It is possible to find a WKB-solution that is regular at the boundaries (see Heading 2013), but here, let us use another common way of matching (6.88) with an approximate *regular* solution to the original equation (6.85) near $\hat{x}_1$ and $\hat{x}_2$. This matching should yield a discrete set of eigenfrequencies ω, as well as the value of φ_0.

In order to find the regular solution near $\hat{x}_1$ and $\hat{x}_2$, we change to the new radial coordinate $\tilde{x} \equiv |\hat{x} - \hat{x}_{1,2}|$ and expand Eq. (6.85) in the main order of the variable $\tilde{x} \ll 1$. Technically, this means that all variables from (6.85) that are non-zero at $\hat{x}_{1,2}$ are set to their exact values at $\hat{x}_{1,2}$. The disc semi-thickness, vanishing at the boundaries, is approximated as $\hat{H} = \hat{H}_{1,2}\tilde{x}^{1/2}$. Here for the constant $\hat{H}_{1,2}$ we get

$$\hat{H}_{1,2} = \delta\hat{x}_{1,2}\left|\frac{\ln\hat{x}_{1,2}}{1+\ln\hat{x}_{1,2}-\hat{x}_{1,2}}\right|^{1/2}$$

We obtain the following near-boundary equation:

$$\tilde{x}\frac{d^2\mathbf{W}}{d\tilde{x}^2} + (n+1/2)\frac{d\mathbf{W}}{d\tilde{x}} + E_{1,2}\mathbf{W} = 0, \tag{6.89}$$

where $E_{1,2} = \dfrac{(2n+1)(-D_{1,2})\,\hat{x}_{1,2}^3}{H_{1,2}^2}$, $D_{1,2}$—are the values of D at points $\hat{x}_{1,2}$.

The regular solution to (6.89) at $\tilde{x} = 0$ has the form:

$$\mathbf{W} = \mathbf{C}_{1,2}\,\tilde{x}^{-(2n-1)/4}\,J_{n-1/2}(\tilde{z}), \tag{6.90}$$

where $\tilde{z} = 2E_{1,2}^{1/2}\,\tilde{x}^{1/2}$.

Note that Eq. (6.89) at $\tilde{x} \to 0$ is equivalent to the boundary condition for the enthalpy perturbation at the free boundary of the flow, which states that the Lagrangian enthalpy perturbation vanishes at the boundary points $\hat{x}_{1,2}$, $\Delta h|_{x_{1,2}} = 0$ (see, for example Glatzel 1987a).

As the denominator $\tilde{z}$ contains the small δ, at some distance from the boundary points $\tilde{z} \gg 1$ yet under the condition $\tilde{x} \ll 1$. In this region, $\mathbf{W}$ is given by the asymptotic of (6.90) for large arguments:

$$\mathbf{W} \approx \mathbf{C}_{1,2}\,\tilde{x}^{-n/2}(4\pi^2 E_{1,2})^{-1/4}\cos\left(2E_{1,2}^{1/2}\,\tilde{x}^{1/2} - n\,\pi/2\right) \tag{6.91}$$

The matching of (6.91) and the WKB solution yields the zero phase $\varphi_0 = -n\pi/2$ in Eq. (6.88) and the following dispersion equation:

$$\int_{\hat{x}_1}^{\hat{x}_2}\left((2n+1)\frac{-D\hat{x}^3}{\hat{H}^2} - \frac{m^2}{\hat{x}^2}\right)^{1/2} d\hat{x} = \pi(n+p), \tag{6.92}$$

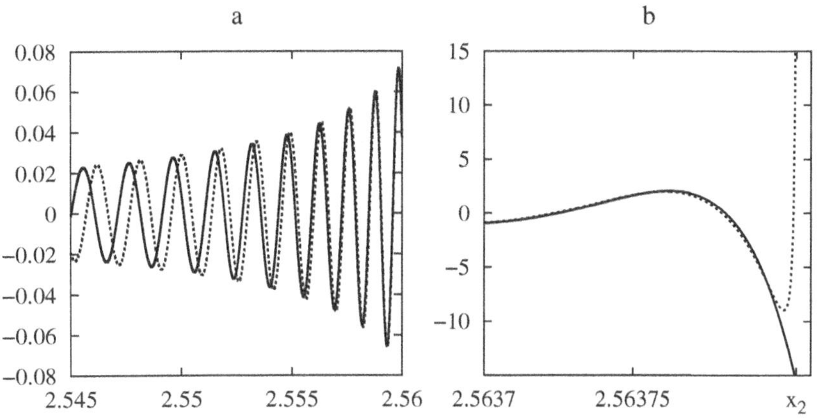

Fig. 6.7 In this figure solutions within different parts of the torus are shown. The solid line denotes a solution (6.90), and the dotted line a WKB-solution. The right panel shows the right borderland of the torus. The WKB solution has an infinite derivative in this region, so the solution (6.90) is regular here. On the left panel a larger region is shown. When the distance from the boundary x_2 is increasing, the differences between the two solutions become more significant. However, in a certain region both of the solutions are very similar. The following parameters were used: $\delta = 0.003$, $n = 1.5$, $m = 10$, $x_d = 2.0$

where p is an integer number. Solving (6.92) for different p yields a discrete set of ω that enters D. This is the sequence of eigenfrequencies of neutral modes that we are interested in.

The modes profiles are given by Eqs. (6.88) and (6.90) with account for the relations between the corresponding constants (see Fig. 6.7):

$$\frac{\mathbf{C}_0}{\mathbf{C}_1} = \left(\frac{\hat{H}_1^{2n+1}}{2\pi\hat{x}_1^{3n-1}(-D_1)}\right)^{1/2}, \quad \frac{\mathbf{C}_2}{\mathbf{C}_1} = (-1)^p \left[\left(\frac{\hat{x}_2}{\hat{x}_1}\right)^{3n-1}\frac{D_2}{D_1}\left(\frac{\hat{H}_1}{\hat{H}_2}\right)^{2n+1}\right]^{1/2} \tag{6.93}$$

After obtaining the profile $\mathbf{W}(\hat{x})$ for a given ω_i, the corresponding complex Fourier harmonics of the Eulerian velocity perturbations $\mathbf{v}_r(\hat{x})$ and $\mathbf{v}_\varphi(\hat{x})$ can be calculated from (6.83) and (6.84). Thus, we find the whole eigenvector $\mathbf{f}_i \equiv \{\mathbf{v}_r, \mathbf{v}_\varphi, \mathbf{W}\}$ of operator $\mathbf{Z}$ corresponding to its eigenvalue $\exp(-\mathrm{i}\omega_i t)$.

6.3.2.4 Optimal Growth

The explicit form of eigenvectors of a dynamical operator allows us to calculate *the optimal growth*, i.e. to find the linear combination of these vectors that demonstrates the maximum increase of the norm at a given time. The optimal growth at the time t is

$$G(t) = \max_{\mathbf{q}(0)} \frac{||\mathbf{q}(t)||^2}{||\mathbf{q}(0)||^2}. \tag{6.94}$$

This is a generalization of (6.46) for the spatially global case.

The inner product of two vectors from the linear span of N eigenvectors of $\mathbf{Z}$ is introduced such that the square of the corresponding norm recovers the acoustic energy of the perturbation (6.15):

$$(\mathbf{f}, \mathbf{g}) = \pi \int_{r_1}^{r_2} \Sigma \left((v_r)_f (v_r)^*_g + (v_\varphi)_f (v_\varphi)^*_g + (n + 1/2) \frac{(W)_f (W)^*_g}{h_*} \right) r dr, \tag{6.95}$$

where the indices 'f' or 'g' indicate the relation of some physical variable to the vector $\mathbf{f}$ or $\mathbf{g}$, respectively. We recall that by v_r, v_φ and W here we mean azimuthal Fourier harmonics of the Eulerian perturbations of the velocity and enthalpy components, respectively.

Now, let us apply the procedure of calculation of the optimal combination of eigenvectors described above. As we have the eigenvectors in analytical form, the matrix $\mathbf{M}$ can be obtained by simple numerical integration of a combination of elementary functions using the inner product formula (6.95):

$$M_{ij} = (\mathbf{f}_i, \mathbf{f}_j) \tag{6.96}$$

Next, we perform the Cholesky decomposition $\mathbf{M} = \mathbf{F}^T \mathbf{F}$ and then the singular value decomposition of the matrix $\mathbf{F}\mathbf{P}\mathbf{F}^{-1}$. Both these procedures are standard in numerical methods of matrix algebra.

In Fig. 6.8 we show an example of the dependence of the maximum possible energy growth, $G(t)$, among all superpositions of 20 neutral modes at time t, on a time scale of the order of the sonic time $t_s \sim (\delta \Omega_0)^{-1}$ and $\sim 10 t_s$, respectively. The left panel of Fig. 6.8 also shows the energy growth of the optimal mode combinations $g(t)$. Clearly, the curves $g(t)$ touch the general optimal growth curve $G(t)$, as must be the case, each at its own optimization time. The optimal growth itself in this model has a quasi-periodic form by reaching maxima at times $\sim t_s$, and the thinner the torus, the higher values g the mode superposition can reach.

6.3.2.5 The Angular Momentum Flux

In Sect. 6.3.2.4 we have shown that some combinations of modes can demonstrate a significant growth in acoustic energy. Consider in more detail what this optimal perturbation is. The perturbation amplitude growth suggests that the main flow

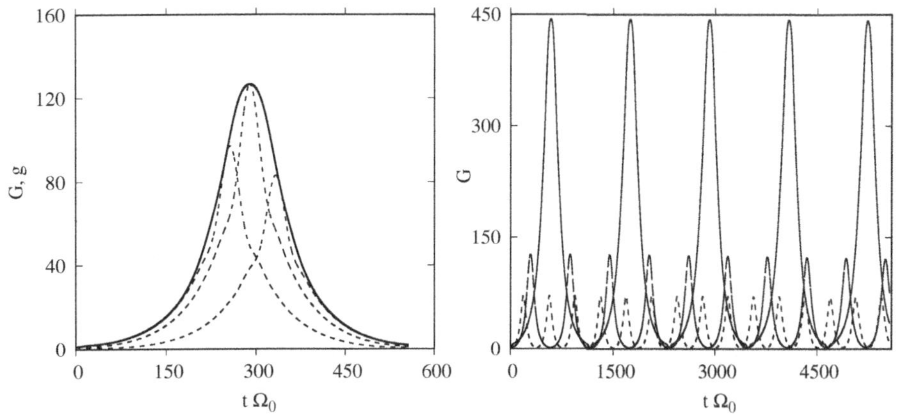

Fig. 6.8 The left panel shows the optimal growth curve $G(t)$ (the solid curve) of a linear combination of slow modes in a thin disc for $\delta = 0.002$. The dashed lines show the growth factor of the total acoustic energy $g(t)$ of the particular optimal perturbations as a function of time. These perturbations are optimal for time intervals $t = 250, 290, 390$ expressed in units of the characteristic Keplerian period $2\pi\Omega_0^{-1}$. The right panel shows curves of $G(t)$ only. The solid, dashed and dotted lines correspond to $\delta = 0.001, 0.002, 0.003$, respectively. The linear combination shown has the dimensionality $N = 20$, parameters are $x_d = 1.0$, $m = 25$, $n = 3/2$ (Figure is quoted from Razdoburdin and Zhuravlev (2012))

transfers energy to perturbations. The first term on the right-hand side of (6.13) is responsible for this, and its integrand sometimes is referred to as the Reynolds stress (which we denote as F_R, see Kojima (1989)). It turns out that F_R is simply related to the density of the specific angular momentum flux excited by perturbations, F: $F_R = -\frac{d\Omega}{d\hat{x}} F$ (see Sections 2.3 and 4 of Savonije and Heemskerk 1990). Clearly, for Keplerian rotation, F_R and F have the same sign: if the perturbation energy increases, $F > 0$, angular momentum flux to the torus periphery takes place, and vice versa.

In order to calculate the evolution of the profile F for the optimal superposition of modes represented by the curve $g(t)$ for $t = 290$ in Fig. 6.8, let us use the following expression for F

$$F = \hat{x}\Sigma < \delta v_r \delta v_\varphi > . \tag{6.97}$$

Figure 6.9 shows how the radial distribution of F changes in the interval $(\hat{x}_1, \hat{x}_2)$. At first, we see that F is radially localized, and its localization region changes with time: during the perturbation growth phase it shifts towards the inner torus boundary, whereas during the perturbation decay phase it moves back to the outer boundary. Therefore, in this case the non-modal growing perturbation is represented by a wave packet containing a set of neutral modes (each of the modes, as we recall, rotates like a solid body with an angular velocity somewhat smaller than the angular velocity of the flow). Initially, this wave packet is localized near the outer boundary

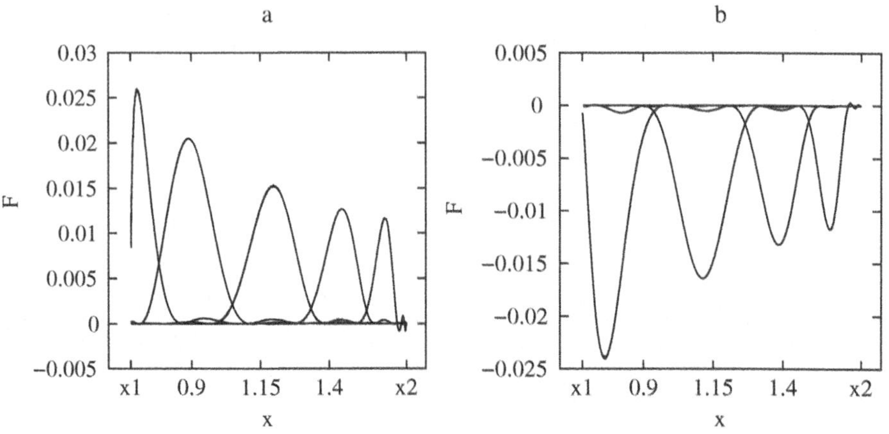

Fig. 6.9 Radial profiles of the azimuthally averaged angular momentum flux density F of the optimal perturbation, which corresponds to $g(t)$ shown in Fig. 6.8a for the optimization time $t = 290$. (**a**) Profiles of F at the instants $t = 50, 100, 150, 200, 240$ before the $g(t)$ maximum. Each profile has one large maximum shifting from the outer disc boundary x_2 to the inner disc boundary x_1 as t increases. (**b**) Profiles of F at the instants $t = 290, 350, 400, 450$ after the $g(t)$ maximum. Correspondingly, each profile has one large minimum shifting from the inner disc boundary x_1 to the outer disc boundary x_2 as t increases further. The linear combination has the dimensionality $N = 20$, and the parameters are $\delta = 0.002$, $x_d = 1.0$, $m = 25$, $n = 3/2$ (Figure from Razdoburdin and Zhuravlev (2012))

of the torus and moves towards the inner boundary. This causes outflow of angular momentum to the disc periphery, since $F > 0$, and its acoustic energy increases. At the moment of reflection from the inner boundary, the sign of F and the direction of motion of the wave packet reverse, which later leads to a decrease in its acoustic energy, and the angular momentum flows back to the inner parts of the torus. As there is no viscous dissipation and the background flow is stationary, obviously, the evolution of the optimal perturbation will continue to repeat itself: the wave packet, after reflecting from the outer boundary, will go back towards the inner boundary. Note also that the obtained shape of $G(t)$ suggests that during the evolution of this particular type of perturbations there are epochs (time intervals counted from the conventional start of the perturbation evolution) when there is no combination of modes to be amplified. These epochs correspond to minima on the curve $G(t)$ (see Fig. 6.8). The reason is that only the wave packets localized near the outer disc boundary can exhibit significant growth. At the same time, the velocity of their radial motion is determined by the sound velocity in the flow, and hence the time intervals 'favourable' for non-modal growth always take the value $\sim \hat{x}_d/\delta$.

If we plot the lines of constant phase of perturbations corresponding to the optimal wave packet on the plane (r, φ), it turns out that at the growth stage it corresponds to a *trailing* spiral. It is opened at the initial time, but while propagating towards the inner boundary it winds up stronger and stronger. Oppositely, after the reflection from the inner boundary, it transforms into a tightly wound *leading* spiral, and while moving back towards the outer boundary the degree of winding gradually decreases. This behavior of the optimal perturbation is similar to the process of enhancement/weakening of the shear density waves by a shear flow we discussed in Sect. 6.2.2 in the context of the spatially local problem.

The rate of spiral twist is controlled by the dynamical time-scale, while the rate of radial drift—by the time-scale of sound. Thus for smaller δ, the spiral manages to be twisted more strongly during the drift from the outer to the inner boundary. For that reason a decrease of δ (see Fig. 6.8) leads to an increase of the maximal factor of transient amplification.

6.3.3 Alternative: A Variational Approach

Singular vectors can be found alternatively by a variational method . This method represents a generalization of the *power iterations* procedure of looking for matrix eigenvalues and eigenvectors in a finite dimensional framework (see, for example, the monograph by Golub and Van Loan (1996)). The variational method requires less computational power than the matrix method (Luchini 2000), and, importantly, it can be applied to non-stationary background flows, as well as used to solve the non-linear problem of transient dynamics of finite-amplitude perturbations. Unlike the matrix method, it does not require discrete representation of the dynamical operator, i.e., for example the decomposition of perturbations by eigenvectors, whose computation in a shear flow faces the known difficulty while bypassing the corotation and Lindblad resonances (see Lin 1955).

As for linear dynamics, the variational method turns out to be *equivalent* to solving the more simple problem of seeking the maximum eigenvalue of the operator $\mathbf{Z}^\dagger\mathbf{Z}$ (see Sect. 6.3.1 and, for example, Andersson et al. (1999) as well). That is why we start with solving exactly that problem, whereas a derivation of the variational method itself, directly from the variational principle, will be given below together with a generalization to the non-linear case.

6.3.3.1 Linear Autonomous Operators

In Sect. 6.3.1, after the singular values were introduced, we discussed that the first singular value is simultaneously the maximum eigenvalue of the composite operator $\mathbf{Z}^\dagger\mathbf{Z}$, and the first right singular vector is the corresponding eigenvector of this operator. First, let us try to understand what the action of $\mathbf{Z}^\dagger\mathbf{Z}$ on the initial state vector $\mathbf{q}(0)$ is equivalent to. Here, the action of the first (right) part of the composite operator is known from its definition (6.54): this is the integration of the equations of perturbation dynamics, for example, the set (6.8)–(6.10), until time t starting from the initial condition $\mathbf{q}(0)$. We symbolically rewrite this as

$$\frac{\partial \mathbf{q}}{\partial t} = \mathbf{A}\mathbf{q}. \tag{6.98}$$

Note that due to the linearity of the problem, the operator $\mathbf{A}$ in (6.98) does not depend on $\mathbf{q}$ itself.

The subsequent action of the operator $\mathbf{Z}^\dagger$ on $\mathbf{q}(t)$ is not difficult to understand if the operator $\mathbf{A}$ is *autonomous*, i.e. time-independent (see Farrell and Ioannou 1996a).

Then, the integration of (6.98) can be written in the operator form: $\mathbf{q}(t) = \mathrm{e}^{\mathbf{A}t}\mathbf{q}(0)$, i.e. $\mathbf{A}$ and $\mathbf{Z}$ are related as

$$\mathbf{Z} = \mathrm{e}^{\mathbf{A}t}. \tag{6.99}$$

The right-hand side of (6.99) is called *the operator exponent* and should be understood as an infinite series $\mathbf{I} + \mathbf{A}t + (\mathbf{A}t)^2/2 + \ldots$.

The operator adjoint to $\mathbf{Z}$ can also be written through the operator exponent $\mathbf{Z}^\dagger = \mathrm{e}^{\mathbf{A}^\dagger \mathbf{t}}$, where $\mathbf{A}^\dagger$ is the operator adjoint to $\mathbf{A}$. $\mathbf{A}^\dagger$ is defined by the Lagrange relation $\left(\mathbf{Aq}, \tilde{\mathbf{q}}\right) = \left(\mathbf{q}, \mathbf{A}^\dagger\tilde{\mathbf{q}}\right)$, where $\mathbf{q}$ and $\tilde{\mathbf{q}}$ are arbitrary vectors. This expression for $\mathbf{Z}^\dagger$ follows from the application of the conjugation operation to the infinite operator series given above. Now consider the inner product:

$$\left(\frac{\partial \mathbf{q}}{\partial t}, \tilde{\mathbf{q}}\right) = \left(\mathbf{Aq}, \tilde{\mathbf{q}}\right) = \left(\mathbf{q}, \mathbf{A}^\dagger\tilde{\mathbf{q}}\right). \tag{6.100}$$

On the other hand,

$$\begin{aligned}\left(\frac{\partial \mathbf{q}}{\partial t}, \tilde{\mathbf{q}}\right) = \frac{\partial}{\partial t}\left(\mathbf{q}, \tilde{\mathbf{q}}\right) - \left(\mathbf{q}, \frac{\partial \tilde{\mathbf{q}}}{\partial t}\right) = \frac{\partial}{\partial t}\left(\mathrm{e}^{\mathbf{A}t}\mathbf{q}(0), \tilde{\mathbf{q}}\right) - \left(\mathbf{q}, \frac{\partial \tilde{\mathbf{q}}}{\partial t}\right) = \\ \left(\mathbf{q}(0), \frac{\partial}{\partial t}\left(\mathrm{e}^{\mathbf{A}^\dagger t}\tilde{\mathbf{q}}\right)\right) - \left(\mathbf{q}, \frac{\partial \tilde{\mathbf{q}}}{\partial t}\right).\end{aligned} \tag{6.101}$$

Combining (6.100) and (6.101) yields the identity:

$$\left(\mathbf{q}(0), \frac{\partial}{\partial t}\left(\mathrm{e}^{\mathbf{A}^\dagger t}\tilde{\mathbf{q}}\right)\right) - \left(\mathbf{q}, \frac{\partial \tilde{\mathbf{q}}}{\partial t}\right) = \left(\mathbf{q}, \mathbf{A}^\dagger\tilde{\mathbf{q}}\right). \tag{6.102}$$

It is easy to see that if $\tilde{\mathbf{q}}$ and $\frac{\partial \tilde{\mathbf{q}}}{\partial t}$ are related as

$$\frac{\partial \tilde{\mathbf{q}}}{\partial t} = -\mathbf{A}^\dagger\tilde{\mathbf{q}}, \tag{6.103}$$

then $\tilde{\mathbf{q}}(t) = \mathrm{e}^{-\mathbf{A}^\dagger \mathbf{t}}\tilde{\mathbf{q}}(0)$ and the identity (6.102) is fulfilled for an arbitrary $\mathbf{q}$.

Thus, the action of operator $\mathbf{Z}^\dagger = \mathrm{e}^{\mathbf{A}^\dagger t}$ is equivalent to integration of Eq. (6.103) *backwards* in time from the instant t with initial condition $\mathbf{q}(t)$ down to the instant $t = 0$. Equation (6.103) is called *the adjoint equation*.

Additionally, note that although the operator $\mathbf{Z}$ can be represented as $\mathbf{Z} = e^{\mathbf{A}t}$ and $\mathbf{Z}^\dagger$ as $\mathbf{Z}^\dagger = e^{\mathbf{A}^\dagger \mathbf{t}}$, the composite operator *cannot* be represented as $\mathbf{Z}^\dagger\mathbf{Z} = e^{(\mathbf{A}+\mathbf{A}^\dagger)\mathbf{t}}$. In order to see this, it is sufficient to employ the series expansion of those operators.

Thus, the action of the composite operator $\mathbf{Z}^\dagger\mathbf{Z}$ on the initial vector $\mathbf{q}(0)$ is equivalent to integration of the original equation (6.98) with the initial condition $\mathbf{q}(0)$ forwards in time up to the instant t, and to subsequent integration of the adjoint equation (6.103) with the initial condition in the form of the vector $\mathbf{q}(t)$ we have just obtained by integrating (6.98)—backwards in time down to $t = 0$.

If the action of the composite operator $\mathbf{Z}^\dagger\mathbf{Z}$ on some vector is equivalent to its multiplication by a constant, this vector is a right singular vector of $\mathbf{Z}$, and the constant is the square of the corresponding singular value: $\mathbf{Z}^\dagger\mathbf{Z}\mathbf{v} = \sigma^2\mathbf{v}$. However, we need only the first, i.e. the largest, right singular vector. In order to obtain it, consider an iteration procedure with one step consisting of action by the composite operator $\mathbf{Z}^\dagger\mathbf{Z}$ with subsequent normalization of the result to unity. To show convergence of iterations to the first singular vector, consider the decomposition of an arbitrary state vector over the singular vectors $\mathbf{q}(0) = \sum\limits_{k=1}^{\infty} q_k \mathbf{v}_k(0)$ and let it be acted on by the iteration operator: $\mathbf{Z}^\dagger\mathbf{Z}\mathbf{q}(0) = \sum\limits_{k=1}^{\infty} \sigma_k^2 q_k \mathbf{v}_k(0)$.

Obviously, the iteration operator increases the weight of each singular vector in proportion to the square of its singular value. Thus, the limit $\left(\mathbf{Z}^\dagger\mathbf{Z}\right)^{p\to\infty}\mathbf{q}(0)$, where p is a natural number, for an arbitrary initial state vector $\mathbf{q}(0)$ is equal to the first right singular vector, since it corresponds to the maximum singular value. The rate of divergence depends on the difference between the singular vectors.

Note that in order to converge exactly to the first singular vector, the initial approximation should not be orthogonal to it, so that in the decomposition of the vector $\mathbf{q}(0)$ the weight of the first singular vector is non-zero, $q_1 \neq 0$. Otherwise, the action of the iteration operator will not increase this weight: $\sigma_1^2 q_1 = 0$. In the latter case, the iteration scheme will converge to the singular vector with the largest singular value among all vectors that have non-zero weight in the initial decomposition.

After all remarks, we would like to stress once again that in order to find the first right singular vector it is necessary to apply an iteration procedure, which includes the integration of the original equation (6.98) forwards in time and of the adjoint equation (6.103)—backwards in time with the subsequent normalization to unity in each iteration step.

6.3.3.2 Linear Non-autonomous Operators

In the case of a time-dependent operator $\mathbf{A}$ (so-called *non-autonomous* operator, see Farrell and Ioannou (1996b)), the action of operator $Z^\dagger$ also corresponds to integration of Eq. (6.103) backwards in time, which can be verified as follows.

For the non-autonomous operator **A**, the action of operator **Z** can be factorized as a product of actions of infinitesimal operators:

$$\mathbf{Z}(\tau) = \lim_{n\to\infty} \prod_{j=1}^{n} \mathrm{e}^{\mathbf{A}(t_j)\delta t}, \tag{6.104}$$

where $\delta t = \tau/n$; $(j-1)\delta t < t_j < j\delta t$, see Farrell and Ioannou (1996b).

The conjugation of the product of operators yields

$$\mathbf{Z}^\dagger(\tau) = \lim_{n\to\infty} \prod_{j=n}^{1} \mathrm{e}^{\mathbf{A}^\dagger(t_j)\delta t}. \tag{6.105}$$

Clearly, at each time interval δt the integration is performed backwards in time, and the intervals themselves are ordered with decreasing j. Therefore the action of $\mathbf{Z}^\dagger$ is again equivalent to integration of (6.103) backwards in time.

Thus, like in the case of autonomous operators, the action of $\mathbf{Z}^\dagger\mathbf{Z}$ is equivalent to consecutive integration of (6.98) forwards in time and of (6.103) backwards in time.

Correspondingly, the iteration procedure to search for the first singular vector presented above is applicable to non-autonomous operators as well.

6.3.3.3 Calculation of Consecutive Singular Vectors

Singular vectors produce an orthogonal set of vectors that can be used as a basis for decomposition of any linear perturbation. Thus, it could be useful to calculate not only the first but also the consecutive singular vectors. Therefore, below we will briefly describe their calculation using the variational method.

In order that the iterations described above converge not to the first singular vector but to a vector with number N, it is sufficient that the domain of the iteration operator completes the subset of linear combinations of previous $N-1$ vectors, or, what is equivalent, the initial approximation is orthogonal to the already obtained singular vectors, i.e. the condition $\big(\mathbf{q}(0), \mathbf{v}_j(0)\big) = 0$ should be satisfied for $j < N$. In this case, the action of the iteration operator will be orthogonal to the obtained singular vectors:

$$\left(\mathbf{Z}^\dagger\mathbf{Z}\mathbf{q}(0), \mathbf{v}_j(0)\right) = \left(\mathbf{Z}^\dagger\mathbf{Z}\sum_{k=N}^{\infty} q^k \mathbf{v}_k(0), \mathbf{v}_j(0)\right) = \left(\sum_{k=N}^{\infty} \sigma_k^2 q^k \mathbf{v}_k(0), \mathbf{v}_j(0)\right) = 0 \tag{6.106}$$

Thus, if we expand some vector over the singular vectors in the form

$$\mathbf{q}(0) = \sum_{k=1}^{\infty} q^k \mathbf{v}_k(0), \tag{6.107}$$

then the change of the initial condition in the iteration procedure by $\mathbf{q}(0) - \sum_{k=1}^{N-1} q^k \mathbf{v}_k(0)$ provides the convergence of power iterations to the singular vector number N. Thus, having the previous $N-1$ singular vectors it is always possible to calculate the next one.

6.3.3.4 Generalization to the Non-linear Case

In the case of non-linear dynamics, the justification for iterative computation of optimal growth presented in the two previous sections becomes invalid. However, in a somewhat generalized form it can be obtained directly from the variational principle, as we will show below.

The problem is formulated as a search for the initial condition demonstrating the maximum growth of the norm at a given time, i.e. it is required to find a vector such that $\mathbf{q}(0)$, for which the functional

$$\mathscr{G}(\tau) = \frac{||\mathbf{q}(t)||^2}{||\mathbf{q}(0)||^2} \tag{6.108}$$

reaches maximum provided that the vector $\mathbf{q}$ satisfies the dynamical equations written in operator form (6.98). To do this, a technique similar to the well-known Lagrange multipliers method of finding conditional extremum of a function is used.

The Lagrangian necessary to find the conditional extremum in this case includes two terms: the functional whose maximum is searched for, and the so-called 'penalty' term, which is non-zero only if $\mathbf{q}$ does not satisfy the dynamical equations (6.98) (see also Corbett and Bottaro (2001), Guégan et al. (2006) and the review Schmid (2007)):

$$\mathscr{L}\left(\mathbf{q}, \tilde{\mathbf{q}}\right) = \mathscr{G}(\mathbf{q}) - \int_0^t \left(\tilde{\mathbf{q}}, \dot{\mathbf{q}} - \mathbf{A}(\mathbf{q})\mathbf{q}\right) d\tau. \tag{6.109}$$

Apparently, the penalty term in (6.109) is written as the inner product of the Lagrange multipliers (entering $\tilde{\mathbf{q}}$) and Eq. (6.98), and additionally integrated over time. Unlike the well-known problem of finding conditional extremum of a function, the Lagrangian in this case is a functional defined for all possible shapes of $\mathbf{q}$, and the Lagrange multipliers themselves are functions rather than numbers.

The extremum of (6.109) is reached when variations of the Lagrangian with respect to $\mathbf{q}$ and $\tilde{\mathbf{q}}$ vanish simultaneously. These variations are defined as (see book Gunzburger (2003))

$$\frac{\partial \mathscr{L}}{\partial \mathbf{q}}\delta\mathbf{q} = \lim_{\epsilon\to 0}\frac{\mathscr{L}\left(\mathbf{q}+\epsilon\delta\mathbf{q},\tilde{\mathbf{q}}\right)-\mathscr{L}\left(\mathbf{q},\tilde{\mathbf{q}}\right)}{\epsilon} \tag{6.110}$$

$$\frac{\partial \mathscr{L}}{\partial \tilde{\mathbf{q}}}\delta\tilde{\mathbf{q}} = \lim_{\epsilon\to 0}\frac{\mathscr{L}\left(\mathbf{q},\tilde{\mathbf{q}}+\epsilon\delta\tilde{\mathbf{q}}\right)-\mathscr{L}\left(\mathbf{q},\tilde{\mathbf{q}}\right)}{\epsilon}, \tag{6.111}$$

where $\delta\mathbf{q}$ and $\delta\tilde{\mathbf{q}}$ are arbitrary functions taken at any time.

Variation with respect to indefinite multipliers clearly reads

$$\frac{\partial \mathscr{L}}{\partial \tilde{\mathbf{q}}}\delta\tilde{\mathbf{q}} = -\lim_{\epsilon\to 0}\frac{1}{\epsilon}\int_0^t\left(\epsilon\delta\tilde{\mathbf{q}},\dot{\mathbf{q}}-\mathbf{A}(\mathbf{q})\mathbf{q}\right)d\tau = -\int_0^t\left(\delta\tilde{\mathbf{q}},\dot{\mathbf{q}}-\mathbf{A}(\mathbf{q})\mathbf{q}\right)d\tau. \tag{6.112}$$

Equating (6.112) to zero we obtain, by arbitrariness of $\delta\tilde{\mathbf{q}}$, Eq. (6.98). To compute variations with respect to the state vectors we use the Lagrange identity: $\left(\tilde{\mathbf{q}},\mathbf{A}\mathbf{q}\right) = \left(\mathbf{A}^{\dagger}\tilde{\mathbf{q}},\mathbf{q}\right)$ (see, for example, Marchuk (1998) for more detail about adjoint operators in non-linear problems) and take the penalty term by parts, after which the Lagrangian can be rewritten as

$$\mathscr{L}\left(\mathbf{q},\tilde{\mathbf{q}}\right) = \mathscr{G}(\mathbf{q}) - \left(\tilde{\mathbf{q}},\mathbf{q}\right)\Big|_0^t + \int_0^t\left(\dot{\tilde{\mathbf{q}}}+\mathbf{A}^{\dagger}(\tilde{\mathbf{q}})\tilde{\mathbf{q}},\mathbf{q}\right)d\tau. \tag{6.113}$$

Taking into account the smallness of ϵ and the real-valued inner product,[9] we then calculate the variation with respect to the state vectors:

$$\frac{\partial \mathscr{L}}{\partial \mathbf{q}}\delta\mathbf{q} = \lim_{\epsilon\to 0}\frac{1}{\epsilon}\left[\frac{||\mathbf{q}(t)+\epsilon\delta\mathbf{q}(t)||^2}{||\mathbf{q}(0)+\epsilon\delta\mathbf{q}(0)||^2} - \frac{||\mathbf{q}(t)||^2}{||\mathbf{q}(0)||^2} - \left(\tilde{\mathbf{q}}(t),\epsilon\delta\mathbf{q}(t)\right) + \right.$$
$$\left. + \left(\tilde{\mathbf{q}}(0),\epsilon\delta\mathbf{q}(0)\right) + \int_0^t\left(\dot{\tilde{\mathbf{q}}}+\mathbf{A}^{\dagger}(\tilde{\mathbf{q}})\tilde{\mathbf{q}},\epsilon\delta\mathbf{q}\right)d\tau\right], \tag{6.114}$$

[9] A real-valued inner product is additionally required only in this section to obtain in simple form the constraints (6.118) and (6.119), see below.

Here the first term can be recast in the form

$$\lim_{\epsilon\to 0}\frac{1}{\epsilon}\frac{||\mathbf{q}(t)+\epsilon\delta\mathbf{q}(t)||^2}{||\mathbf{q}(0)+\epsilon\delta\mathbf{q}(0)||^2}=\lim_{\epsilon\to 0}\frac{1}{\epsilon}\frac{||\mathbf{q}(t)||^2+\epsilon(\delta\mathbf{q}(t),\mathbf{q}(t))+\epsilon(\mathbf{q}(t),\delta\mathbf{q}(t))}{||\mathbf{q}(0)||^2+\epsilon(\delta\mathbf{q}(0),\mathbf{q}(0))+\epsilon(\mathbf{q}(0),\delta\mathbf{q}(0))}. \tag{6.115}$$

As the inner product is real-valued, we have $(\delta\mathbf{q}(t),\mathbf{q}(t)) = (\mathbf{q}(t),\delta\mathbf{q}(t))$, and so the transformation can be continued:

$$\begin{aligned}
&\lim_{\epsilon\to 0}\frac{1}{\epsilon}\left[\frac{||\mathbf{q}(t)+\epsilon\delta\mathbf{q}(t)||^2}{||\mathbf{q}(0)+\epsilon\delta\mathbf{q}(0)||^2}-\frac{||\mathbf{q}(t)||^2}{||\mathbf{q}(0)||^2}\right]\\
&=\lim_{\epsilon\to 0}\frac{1}{\epsilon}\left[\frac{||\mathbf{q}(t)||^2+2\epsilon(\delta\mathbf{q}(t),\mathbf{q}(t))}{||\mathbf{q}(0)||^2+2\epsilon(\delta\mathbf{q}(0),\mathbf{q}(0))}-\frac{||\mathbf{q}(t)||^2}{||\mathbf{q}(0)||^2}\right]=\\
&=\lim_{\epsilon\to 0}\frac{1}{\epsilon}\left[\frac{2\epsilon(\delta\mathbf{q}(t),\mathbf{q}(t))}{||\mathbf{q}(0)||^2}-\frac{2\epsilon(\delta\mathbf{q}(0),\mathbf{q}(0))||\mathbf{q}(t)||^2}{||\mathbf{q}(0)||^4}\right]=\\
&=\frac{2(\delta\mathbf{q}(t),\mathbf{q}(t))}{||\mathbf{q}(0)||^2}-2(\delta\mathbf{q}(0),\mathbf{q}(0))\frac{||\mathbf{q}(t)||^2}{||\mathbf{q}(0)||^4},
\end{aligned} \tag{6.116}$$

which ultimately gives the variation:

$$\begin{aligned}
\frac{\partial\mathscr{L}}{\partial\mathbf{q}}\delta\mathbf{q}=&\frac{2(\delta\mathbf{q}(t),\mathbf{q}(t))}{||\mathbf{q}(0)||^2}-2(\delta\mathbf{q}(0),\mathbf{q}(0))\frac{||\mathbf{q}(t)||^2}{||\mathbf{q}(0)||^4}-\big(\tilde{\mathbf{q}}(t),\delta\mathbf{q}(t)\big)+\\
&+\big(\tilde{\mathbf{q}}(0),\delta\mathbf{q}(0)\big)+\int_0^t\Big(\dot{\tilde{\mathbf{q}}}+\mathbf{A}^\dagger(\tilde{\mathbf{q}})\tilde{\mathbf{q}},\delta\mathbf{q}\Big)\,d\tau.
\end{aligned} \tag{6.117}$$

Since the variations of $\delta\mathbf{q}$ taken at different instants are independent from each other, equating (6.117) to zero yields the equation for indefinite multipliers (6.103), which provides the vanishing of the Lagrangian variation in the interval $0 < \tau < t$, as well as the relations between $\mathbf{q}$ and $\tilde{\mathbf{q}}$, which are necessary for the Lagrangian variations to vanish at moments $\tau = 0$ and $\tau = t$:

$$\tilde{\mathbf{q}}(t)=\frac{2}{||\mathbf{q}(0)||^2}\mathbf{q}(t) \tag{6.118}$$

$$\mathbf{q}(0)=\frac{||\mathbf{q}(0)||^4}{2||\mathbf{q}(t)||^2}\tilde{\mathbf{q}}(0). \tag{6.119}$$

The vectors $\mathbf{q}$ and $\tilde{\mathbf{q}}$ satisfying Eqs. (6.98) and (6.103) and the constraints (6.118) and (6.119) turn the Lagrangian variations to zero, and hence precisely for them the functional (6.108) reaches extremum.

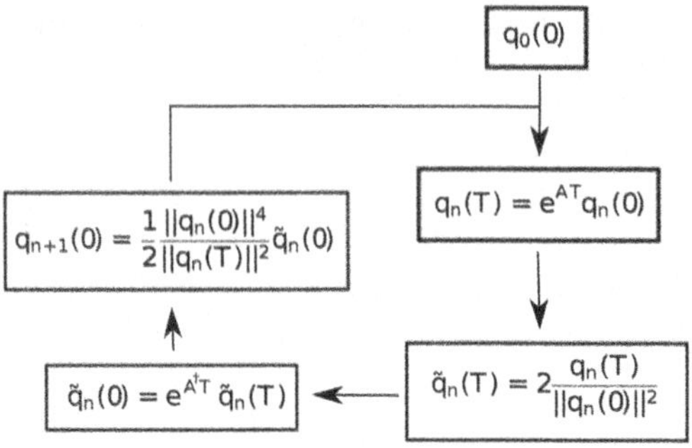

Fig. 6.10 Schematics of the iteration loop to search for the optimal perturbation at time T satisfying the general set (6.98) (see the review Schmid (2007))

As for linear systems, the joint solution of equations can be found using the power iteration method schematically shown in Fig. 6.10. This issue is further discussed in papers Cherubini et al. (2010, 2011), Cherubini and De Palma (2013). Note once again that for linear perturbations the optimization of the functional (6.108) is reduced to looking for the maximal eigenvalue of the composite operator $\mathbf{Z}^\dagger\mathbf{Z}$.

6.3.4 Adjoint Equations

6.3.4.1 Derivation of Adjoint Equations

In order to obtain an explicit form of adjoint equations to the set (6.8)–(6.10), we will use the norm identical to the total acoustic energy of the perturbations (6.15). The inner product corresponding to this norm is given by Eq. (6.95), which we have already used. We represent $(\tilde{\mathbf{q}}, \mathbf{A}\mathbf{q})$ as

$$
\begin{aligned}
(\tilde{\mathbf{q}}, \mathbf{A}\mathbf{q}) = \pi \int_{r_{in}}^{r_{out}} \Sigma \Bigg[& \delta\tilde{v}_r \left(im\Omega\delta v_r^* + 2\Omega\delta v_\varphi^* - \frac{\partial \delta h^*}{\partial r} \right) \\
& + \delta\tilde{v}_\varphi \left(-\frac{\kappa^2}{2\Omega}\delta v_r^* + im\Omega\delta v_\varphi^* + \frac{im\delta h^*}{r} \right) + \\
& + \frac{\delta\tilde{h}}{a_*^2} \left(-\frac{a_*^2}{\Sigma r}\frac{\partial}{\partial r}\left(r\Sigma\delta v_r^*\right) + \frac{ima_{*^2}}{r}\delta v_\varphi^* + im\Omega\delta h^* \right) \Bigg] r dr.
\end{aligned}
\tag{6.120}
$$

Now, using the Lagrange identity $(\tilde{\mathbf{q}}, \mathbf{A}\mathbf{q}) = (\mathbf{A}^\dagger\tilde{\mathbf{q}}, \mathbf{q})$ and Eq. (6.103) in the left part of this expression, we represent the inner product according to (6.95). The right-hand side can be rearranged in a way to get the components of $\delta\mathbf{q}$ in the form of multipliers. Here, the spatial derivatives are rearranged using integration by parts.

We obtain

$$
\begin{aligned}
&\pi \int\limits_{r_{in}}^{r_{out}} \Sigma r dr \left[-\delta v_r^* \frac{\partial \tilde{v}_r}{\partial t} - \delta v_\varphi^* \frac{\partial \tilde{v}_\varphi}{\partial t} - \delta h^* \frac{\partial \tilde{h}}{\partial t} \right] = \\
&= \pi \int\limits_{r_{in}}^{r_{out}} \Sigma r dr \left[\delta v_r^* \left(i m \Omega \delta \tilde{v}_r - \frac{\kappa^2}{2\Omega} \delta \tilde{v}_\varphi + \frac{\partial \delta \tilde{h}}{\partial r} \right) \right. \\
&+ \delta v_\varphi^* \left(2\Omega \delta \tilde{v}_r + i m \Omega \delta \tilde{v}_\varphi + \frac{im}{r} \delta \tilde{h} \right) + \\
&\left. + \delta h^* \left(\frac{1}{r\Sigma} \frac{\partial}{\partial r} (r \Sigma \delta \tilde{v}_r) + \frac{im}{r} \delta \tilde{v}_\varphi + \frac{i m \Omega}{a_*^2} \delta \tilde{h} \right) \right] \\
&- \pi r \Sigma \delta \tilde{h} \delta v_r^* \Big|_{r_{in}}^{r_{out}} - \pi r \Sigma \delta \tilde{v}_r \delta h^* \Big|_{r_{in}}^{r_{out}} .
\end{aligned}
\tag{6.121}
$$

The substitutions on the right-hand side of (6.121) vanish since $\Sigma \to 0$ at the boundaries.

The components of variation $\delta \mathbf{q}$ are arbitrary and independent, so (6.121) is transformed into three independent equalities each corresponding to a certain component of $\delta \mathbf{q}$. These equalities result in a set of adjoint equations:

$$\frac{\partial \delta \tilde{v}_r}{\partial t} = -im\Omega\, \delta \tilde{v}_r + \frac{\kappa^2}{2\Omega} \delta \tilde{v}_\varphi - \frac{\partial \delta \tilde{h}}{\partial r}, \tag{6.122}$$

$$\frac{\partial \delta \tilde{v}_\varphi}{\partial t} = -2\Omega \delta \tilde{v}_r - im\Omega\, \delta \tilde{v}_\varphi - \frac{im}{r} \delta \tilde{h}, \tag{6.123}$$

$$\frac{\partial \delta \tilde{h}}{\partial t} = -\frac{a_*^2}{r\Sigma} \frac{\partial}{\partial r} (r \Sigma \delta \tilde{v}_r) - \frac{i m a_*^2}{r} \delta \tilde{v}_\varphi - im\Omega\, \delta \tilde{h}. \tag{6.124}$$

Changing to the local space limit in (6.122)–(6.124) (as we did in Sect. 6.2.2 to obtain the set (6.17)–(6.19) from Eqs. (6.8)–(6.10)) we get an explicit form of equations adjoint to the set (6.17)–(6.19):

$$\left(\frac{\partial}{\partial t} - q\Omega_0 x \frac{\partial}{\partial y} \right) \tilde{u}_x - (2-q)\Omega_0 \tilde{u}_y = -\frac{\partial \tilde{W}}{\partial x}, \tag{6.125}$$

$$\left(\frac{\partial}{\partial t} - q\Omega_0 x \frac{\partial}{\partial y} \right) \tilde{u}_y + 2\Omega_0 \tilde{u}_x = -\frac{\partial \tilde{W}}{\partial y}, \tag{6.126}$$

$$\left(\frac{\partial}{\partial t} - q\Omega_0 x \frac{\partial}{\partial y} \right) \tilde{W} + a_*^2 \left(\frac{\partial \tilde{u}_x}{\partial x} + \frac{\partial \tilde{u}_y}{\partial y} \right) = 0, \tag{6.127}$$

where tildes above u_x, u_y and W means that these quantities compose an adjoint state vector.

Finally, changing to the co-moving reference frame in (6.125)–(6.127) yields adjoint equations for a particular SFH:

$$\frac{d\hat{\tilde{u}}_x}{dt} = (2-q)\hat{\tilde{u}}_y - \mathrm{i}\,\tilde{k}_x(t)\hat{\tilde{W}}, \tag{6.128}$$

$$\frac{d\hat{\tilde{u}}_y}{dt} = -2\hat{\tilde{u}}_x - \mathrm{i}\,k_y\hat{\tilde{W}}, \tag{6.129}$$

$$\frac{d\hat{\tilde{W}}}{dt} = -\mathrm{i}\,(\,\tilde{k}_x(t)\hat{\tilde{u}}_x + k_y\hat{\tilde{u}}_y\,). \tag{6.130}$$

Applying the power iteration method jointly to (6.8)–(6.10) and (6.122)–(6.124) for global azimuthal Fourier harmonics of two-dimensional perturbations, or to the sets (6.23)–(6.25) and (6.128)–(6.130) for the local SFH, we automatically arrive at the optimal initial profiles of the enthalpy and velocity component perturbations that maximize the total acoustic energy growth at a given time interval. This problem in application to Keplerian flows was solved by Zhuravlev and Razdoburdin (2014).

To illustrate this iteration algorithm, let us consider convergence of two different initial conditions to an optimal perturbation in a global approach (i.e. with usage of the sets (6.8–6.10) and (6.122–6.124)). In Fig. 6.11 a change of the perturbation profile at $t = 0$ during the iteration procedure is shown. It is easy to see that neither the profile nor the amplification factor of the resulting perturbations depends on the initial profile. However, the initial profile have influence on the convergence rate.

6.3.4.2 Non-normality Condition for Z

Here we show that non-normality of the dynamical operator determined by the set of Eqs. (6.8)–(6.10) is a direct consequence of the angular velocity gradient in the flow. We already discussed this in Sect. 6.3.1, where we introduced the notion of singular vectors. Now we can prove this rigorously in a rather general case, since the explicit form of the operator $\mathbf{A}^\dagger$, defined by the set (6.122)–(6.124), is known. First, let us calculate the commutator of $\mathbf{A}$ and $\mathbf{A}^\dagger$:

$$\left[\mathbf{A}, \mathbf{A}^\dagger\right] = \begin{pmatrix} \frac{16\Omega^4-\kappa^4}{4\Omega^2} & 0 & \frac{im}{2r\Omega}\left(4\Omega^2-\kappa^2\right) \\ 0 & \frac{\kappa^4-16\Omega^4}{4\Omega^2} & \frac{4\Omega^2-\kappa^2}{2\Omega}\frac{\partial}{\partial r} \\ \frac{ima_*^2}{2r\Omega}\left(\kappa^2-4\Omega^2\right) & \frac{a_*^2}{r\Sigma}\frac{\partial}{\partial r}\left(\frac{r\Sigma}{2\Omega}\left(\kappa^2-4\Omega^2\right)\right)+\frac{a_*^2}{2\Omega}\left(\kappa^2-4\Omega^2\right)\frac{\partial}{\partial r} & 0 \end{pmatrix}. \tag{6.131}$$

It is not difficult to see that $\left[\mathbf{A}, \mathbf{A}^\dagger\right]$ vanishes for $\kappa = 2\Omega$, which corresponds to solid-body rotation. In this case the commutator $\left[\mathbf{Z}, \mathbf{Z}^\dagger\right] = \left[\mathrm{e}^{\mathbf{A}t}, \mathrm{e}^{\mathbf{A}^\dagger t}\right]$ can easily

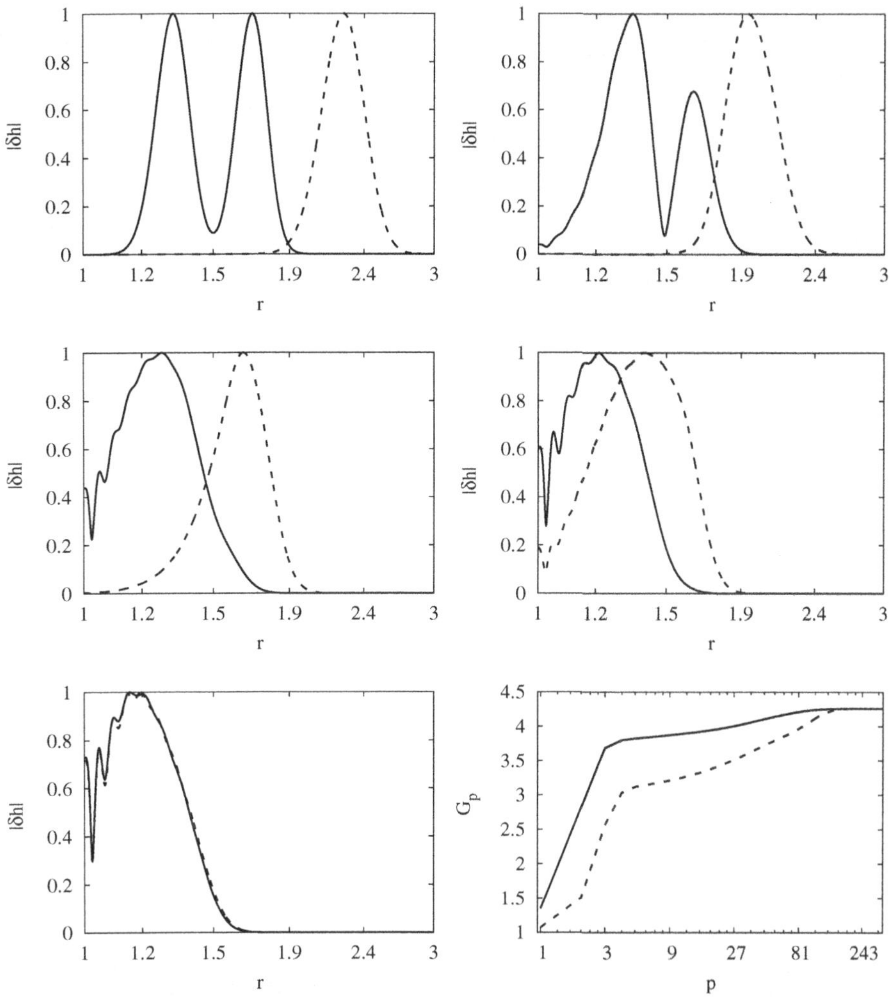

Fig. 6.11 Illustration of iteration convergence for two different initial conditions. In the top left panel initial radial profiles of $|\delta h|$ are shown. In the top right, middle left, middle right and bottom left panels radial profiles of $|\delta h|$ are shown after 25, 80, 120 and 350 iterations respectively. In the bottom right panel the amplification factor G_p as a function of iteration number p is shown. Solid lines denote the radial disturbance of $|\delta h|$ for an initial profile in the form of a'double gaussian function' (the sum of two gaussian distributions with different mean). Dotted lines denote the initial profile in the form of a single gaussian function. For both initial conditions, initial velocities are equal to zero before iterations start. A Shakura-Sunyaev disc was used as background flow. The azimuthal number was set to $m = 5$, the polytropic index $n = 3/2$, the optimization time $T = 3$ and $\delta = 0.05$ (see paper Zhuravlev and Razdoburdin (2014) for a detailed description of the background flow)

be found, since for commuting operators the product of their operator exponents is equal to the exponent of their sum, which can be easily verified by writing the operator exponents as the corresponding infinite series

$$\left[e^{\mathbf{A}t}, e^{\mathbf{A}^\dagger t}\right] = e^{\mathbf{A}t}e^{\mathbf{A}^\dagger t} - e^{\mathbf{A}^\dagger t}e^{\mathbf{A}t} = e^{(\mathbf{A}+\mathbf{A}^\dagger)t} - e^{(\mathbf{A}^\dagger+\mathbf{A})t} = 0. \tag{6.132}$$

Thus, the operator $\mathbf{Z}$ becomes normal for solid-body rotation.

The inverse statement is also valid: if $\mathbf{Z}$ is normal at any time t, the rotation is solid-body. To see this, use the Campbell-Baker-Hausdorff formula (Richtmyer 1981, Ch. 25) to represent the composite operators $\mathbf{Z}\mathbf{Z}^\dagger$ and $\mathbf{Z}^\dagger\mathbf{Z}$:

$$e^{\mathbf{A}t}e^{\mathbf{A}^\dagger t} = \exp\left(\left(\mathbf{A}+\mathbf{A}^\dagger\right)t + \frac{t^2}{2}\left[\mathbf{A}, \mathbf{A}^\dagger\right] + \frac{t^3}{12}\left[\mathbf{A}, \left[\mathbf{A}, \mathbf{A}^\dagger\right]\right] - \frac{t^3}{12}\left[\mathbf{A}^\dagger, \left[\mathbf{A}, \mathbf{A}^\dagger\right]\right] + \dots\right) \tag{6.133}$$

$$\begin{aligned} e^{\mathbf{A}^\dagger t}e^{\mathbf{A}t} &= \exp\left(\left(\mathbf{A}^\dagger+\mathbf{A}\right)t + \frac{t^2}{2}\left[\mathbf{A}^\dagger, \mathbf{A}\right] + \frac{t^3}{12}\left[\mathbf{A}^\dagger, \left[\mathbf{A}^\dagger, \mathbf{A}\right]\right] - \frac{t^3}{12}\left[\mathbf{A}, \left[\mathbf{A}^\dagger, \mathbf{A}\right]\right] + \dots\right) = \\ &= \exp\left(\left(\mathbf{A}^\dagger+\mathbf{A}\right)t - \frac{t^2}{2}\left[\mathbf{A}, \mathbf{A}^\dagger\right] + \frac{t^3}{12}\left[\mathbf{A}, \left[\mathbf{A}, \mathbf{A}^\dagger\right]\right] - \frac{t^3}{12}\left[\mathbf{A}^\dagger, \left[\mathbf{A}, \mathbf{A}^\dagger\right]\right] + \dots\right). \end{aligned} \tag{6.134}$$

The equality $\left[e^{\mathbf{A}t}, e^{\mathbf{A}^\dagger t}\right] = 0$ is fulfilled for any t, therefore the terms with the same powers of t must be independently equal to zero, which is possible only if the commutator $\left[\mathbf{A}, \mathbf{A}^\dagger\right] = 0$. The last equality is valid for solid-body rotation only.

This implies that solid-body rotation is necessary and sufficient for the dynamical operator $\mathbf{Z}$ of the set (6.8)–(6.10) to be normal. Thus, any deviation from solid-body rotation, for example the appearance of an angular velocity gradient in astrophysical discs, makes the dynamical operator non-normal and perturbation modes non-orthogonal to each other.

6.4 Optimal Perturbations in Keplerian Discs

In the concluding section of this chapter we would like to briefly discuss the use of the variational method to search for optimal perturbations in astrophysical discs. We consider geometrically thin discs with an almost Keplerian azimuthal velocity profile in the background flow. In the numerical calculations we are going to consider a radially infinite disc with only an inner (free) boundary and a thin quasi-Keplerian torus with inner and outer radial boundaries. The latter configuration was used in Sect. 6.3.2 for the analysis of superposition of neutral modes to illustrate the matrix method of optimization. However, for the sake of methodology, we start with the simplest analytically tractable problem of transient growth of local short-wave perturbations with $k_y \gg 1$ which we discussed in detail in Sect. 6.2.3.

6.4.1 Local Approximation

Indeed, let us apply the power iteration method to the sets (6.23)–(6.25), (6.128)–(6.130) in the limit $k_y \gg 1$ corresponding to an incompressible fluid. In this limit, the set (6.23)–(6.25) can be reduced to one equation for $\hat{u}_x$:

$$\frac{d\hat{u}_x}{dt} + 2qk_y \frac{\tilde{k}_x}{k_y^2 + \tilde{k}_x^2} \hat{u}_x = 0, \tag{6.135}$$

giving the analytical solution

$$\hat{u}_x(t) = \hat{u}_x(0) \frac{k_x^2 + k_y^2}{\tilde{k}_x^2 + k_y^2}, \tag{6.136}$$

which, of course, repeat (6.37) for $k_y \gg 1$.

At the same time, the adjoint equations (6.128)–(6.130) in the limit of an incompressible fluid suggest that the quantity $\hat{\tilde{u}}_x$ conjugate to $\hat{u}_x$ is *conserved*[10]:

$$\frac{d\hat{\tilde{u}}_x}{dt} = 0. \tag{6.137}$$

Obviously, after p iterations of the arbitrary initial profile $\hat{u}_x^{in}(k_x, k_y, t = 0)$ we obtain that it is multiplied by the factor:

$$\left[\frac{k_x^2 + k_y^2}{(\tilde{k}_x(t)^2 + k_y^2} \right]^p . \tag{6.138}$$

With account for renormalization of the solution at each iteration, while $p \to \infty$, the factor (6.138) suppresses all SFH composing $\hat{u}_x^{in}(k_x, k_y, t = 0)$ except the

[10]It can be verified that the value of I, which is conserved in the direct equations (6.23)–(6.25), becomes time-dependent in the adjoint equations (6.128)–(6.130) (see the Appendix in paper Zhuravlev and Razdoburdin (2014)).

optimal SFH corresponding to a maximum of (6.138) as a function of k_x. For a fixed time interval t this k_x takes the value

$$k_x = 1/2k_y\,(-qt - ((qt)^2 + 4)^{1/2}). \qquad (6.139)$$

Plugging (6.139) into the SFH growth factor (6.44) yields the sought for optimal growth G, which for the local problem is defined as (6.46):

$$G(t) = \frac{(qt)^2 + qt[(qt)^2 + 4]^{1/2} + 4}{(qt)^2 - qt[(qt)^2 + 4]^{1/2} + 4}. \qquad (6.140)$$

Expression (6.140) represents the first singular value which the iteration loop for short-wave local vortices converges to. Comparison of optimal growth factor given by expression (6.140) with growth rate of individual perturbations can be found in Fig. 6.12. Apparently, for large time intervals, $qt \gg 1$, it gives $G \approx (qt)^2$, which reproduces the approximate estimate of G according to formula (6.48).

Also note that an exact result (6.140) could be obtained in this simple example directly from the expression for the growth factor (6.44) by calculating the maximum of g as a function of k_x at a fixed t.

For arbitrary k_y the optimal growth can be obtained by numerical forwards-backwards integration of the full set of direct and adjoint equations, which are ordinary differential equations for SFH.

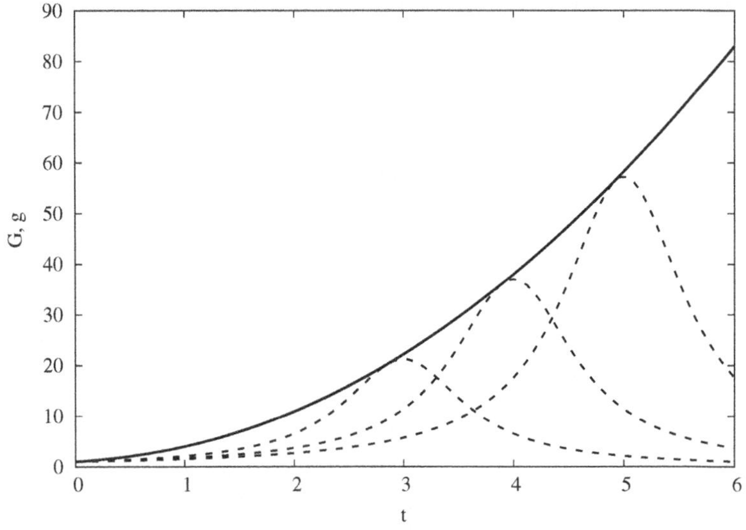

Fig. 6.12 Comparison of growth rate of individual perturbations, acquired with the help of Eq. (6.44) (dotted lines) for $k_y = 100$, $q = 1.5$, with optimal growth rate from Eq. (6.140) (solid line). Initial k_x for individual perturbations are chosen to provide swing at moments $t = 3, 4, 5$. It is easy to see that for all moments $G(t) \geq g(t)$

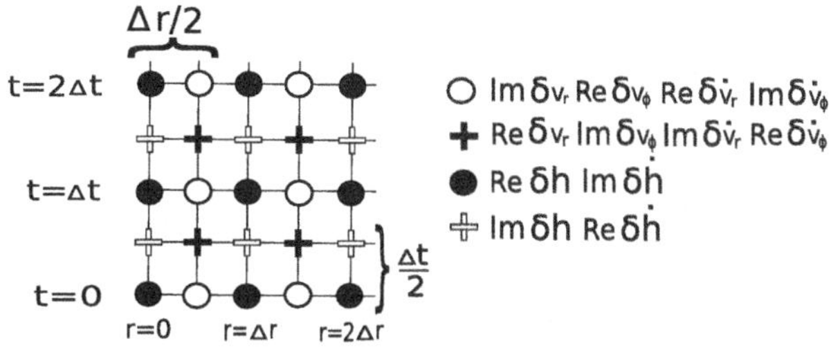

Fig. 6.13 Illustration to the numerical scheme of integration of Eqs. (6.8)–(6.10) and (6.122)–(6.124)

6.4.2 *Global Problem*

In the case where the azimuthal scale of perturbations is comparable to the horizontal disc scale it is necessary to numerically solve the set of partial differential equations (6.8)–(6.10) and (6.122)–(6.124), which was done in paper Zhuravlev and Razdoburdin (2014), using a second-order explicit difference scheme (leap-frog) (see, for example, Frank and Robertson 1988).

In this difference scheme, each equation is separated into real and imaginary parts and on the plane (r, t) 4 grids are introduced. Unknown variables are calculated in the nodes of these grids using the corresponding differences (Fig. 6.13). The nodes are shifted with respect to each other by half a time step Δt and/or by half a radial step Δr. This allows the use of a central approximation to the calculated derivatives with respect to r and t, which provides an accuracy of the order of $(\Delta r)^2$ and $(\Delta t)^2$. The time step is determined using the radial step and the Courant condition that follows from the local dispersion relation, which can be obtained from the equations being integrated.

6.4.2.1 Comparison of the Transient Growth of Vortices in Global and Local Space Limits

As a background flow, consider an infinite Keplerian disc that has only an inner boundary at $r = r_1$. To see how the cylindrical geometry of the disc and, mainly, the accurate profile of the Keplerian angular velocity, $\Omega = \Omega(r_1)(r/r_1)^{-3/2}$, affect the transient growth, we assume for simplicity that all other values in the equations for the perturbations are constant:

$$\Sigma = const, \quad a_{eq} = (\delta/\sqrt{2n})(\Omega r)|_{r_1}. \tag{6.141}$$

As shown in paper Zhuravlev and Razdoburdin (2014), an account for a more realistic distribution of Σ and a_{eq} (for example, as in standard accretion discs) does not change the qualitative conclusions presented below. The results of local and global calculations of optimal perturbations using the variational method are shown in Fig. 6.14.

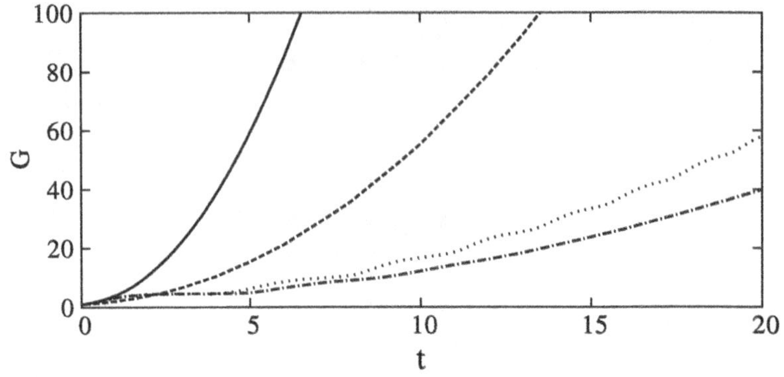

Fig. 6.14 Comparison of the optimal growth for small-scale and large-scale vortices (see Sect. 6.2.3) in the global and local space limits. The solid and dotted curves are calculated for local SFH using formula (6.46) using an iteration loop for Eqs. (6.23)–(6.25) and (6.128)–(6.130) for $k_y = 12.5$ and $k_y = 0.125$, respectively. Harmonics with $m = 5$ are taken as global perturbations. The optimal perturbations are calculated using formula (6.94) using an iteration loop for Eqs. (6.8)–(6.10) and (6.122)–(6.124) with the polytropic index $n = 3/2$. Using the relation $(m/r)H \sim k_y$, for similar large-scale and small-scales vortices, a disc with $\delta = 0.05$ (the dashed-dotted line) and a formally thick disc with $\delta = 5$ (the dashed line) were considered, respectively. In both cases time is expressed in units of $\Omega(r_1)^{-1}$

Here we compare the transient growth of vortices with azimuthal scale both smaller and larger than the disc thickness. The main qualitative conclusion is that the growth rate of small-scale vortices ($\lambda_\varphi < H$) decreases much faster, as one proceeds from the shearing sheet approximation to the scales of order of the disc radial size (i.e. from $m = \infty$ to $m \sim 1$), than that of large-scale vortices ($\lambda_\varphi > H$). It can be verified that in the limiting case of global perturbations with $m = 1$, the value of G for small-scale and large-scale vortices differ to within a factor of 1.5–2 only, for the given parameters and for time intervals up to $t \sim 20$. At the same time, for local vortices, the value of G for $\lambda_\varphi < H$ and $\lambda_\varphi > H$ differs by several orders of magnitude. This suggests that global large-scale vortices in thin Keplerian discs can also exhibit a growth of dozens of times on quite short time scales of the order of a few Keplerian periods at the inner disc boundary. In turn, this may imply importance of transient growth of perturbations for angular momentum transfer on scales much larger than the disc thickness.

6.4.2.2 Transient Spirals and Modes in a Quasi-Keplerian Torus

Finally, let us return to the disc model considered above for illustration of the matrix method (see Sect. 6.3.2). As is well known (see, for example, Glatzel (1987a,b), Glatzel (1988), as well as Zhuravlev and Shakura (2007b)), this flow demonstrates a weak spectral instability, since there are exponentially growing inertial-acoustic modes present. As we have already mentioned in Sect. 6.2.1, their increments rapidly decrease with decreasing relative geometrical thickness of the torus, i.e. with Ω approaching a Keplerian profile. Then, perturbations can grow only due to the transient mechanism of shortening of leading spirals by the shear flow (see the discussion in Sect. 6.2.3), which occurs on short time scales of the order of a few Keplerian periods in the flow. However, in the intermediate case, where the pressure gradient in the torus is sufficiently high, both non-modal and modal perturbation growth can occur simultaneously but on different time scales. The exponential growth of modes will always dominate over the transient growth starting from some large time intervals. Interestingly, essentially this means that as calculating the first singular value of the dynamical operator employing the variational method, starting from some t the curve $G(t)$ should become exponential corresponding to the most unstable mode. At the same time, the iteration loop, which always converges to the optimal initial perturbation vector $\mathbf{q}(t = 0)$, must then give not a leading spiral, but a *mode*. Whereas the spiral starts being shrunk by the shear flow and enhanced due to the perimeter shortening (see the discussion in Sect. 6.2.1) at the time $t > 0$, the mode rotates like a solid body with angular velocity equal to Ω at the corotation radius inside the flow, since its amplitude increases due to the resonance energy exchange with the flow at this radius. Thus, the method of optimization of perturbations can be applied both to study the transient growth of perturbations and to find the profiles and increments of the most unstable modes in arbitrary complex shear flows, i.e. to solve the spectral problem as well.

An example of the calculation of a transient spiral and of an unstable mode in the same toroidal flow by joint solution of the set (6.8)–(6.10) and (6.122)–(6.124), employing the variational method, was presented in Figs. 6.1 and 6.2 in the Introduction. As we see, even for $\delta = 0.3$ the maximal increment is very low, and it takes $\sim 10^3$ Keplerian periods for the most unstable mode to at least double its amplitude. At the same time, the transient growing spiral increases by a factor of 6 already after a few rotational periods at the inner disc boundary.

6.5 Conclusions

This chapter is devoted to the transient dynamics of perturbations, which is of special interest in the theory of astrophysical discs, in particular accretion discs. Exponentially growing perturbations do not exist in a homogeneous inviscid Keplerian flow, provided that there are no conditions for the magneto-rotational instability. Nevertheless, observations suggest that also in this case angular momentum should be somehow transported outwards. At least, this implies that there should be some mechanism of energy transfer from the regular rotational motion to hydrodynamical perturbations. In spectrally stable flows the transient growth mechanism is responsible for this. Here it was introduced by a simple example of two-dimensional vortices and it was discussed that the reason for their growth is the shortening of the length of leading spirals by the differential rotation of the flow (see Figs. 6.2 and 6.3). Notwithstanding their seeming simplicity, these (quasi-)columnar structures exhibit the strongest ability to extract energy from spectrally stable differentially rotating flows (see Maretzke et al. 2014). Physically, the energy growth of vortices takes place due to their own angular momentum conservation, which in the local limit is expressed by the conservation of their potential vorticity and the existence of the invariant I (see Sect. 6.2.2). Here we considered both small-scale ($k_y \gg 1$) and large-scale ($k_y \ll 1$) vortices and compared their optimal growth with account for non-zero effective viscosity in the disc (see Fig. 6.4). Importantly, the transient growth of large-scale vortices strongly increases for super-Keplerian rotation, which can be significant in relativistic discs where $q > 3/2$. In this chapter, special attention was given to the mathematical aspects of non-modal analysis and to methods of optimal perturbations computation. We have discussed in detail that the transient growth is a consequence of non-normality of the governing dynamical operator of the problem and non-orthogonality of its eigenvectors, i.e. modes of perturbations (see Figs. 6.5 and 6.6). Therefore, the growth of arbitrary perturbations can be adequately studied by calculating not eigenvectors but singular vectors of this operator. We have considered two methods: a matrix and variational one and applied them to the particular problems (see the corresponding results in Figs. 6.8 and 6.14). The matrix method requires a discrete representation of the dynamical operator, for example, in the basis of its eigenvectors. The variational method is reduced to iterative integration of the set of direct and adjoint equations forwards and backwards in time, respectively. We have emphasized that the variational method is more universal and can be applied to the study of non-modal dynamics of perturbations in non-stationary flows, as well as to non-linear problems.

As was discussed, the transient growth of perturbations is used in the concept of bypass transition to turbulence in laminar flows. It can be also important as a mechanism of enhanced angular momentum transfer and stimulation of the accretion rate in weakly turbulized discs. Note that turbulence emerging due to the bypass mechanism is fundamentally different from 'classical' turbulence, in which the energy transfer from the background flow is mediated by modes exponentially growing on large spatial scales, whereas the non-linear interactions nothing but redistribute this energy between modes with other wave vectors k (the so-called direct or inverse cascade). This means that in phase space an energy flux $\epsilon_T(k)$ arises which brings (in the case of direct cascade) the kinetic energy of perturbations to small scales where viscous dissipation occurs. In this picture, the mode distribution over the directions of $\mathbf{k}$ in phase space is of minor importance, and ϵ_T can be non-zero only along the direction of change of the module $\mathbf{k}$. A completely different situation should arise if the transient growth of perturbations is responsible for the energy transfer from the background flow. This linear mechanism appears as leading spirals in the disc, i.e. spatial Fourier harmonics corresponding to only such values $\mathbf{k}$ that $k_x/k_y < 0$. In a spectrally stable flow, where there is no energy supply to the leading spirals, initial perturbations inevitably decay as the leading spirals turn into trailing ones. Thus, the turbulent state here is possible only due to positive non-linear feedback, which can exist only in the appearance of non-zero ϵ_T also in the direction of positional change of the vector $\mathbf{k}$, i.e. in the phase space angles, when the trailing spirals return a part of their energy to the leading spirals, sufficient to sustain transient growth. Simultaneously, the rest of the energy stored in the trailing spirals dissipates into heat due to their ultimate transition to higher k. Here, heat dissipation can be not due to a direct cascade, but to a purely linear winding up of the trailing spiral by the flow, i.e. to the increase in time of the ratio $k_x/k_y > 0$ at $k_y = const$. As we see, the *transverse* cascade is an essential part of the alternative picture of turbulence in a shear flow, which is the angular redistribution of spatial Fourier harmonics of perturbations (see, for example, the appendix in Chagelishvili et al. (2003)). The maintenance of transient growth of small perturbations by the transverse cascade was studied in detail in Horton et al. (2010) for a two-dimensional Couette flow. These profound changes in the concept of the possible structure of turbulent flows should affect both analytical estimates of the turbulent viscosity coefficient (see, for example, Canuto et al. 1984), and numerical simulations of turbulence in astrophysical discs (see, for example, Simon et al. (2009), Davis et al. (2010), where spectral properties of turbulence averaged over the directions of $\mathbf{k}$ were mostly studied). Note that we deliberately cited here numerical simulations in discs containing magnetic fields, in which modal growth of perturbations due to the magneto-rotational instability takes place. The point is that recent studies Squire and Bhattacharjee (2014a,b) show that even in Keplerian flows, where the magneto-rotational instability operates, the optimal transiently growing perturbations dominate over exponentially growing modes on short time-scales. Like in an unmagnetized flow, these transient perturbations are locally represented by shear harmonics. Thus, non-modal dynamics of perturbations can be an essential tool in extracting energy from the background flow in MHD-turbulent accretion discs as well. Another hint of this is the recent paper Mamatsashvili et al. (2014), which studied numerically (similar to Horton et al. 2010) the transverse cascade of shear harmonics in a spectrally stable plane-parallel magnetized flow and demonstrated that two-dimensional turbulence arises due to a positive feedback with linear transient growth of shear harmonics. The plane Poiseuille flow provides another example of a shear flow in which the bypass transition to turbulence turns out to be more preferable than the 'classical' mechanism, despite the presence of growing modes. Here we mention Schmid et al. (1996) and Reddy et al. (1998), who numerically studied not the developed turbulence (as is usually done in most of papers on MHD-turbulence in Keplerian flows), but some scenarios of the transition to turbulence from regular initial small perturbations of different types (see also Ch. 9 of book Schmid and Henningson (2001)). It turned out that the previously accepted scenario of a transition due to secondary instability of saturated modes requires much more time and/or significantly higher initial perturbation amplitudes than a transition due to secondary instability of the so-called streaks grown due to the transient mechanism. For the sake of clarity, we stress that streaks in the 3D model of a plane Poiselle flow grow from the so called vortex *rolls* due to the lift-up mechanism, which is also a variant of transient growth, but differs from the (swing) amplification of 2D vortices considered throughout this work. Anyhow, as follows from Fig. 1 of Schmid et al. (1996), the time of turbulence development from regular initial perturbations strongly depends on their amplitudes. This is not surprising, since vortex rolls (just like the 2D spatial Fourier harmonics studied above) of smaller amplitude require more time to saturate, after which secondary instability comes into play leading directly to a breakdown to turbulence. Clearly, the time for such a transition can be as long as hundreds of characteristic shear times, and nevertheless this does not affect later the properties and power of turbulent motions. Although we at present have results only from studies of plane-parallel flows, in the future results may be obtained in a similar way for quasi-Keplerian flows with high Reynolds numbers, since locally such flows differ from plane-parallel flows only by the presence of the Coriolis force stabilizing the flow. At last, additional useful evidence presented here are the simplified finite-dimensional dynamical models of non-normal systems with positive feedback that recover basic properties of transition to turbulence in spectrally stable shear flows (see Trefethen et al. 1993; Waleffe 1995). For example, in Fig. 10 of Trefethen et al. (1993) it can be seen that the time for such a simplified model to reach one and the same 'turbulent' state increases with a decrease of the initial perturbation amplitude and ultimately becomes infinite.

To conclude, we stress once again that here we have not discussed the aspects of three-dimensional perturbation dynamics. Meanwhile, there are indications that taking into account the natural inhomogeneity of the disc due to vertical density and pressure gradients gives a qualitatively new picture of both the transient growth of perturbations and the subsequent transition to turbulence (see Lominadze 2011). Here, the perturbation dynamics is essentially three-dimensional, and it can be shown that for three-dimensional transient vortices there is a time-conserved analogue of the invariant of motion I (see Tevzadze et al. 2003, 2008). The new numerical calculations carried out in Marcus et al. (2014) also point out that taking into account the disc vertical inhomogeneity can result in its destabilization in the subcritical regime at high Reynolds numbers, unlike the case observed in a homogeneous flow (see Shen et al. 2006).

References

Afshordi N, Mukhopadhyay B, Narayan R (2005) Bypass to turbulence in hydrodynamic accretion: Lagrangian analysis of energy growth. Astrophys J 629:373–382. https://doi.org/10.1086/431418

Andersson P, Berggren M, Henningson DS (1999) Optimal disturbances and bypass transition in boundary layers. Phys Fluids 11:134–150. https://doi.org/10.1063/1.869908

Armitage PJ (2009) Astrophysics of planet formation. Cambridge University Press, Cambridge

Avila M (2012) Stability and angular-momentum transport of fluid flows between corotating cylinders. Phys Rev Lett 108(12):124501. https://doi.org/10.1103/PhysRevLett.108.124501. ArXiv:1203.4923

Balbus SA (2003) Enhanced angular momentum transport in accretion disks. Annu Rev Astron Astrophys 41:555–597. https://doi.org/10.1146/annurev.astro.41.081401.155207. ArXiv:astro-ph/0306208

Balbus SA, Hawley JF (1991) A powerful local shear instability in weakly magnetized disks. i - linear analysis. ii - nonlinear evolution. Astrophys J 376:214–233. https://doi.org/10.1086/170270

Balbus SA, Hawley JF (1998) Instability, turbulence, and enhanced transport in accretion disks. Rev Mod Phys 70:1–53. https://doi.org/10.1103/RevModPhys.70.1

Balbus SA, Hawley JF, Stone JM (1996) Nonlinear stability, hydrodynamical turbulence, and transport in disks. Astrophys J 467:76

Binney J, Tremaine S (2008) Galactic dynamics, 2nd edn. Princeton University Press, Princeton

Blaes OM, Glatzel W (1986) On the stability of incompressible constant angular momentum cylinders. Mon Not R Astron Soc 220:253–258

Bodo G, Chagelishvili G, Murante G, Tevzadze A, Rossi P, Ferrari A (2005) Spiral density wave generation by vortices in keplerian flows. Astron Astrophys 437:9–22. https://doi.org/10.1051/0004-6361:20041046

Brekhovskikh LM, Goncharov V (1985) Mechanics of continua and wave dynamics. Berlin: Springer

Butler KM, Farrell BF (1992) Three-dimensional optimal perturbations in viscous shear flow. Phys Fluids A 4(8):1637–1650. https://doi.org/10.1063/1.858386

Canuto VM, Goldman I, Hubickyj O (1984) A formula for the shakura–sunyaev turbulent viscosity parameter. Astrophys J Lett 280:L55

Chagelishvili GD, Rogava AD, Segal IN (1994) Hydrodynamic stability of compressible plane couette flow. Phys Rev E 50:4283. https://doi.org/10.1103/PhysRevE.50.R4283

Chagelishvili GD, Chanishvili RG, Lominadze DG (1996) Physics of the amplification of vortex disturbances in shear flows. JETP Lett 7:543–549. https://doi.org/10.1134/1.567072

Chagelishvili GD, Tevzadze AG, Bodo G, Moiseev SS (1997) Linear mechanism of wave emergence from vortices in smooth shear flows. Phys Rev Lett 79:3178–3181. https://doi.org/10.1103/PhysRevLett.79.3178

Chagelishvili GD, Zahn JP, Tevzadze AG, Lominadze JG (2003) On hydrodynamic shear turbulence in keplerian disks: via transient growth to bypass transition. Astron Astrophys 402:401–407. https://doi.org/10.1051/0004-6361:20030269

Charney JG, Fjörtoft R, Von Neuman J (1950) Numerical integration of the barotropic vorticity equation. Tellus 2:237–254

Cherubini S, De Palma P (2013) Nonlinear optimal perturbations in a couette flow: bursting and transition. J Fluid Mech 716:251–279. https://doi.org/10.1017/jfm.2012.544

Cherubini S, Robinet JC, Bottaro A, de Palma P (2010) Optimal wave packets in a boundary layer and initial phases of a turbulent spot. J Fluid Mech 656:231–259. https://doi.org/10.1017/S002211201000114X

Cherubini S, de Palma P, Robinet JC, Bottaro A (2011) The minimal seed of turbulent transition in the boundary layer. J Fluid Mech 689:221–253. https://doi.org/10.1017/jfm.2011.412

Churilov SM, Shuhman IG (1981) On the relation between volume and surface adiabatic indices for gaseous subsystems of flat galaxies. Astron Tsirk 1157:1–2

Coles D (1965) Transition in circular couette flow. J Fluid Mech 21:385–425. https://doi.org/10.1017/S0022112065000241

Corbett P, Bottaro A (2001) Optimal linear growth in swept boundary layers. J Fluid Mech 435:1–23

Davis SW, Stone JM, Pessah ME (2010) Sustained magnetorotational turbulence in local simulations of stratified disks with zero net magnetic flux. Astrophys J 713:52–65

Drazin PG, Reid WH (1981) Hydrodynamic stability. NASA STI/Recon Technical Report A 82:17950

Drury LO (1985) Acoustic amplification in discs and tori. Mon Not R Astron Soc 217:821–829

Edlund EM, Ji H (2014) Nonlinear stability of laboratory quasi-keplerian flows. Phys Rev Lett 89(2):021004. https://doi.org/10.1103/PhysRevE.89.021004. ArXiv:1401.6183

Farrell BF (1988) Optimal excitation of perturbations in viscous shear flow. Phys Fluids 31:2093–2102. https://doi.org/10.1063/1.866609

Farrell BF, Ioannou PJ (1996a) Generalized stability theory. Part i: autonomous operators. J Atmos Sci 53:2025–2040. https://doi.org/10.1175/1520-0469(1996)053<2025:GSTPIA>2.0.CO;2

Farrell BF, Ioannou PJ (1996b) Generalized stability theory. Part ii: nonautonomous operators. J Atmos Sci 53:2041–2053. https://doi.org/10.1175/1520-0469(1996)053<2041:GSTPIN>2.0.CO;2
Frank J, Robertson JA (1988) Numerical studies of the dynamical stability of differentially rotating tori. Mon Not R Astron Soc 232:1–33
Fridman AM (1989) On the dynamics of a viscous differentially rotating gravitating medium. Sov Astron Lett 15:487
Fridman AM, Bisikalo DV (2008) The at of accretion disks of close binary stars: overreflection instability and developed turbulence. Phys Usp 51:551–576
Friedman JL, Schutz BF (1978) Lagrangian perturbation theory of nonrelativistic fluids. Astrophys J 221:937–957. https://doi.org/10.1086/156098
Glatzel W (1987a) On the stability of compressible differentially rotating cylinders. Mon Not R Astron Soc 225:227–255
Glatzel W (1987b) On the stability of compressible differentially rotating cylinders. ii. Mon Not R Astron Soc 228:77–100
Glatzel W (1988) Sonic instabilities in supersonic shear flows. Mon Not R Astron Soc 231:795–821
Goldreich P, Lynden-Bell D (1965) II. Spiral arms as sheared gravitational instabilities. Mon Not R Astron Soc 130:125
Goldreich P, Narayan R (1985) Non-axisymmetric instability in thin discs. Mon Not R Astron Soc 213:7P–10P
Goldreich P, Goodman J, Narayan R (1986) The stability of accretion tori. i - long-wavelength modes of slender tori. Mon Not R Astron Soc 221:339–364
Golub GH, Reinsch C (1970) Singular value decomposition and least squares solutions. Numer Math 14:403–420. https://doi.org/10.1007/BF02163027
Golub GH, Van Loan CF (1996) Matrix computations. Johns Hopkins University Press, Baltimore
Guégan A, Schmid PJ, Huerre P (2006) Optimal energy growth and optimal control in swept hiemenz flow. J Fluid Mech 566:11–45. https://doi.org/10.1017/S0022112006001303
Gunzburger MD (2003) Perspectives in flow control and optimization. J Soc Ind Appl Math. https://doi.org/10.1137/1.9780898718720
Hanifi A, Schmid PJ, Henningson DS (1996) Transient growth in compressible boundary layer flow. Phys Fluids 8:826–837. https://doi.org/10.1063/1.868864
Hawley JF, Gammie CF, Balbus SA (1995) Local three-dimensional magnetohydrodynamic simulations of accretion disks. Astrophys J 440:742. https://doi.org/10.1086/175311
Hawley JF, Balbus SA, Winters WF (1999) Local hydrodynamic stability of accretion disks. Astrophys J 518:394–404. https://doi.org/10.1086/307282. ArXiv:astro-ph/9811057
Heading JA (2013) Introduction to phase-integral methods. Dover Publications, Mineola
Heinemann T, Papaloizou JCB (2009a) The excitation of spiral density waves through turbulent fluctuations in accretion discs - I. WKBJ theory. Mon Not R Astron Soc 397:52–63. https://doi.org/10.1111/j.1365-2966.2009.14799.x. ArXiv:0812.2068
Heinemann T, Papaloizou JCB (2009b) The excitation of spiral density waves through turbulent fluctuations in accretion discs - ii. numerical simulations with MRI-driven turbulence. Mon Not R Astron Soc 397:64–74. https://doi.org/10.1111/j.1365-2966.2009.14800.x. ArXiv:0812.2471
Henningson DS, Reddy SC (1994) On the role of linear mechanisms in transition to turbulence. Phys Fluids 6:1396–1398. https://doi.org/10.1063/1.868251
Hill GW (1878) Researches in the lunar theory. Am J Math 1:5–26
Horton W, Kim JH, Chagelishvili GD, Bowman JC, Lominadze JG (2010) Angular redistribution of nonlinear perturbations: a universal feature of nonuniform flows. Phys Rev E 81(6):066304. https://doi.org/10.1103/PhysRevE.81.066304
Ioannou PJ, Kakouris A (2001) Stochastic dynamics of keplerian accretion disks. Astrophys J 550:931–943. https://doi.org/10.1086/319791

Ji H, Burin M, Schartman E, Goodman J (2006) Hydrodynamic turbulence cannot transport angular momentum effectively in astrophysical disks. Nature 444:343–346, https://doi.org/10.1038/nature05323, ArXiv:astro-ph/0611481

Johnson BM, Gammie CF (2005) Linear theory of thin, radially stratified disks. Astrophys J 626:978–990. https://doi.org/10.1086/430081

Joseph DD (1976) Stability of fluid motions. i, ii. NASA STI/Recon Technical Report A 77:287

Julian WH, Toomre A (1966) Non-axisymmetric responses of differentially rotating disks of stars. Astrophys J 146:810. https://doi.org/10.1086/148957

Kato S (1987) Instability of isentropic geometrically thin disks due to corotation resonance. Astron Soc Jpn 39(4):645–666

Kato S (2001) Basic properties of thin-disk oscillations. Publ Astron Soc Jpn 53(1):1–24

Kelvin L (1887) On the stability of steady and of periodic fluid motion. Philos Mag 23:459–539

Klahr H, Hubbard A (2014) Convective overstability in radially stratified accretion disks under thermal relaxation. Astrophys J 788:8. https://doi.org/10.1088/0004-637X/788/1/21

Kojima Y (1986) The dynamical stability of a fat disk with constant specific angular momentum. Prog Theor Phys 75:251–261.https://doi.org/10.1143/PTP.75.251

Kojima Y (1989) Non-axisymmetric unstable modes of a differentially rotating torus. Mon Not R Astron Soc 236:589–602

Kojima Y, Miyama SM, Kubotani H (1989) Effects of entropy distributions on non-axisymmetric unstable modes in differentially rotating tori and cylinders. Mon Not R Astron Soc 238:753–768

Korn GA, Korn TM (1968) Mathematical Handbook for Scientists and Engineers. McGraw-Hill, New York

Kurbatov EP, Bisikalo DV, Kaygorodov PV (2014) On the possible turbulence mechanism in accretion disks in nonmagnetic binary stars. Phys Usp 57:787–198

Landau LD, Lifshitz EM (1980) Statistical physics, vol 1. Pegramon Press, Oxford

Landau LD, Lifshitz EM (1987) Fluid mechanics. Pergamon Press, Oxford

Lesur G, Longaretti PY (2005) On the relevance of subcritical hydrodynamic turbulence to accretion disk transport. Astron Astrophys 444:25–44. https://doi.org/10.1051/0004-6361:20053683

Lin C (1955) The theory of hydrodynamic stability. Cambridge University Press, Cambridge

Lominadze JG (2011) Development of the theory of instabilities of differentially rotating plasma with astrophysical applications. In: Bonanno A, de Gouveia Dal Pino E, Kosovichev AG (eds) Proceedings of the International Astronomical Union, IAU symposium: advances in plasma astrophysics, IAU symposium, vol 274, pp 318–324. https://doi.org/10.1017/S1743921311007216

Lominadze DG, Chagelishvili GD, Chanishvili RG (1988) The evolution of nonaxisymmetric shear perturbations in accretion disks. Sov Astron Lett 14(5):364

Longaretti PY (2002) On the phenomenology of hydrodynamic shear turbulence. Astrophys J 576:587–598. https://doi.org/10.1086/341630. ArXiv:astro-ph/0205430

Lovelace RVE, Li H, Colgate SA, Nelson AF (1999) Rossby wave instability of keplerian accretion disks. Astrophys J 513:805–810. https://doi.org/10.1086/306900. ArXiv:astro-ph/9809321

Luchini P (2000) Reynolds-number-independent instability of the boundary layer over a flat surface:optimal perturbations. J Fluid Mech 404:289–309. https://doi.org/10.1017/S0022112099007259

Malik M, Alam M, Dey J (2006) Nonmodal energy growth and optimal perturbations in compressible plane couette flow. Phys Fluids 18(3):034103–034103-14. https://doi.org/10.1063/1.2186671. ArXiv:0804.0065

Mamatsashvili GR, Gogichaishvili DZ, Chagelishvili GD, Horton W (2014) Nonlinear transverse cascade and two-dimensional magnetohydrodynamic subcritical turbulence in plane shear flows. Phys Rev E 89(4):043101. https://doi.org/10.1103/PhysRevE.89.043101. ArXiv:1409.8543

Marchuk GI (1998) Construction of adjoint operators in non-linear problems of mathematical physics. Sbornik Math 189:1505

Marcus P, Pei S, Jiang CH, Barranco J, Hassanzadeh P, Lecoanet D (2014) Zombie vortex instability i: a purely hydrodynamic instability to resurrect the dead zones of protoplanetary disks. ArXiv e-prints1410.8143
Maretzke S, Hof B, Avila M (2014) Transient growth in linearly stable Taylor–Couette flows. J Fluid Mech 742:254–290. https://doi.org/10.1017/jfm.2014.12. ArXiv:1304.7032
Menou K (2000) Viscosity mechanisms in accretion disks. Science 288:2022–2024. https://doi.org/10.1126/science.288.5473.2022. ArXiv:astro-ph/0009022
Meseguer Á (2002) Energy transient growth in the taylor-couette problem. Phys Fluids 14:1655–1660. https://doi.org/10.1063/1.1464851
Monin AS, Yaglom AM (1971) Statistical fluid mechanics, vol 1. MIT Press, Cambridge
Mukhopadhyay B, Afshordi N, Narayan R (2005) Bypass to turbulence in hydrodynamic accretion disks: an eigenvalue approach. Astrophys J 629:383–396. https://doi.org/10.1086/431419
Narayan R (1991) Instabilities in thick disks. In: Bertout C, Collin-Souffrin S, Lasota JP (eds) Proceedings of IAU Colloq. 129, the 6th Institute d'Astrophysique de Paris (IAP) meeting: structure and emission properties of accretion disks, p 153. Editions Frontieres, Gif-sur-Yvette
Narayan R, Goodman J (1989) Non-axisymmetric shear instabilities in thick accretion disks. In: Meyer F (ed) Proceedings of a NATO advanced research workshop: theory of accretion disks, vol 290. NATO Advanced Science Institutes (ASI), Series C, Garching, p 231
Narayan R, Goldreich P, Goodman J (1987) Physics of modes in a differentially rotating system - analysis of the shearing sheet. Mon Not R Astron Soc 228:1–41
Okazaki AT, Kato S, Fukue J (1987) Global trapped oscillations of relativistic accretion disks. Astron Soc Jpn 39:457–473
Orr WM (1907a) The stability or instability of the steady motions of a liquid i. Proc R Ir Acad A 27:9–68
Orr WM (1907b) The stability or instability of the steady motions of a perfect liquid and of a viscous liquid ii. Proc R Ir Acad A 27:69–138
Paoletti MS, Lathrop DP (2011) Angular momentum transport in turbulent flow between independently rotating cylinders. Phys Rev Lett 106(2):024501. https://doi.org/10.1103/PhysRevLett.106.024501. ArXiv:1011.3475
Paoletti MS, van Gils DPM, Dubrulle B, Sun C, Lohse D, Lathrop DP (2012) Angular momentum transport and turbulence in laboratory models of keplerian flows. Astron Astrophys 547:A64. https://doi.org/10.1051/0004-6361/201118511. ArXiv:1111.6915
Papaloizou JCB, Pringle JE (1984) The dynamical stability of differentially rotating discs with constant specific angular momentum. Mon Not R Astron Soc 208:721–750
Papaloizou JCB, Pringle JE (1985) The dynamical stability of differentially rotating discs. ii. Mon Not R Astron Soc 213:799–820
Papaloizou JCB, Pringle JE (1987) The dynamical stability of differentially rotating discs. iii. Mon Not R Astron Soc 225:267–283
Pringle JE, King A (2007) Astrophysical flows. Cambridge University Press, Cambridge
Rayleigh L (1880) On the stability or instability of certain fluid motions. Sci Papers 1:474–484
Rayleigh L (1916) On the dynamics of revolving fluids. Sci Papers 6:447–453
Razdoburdin DN, Zhuravlev VV (2012) Optimal growth of small perturbations in thin gaseous disks. Astron Lett 38:117–127. https://doi.org/10.1134/S1063773712010069
Razdoburdin DN, Zhuravlev VV (2017) Transient growth of perturbations on scales beyond the accretion disc thickness. Mon Not R Astron Soc 467:849–872. https://doi.org/10.1093/mnras/stx050. ArXiv:1701.02535
Rebusco P, Umurhan OM, Klúzniak W, Regev O (2009) Global transient dynamics of three-dimensional hydrodynamical disturbances in a thin viscous accretion disk. Phys Fluids 21:076601. https://doi.org/10.1063/1.3167411
Reddy SC, Henningson DS (1993) Energy growth in viscous channel flows. J Fluid Mech 252:209–238
Reddy SC, Schmid PJ, Henningson DS (1993) Pseudospectra of the Orr–sommerfeld operator. SIAM J Appl Math 53(01):15–47. https://doi.org/10.1137/0153002

Reddy SC, Schmid PJ, Baggett JS, Henningson DS (1998) On stability of streamwise streaks and transition thresholds in plane channel flows. J Fluid Mech 365:269–303
Regev O, Umurhan OM (2008) On the viability of the shearing box approximation for numerical studies of MHD turbulence in accretion disks. Astron Astrophys 481:21–32. https://doi.org/10.1051/0004-6361:20078413. ArXiv:0711.0794
Remillard RA, McClintock JE (2006) X-ray properties of black-hole binaries. Annu Rev Astron Astrophys 44:49–92. https://doi.org/10.1146/annurev.astro.44.051905.092532. ArXiv:astro-ph/0606352
Richard D, Zahn JP (1999) Turbulence in differentially rotating flows. what can be learned from the Couette–Taylor experiment. Astron Astrophys 347:734–738. ArXiv:astro-ph/9903374
Richtmyer RD (1981) Principles of advanced mathematical physics, vol 2. Springer, New York
Salhi A, Pieri AB (2014) Wave-vortex mode coupling in neutrally stable baroclinic flows. Phys Rev E 90(4):043003. https://doi.org/10.1103/PhysRevE.90.043003
Savonije GJ, Heemskerk MHM (1990) Non-axisymmetric unstable modes in a thin differentialy rotating gaseous disk. Astron Astrophys 240(1):191–202
Schartman E, Ji H, Burin MJ (2009) Development of a Couette–Taylor flow device with active minimization of secondary circulation. Rev Sci Instrum 80(2):024501. https://doi.org/10.1063/1.3077942
Schartman E, Ji H, Burin MJ, Goodman J (2012) Stability of Quasi–Keplerian shear flow in a laboratory experiment. Astron Astrophys 543:A94. https://doi.org/10.1051/0004-6361/201016252. ArXiv:1102.3725
Schmid PJ (2007) Nonmodal stability theory. Annu Rev Fluid Mech 39:129–162. https://doi.org/10.1146/annurev.fluid.38.050304.092139
Schmid PJ, Henningson DS (2001) Stability and transition in shear flows. Springer, New York
Schmid PJ, Reddy SC, Henningson DS (1996) Transition thresholds in boundary layer and channel flows. In: Gavrilakis S, Machiels L, Monkewitz PA (eds) Advances in turbulence VI. Kluwer Academic Publishers, Dordrecht, pp 381–384
Sekiya M, Miyama SM (1988) The stability of a differentially rotating cylinder of an incompressible perfect fluid. Mon Not R Astron Soc 234:107–114
Shen Y, Stone JM, Gardiner TA (2006) Three-dimensional compressible hydrodynamic simulations of vortices in disks. Astrophys J 653:513–524. https://doi.org/10.1086/508980
Simon JB, Hawley JF, Beckwith K (2009) Nonlinear stability, hydrodynamical turbulence, and transport in disks. Astrophys J 690:974–997
Squire J, Bhattacharjee A (2014a) Magnetorotational instability: nonmodal growth and the relationship of global modes to the shearing box. Astrophys J 797:15. https://doi.org/10.1088/0004-637X/797/1/67
Squire J, Bhattacharjee A (2014b) Nonmodal growth of the magnetorotational instability. Phys Rev Lett 113(2):025006. https://doi.org/10.1103/PhysRevLett.113.025006. ArXiv:1406.6582
Stepanyants YA, Fabrikant AL (1989) Sov Phys Usp 32:783 (1989)
Stone JM, Hawley JF, Gammie CF, Balbus SA (1996) Three-dimensional magnetohydrodynamical simulations of vertically stratified accretion disks. Astrophys J 463:656. https://doi.org/10.1086/177280
Tassoul JL (1978) Theory of rotating stars. Princeton University Press, Princeton
Taylor GI (1936) Fluid friction between rotating cylinders. i. Torque measurements. R Soc Lond Proc Ser A 157:546–564. https://doi.org/10.1098/rspa.1936.0215
Tevzadze AG, Chagelishvili GD, Zahn JP, Chanishvili RG, Lominadze JG (2003) On hydrodynamic shear turbulence in stratified keplerian disks: transient growth of small-scale 3d vortex mode perturbations. Astron Astrophys 407:779–786. https://doi.org/10.1051/0004-6361:20030867
Tevzadze AG, Chagelishvili GD, Zahn JP (2008) Hydrodynamic stability and mode coupling in keplerian flows: local strato-rotational analysis. Astron Astrophys 478:9–15. https://doi.org/10.1051/0004-6361:20078386

Tevzadze AG, Chagelishvili GD, Bodo G, Rossi P (2010) Linear coupling of modes in two-dimensional radially stratified astrophysical discs. Mon Not R Astron Soc 401:901–912. https://doi.org/10.1111/j.1365-2966.2009.15723.x

Trefethen LN, Trefethen AE, Reddy SC, Driscoll TA (1993) Hydrodynamic stability without eigenvalues. Science 261(5121):578–584. https://doi.org/10.1126/science.261.5121.578

Umurhan OM, Nemirovsky A, Regev O, Shaviv G (2006) Global axisymmetric dynamics of thin viscous accretion disks. Astron Astrophys 446:1–18. https://doi.org/10.1051/0004-6361:20053960

Vilenkin NY, Gorin EA, Kostyuchenko AG, Krasnosel'skii MA, Maslov VP, Mityagin BS, Petunin YI, Rutitskii YB, Sobolev VI, Ya SV, Fadeev LD, Tsitlanadze ES (1972) Functional analysis. Wolters-Noordhoff, Groningen

Volponi F (2010) Linear transport in fully stratified discs. Mon Not R Astron Soc 406:551–667. https://doi.org/10.1111/j.1365-2966.2010.16688.x

Waleffe F (1995) Transition in shear flows. nonlinear normality versus non-normal linearity. Phys Fluids 7:3060–3066

Wendt G (1933) Potentialtheoretische behandlung des wehneltzylinders. Ann Phys 409:445–459. https://doi.org/10.1002/andp.19334090408

Yecko PA (2004) Accretion disk instability revisited. transient dynamics of rotating shear flow. Astron Astrophys 425:385–393. https://doi.org/10.1051/0004-6361:20041273

Youdin AN, Kenyon SJ (2013) From disks to planets. Springer Science, Dordrecht. https://doi.org/10.1007/978-94-007-5606-9_1

Zeldovich YB (1981) On the friction of fluids between rotating cylinders. R Soc Lond Proc Ser A 374:299–312. https://doi.org/10.1098/rspa.1981.0024

Zhuravlev VV, Razdoburdin DN (2014) A study of the transient dynamics of perturbations in keplerian discs using a variational approach. Mon Not R Astron Soc 442:870–890. https://doi.org/10.1093/mnras/stu848

Zhuravlev VV, Shakura NI (2007a) Dynamical instability of laminar axisymmetric flows of ideal incompressible fluid. Astron Lett 33:536

Zhuravlev VV, Shakura NI (2007b) Dynamical instability of laminar axisymmetric flows of ideal fluid with stratification. Astron Lett 33:673. https://doi.org/10.1134/S1063773707110059

Zhuravlev VV, Shakura NI (2009) Temporal behavior of global perturbations in compressible axisymmetric flows with free boundaries. Astron Nachr 330(88):84–91. https://doi.org/10.1002/asna.200811128

Chapter 7
Quasi-Spherical Subsonic Accretion onto Magnetized Neutron Stars

Nikolay Shakura, Konstantin Postnov, Alexandra Kochetkova, and Linnea Hjalmarsdotter

Abstract A theory of quasi-spherical subsonic accretion onto slowly rotating magnetized neutron stars is presented. In this regime, the accreted matter settles with subsonic velocities onto the rotating magnetosphere forming an extended quasi-spherical shell. The accretion rate in the shell is determined by the ability of the plasma to enter the magnetosphere due to the Rayleigh-Taylor instability with account for cooling. This accretion regime may be established for moderate X-ray luminosities, corresponding to accretion rates $\dot{M} < \dot{M}^{\dagger} \simeq 4 \times 10^{16}\,\mathrm{g\,s^{-1}}$. For higher accretion rates a free-fall gap appears, due to strong Compton cooling of the flow above the magnetosphere, and accretion becomes highly non-stationary. Observations of spin-up and spin-down in equilibrium wind-fed X-ray pulsars with known orbital periods (like GX 301-2 and Vela X-1) enable the determination of the basic dimensionless model parameters and estimation of the neutron star magnetic field. In equilibrium pulsars with independently measured magnetic fields, the model enables the stellar wind velocity to be independently estimated. For non-equilibrium pulsars, there exists a maximum spin-down rate of the accreting neutron star. The model can also explain bright flares in Supergiant Fast X-ray Transients if stellar winds of the O-supergiant companions are magnetized.

N. Shakura
Sternberg Astronomical Institute, Lomonosov Moscow State University, Moscow, Russia

Kazan Federal University, Kazan, Russia

K. Postnov (✉)
Sternberg Astronomical Institute, Lomonosov Moscow State University, Moscow, Russia

National Research University Higher School of Economics, Moscow, Russia

A. Kochetkova · L. Hjalmarsdotter
Sternberg Astronomical Institute, Lomonosov Moscow State University, Moscow, Russia
e-mail: astrogirl@telia.com

N. Shakura (ed.), *Accretion Flows in Astrophysics*, Astrophysics and Space Science Library 454, https://doi.org/10.1007/978-3-319-93009-1_7

7.1 Introduction

The X-ray pulsar phenomenon appears during accretion of matter onto rotating strongly magnetized neutron stars (NSs) in binary systems. If the secondary companion of the binary (the optical star) fills its Roche lobe, an accretion disc is formed around the NS. If the secondary companion is an early type massive star, the NS may accrete from its powerful stellar wind. In this case, depending on the stellar wind parameters, either an accretion disc forms around the NS magnetosphere or accretion proceeds quasi-spherically. The strong magnetic field of the NS (of the order of 10^{12}–10^{13} G) alters the accretion flow near the NS magnetosphere that forms at a certain distance from the NS. The plasma flow gets frozen into the magnetic field and is canalized towards the polar cap region onto the NS surface, where hot spots or accretion columns are produced. If the magnetic dipole axis is misaligned with the NS rotation, pulsating X-ray emission may be observed. Most X-ray pulsars exhibit stochastic variations of the NS spin frequency and X-ray flux. Many sources also show long-term variability in the NS spin frequency, when the latter increases (spins-up) or decreases (spins-down) on average, as well as switches between spin-up and spin-down (so-called spin reversals) (see Bildsten et al. (1997) for a detailed review and references).

The most studied case is accretion through a geometrically thin disc onto relativistic compact stars (Shakura and Sunyaev 1973). Here, the spin-up torque acting on the NS can be written as Pringle and Rees (1972) $K_{su} \approx \dot{M}\sqrt{GMR_A}$. The inner radius of the disc around an X-ray pulsar is determined by the Alfvén surface R_A located at distance $R_A \sim \dot{M}^{-2/7}$, therefore $K_{su} \sim \dot{M}^{6/7}$, i.e. for disc accretion, the spin-up torque is almost linearly dependent on the mass accretion rate (X-ray luminosity). The spin-down torque for disc accretion is, in the first approximation, independent of $\dot{M}$: $K_{sd} \sim -\mu^2/R_c^3$, where $R_c = (GM/(\omega^*)^2)^{1/3}$ is the corotation radius, ω^* is the NS rotational frequency and μ is the NS dipole magnetic moment. In fact, the torques in disc accretion are determined by complex disc-magnetospheric interaction (see, for example, Ghosh and Lamb (1979), Lovelace et al. (1995) and the discussion in Kluźniak and Rappaport (2007)) and may thus have a more complicated dependence on the mass accretion rate and other parameters.

Measurements of NS spin-up/spin-down in X-ray pulsars can be used to estimate a very important NS characteristic – its magnetic field. The NS spin period in X-ray pulsars is usually close to the equilibrium value P_{eq}, at which the total torque applied to the NS vanishes, $K = K_{su} + K_{sd} = 0$. Therefore, by assuming that the observed NS spin frequency is $\omega^* = 2\pi/P_{eq}$, the equation for the equilibrium period with known $\dot{M}$ enables the NS magnetic field to be estimated.

In the case of quasi-spherical accretion, which may take place in binary systems in which the optical star does not fill its Roche lobe and no disc is formed, the situation turns out to be more complicated. Clearly, to spin-up or spin down a NS in this regime, the amount and sign of the angular momentum of matter gravitationally captured from the stellar wind are important.To within a factor of the order of one (which can be positive or negative, see, for example, numerical simulations (Fryxell and Taam 1988; Ruffert 1997, 1999)), the torque applied to the NS in this case is proportional to $\dot{M}\omega_B R_B^2$, where $\omega_B = 2\pi/P_B$ is the orbital angular frequency, $R_B = 2GM/(V_w^2 + v_{orb}^2)$ is the Bondi gravitational capture radius, V_w is the stellar wind velocity near the NS, and v_{orb} is its orbital velocity. In reality, the orbital eccentricity in high-mass X-ray binaries (HMXBs) is usually non-zero, and the stellar wind can be variable and inhomogeneous. Therefore, the spin-up torque K_{su} can be a complicated function of time. The spin-down torque in this case is even more uncertain since it is no longer possible to write down a simple relation like $-\mu^2/R_c^3$ (the corotation radius R_c does not have a physical meaning in quasi-spherical accretion; in slowly rotating X-ray pulsars it is much larger than the Alfvén radius at which in fact the angular momentum transfer from the accreting matter to the NS magnetosphere occurs). For example, the use of the braking torque in the form $-\mu^2/R_c^3$ formally leads to very high magnetic fields for long-period X-ray pulsars (of the order of 10^{14} G and even higher). This is apparently a result of an underestimation of the spin-down torque applied to the NS magnetosphere in the quasi-spherical accretion regime.

The matter captured from the stellar wind may accrete onto the NS in different ways. Indeed, if the X-ray flux from the accreting NS is high enough, the stellar wind matter, heated downstream the bow shock, rapidly cools down by the radiation via Compton cooling and falls freely onto the magnetosphere. The velocity of the freely falling matter rapidly exceeds the sound speed, and a shock appears above the magnetosphere. This regime of accretion was studied in Burnard et al. (1983). Depending on the specific angular momentum vector direction (along or opposite to the orbital angular momentum), the NS can spin-up or spin-down. However, if the X-ray flux (more precisely, the energy density of photons) is below a certain value, the plasma heated behind the Bondi radius has no time to cool down, and the fall of matter towards the magnetosphere may proceed subsonically (the settling accretion regime). In the last case, a hot quasi-spherical shell arises around the magnetosphere (Davies and Pringle 1981) (see Fig. 7.1). Due to additional energy release (especially close to the base of the shell), the temperature gradient across the shell becomes superadiabatic, giving rise to large-scale convective motions in the shell. The convection generates turbulence, and thus the motion of a fluid parcel in such a shell is very intricate. If the plasma is able to enter the magnetosphere and then fall onto the NS, the accretion rate in the entire shell will be determined by the magnetosphere. For example, under certain conditions, a shell may be present, but the accretion rate through it may be very small. Thus, in the shell, there may be slow subsonic settling of matter on top of large-scale convective motions. This accretion regime is possible at relatively low X-ray luminosities, $L_x < 4 \times 10^{36}\,\mathrm{erg\,s^{-1}}$ (see below), and is radically different from the numerical calculations of quasi-spherical accretion onto NSs mentioned above. When a quasi-spherical shell is present, its interaction with the rotating NS magnetosphere will spin-up or spin-down the NS depending on the sign of the difference between the angular velocity of the accreting matter and the magnetospheric boundary. Therefore, in the settling accretion regime,

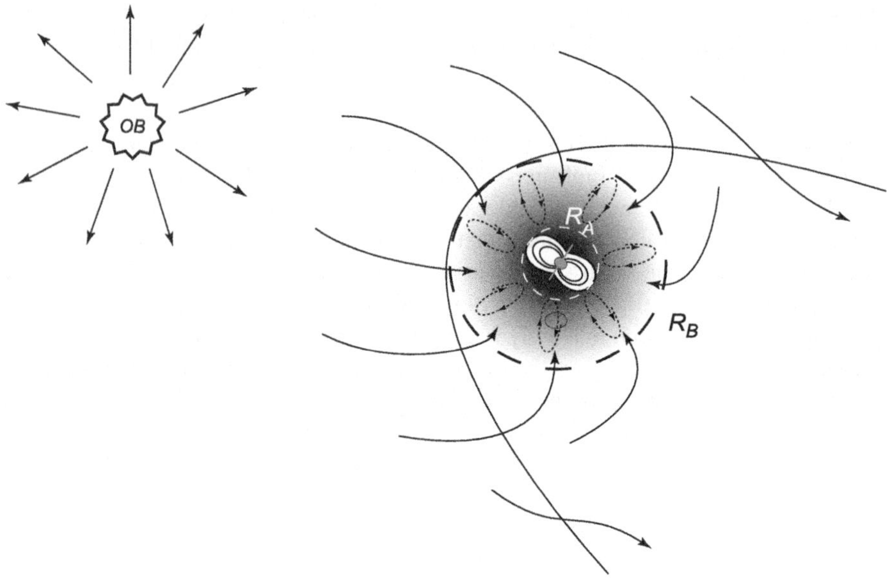

Fig. 7.1 Schematics of quasi-spherical accretion from the stellar wind of the optical component of a binary system (to the left) onto a magnetized NS (to the right). In the subsonic settling regime, a quasi-spherical shell (dark region) is formed between the bow-shock (the parabolic curve) and the NS magnetosphere with radius R_A, in which large-scale convective motions appear that may remove angular momentum from the magnetosphere. The outer radius of the shell is determined by the gravitational capture Bondi radius R_B

both spin-up and spin-down of the NS is possible, even if the specific angular momentum of the captured stellar wind matter is aligned with the orbital angular momentum. Here the angular momentum may flow towards the rotating neutron star or outwards through the shell.

In the literature, there are several models (see especially Illarionov and Kompaneets (1990) and Bisnovatyi-Kogan (1991)), in which the spin-down torque applied to the NS magnetosphere in the case of quasi-spherical accretion is written as $K_{sd} \sim -\dot{M} R_A^2 \omega^*$. With account for the standard definition of the Alfvén radius, $R_A \sim \dot{M}^{-2/7} \mu^{4/7}$, this torque is proportional to $K_{sd} \sim -\mu^{8/7} \dot{M}^{3/7}$. In our model, the matter of the shell settles down with a subsonic velocity as it cools down near the magnetospheric boundary, and the Alfvén radius is determined differently: $R_A \sim \dot{M}^{-2/11} \mu^{6/11}$ (see below).

There can be two different mechanisms of angular momentum removal from the rotating magnetosphere outwards the shell. In the first case (we call it *moderate coupling*), the angular momentum transfer is mediated by convective motions in the shell, and the spin-down torque in the settling accretion regime depends on the accretion rate as $K_{sd} \sim -\dot{M}^{3/11}$. In this case the characteristic velocity of convective motions is subsonic. In addition, a settling regime is possible in which the angular momentum transfer is due to the shear turbulence near the magnetosphere (the case of *weak coupling*). In this case, the characteristic velocity of the shear flow near the magnetosphere is of the order of its linear rotational velocity. Then $K_{sd} \sim \mu^2/R_c^3 \sim \mu^2\omega^{*2}/(GM)$, i.e. in the weak coupling regime the spin-down torque does not depend on the accretion rate at all.

To stress the difference between the two possible subsonic accretion regimes (the moderate and weak coupling), we may rewrite the spin-down torque due to convection (the moderate coupling) using the corotation radius and the Alfvén radius in the form $K_{sd} \sim -\mu^2/\sqrt{R_c^3 R_A^3} \sim -(\mu^2/R_c^3)(R_c/R_A)^{3/2}$ (see below in Sect. 7.11). As the factor $(R_c/R_A)^{3/2} \sim (\omega_K(R_A)/\omega^*)$ in reality may be of the order of a factor 10 or larger, the use of the spin-down torque in the 'traditional' form μ^2/R_c^3 may strongly overestimate the NS magnetic field.

The dependence of the spin-down torque on the accretion rate in the quasi-spherical accretion regime suggests that variations in the mass accretion rate (and hence in the X-ray luminosity) should result in a change from spin-up (at high luminosities) to spin-down (at low luminosities) at some critical value of the mass accretion rate $\dot{M}$ (or R_A), which will be different for different sources. This phenomenon (also known as 'torque reversal') is indeed observed in some X-ray pulsars with quasi-spherical accretion, for example, in Vela X-1, GX 301-2 and GX 1+4, and below we consider these objects in more detail.

7.2 Two Regimes of Quasi-Spherical Wind Accretion

We start with the basic physical picture of quasi-spherical wind accretion onto NSs in binary systems. Quasi-spherical accretion is most likely to occur in wind-fed high-mass X-ray binaries when the optical star of early spectral class (OB) does not fill its Roche lobe, but experiences significant mass loss via stellar wind. We shall discuss the wind accretion regime, in which a bow shock forms in the stellar wind around the compact star. The characteristic distance at which the bow shock forms is approximately equal to the gravitational capture (Bondi) radius

$$R_B = 2GM/(v_w^2 + v_{orb}^2)\,, \tag{7.1}$$

where v_w is the wind velocity (typically 100–1000 km/s), v_{orb} is the orbital velocity of the NS, which is usually much smaller than v_w, so below we will neglect it. The rate of gravitational capture of mass from a wind with density ρ_w near the orbital position of the NS is the Bondi-Hoyle-Littleton mass accretion rate:

$$\dot{M}_B \simeq \rho_w R_B^2 v_w \propto \rho_w v_w^{-3}\,. \tag{7.2}$$

7.2.1 Supersonic (Bondi-Hoyle-Littleton) Accretion

As noted in the Introduction, there can be two different cases of quasi-spherical accretion. The classical Bondi-Hoyle-Littleton accretion takes place when the shocked matter cools down rapidly, and falls freely towards the NS magnetosphere (see Fig. 7.2) by forming a shock at some distance above the magnetosphere. Here the shocked matter cools down (mainly by Compton processes) and enters the magnetosphere via the Rayleigh-Taylor instability (Arons and Lea 1976). The magnetospheric boundary is characterized by the Alfvén radius R_A, which may be calculated from the balance of the ram pressure of the infalling matter and the magnetic field pressure at the magnetospheric boundary: $\rho v_{ff}^2(R_A) = B^2/8\pi$. Making use of the mass continuity equation in the shell, $\dot{M} = 4\pi R^2 \rho(R) v_{ff}(R)$, and assuming a dipole NS magnetic field, the standard result (Davidson and Ostriker 1973) is obtained:

$$R_A = \left(\frac{\mu^2}{\dot{M}\sqrt{2GM}} \right)^{2/7} . \quad (7.3)$$

The captured matter from the wind carries a specific angular momentum $j_w \sim \omega_B R_B^2$ (Illarionov and Sunyaev 1975). Depending on the sign of j_w (prograde or retrograde), the NS can spin-up or spin-down. This regime of quasi-spherical accretion occurs in bright X-ray pulsars with $L_x > 4 \times 10^{36}$ erg s^{-1} (Burnard et al. 1983; Shakura et al. 2012).

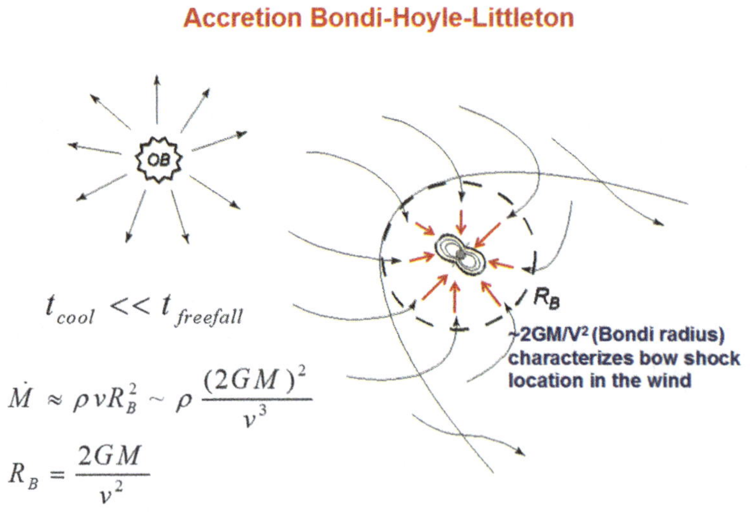

Fig. 7.2 Supersonic (Bondi-Hoyle-Littleton) accretion onto a magnetized NS

7.2.2 Subsonic (Settling) Accretion

If the captured wind matter behind the bow shock at R_B remains hot (when the plasma cooling time is much longer than the free-fall time, $t_{cool} \gg t_{ff}$), a hot quasi-static shell forms around the magnetosphere, and subsonic (settling) accretion sets in (see Fig. 7.3). In this case, both spin-up and spin-down of the NS is possible, even if the sign of j_w is positive (prograde). The shell mediates the angular momentum transfer from the NS magnetosphere via viscous stresses due to convection and turbulence. In this regime, the mean radial velocity of matter in the shell u_r is smaller than the free-fall velocity u_{ff}: $u_r = f(u)u_{ff}$, $f(u) < 1$, and is determined by the plasma cooling rate near the magnetosphere (due to Compton or radiative cooling):

$$f(u) \sim [t_{ff}(R_A)/t_{cool}(R_A)]^{1/3}. \tag{7.4}$$

In the settling accretion regime, the actual mass accretion rate onto the NS may be significantly lower than the Bondi mass accretion rate,

$$\dot{M} = f(u)\dot{M}_B\,. \tag{7.5}$$

Settling accretion occurs for $L_x < 4 \times 10^{36}\,\mathrm{erg\,s^{-1}}$ (Shakura et al. 2012).

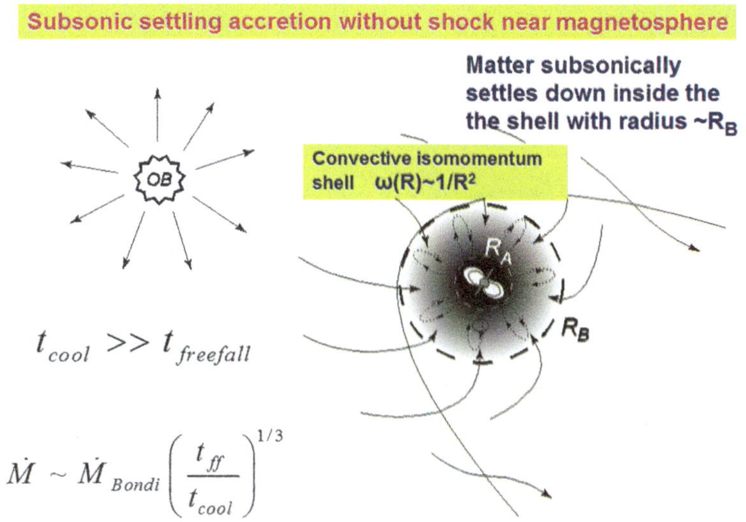

Fig. 7.3 Subsonic settling accretion onto a magnetized NS

7.2.3 The Structure of a Subsonic Shell Around a Neutron Star Magnetosphere

Let us consider the torques acting on a magnetized NS quasi-spherically accreting from the stellar wind of an optical star. The wind matter is gravitationally captured by the moving NS, and a bow-shock is formed near the Bondi radius $R \sim R_B$. Suppose a quasi-spherical shell is formed around the NS magnetosphere as described above. In such a shell, the temperature remains high (of the order of the virial temperature, see Davies and Pringle (1981)), and the key question arises as to whether the hot plasma may enter the magnetosphere. Two-dimensional calculations (Elsner and Lamb 1977) showed that hot monoatomic ideal gas is stable against the Rayleigh-Taylor instability at the magnetospheric boundary, suggesting that plasma cooling is needed for the plasma to enter the magnetosphere. However, a careful consideration of 3D-calculations (Arons and Lea 1976) shows that the hot plasma is marginally stable at the magnetospheric boundary (to within a 5% accuracy of these calculations). Compton cooling and possible dissipative processes (magnetic reconnection, etc.) facilitate the plasma entering the magnetosphere. Below we show that subsonic settling accretion from a hot shell enables NS spin-down.

In the zero approximation, we may neglect both rotation and radial motion of matter in the shell and consider its structure in hydrostatic equilibrium. The velocity of radial motion of matter in the shell u_r is below the sound speed c_s. Under these assumptions, the characteristic heating/cooling time of the plasma should be shorter than the characteristic free-fall time.

In the general case, there is gas pressure and anisotropic turbulent motion in the shell, so Pascal's law is violated. Then the hydrostatic equilibrium equation may be derived from the equations of motion (7.54) with the stress tensor components (7.57)–(7.59) and zero viscosity (see Sect. 7.3 for more detail):

$$-\frac{1}{\rho}\frac{dP_g}{dR} - \frac{1}{\rho R^2}\frac{d(P_{\|}^t R^2)}{dR} + \frac{2P_{\perp}^t}{\rho R} - \frac{GM}{R^2} = 0 \tag{7.6}$$

Here $P_g = \rho c_s^2/\gamma$ is the gas pressure and P^t stands for the contribution due to turbulent motions:

$$P_{\|}^t = \rho < u_{\|}^2 >= \rho m_{\|}^2 c_s^2 = \gamma P_g m_{\|}^2 \tag{7.7}$$

$$P_{\perp}^t = \rho < u_{\perp}^2 >= \rho m_{\perp}^2 c_s^2 = \gamma P_g m_{\perp}^2 \tag{7.8}$$

($< u_t^2 >=< u_{\|}^2 > +2 < u_{\perp}^2 >$ is the turbulent velocity dispersion while $m_{\|}^2$ and $m_{\perp}^2$ are the radial and tangential turbulent Mach numbers squared. For example, for isotropic turbulence we have $m_{\|}^2 = m_{\perp}^2 = (1/3)m_t^2$, where m_t is the turbulent Mach number). The total pressure is given by the sum of the gas and turbulent terms: $P_g + P_t = P_g(1 + \gamma m_t^2)$. Generally, the turbulent Mach numbers in the shell may depend on radius, however in our model we will treat them as constants. Moreover, the turbulent heating (important from the dynamical point of view, see Sect. 7.7) changes the physical parameters in real X-ray pulsars by less than a factor two.

In the first approximation, we assume that the entropy S in the shell constant. For an ideal gas with the adiabatic exponent γ and the equation of state $P = Ke^{S/cv}\rho^\gamma$, the density may be expressed as a function of temperature $\rho \sim T^{1/(\gamma-1)}$. Integrating these hydrostatic equilibrium equation (7.6) yields:

$$\frac{\mathcal{R}T}{\mu_m} = \left(\frac{\gamma-1}{\gamma}\right)\frac{GM}{R}\left(\frac{1}{1+\gamma m_\parallel^2 - 2(\gamma-1)(m_\parallel^2 - m_\perp^2)}\right) = \frac{\gamma-1}{\gamma}\frac{GM}{R}\psi(\gamma, m_t)\,. \tag{7.9}$$

(In this solution we have neglected the integration constant which is not important deep inside the shell. It could be important in the outer parts of the shell, but as these parts are located close to the shock and near $\sim R_B$ and are not spherically symmetric, their structure should be calculated numerically.)

Note that including turbulence somewhat decreases the temperature in the shell. However, the most essential here is that the turbulent anisotropy, due to convection in the stationary case changes the radial distribution of the angular velocity. Below we will show that in the case of isotropic turbulence, the rotation of the shell can be close to quasi-Keplerian, $\omega(R) \sim R^{-3/2}$. In the case of strongly anisotropic turbulence with $m_\parallel^2 \gg m_\perp^2$, a distribution of the angular velocity with constant specific angular momentum (isomomentum law), $\omega(R) \sim R^{-2}$, may be established. Below we will see that shells around real X-ray pulsars most likely have an isomomentum angular velocity distribution.

Let us now determine how the density changes in a quasi-static shell in the inner layers with $R \ll R_B$. For a fully ionized gas with $\gamma = 5/3$, we find the density changes as:

$$\rho(R) = \rho(R_A)\left(\frac{R_A}{R}\right)^{3/2} \tag{7.10}$$

and the gas pressure changes as:

$$P(R) = P(R_A)\left(\frac{R_A}{R}\right)^{5/2}\,. \tag{7.11}$$

These equations describe the structure of an ideal static adiabatic shell above the magnetosphere. Of course, for $R \sim R_B$ the problem becomes significantly non-spherically symmetric, and to calculate the structure of the outer parts of the shell numerical methods are required.

Corrections to the adiabatic temperature gradient due to convective transport of energy in the shell are calculated in Sect. 7.6.

7.2.4 The Alfvén Surface

By definition, at the magnetospheric boundary (the Alfvén surface), the total pressure (including isotropic gas pressure and the possibly anisotropic turbulent pressure) is balanced by the magnetic field pressure $B^2/(8\pi)$:

$$P_g + P_t = P_g(R_A)(1+\gamma m_t^2) = \frac{B^2(R_A)}{8\pi}. \tag{7.12}$$

The magnetic field at the Alfvén radius is determined by the dipole magnetic field of the neutron star and the field produced by currents flowing over the magnetosphere (in the magnetopause):

$$P_g(R_A) = \frac{K_2}{(1+\gamma m_t^2)} \frac{B_0^2}{8\pi} \left(\frac{R_0}{R_A}\right)^6 = \frac{\rho \mathcal{R} T}{\mu_m} \tag{7.13}$$

where the dimensionless coefficient K_2 takes into account the contribution from the magnetospheric currents, and the factor $1/(1+\gamma m_t^2)$ appears due to turbulent pressure. For example, in the model by Arons and Lea (1976) (see Eq. (31) in that paper), $K_2 = (2.75)^2 \approx 7.56$. Near the magnetospheric cusp (where the curvature of the magnetic field lines is maximal), the size of the Alfvén surface is about 0.51 times the equatorial size (Arons and Lea 1976). Everywhere below we will assume that R_A is the equatorial Alfvén surface unless stated otherwise.

The plasma enters the magnetosphere mainly via the Rayleigh-Taylor instability. In the stationary case, we may introduce a constant accretion rate $\dot{M}$ onto the neutron star. From the continuity equation in the shell we find

$$\rho(R_A) = \frac{\dot{M}}{4\pi u_r(R_A) R_A^2}. \tag{7.14}$$

Clearly, the radial velocity of the matter entering the magnetosphere is below the free-fall velocity, therefore we introduce the dimensionless coefficient $f(u) = u_r/\sqrt{2GM/R} < 1$. Then the density at the magnetospheric boundary reads

$$\rho(R_A) = \frac{\dot{M}}{4\pi f(u)\sqrt{2GM/R_A} R_A^2}. \tag{7.15}$$

For example, in the model Arons and Lea (1976) $f(u) \approx 0.1$. In our case, at high X-ray luminosities the dimensionless factor $f(u)$ may be as high as $\approx$0.5 (however, still at luminosities below the critical value $\sim 4\times10^{36}$ erg s^{-1}, for which settling accretion may occur).

If we imagine that the magnetosphere is fully impenetrable and the accretion rate in the shell $\dot{M} \to 0$, then $u_r \to 0$ and $f(u) \to 0$. However, the density near the magnetospheric boundary remains finite. In some sense, the matter leaks through the magnetosphere onto the neutron star, and the leakage rate can be either very low ($\dot{M} \to 0$) or have a finite non-zero value ($\dot{M} \neq 0$).

Eliminating density from Eq. (7.13) with the help of the continuity equation and making use of (7.9) together with the definition of the dipole magnetic moment,

$$\mu = \frac{1}{2} B_0 R_0^3$$

(where R_0 is the neutron star radius and B_0 is the magnetic field at the NS pole, which in the 3D-case is two times as high as the equatorial field), we find the expression for the Alfvén radius at the quasi-spherical accretion stage:

$$R_A = \left[\frac{4\gamma}{(\gamma - 1)} \frac{f(u)K_2}{\psi(\gamma, m_t)(1 + \gamma m_t^2)} \frac{\mu^2}{\dot{M}\sqrt{2GM}} \right]^{2/7} . \tag{7.16}$$

It should be emphasized that when a hot shell is present, the Alfvén radius is determined by the static gas pressure (and a possible contribution from turbulent motions) at the magnetospheric boundary and has a certain value even for a zero mass accretion rate through the shell. The dependence of the factor $f(u)$ on $\dot{M}$ in the settling accretion regime with account for cooling will be obtained below (see Eq. (7.29)). In the supersonic accretion (Bondi) regime, evidently, $f(u) = 1$. Note that in the Bondi regime (Bondi 1952) a subsonic flow can be formed, but with a smaller (compared to the maximum possible) accretion rate $\dot{M}$. In the Bondi regime (i.e. in the adiabatic regime without heating/cooling of the gas), the choice of solution is determined by the boundary conditions.

7.2.5 The Mean Rate of the Flow of Matter Through the Magnetosphere

In this Section, we will consider the case of isotropic turbulence, i.e., we will set the factor $K_t = \psi(\gamma, m_t)(1 + \gamma m_t^2)\psi(\gamma, m_t)(1 + \gamma m_t^2)$ that enters the Alfvén radius definition (7.16) equal to one.

As noted above, the plasma enters the magnetosphere of the slowly rotating neutron star due to the Rayleigh-Taylor instability. The boundary between the plasma and the magnetosphere is stable at high temperatures $T > T_{cr}$, but becomes unstable at $T < T_{cr}$, and remains in a neutral equilibrium at $T = T_{cr}$ (Elsner and Lamb 1977). The critical temperature is:

$$\mathcal{R} T_{cr} = \frac{1}{2} \frac{\cos\chi}{\kappa R_A} \frac{\mu_m GM}{R_A} . \tag{7.17}$$

Here κ is the local curvature of the magnetosphere, χ is the angle that the outer normal to the magnetospheric surface makes with the radius-vector at a given point. The effective gravity acceleration can be written as

$$g_{eff} = \frac{GM}{R_A^2} \cos\chi \left(1 - \frac{T}{T_{cr}}\right) . \tag{7.18}$$

The temperature in the quasi-static shell is given by (7.9), and the condition for magnetospheric instability can thus be rewritten as:

$$\frac{T}{T_{cr}} = \frac{2(\gamma - 1)}{\gamma} \frac{\kappa R_A}{\cos\chi} < 1 . \tag{7.19}$$

Consider, for example, the development of the interchange instability when cooling (predominantly Compton cooling) is present. The temperature changes as (Kompaneets 1957; Weymann 1965):

$$\frac{dT}{dt} = -\frac{T - T_x}{t_C} , \tag{7.20}$$

where the Compton cooling time is

$$t_C = \frac{3}{2\mu_m} \frac{\pi R_A^2 m_e c^2}{\sigma_T L_x} \approx 10.6[\text{s}] R_9^2 \dot{M}_{16}^{-1} . \tag{7.21}$$

Here m_e is the electron mass, σ_T is the Thomson cross section, $L_x = 0.1\dot{M}c^2$ is the X-ray luminosity, T is the electron temperature (which is equal to the ion temperature since the timescale of electron-ion energy exchange here is the shortest possible), T_x is the X-ray temperature and $\mu_m = 0.6$ is the molecular weight. The photon temperature is $T_x = (1/4)T_{cut}$ for a bremsstrahlung spectrum with an exponential cut-off at T_{cut}, typically $T_x = 3$–5 keV. The solution to Eq. (7.20) reads:

$$T = T_x + (T_{cr} - T_x)e^{-t/t_C} . \tag{7.22}$$

We note that $T_{cr} \sim 30$ keV $\gg T_x \sim 3$ keV, and see that for $t \approx 2t_C$ the temperature decreases down to T_x. In the linear approximation the temperature changes as:

$$T \approx T_{cr}(1 - t/t_C) . \tag{7.23}$$

Plugging this expression into (7.18), we find that the effective gravity acceleration increases linearly with time as:

$$g_{eff} \approx \frac{GM}{R_A^2} \frac{t}{t_C} \cos\chi . \tag{7.24}$$

Correspondingly, the velocity of matter due to the instability growth increases with time as:

$$u_i = \int_0^t g_{eff} dt = \frac{GM}{R_A^2}\frac{t^2}{2t_C}\cos\chi\,. \tag{7.25}$$

Let us introduce the mean rate of the instability growth

$$< u_i >= \frac{\int u dt}{t} = \frac{1}{6}\frac{GM}{R_A^2}\frac{t^2}{t_C} = \frac{1}{6}\frac{GM}{R_A^2 t_C}\left(\frac{\zeta R_A}{< u_i >}\right)^2 \cos\chi\,. \tag{7.26}$$

Here $\zeta \lesssim 1$ and ζR_A is the characteristic scale of the instability that grows at the rate $< u_i >$. Therefore, for the mean instability growth rate at the linear stage, we find

$$< u_i >= \left(\frac{\zeta^2 GM}{6t_C}\right)^{1/3} = \frac{\zeta^{2/3}}{12^{1/3}}\sqrt{\frac{2GM}{R_A}}\left(\frac{t_{ff}}{t_C}\right)^{1/3}\cos\chi\,. \tag{7.27}$$

As the factor $\cos\chi \simeq 1$, we will omit it below. Here we have introduced the characteristic time as

$$t_{ff} = \frac{R_A^{3/2}}{\sqrt{2GM}}, \tag{7.28}$$

which is close to the free-fall time at a given radius. Therefore, the factor $f(u)$ becomes:

$$f(u) = \frac{< u_i >}{u_{ff}(R_A)}\,. \tag{7.29}$$

Substituting (7.27) and (7.29) into (7.16), we find for the Alfvén radius in this regime:

$$R_A^{(C)} \approx 1.37 \times 10^9[\mathrm{cm}]\left(\zeta\frac{\mu_{30}^3}{\dot{M}_{16}}\right)^{2/11}\,. \tag{7.30}$$

Plugging (7.30) into (7.29), we obtain the explicit expression for $f(u)$ in the Compton coolingregime:

$$f(u)_C \approx 0.22\zeta^{7/11}\dot{M}_{16}^{4/11}\mu_{30}^{-1/11}\,. \tag{7.31}$$

In the radiation cooling regime, the cooling time is

$$t_{cool}^{(rad)} = \frac{3kT}{2\mu_m n_e \Lambda(T)} = \sqrt{T}/K_{rad}, \tag{7.32}$$

where $\Lambda(T) \approx 2.5 \times 10^{-27}\sqrt{T}$ (in CGS units) is the radiation cooling factor (here we take into account the Gaunt-factor and that the real cooling function at high temperatures goes slightly higher than for a pure free-free emission). With this cooling time, the temperature decreases as

$$\frac{dT}{dt} = -K_{rad}\sqrt{T}\,, \tag{7.33}$$

yielding a non-exponential temperature decay with time

$$\frac{T}{T_0} = \left(1 - \frac{1}{2}\frac{K_{rad}t}{\sqrt{T_0}}\right)^2 . \tag{7.34}$$

In the linear approximation, when $t \ll t_{cool}^{(rad)}$, we get for the radiation cooling law

$$\frac{T}{T_{cr}} = 1 - \frac{t}{t_{cool}^{(rad)}}\,, \tag{7.35}$$

similarly to (7.23) for Compton cooling, and find that

$$R_A^{(rad)} \approx 1.05 \times 10^9[\mathrm{cm}]\zeta^{4/81}\mu_{30}^{16/27}\dot{M}_{16}^{-6/27}\,, \tag{7.36}$$

$$f(u)_{rad} \approx 0.1\zeta^{14/81}\mu_{30}^{2/27}\dot{M}_{16}^{6/27}\,. \tag{7.37}$$

A necessary condition for angular momentum removal from the magnetosphere by convection in the shell is the subsonic settling of matter (the Mach number for the settling velocity $\mathcal{M} \equiv u_r/u_s < 1$), a condition which for $\gamma = 5/3$ reduces to the inequality $f(u) < 1/\sqrt{3}$. Clearly, for accretion rates about 10^{16} g s^{-1} and below this condition is satisfied. It is also important to note that convection and removal of angular momentum in the shell almost cease when the mean radial settling velocity of matter u_r becomes higher than the convective velocity u_c, i.e., when the convective Mach number $m_c = u_c/c_s \sim m_t$ is smaller than the usual Mach number $\mathcal{M} = u_r/c_s$. Oppositely, when the Mach number of the radial flow is smaller than the turbulent Mach number $\mathcal{M} < m_t \sim m_c$, angular momentum can be removed from the rotating magnetosphere through the settling shell.

When the accretion rate in the shell exceeds some critical value, $\dot{M} > \dot{M}^{\dagger}$, the velocity of the accreting matter near the Alfvén surface may exceed the speed of sound, and a supersonic flow with free-fall velocity of the matter appears in some layer above the magnetosphere. This prevents the angular momentum transfer outwards from the rotating magnetosphere. In this case, the settling accretion regime cannot be realized: a shock emerges above the magnetosphere, and interaction of the plasma with the magnetosphere must be treated following the scheme considered, for example, in Burnard et al. (1983). Depending on the character of the inhomogeneities in the captured stellar wind, the specific angular momentum of the matter may be both positive and negative. Therefore, in the supersonic regime, intermittent episodes of both NS spin-up and spin-down are possible.

By assuming a limiting value of the dimensionless settling velocity $f(u) = 0.5$ (below which it is still possible to remove angular momentum from the rotating NS magnetosphere, see Sect. 7.7 for more detail), from Eq. (7.29) we obtain a maximum possible mass accretion rate in the settling regime with angular momentum removal:

$$\dot{M}_{16}^{\dagger} \approx 3.7\zeta^{-7/4}\mu_{30}^{1/4} . \tag{7.38}$$

Note that a similar value for the critical accretion rate in the settling regime will be obtained from a comparison of the characteristic Compton cooling time with the time of convective motions near the Alfvén radius.

To conclude this section, we note that low-luminosity X-ray pulsars may enter the regime of plasma entry into the magnetosphere, due to radiative cooling, via a change in the X-ray radiation beam structure when most of the X-ray photons form a pencil-beam emission diagram illuminating the magnetospheric cusp. This makes it possible to explain the temporal appearance of 'switched-off' states (with low X-ray luminosity) in Vela X-1 and other X-ray pulsars, which display a phase jump in their X-ray pulse profiles (Doroshenko et al. 2011).

7.3 The Structure of a Quasi-Spherical Rotating Shell with Subsonic Accretion

In this section, we will give a detailed derivation of the equations describing the structure of a quasi-spherical shell in the settling accretion regime. We will use the tensor components as measured by a physical observer (the velocity u_a and the viscous stress tensor components W_{ab}).[1]

7.3.1 Basic Equations

We start with writing down the Navier-Stokes equation in spherical coordinates R, θ, ϕ.

Due to the huge values of the Reynolds number in the shell ($\sim 10^{15}$–10^{16} for typical accretion rates $10^{17}\,\mathrm{g\,s^{-1}}$ and magnetospheric radii $\sim 10^{8}\,\mathrm{cm}$), strong turbulence develops in the shell. In this case, the Navier-Stokes equations are referred to as Reynolds equations. In the general case, the turbulent viscosity

[1] Not to be confused with the covariant components in curvilinear coordinates.

depends on the coordinates. Therefore the hydrodynamical equations take the form:

1. The continuity equation:

$$\frac{\partial \rho}{\partial t}+\frac{1}{R^2}\frac{\partial}{\partial R}\left(R^2 \rho u_r\right)+\frac{1}{R\sin\theta}\frac{\partial}{\partial\theta}(\sin\theta\,\rho u_\theta)+\frac{1}{R\sin\theta}\frac{\partial\rho u_\phi}{\partial\phi}=0. \qquad (7.39)$$

2. The R-component of the equation of motion:

$$\frac{\partial u_r}{\partial t}+u_r\frac{\partial u_r}{\partial R}+\frac{u_\theta}{R}\frac{\partial u_r}{\partial\theta}+\frac{u_\phi}{R\sin\theta}\frac{\partial u_r}{\partial\phi}-\frac{u_\phi^2+u_\theta^2}{R}=-\frac{GM}{R^2}+N_R \qquad (7.40)$$

3. The θ-component of the equation of motion:

$$\frac{\partial u_\theta}{\partial t}+u_r\frac{\partial u_\theta}{\partial R}+\frac{u_\theta}{R}\frac{\partial u_\theta}{\partial\theta}+\frac{u_\phi}{R\sin\theta}\frac{\partial u_\theta}{\partial\phi}+\frac{u_r u_\theta-u_\phi^2\cot\theta}{R}=N_\theta \qquad (7.41)$$

4. The ϕ-component of the equation of motion:

$$\frac{\partial u_\phi}{\partial t}+u_r\frac{\partial u_\phi}{\partial R}+\frac{u_\theta}{R}\frac{\partial u_\phi}{\partial\theta}+\frac{u_\phi}{R\sin\theta}\frac{\partial u_\phi}{\partial\phi}+\frac{u_r u_\phi+u_\phi u_\theta\cot\theta}{R}=N_\phi \qquad (7.42)$$

Here, the physically measurable components of forces (including the viscous force and pressure gradient) are written in the form:

$$\rho N_R=\frac{1}{R^2}\frac{\partial}{\partial R}\left(R^2 W_{RR}\right)+\frac{1}{\sin\theta\, R}\frac{\partial}{\partial\theta}(W_{R\theta}\sin\theta)+\frac{1}{\sin\theta\, R}\frac{\partial}{\partial\phi}W_{R\phi}-\frac{W_{\theta\theta}}{R}-\frac{W_{\phi\phi}}{R} \qquad (7.43)$$

$$\rho N_\theta=\frac{1}{R^2}\frac{\partial}{\partial R}\left(R^2 W_{\theta R}\right)+\frac{1}{\sin\theta\, R}\frac{\partial}{\partial\theta}(W_{\theta\theta}\sin\theta)+\frac{1}{\sin\theta\, R}\frac{\partial}{\partial\phi}W_{\theta\phi}$$
$$-\cot\theta\frac{W_{\theta\theta}}{R}+\cot\theta\frac{W_{\phi\phi}}{R} \qquad (7.44)$$

$$\rho N_\phi=\frac{1}{R^3}\frac{\partial}{\partial R}\left(R^3 W_{\phi R}\right)+\frac{1}{\sin\theta\, R}\frac{\partial}{\partial\theta}\left(W_{\phi\theta}\sin\theta\right)+\frac{1}{\sin\theta\, R}\frac{\partial}{\partial\phi}W_{\phi\phi} \qquad (7.45)$$

The physically measurable components of the stress tensors include both the gas pressure P_g (we assume it to be isotropic) and the pressure due to turbulent degrees of freedom P^t (generally, anisotropic). These components are determined following the classical Landau and Lifshitz (1987) treatment, however, with account for the anisotropic turbulent pressure:

$$W_{RR}=-P_g-P^t_{RR}+2\rho\nu_t\frac{\partial u_r}{\partial R}-\frac{2}{3}\rho\nu_t\mathrm{div}\mathbf{u} \qquad (7.46)$$

$$W_{\theta\theta}=-P_g-P^t_{\theta\theta}+2\rho\nu_t\left(\frac{1}{R}\frac{\partial u_\theta}{\partial\theta}+\frac{u_r}{R}\right)-\frac{2}{3}\rho\nu_t\mathrm{div}\mathbf{u} \qquad (7.47)$$

$$W_{\phi\phi} = -P_g - P^t_{\phi\phi} + 2\rho\nu_t \left(\frac{1}{R\sin\theta}\frac{\partial u_\phi}{\partial\phi} + \frac{u_r}{R} + \frac{u_\theta\cot\theta}{R} \right) - \frac{2}{3}\rho\nu_t \mathrm{div}\mathbf{u} \tag{7.48}$$

$$W_{R\theta} = \rho\nu_t \left(\frac{1}{R}\frac{\partial u_r}{\partial\theta} + \frac{\partial u_\theta}{\partial R} - \frac{u_\theta}{R} \right) \tag{7.49}$$

$$W_{\theta\phi} = \rho\nu_t \left(\frac{1}{R\sin\theta}\frac{\partial u_\theta}{\partial\phi} + \frac{1}{R}\frac{\partial u_\phi}{\partial\theta} - \frac{u_\phi\cot\theta}{R} \right) \tag{7.50}$$

$$W_{R\phi} = \rho\nu_t \left(\frac{1}{R\sin\theta}\frac{\partial u_r}{\partial\phi} + \frac{\partial u_\phi}{\partial R} - \frac{u_\phi}{R} \right) . \tag{7.51}$$

In the problem considered here, the turbulence is such that $P^t_{RR} = P^t_{\parallel}$, $P^t_{\theta\theta} = P^t_{\phi\phi} = P^t_{\perp}$. The turbulent pressure components can be expressed via turbulent Mach numbers and will be presented in Sect. 7.6.

In spherical coordinates, the velocity divergence div**u** reads:

$$\mathrm{div}\mathbf{u} = \frac{1}{R^2}\frac{\partial}{\partial R}\left(R^2 u_r\right) + \frac{1}{R\sin\theta}\frac{\partial}{\partial\theta}\left(\sin\theta\, u_\theta\right) + \frac{1}{R\sin\theta}\frac{\partial u_\phi}{\partial\phi}. \tag{7.52}$$

7.3.2 Symmetries of the Problem

Let us consider a spherically symmetric ($\frac{\partial}{\partial\phi} = 0$), stationary($\frac{\partial}{\partial t} = 0$) and purely radial gas accretion ($u_\theta = 0$). For such a problem setup, the continuity equation (7.39) is:

$$\dot{M} = 4\pi R^2 \rho u_r = const\, . \tag{7.53}$$

The constant in this expression is determined by the conditions of plasma entering the magnetosphere.

The Reynolds equations under these assumptions read as follows. The R-component of the equation of motion (7.40) is:

$$\rho\left(u_r\frac{\partial u_r}{\partial R} - \frac{u_\phi^2}{R}\right) = -\rho\frac{GM}{R^2} + \frac{1}{R^2}\frac{\partial}{\partial R}\left(R^2 W_{RR}\right)$$
$$+ \frac{1}{\sin\theta\, R}\frac{\partial}{\partial\theta}\left(W_{R\theta}\sin\theta\right) - \frac{W_{\theta\theta}}{R} - \frac{W_{\phi\phi}}{R} . \tag{7.54}$$

The θ-component of the equation of motion (7.40) is:

$$-\rho\frac{u_\phi^2\cot\theta}{R}=\frac{1}{R^2}\frac{\partial}{\partial R}\left(R^2W_{\theta R}\right)+\frac{1}{\sin\theta\, R}\frac{\partial}{\partial\theta}\left(W_{\theta\theta}\sin\theta\right)-\cot\theta\frac{W_{\theta\theta}}{R}\,. \tag{7.55}$$

The ϕ-component of the equation of motion (7.40) is:

$$\rho\left(u_r\frac{\partial u_\phi}{\partial R}+\frac{u_r u_\phi}{R}\right)=\frac{1}{R^3}\frac{\partial}{\partial R}\left(R^3W_{\phi R}\right)+\frac{1}{\sin\theta\, R}\frac{\partial}{\partial\theta}\left(W_{\phi\theta}\sin\theta\right) \tag{7.56}$$

The components of the viscous stress tensor take the form:

$$W_{RR}=-P_g-P_\parallel^t-\frac{4}{3}\rho\nu_t\left(\frac{u_r}{R}-\frac{\partial u_r}{\partial R}\right) \tag{7.57}$$

$$W_{\theta\theta}=-P_g-P_\perp^t+\frac{2}{3}\rho\nu_t\left(\frac{u_r}{R}-\frac{\partial u_r}{\partial R}\right) \tag{7.58}$$

$$W_{\phi\phi}=-P_g-P_\perp^t+\frac{2}{3}\rho\nu_t\left(\frac{u_r}{R}-\frac{\partial u_r}{\partial R}\right) \tag{7.59}$$

$$W_{R\theta}=\rho\nu_t\frac{1}{R}\frac{\partial u_r}{\partial\theta} \tag{7.60}$$

$$W_{\theta\phi}=\rho\nu_t\left(\frac{1}{R}\frac{\partial u_\phi}{\partial\theta}-\frac{u_\phi\cot\theta}{R}\right) \tag{7.61}$$

$$W_{R\phi}=\rho\nu_t\left(\frac{\partial u_\phi}{\partial R}-\frac{u_\phi}{R}\right) \tag{7.62}$$

Equations (7.57)–(7.62) relate the components of the stress tensor W_{ik} to the strain tensor (the rate of the shear tensor) $S_{ik}=(1/2)(\partial u_i/\partial x_k+\partial u_k/\partial x_i)$:

$$W_{ik}=-2\rho\nu_t S_{ik} \tag{7.63}$$

Here

$$\nu_t\simeq\frac{1}{3}u_t l_t \tag{7.64}$$

is the turbulent viscosity coefficient, l_t is the size of the largest eddies, and u_t is the magnitude of velocity pulsations on the scale of these eddies (see, for example, Thorne and Blandford 2017, chapter 15). The coefficient ν_t is determined by the properties of the turbulent flow itself, which is different from the kinematic viscosity coefficient, which is determined by the intrinsic property of the fluid. To describe the

Reynold stresses for an anisotropic eddy viscosity, instead of one turbulent viscosity coefficient, 81 independent parameters (components of a 4th-rank tensor) appear (see, for example, Monin and I'Aglom 1971). Unfortunately, there is no theory of turbulence. To describe turbulent stresses, empirical formulas that can be tested experimentally are employed.

L. Prandtl in his studies of plane-parallel flows (along the x-axis, for example) introduced relations between the turbulent mixing length l_t, the velocity of turbulent pulsations u_t and the characteristic velocity shear in the direction perpendicular to the mean flow (z):

$$u_t = l_t \left| \frac{du}{dz} \right| . \tag{7.65}$$

Then the turbulent viscosity coefficient reads:

$$\nu_t = C_0 l_t^2 \left| \frac{du}{dz} \right| \tag{7.66}$$

where $C_0 \sim 1$ is a universal dimensionless constant, the precise value of which should be determined from the presently non-existent theory of turbulence. Thus, the turbulent stresses depend quadratically on the shear amplitude,

$$W_{zx} = \rho C_0 l_t^2 \left(\frac{du}{dz} \right)^2 , \tag{7.67}$$

and a non-linearity appears that in the general case strongly complicates the problem. Here we should stress that in fact expression (7.67) for W_{zx} is not a component of some tensor any more, and can be applicable only in a particular coordinate system.

Consider a possible generalization of Prandtl's law for turbulent viscosity in the case of an axially symmetric flow. If turbulence is strongly anisotropic, there exists one more empirical law for turbulent viscosity, Wasiutynski's law (see below), which does not reduce to Prandtl's law for isotropic turbulence. This more general case for anisotropic turbulence will be discussed separately in Sect. 7.5.

7.4 Structure of the Shell for Prandtl's Turbulent Viscosity

7.4.1 The Empirical Prandtl Law for Turbulent Viscosity in Axially Symmetric Flows

Consider an axially symmetric flow with a very high Reynolds number. By generalizing Prandtl's law for turbulent velocity derived for plane-parallel flows, we can write the turbulent velocity scaling in the form: $u_t \sim l_t R(\partial\omega/\partial R)$. Using gas-dynamical similarity laws, we assume $l_t \sim R$, so that

$$u_t = C_1 R^2 \left|\frac{\partial\omega}{\partial R}\right| . \tag{7.68}$$

Note that in our case the turbulent velocity is determined by convection, and $u_t \lesssim 0.5u_{ff}$ (see Sect. 7.7). This means that the constant C_1

$$C_1 \sim u_t/\langle u_\phi \rangle, \tag{7.69}$$

can be very large since $\langle u_\phi \rangle \ll u_t$. Therefore, the turbulent viscosity coefficient is equal to

$$\nu_t = \langle u_t l_t \rangle = C_2 C_1 R^3 \left|\frac{\partial\omega}{\partial R}\right| \tag{7.70}$$

Here $C_2 \approx 1/3$ is a numerical factor arising from statistical averaging. Below we will use the new coefficient $C = C_1 C_2$ which can be much larger than one.

For this viscosity law the turbulent stresses $W_{R\phi}$ are:

$$W_{R\phi} = \rho\nu_t R\frac{\partial\omega}{\partial R} = \rho C R^4 \left(\frac{\partial\omega}{\partial R}\right)^2 . \tag{7.71}$$

Note that in axially symmetric flows with angular momentum increasing outwards (in particular, in Keplerian flows), a stabilization of the flow appears (Zeldovich 1981) which is absent in plane-parallel flows. The stabilization criterion can be characterized by the dimensionless ratio of two energies, the 'Taylor number', according to Zeldovich, Ty $= E_s/E_t$. Here E_s is a measure of the flux stabilization, the kinetic energy needed to interchange two adjacent parcels of gas of mass m, a and b, by a distance $\Delta r \ll r$, while keeping the specific angular momentum $j = \omega r^2$ constant,

$$E_s = \frac{1}{2}m(u_a'^2 - u_a^2) + \frac{1}{2}m(u_b'^2 - u_b^2) \simeq 2m\frac{\omega}{r}\frac{d(\omega r^2}{dr}(\Delta r)^2 \tag{7.72}$$

(here $u_{a,b} = j_{a,b}/r_{a,b}$ and $u'_{a,b} = j_{a,b}/r_{b,a}$ are the velocities before and after the change in positions). The energy E_t is the energy dissipated in, for example, chaotic turbulent motions (see Zeldovich (1981) for more detail):

$$E_t = \frac{1}{2}m(u_a^2 - (u_a{\prime\prime})^2 + \frac{1}{2}m(u_b^2 - (u_b{\prime\prime})^2 \tag{7.73}$$

where u and $u{\prime\prime}$ are velocities after ω smoothing, which are defined by

$$j = m(r_a^2 + r_b^2)\omega = mr_a^2\omega_a + mr_b^2\omega_b . \tag{7.74}$$

Finally, we obtain[2]

$$\mathrm{Ty} = \frac{E_s}{E_t} = 4\frac{d(\omega r^2)^2/dr}{r^5(d\omega/dr)^2} . \tag{7.75}$$

The Taylor number is analogous to the gradient Richardson number, which is

$$\mathrm{Ri} = \frac{N^2}{(du/dr)^2} , \tag{7.76}$$

where N is the Brunt-Väisälä frequency,

$$N^2 = \frac{g}{T}\left(\frac{\partial T}{\partial r} + \frac{g}{c_p}\right) , \tag{7.77}$$

and du/dr is the mean vorticity. Following Bradshaw (1969), it is possible to define another analogue to the Richardson number as

$$\text{`Ri'} = \frac{\varkappa^2}{(du/dr)^2} , \tag{7.78}$$

where

$$\varkappa^2 = \frac{1}{r^3}\frac{d(\omega r^2)^2}{dr} \tag{7.79}$$

is the epicyclic frequency. Clearly, 'Ri' $= 1/4\mathrm{Ty}$.

7.4.2 The Angular Momentum Transfer Equation

A similar problem (rotation of a sphere in a viscous fluid) is solved in Landau and Lifshitz (1987). It is shown there that in this problem the variables may be separated, and one can write $u_\phi(R,\theta) = u_\phi(R)\sin\theta$. Note that the angular velocity $\omega(R) = u_\phi(R)/R$ is independent of the polar angle θ. Our problem setup is different from the rotation of a sphere in a viscous fluid in several aspects: (1) there is a gravity force, (2) the turbulent viscosity changes with the distance R and, in general, may depend on the angle θ, and (3) there is radial motion of matter (accretion). These differences lead, as we will show below, to a radial dependence of the rotational velocity $u_\phi(R) \propto R^{-1/2}$. (Recall that $u_\phi \propto R^{-2}$ in the case of a rotating sphere in a viscous fluid).

[2]Not to be confused with the standard definition of the Taylor number in viscous Taylor-Couette flows, $\mathrm{Ta} = 4\frac{\omega^2 r^4}{\nu}$.

We start by solving Eq. (7.56). At first, note that in order to have $u_\phi(\theta) \sim \sin\theta$, according to (7.61), we must have $W_{\theta\phi} = 0$. Next, using the continuity equation (7.53) and the definition of the angular velocity, we can write Eq. (7.56) in the form of the equation of angular momentum transfer by viscous forces:

$$\frac{\dot{M}}{R}\frac{\partial}{\partial R}\omega R^2 = \frac{4\pi}{R}\frac{\partial}{\partial R}R^3 W_{R\phi}\,. \tag{7.80}$$

We may integrate Eq. (7.80) over R to obtain

$$\dot{M}\omega R^2 = 4\pi R^3 W_{R\phi} + D\,, \tag{7.81}$$

were D is an integration constant. We then rewrite Eq. (7.62) using the derivative of the angular velocity:

$$W_{R\phi} = \rho\nu_t R\frac{\partial\omega}{\partial R}\,. \tag{7.82}$$

Substituting this expression into (7.81) yields

$$\dot{M}\omega R^2 = 4\pi\rho\nu_t R^4\frac{\partial\omega}{\partial R} + D\,. \tag{7.83}$$

This equation for the viscous angular momentum transfer is similar to the one for accretion discs (Shakura and Sunyaev 1973) but, however, differs due to the spherical symmetry of the problem under consideration.

The left-hand side of Eq. (7.83) describes the advective transfer of the angular momentum averaged over the sphere ($1/2\int_0^\pi \omega R^2 \sin^2\theta \sin\theta d\theta = (1/3)\omega R^2$) during the mean motion towards the gravitating centre (accretion). The accretion rate $\dot{M}$ here is negative, as well as the value of the derivative $\frac{\partial\omega}{\partial R}$. The first term on the right-hand side describes the angular momentum transfer outwards by the turbulent viscosity force.

The constant D is determined from the equation

$$D = K_1 K_2\left(\frac{u_c}{u_{ff}}\right)\frac{\mu^2}{R_A^3}\frac{\omega_m - \omega^*}{\omega_K(R_A)} \tag{7.84}$$

(see Eq. (7.185)). Here we consider accretion onto a magnetized neutron star. For $D < 0$, the advective term on the left-hand side of (7.83) dominates over the viscous angular momentum transport outwards. Oppositely, for $D > 0$, the viscous term in (7.83) dominates. In the case $\dot{M} = 0$ (when no plasma enters the magnetosphere), there is only viscous angular momentum transfer.

Now we rewrite (7.84) in the form

$$D = K_1 K_2\left(\frac{u_c}{u_{ff}}\right)\frac{\mu^2}{R_A^6}R_A^3\frac{\omega_m - \omega^*}{\omega_K(R_A)} \tag{7.85}$$

and use the pressure balance condition

$$P(R_A) = P_g(R_A)(1+\gamma m_t^2) = \frac{B^2(R_A)}{8\pi} = \frac{K_2}{2\pi}\frac{\mu^2}{R_A^6}\,. \tag{7.86}$$

Using the continuity equation in the form

$$|\dot{M}| = 4\pi R^2 \rho f(u)\sqrt{GM/R}\,,$$

and the gas pressure Eq. (7.13), we can recast the integration constant $D/|\dot{M}|$ to the form

$$\frac{D}{|\dot{M}|} = K_1 K_2 \left(\frac{u_c}{u_{ff}}\right)\frac{(\gamma-1)}{\gamma}\psi(\gamma, m_t)\frac{(\omega_m-\omega^*)R_A^2}{2\sqrt{2}f(u)}(1+\gamma m_t^2)\,. \tag{7.87}$$

Consider the rotation of a neutron star near equilibrium with $\dot{\omega}^* = 0$. In this case, from Eq. (7.188) we find

$$\omega_m - \omega^* = -\frac{z}{Z}\omega^*\,, \tag{7.88}$$

hence using the expression for Z (7.187) we obtain:

$$\frac{D}{|\dot{M}|} = -zR_A^2\omega^*\,. \tag{7.89}$$

We stress that for equilibrium neutron star rotation the value of the constant D is fully determined by the dimensionless specific angular momentum of matter near the Alfvén surface z.

7.4.3 The Rotation Law in the Shell

Equation (7.83) can be used to find the rotation law in the shell $\omega(R)$. At large distances, $R \gg R_A$ (remember that R_A determines the location of the shell base), the constant D is small compared to other terms, and we can set $D \approx 0$. Thus, when deriving the rotation law, we will neglect this constant in the right-hand side of Eq. (7.83). Next, we substitute (7.70) and the density distribution (which, as we show below, does not differ from the hydrostatic distribution)

$$\rho(R) = \rho(R_A)\left(\frac{R_A}{R}\right)^{3/2} \tag{7.90}$$

into Eq. (7.83). We thus obtain:

$$\left|\dot{M}\right| \omega R^2 = 4\pi\rho(R_A)\left(\frac{R_A}{R}\right)^{3/2} C R^7 \left(\frac{\partial \omega}{\partial R}\right)^2 . \tag{7.91}$$

After integration of this non-linear equation, we find

$$2\omega^{1/2} = \pm\frac{4}{3}\frac{K^{1/2}}{R^{3/4}} + D_1 , \tag{7.92}$$

where

$$K = \frac{|\dot{M}|}{4\pi\rho(R_A) C R_A^{3/2}} \tag{7.93}$$

and D_1 is the integration constant. In Eq. (7.92) we choose only the positive solution (the sign minus with constant $D_1 > 0$ corresponds to a solution with angular velocity increasing outwards, which is possible if the neutron star spin period is very large). If $D_1 \neq 0$, at large distances $R \gg R_A$ (near the external bow shock) solid-body rotation would lead to $\omega \to const \approx \omega_B$. (However, we remind the reader that our treatment is not applicable to the external shock region.) At small distances from the Alfvén surface, the influence of this constant is insignificant, and we will neglect it below. Then we find

$$\omega(R) = \frac{4}{9}\frac{|\dot{M}|}{4\pi\rho(R_A) C R_A^3}\left(\frac{R_A}{R}\right)^{3/2} = \omega_m (R_A/R)^{3/2} \tag{7.94}$$

i.e. a quasi-Keplerian rotation law, $\omega(R) \sim R^{-3/2}$. The constant ω_m in Eq. (7.94) is obtained after plugging $\dot{M}$ from the continuity equation at $R = R_A$ into Eq. (7.94):

$$\omega_m \equiv \tilde{\omega}\omega(R_A) = \frac{4}{9}\tilde{\omega}\frac{|u_r(R_A)|}{C R_A} . \tag{7.95}$$

(Here the correction factor $\tilde{\omega} > 1$ is introduced to account for the deviation of the exact solution from the quasi-Keplerian law near R_A.)

As the radial velocity of the matter in the settling accretion regime $u_r(R_A)$ is smaller than the free-fall velocity, the above equation implies that $\omega_m < \omega_K(R_A)$, i.e. smaller than the Keplerian angular rotation velocity. For a self-consistent solution, the coefficient C in Prandtl's law should be determined, according to Eq. (7.95), from the ratio of the radial velocity of the matter u_r to the linear rotational velocity u_ϕ:

$$C = \frac{4}{9}\tilde{\omega}\frac{|u_r(R_A)|}{\omega_m R_A} = \frac{4}{9}\tilde{\omega}\frac{|u_r(R_A)|}{u_\phi(R_A)} . \tag{7.96}$$

Note that this ratio is independent of the radius R and remains constant along the shell radius. Indeed, the radial dependence of the velocity u_r follows from the continuity equation with account for the density distribution (7.90)

$$u_r(R) = u_r(R_A)\left(\frac{R_A}{R}\right)^{1/2} . \tag{7.97}$$

For quasi-Keplerian rotation $u_\phi(R) \sim 1/R^{1/2}$, and thus the ratio u_r/u_ϕ remains constant.

Finally, the angular velocity of the shell near the magnetosphere ω_m is related to the angular velocity near the external shock as

$$\omega_m = \tilde{\omega}\omega_B\left(\frac{R_B}{R_A}\right)^{3/2} . \tag{7.98}$$

In reality, when approaching R_A, the integration constant D (that we have neglected at large distances $R \gg R_A$) should be taken into account. Therefore, the rotation law near the magnetosphere should be somewhat different from the quasi-Keplerian one.

We stress the principal difference of the accretion regime we consider from disc accretion. In disc accretion, the radial velocity of the matter is much smaller than the velocity of turbulent motions, and the tangential velocity is almost Keplerian and is much larger than the turbulent velocity. In quasi-spherical subsonic accretion, the radial velocity of the matter is not determined by the rate of angular momentum transfer. Instead, the radial velocity depends solely on the 'permeability' of the neutron star magnetospheric boundary to infalling plasma. In our case, it turns out to be of the order of the velocity of convective motions in the shell. The tangential velocity in the quasi-Keplerian law obtained above is much smaller than the convective velocities in the shell. Also note that in disc accretion the turbulence can be effectively described by the single dimensionless parameter, $\alpha \approx u_t^2/u_s^2$ with $0 < \alpha < 1$ (Shakura and Sunyaev 1973). The gas in an accretion disc rotates differentially with a supersonic (almost Keplerian) velocity, whereas in our case the shell rotates differentially with a significantly subsonic velocity at any radius, and the turbulence in the shell is subsonic. It is also evident that our case is significantly different from free-fall accretion onto the magnetosphere with the formation of a shock, which was considered, for example, in Arons and Lea (1976).

7.4.4 The Case Without Accretion

We now consider the situation where the plasma cannot penetrate the magnetosphere and there is no accretion onto the neutron star. This case is similar to the subsonic propeller considered in Davies and Pringle (1981). Then Eq. (7.83) takes the form:

$$0 = 4\pi\rho\nu_t R^4 \frac{\partial\omega}{\partial R} + D . \tag{7.99}$$

(Remember that the constant D is determined by the neutron star spin-down rate, $D = I\dot{\omega}^* < 0$.) Solving this equation as before, we obtain the rotational law in the shell for the case without accretion:

$$\omega(R) = \omega_m \left(\frac{R_A}{R}\right)^{7/4}, \tag{7.100}$$

where

$$\omega_m = \frac{I|\dot{\omega}^*|}{7\pi\rho(R_A)\nu_t(R_A)R_A^3}. \tag{7.101}$$

From (7.70) we find:

$$\nu_t(R_A) = \frac{7}{4}C\omega_m R_A^2, \tag{7.102}$$

therefore for ω_m we get:

$$\omega_m = \frac{2}{7}\left(\frac{I|\dot{\omega}^*|}{\pi C\rho(R_A)R_A^5}\right)^{1/2}. \tag{7.103}$$

On the other hand, the angular velocity of matter near the magnetosphere ω_m is related to the parameters near the external shock as:

$$\omega_m = \omega_B \left(\frac{R_B}{R_A}\right)^{7/4}. \tag{7.104}$$

A further discussion of this case and applications to possible observational evidence of hot shells around non-accreting magnetized neutron stars can be found in Postnov et al. (2017).

7.5 The Structure of the Shell and the Rotation Law for Wasiutynski's Turbulent Viscosity Law

Parndtl's rule for turbulent viscosity used above relates the scale and velocity of turbulent pulsations to the mean angular rotational velocity of matter, and is successfully applied in cases where the turbulence is generated by the shear flow itself. In our problem, the turbulence arises due to large-scale convective motions in the gravitational field. During radial convection, strongly anisotropic turbulent motions may appear (the radial dispersion of the chaotic motions may be much larger than the dispersion in the tangential direction), and Prandtl's law may be inapplicable. The anisotropic turbulence is much more complicated and poorly understood.

In the first approximation, we may use the empirical expression for the component $W_{R\phi}$, derived by Wasiutynski (1946)[3]:

$$W_{R\phi} = \rho\left(\nu_t R\frac{d\omega}{dR} + (\nu_r - \nu_t)\frac{1}{R}\frac{d\omega R^2}{dR}\right), \tag{7.105}$$

or

$$W_{R\phi} = 2\rho(\nu_r - \nu_t)\omega + \nu_r \rho R\frac{d\omega}{dR}, \tag{7.106}$$

where the radial and tangential kinematic viscosity coefficients are

$$\nu_r = C_\parallel \langle |u^t_\parallel|\rangle R$$

and

$$\nu_t = C_\perp \langle |u^t_\perp|\rangle R\,,$$

respectively. The dimensionless constants $C_\parallel$ and $C_\perp$ are of the order of one. In the isotropic case, $\nu_r = \nu_t$, $W_{R\phi} \sim d\omega/dR$, and in the strongly anisotropic case, $\nu_r \gg \nu_t$, $W_{R\phi} \sim d(\omega R^2)/dR$. Using these definitions and plugging (7.106) into (7.81), we obtain:

$$\omega R^2\left(1 - \frac{2C_\perp\langle|u^t_\perp|\rangle}{|u_r|}\right) = C_\parallel\frac{\langle|u^t_\parallel|\rangle}{|u_r|}\frac{Rd(\omega R^2)}{dR} - \frac{D}{|\dot{M}|}. \tag{7.107}$$

Note that due to the self-similar structure of the shell $u^t_\parallel \sim u^t_\perp \sim u_r \sim R^{-1/2}$, and therefore the ratios $\langle|u^t_\parallel|\rangle/u_r$ and $\langle|u^t_\perp|\rangle/u_r$ are constants. The above equation has an obvious solution:

$$\omega R^2 + \frac{D}{|\dot{M}|}\frac{1}{1 - 2C_\perp\frac{\langle|u^t_\perp|\rangle}{|u_r|}}$$
$$= \left[\omega_B R_B^2 + \frac{D}{|\dot{M}|}\frac{1}{1 - 2C_\perp\frac{\langle|u^t_\perp|\rangle}{|u_r|}}\right]\left(\frac{R_B}{R}\right)^{\frac{|u_r|}{C_\parallel\langle|u^t_\parallel|\rangle}\left(1-2C_\perp\frac{\langle|u^t_\perp|\rangle}{|u_r|}\right)} \tag{7.108}$$

(here the integration constant is defined such that $\omega(R_B) = \omega_B$).

[3] In Wasiutynski's paper, all equations are written for covariant components of the stress tensor $\tau_{R\phi}$, while here we, following Landau and Lifshits, write all values for physically measurable components, i.e. $W_{R\phi} = \tau_{R\phi}/(R\sin\theta)$, etc.

Now consider the equilibrium situation with $\dot{\omega}^* = 0$. Then, as we remember,

$$\frac{D}{|\dot{M}|} = -z\omega^* R_A^2 \, , \omega_m = (1 - z/Z)\omega^* \, .$$

At first, consider the case of strongly anisotropic almost radial turbulence for which $\langle |u_\perp^t|\rangle = 0$. Here the specific angular momentum at the Alfvén radius is

$$\omega_m R_A^2 \left[1 + \frac{z}{1 - z/Z} \left(\left(\frac{R_B}{R_A} \right)^{\frac{|u_r|}{C_\parallel \langle |u_\parallel^t|\rangle}} - 1 \right) \right] = \omega_B R_B^2 \left(\frac{R_B}{R_A} \right)^{\frac{|u_r|}{C_\parallel \langle |u_\parallel^t|\rangle}} \, . \tag{7.109}$$

We see that for very weak accretion (or in the limit when there is no accretion at all) $|u_r| \ll C_\parallel \langle |u_\parallel^t|\rangle$, i.e. virtually isomomentum rotation is established in the shell.

The next case is when the anisotropy is such that $C_\perp \langle |u_\perp^t|\rangle / |u_r| = 1/2$. Then a strictly isomomentum angular momentum distribution is established in the shell: $\omega_m R_A^2 = \omega_B R_B^2$.

If the turbulence is fully isotropic, then $C_\perp \langle |u_\perp^t|\rangle = C_\parallel \langle |u_\parallel^t|\rangle = \tilde{C} \langle |u^t|\rangle$. By denoting $\epsilon = |u_r| / (\tilde{C} \langle |u^t|\rangle)$, we find:

$$\omega_m R_A^2 \left[1 + \left(\frac{z}{1 - z/Z} \right) \left(\frac{1}{2/\epsilon - 1} \right) \left(1 - \left(\frac{R_A}{R_B} \right)^{2-\epsilon} \right) \right] = \omega_B R_B^2 \left(\frac{R_A}{R_B} \right)^{2-\epsilon} \, . \tag{7.110}$$

Note that for $\epsilon \to 0$ (no accretion through the magnetosphere) $\omega_m \to \omega_B$, i.e., we get solid-body rotation without accretion (cf. the first case above). For $\epsilon = 3/2$, the rotation is almost quasi-Keplerian. Recall that quasi-Keplerian rotation was obtained above using Prandtl's turbulent viscosity prescription. Then it was the unique solution. For anisotropic turbulence, in contrast, quasi-Keplerian rotation is a particular case of a more general solution that is obtained using Wasiutynski's turbulent viscosity law.

As shown in Shakura et al. (2012), the quasi-Keplerian rotation in the shell is less favored by observations. Therefore, we conclude that in quasi-spherical shells at the stage of subsonic settling accretion an almost isomomentum rotation, caused by the anisotropic turbulence due to convection, is most likely.

7.6 Corrections to the Radial Temperature Gradient

Let us estimate how strongly the temperature gradient in the shell differs from the adiabatic law when convective motions are present. Multiplying Eq. (7.186) by $(1/2)(\omega_m - \omega^*)$ yields the convection heating rate caused by the shell-magnetosphere interaction:

$$L_c = \frac{1}{2} Z \dot{M} R_A^2 (\omega_m - \omega^*)^2 \, . \tag{7.111}$$

Multiplying the same Eq. (7.186) by ω^* yields the rate of change of the neutron star's mechanical energy:

$$L_k = Z\dot{M}R_A^2\omega^*(\omega_m - \omega^*)\,. \tag{7.112}$$

Thus, the energy balance equation can be presented in the form:

$$L_t = L_c + L_k = \frac{1}{2}Z\dot{M}R_A^2(\omega_m^2 - \omega^{*2})\,. \tag{7.113}$$

Note that the formula obtained for L_c looks like the equation describing the energy release in the boundary layer of an accretion disc (Koh et al. 1997; Shakura and Sunyaev 1988).

The convective energy flux is

$$q_c = \frac{L_c}{4\pi R^2} = \frac{Z\dot{M}R_A^2(\omega_m - \omega^*)^2}{8\pi R^2}\,. \tag{7.114}$$

On the other hand, the convective energy flux can be related to the entropy gradient (see Shakura et al. 1978):

$$q_c = -\rho\nu_c T\frac{dS}{dR}\,, \tag{7.115}$$

where S is the specific entropy (per gram). Here ν_c is the radial thermal conductivity coefficient,

$$\nu_c =< u_c l_c >= C_h u_c R\,, \tag{7.116}$$

where the characteristic scale of convection is $l_c \sim R$, the velocity of the convective motions $u_c \sim c_s \sim R^{-1/2}$, and C_h is a numerical factor of the order of one. Thus

$$\nu_c = \nu_c(R_A)\left(\frac{R}{R_A}\right)^{1/2}\,. \tag{7.117}$$

Next, make use of the thermodynamic identity for the specific enthalpy H:

$$\frac{dH}{dR} = \frac{1}{\rho}\frac{dP_g}{dR} + T\frac{dS}{dR}\,. \tag{7.118}$$

Recall that the enthalpy can be written in the form

$$dH = c_p dT\,,$$

where

$$c_p = T\left(\frac{\partial S}{\partial T}\right)_p = \frac{\gamma}{\gamma - 1}\frac{\mathscr{R}}{\mu_m}$$

is the specific thermal capacity at constant pressure. We now express $T(dS/dR)$ from Eq. (7.115) and use the hydrostatic equilibrium equation (7.9) in the form

$$\frac{dP_g/\rho}{dR} = -\frac{\mathcal{R}}{\mu_m c_p}\frac{GM}{R^2}\psi(\gamma, m_t)$$

to cast the identity (7.118) into the form

$$\frac{dT}{dR} = -\frac{1}{c_p}\left[\frac{GM}{R^2}\psi(\gamma, m_t) - \frac{Zu_r(R_A)}{2\nu_c(R_A)}\left(\frac{R_A}{R}\right)R_A^2(\omega_m - \omega^*)^2\right]. \tag{7.119}$$

By definition, the adiabatic temperature gradient is determined by the first term on the right-hand side, $(dT/dR)_{ad} = g/c_p$. Equation (7.119) can be integrated to find the real dependence of the temperature on radius in a convective shell:

$$T = \frac{1}{c_p}\left[\frac{GM}{R}\psi(\gamma, m_t) - \frac{Zu_r(R_A)}{2\nu_c(R_A)}R_A^3(\omega_m - \omega^*)^2 \ln\left(\frac{R}{R_A}\right)\right]. \tag{7.120}$$

Near the equilibrium ($I\dot{\omega}^* = 0$) we may use Eq. (7.88) to obtain

$$T = \frac{1}{c_p}\left[\frac{GM}{R}\psi(\gamma, m_t) - \frac{|u_r(R_A)|}{2C_h u_c(R_A)}\omega^{*2}R_A^2\frac{z^2}{Z}\ln\left(\frac{R}{R_A}\right)\right]. \tag{7.121}$$

This solution shows that in the region between R_A and R_B in the shells around slowly rotating X-ray pulsars (i.e. those in which $\omega_m \ll \omega_K(R_A)$), the temperature distribution is similar to the adiabatic law with a temperature gradient close to (but still steeper, enabling convection) the adiabatic one (7.9):

$$T \approx \frac{\gamma - 1}{\gamma}\frac{GM}{\mathcal{R}R}\psi(\gamma, m_t). \tag{7.122}$$

Here we have taken into account only the energy release caused by the angular velocity difference near the magnetosphere. In fact, there can be additional energy sources in the shell, for example, heating of plasma due to magnetic reconnection and turbulence (see Sect. 7.7), etc.

7.7 Dynamics of Static Spherically-Symmetric Gas Flow

In this section, we will consider the gas-dynamic equations for a spherically symmetric flow of an ideal gas onto a Newtonian gravitating center. This problem was first solved in the classical paper by Bondi (1952) in the case of an adiabatic accretion. The adiabatic gas outflow from stars (stellar wind) was later studied by Parker (1963). A detailed and thorough discussion of the problem can be found in the monograph by Beskin (2010). Here we will focus on the role of gas cooling/heating near the Alfvén surface, including the effect of turbulence/convection (generally, anisotropic) in the gas flow through a shell around the magnetosphere around a rotating neutron star. As discussed above, at low X-ray luminosities the quasi-static gas shell can remove angular momentum from the rotating neutron star magnetosphere via convective turbulent motions of the gas. If the mass accretion rate through the shell exceeds some critical value, strong Compton cooling gives rise to the appearance of a free-fall zone above the magnetosphere, and the angular momentum cannot be transferred upstream such a flow.

The equation of motion (7.54) in the absence of viscosity reads:

$$u_r\frac{du_r}{dR}=-\frac{1}{\rho}\frac{dP_g}{dR}-\frac{1}{\rho}\frac{dP_{\parallel}^t}{dR}-\frac{2(P_{\parallel}^t-P_{\perp}^t)}{\rho R}-\frac{GM}{R^2} \tag{7.123}$$

Here $P_g=\rho c_s^2/\gamma$ is the gas pressure and P^t is the pressure due to turbulent pulsations, which in general may be anisotropic:

$$P_{\parallel}^t=\rho<u_{\parallel}^2>=\rho m_{\parallel}^2c_s^2=\gamma P_g m_{\parallel}^2 \tag{7.124}$$

$$P_{\perp}^t=2\rho<u_{\perp}^2>=2\rho m_{\perp}^2c_s^2=2\gamma P_g m_{\perp}^2 \tag{7.125}$$

(here $<u_t^2>=<u_{\parallel}^2>+2<u_{\perp}^2>$ is the turbulent velocity dispersion, $m_{\parallel}^2$ and $m_{\perp}^2$ are radial and tangential Mach numbers squared).

From the first law of thermodynamics we find:

$$\frac{dE}{dR}=\frac{P_g}{\rho^2}\frac{d\rho}{dR}+T\frac{dS}{dR}, \tag{7.126}$$

where the specific internal energy (per gram) is

$$E=c_VT=\frac{c_s^2}{\gamma(\gamma-1)}, \tag{7.127}$$

and the specific thermal capacity (per gram) is

$$c_V=\frac{\mathcal{R}}{\mu_m}\frac{1}{\gamma-1}. \tag{7.128}$$

From the second law of thermodynamics the change in the specific entropy of the gas can be written through the rate of change of the specific heat dQ/dt [erg s^{-1} g^{-1}] as

$$T\frac{dS}{dR}=\frac{dQ}{dR}=\frac{dQ/dt}{u_r}. \tag{7.129}$$

Using the continuity equation

$$\dot{M} = 4\pi R^2 \rho u_r\,, \tag{7.130}$$

we find

$$\frac{1}{\rho}\frac{d\rho}{dR} = -\frac{2}{R} - \frac{1}{2u_r^2}\frac{du_r^2}{dR}\,. \tag{7.131}$$

Making use of the relation $c_s^2 = \gamma \mathscr{R} T$, we finally obtain:

$$\frac{1}{c_s^2}\frac{dc_s^2}{dR} = (\gamma-1)\left[-\frac{2}{R} - \frac{1}{2u_r^2}\frac{du_r^2}{dR}\right] + \frac{dQ/dt}{u_r c_V T}\,. \tag{7.132}$$

Note that this equation can be also derived from the equation of state of an ideal gas written in the form

$$P_g = K e^{S/c_V} \rho^\gamma\,, \tag{7.133}$$

where K is some constant.

Using Eq. (7.132), the gas pressure gradient can be presented in the form:

$$\frac{1}{\rho}\frac{dP_g}{dR} = \frac{c_s^2}{c_P u_r}\frac{dQ/dt}{T} + c_s^2\left[-\frac{2}{R} - \frac{1}{2u_r^2}\frac{du_r^2}{dR}\right] \tag{7.134}$$

Plugging (7.134) into the equation of motion finally yields:

$$\frac{1}{2}\frac{1}{u_r^2}\frac{du_r^2}{dR} = \left[c_s^2(1+\gamma m_\parallel^2)\left(\frac{2}{R} - \frac{dQ/dt}{c_P u_r T}\right) - 2c_s^2\frac{(m_\parallel^2 - m_\perp^2)}{R} - \frac{GM}{R^2}\right] / \left[u_r^2 - c_s^2(1+\gamma m_\parallel^2)\right]. \tag{7.135}$$

Also, note that in a strongly anisotropic case where $m_\parallel^2 = m_t^2 \gg m_\perp^2$, the role of the turbulence increases compared to the isotropic case where $m_\parallel^2 = m_\perp^2 = (1/3)m_t^2$.

We may also introduce the Mach number in the flow $\mathscr{M} \equiv u_r/c_s$. Then from Eqs. (7.132) and (7.135) we can derive the equation for the Mach number:

$$\frac{[\mathscr{M}^2-(1+\gamma m_\parallel^2)]}{\mathscr{M}^2}\frac{d\mathscr{M}^2}{dR} = \left\{\frac{2\left[(\gamma-1)\mathscr{M}^2-(\gamma+1)(m_\parallel^2-m_\perp^2)\right]}{R} - \frac{\left[\mathscr{M}^2+\gamma(1+\gamma m_\parallel^2)\right]}{c_P T}\frac{dQ}{dR} - \frac{(\gamma+1)GM}{R^2 c_s^2}\right\}, \tag{7.136}$$

where we have used the substitution $(dQ/dt) = u_r(dQ/dR)$. Equations (7.132), (7.135) and (7.136) can be applied to describe the dynamics of the accreting flow in terms of pairs of independent variables (u_r, c_s), $(u_r, \mathcal{M})$ or $(c_s, \mathcal{M})$. Here we will consider the behavior of the flow near the singular point only. To this goal, we can use Eq. (7.135).

Equation (7.135) has one singular saddle point in which the denominator in the right-hand side vanishes:

$$u_r^2 = c_s^2(1 + \gamma m_\parallel^2) \,. \tag{7.137}$$

For a solution to exist in this point, the numerator in the right-hand side must also vanish, which yields the quadratic equation for the flow velocity in the singular point:

$$u_r^2 \frac{2}{R} \left(\frac{1 + (\gamma - 1)m_\parallel^2 + m_\perp^2}{1 + \gamma m_\parallel^2} \right) - u_r \left(\frac{dQ/dt}{c_P T} \right) - \frac{GM}{R^2} = 0 \,. \tag{7.138}$$

Recall that in the adiabatic case $(dQ/dt = 0)$ without turbulence, in the singular point we would simply have

$$u_r^2 = c_s^2 = \frac{GM}{2R} \,. \tag{7.139}$$

We stress that in the presence of turbulence, the velocity in the singular point increases. For example, for $\gamma = 5/3$ and a strong anisotropic turbulence we find $u_r^2 = c_s^2(1 + (5/3)m_\parallel^2)$; for an isotropic turbulence the correction is smaller: $u_r^2 = c_s^2(1 + (5/9)m_t^2)$. Due to turbulence, the transition through the sound velocity (the sound point where $u_r^2 = c_s^2$) occurs above the saddle point, and there is no singularity at the sound point.

The turbulent heating rate in the quasi-static shell c $(dQ/dt)_t^+$ can be determined as:

$$\left(\frac{dQ}{dt} \right)_t^+ = \frac{1}{2} \frac{< u_t^2 >}{t_t} \,, \tag{7.140}$$

where the characteristic turbulent heating time is

$$t_t = \alpha_t \frac{R}{u_t} = \alpha_t \frac{R}{m_t c_s} \,. \tag{7.141}$$

Here α_t is a dimensionless constant determining the turbulent energy dissipation rate, and the turbulent Mach number is $m_t^2 \equiv m_\parallel^2 + 2m_\perp^2$. Thus, the turbulent heating rate can be written in the form

$$\left(\frac{dQ}{dt} \right)_t^+ = \frac{c_s^3}{2\alpha_t R} m_t^3 \,. \tag{7.142}$$

In the case of Compton cooling we have

$$\left(\frac{dQ}{dt}\right)^{-}_{C} = -\frac{c_V(T - T_x)}{t_C}, \tag{7.143}$$

were t_C is the characteristic Compton cooling time (7.21).

Equation (7.138) can now be recast to the form

$$u_r^2 \frac{2}{R}\left(\frac{1 + (\gamma - 1)m_\parallel^2 + m_\perp^2}{1 + \gamma m_\parallel^2}\right) - u_r^2 \frac{c_s}{u_r}\frac{\gamma(\gamma - 1)m_t^3}{2\alpha_t R} + \frac{u_r(1 - T_x/T)}{\gamma t_C} - \frac{GM}{R^2} = 0. \tag{7.144}$$

As we are studying an accretion processes, the sign of the velocity $u_r = dR/dt$ is negative, $u_r = -|u_r|$. Then, the absolute value of the flow velocity at the singular point where the sound velocity is $c_s/|u_r| = -1/(1 + \gamma m_\parallel^2)^{1/2}$ can be found from the quadratic equation:

$$u_r^2 \frac{2}{R}\left(\frac{1 + (\gamma - 1)m_\parallel^2 + m_\perp^2}{1 + \gamma m_\parallel^2}\right) + u_r^2 \frac{1}{(1 + \gamma m_\parallel^2)^{1/2}}\frac{(\gamma - 1)m_t^3}{2\alpha_t R}$$
$$- \frac{|u_r|(1 - T_x/T)}{\gamma t_C} - \frac{GM}{R^2} = 0. \tag{7.145}$$

In this case, the solution to Eq. (7.138) reads:

$$|u_r| = \frac{R(1 - T_x/T)}{4\gamma t_C A} + \sqrt{\frac{2GM}{R}}\left[\frac{1}{4A} + \frac{R}{2GM}\frac{R^2(1 - T_x/T)^2}{16\gamma^2 t_C^2 A^2}\right]^{1/2}, \tag{7.146}$$

where we have introduced the dimensionless factor

$$A = \frac{1 + (\gamma - 1)m_\parallel^2 + m_\perp^2}{1 + \gamma m_\parallel^2} + \frac{(\gamma - 1)(m_\parallel^2 + 2m_\perp^2)^{3/2}}{4\alpha_t(1 + \gamma m_\parallel^2)^{1/2}}. \tag{7.147}$$

For isotropic turbulence with $m_\parallel = m_\perp = 1/\sqrt{3}$, $m_t = 1$, for $\gamma = 5/3$ this factor is $A \approx 1.23$, and for strongly anisotropic turbulence when $m_\parallel = 1, m_\perp = 0, m_t = 1$, this factor is $A \approx 0.8$.

In units of the free-fall velocity, the solution to Eq. (7.146) has the form:

$$f(u) = \frac{|u_r|}{u_{ff}} = \frac{(1 - T_x/T)}{4\gamma A}\left(\frac{t_{ff}}{t_C}\right) + \frac{1}{2}\left[\frac{1}{A} + \frac{(1 - T_x/T)^2}{4\gamma^2 A^2}\left(\frac{t_{ff}}{t_C}\right)^2\right]^{1/2}. \tag{7.148}$$

With the onset of Compton cooling the temperature changes exponentially:

$$T = T_x + (T_{cr} - T_x)e^{-t/t_C}. \tag{7.149}$$

When cooling is slow, $t_{ff}/t_C \ll 1$, the critical point lies under the Alfvén surface, and through the flow down to the magnetosphere no transition through the sound speed occurs. It is under such conditions that slow settling accretion can be realized. If the critical point lies above the Alfvén surface, a supersonic transition in the flow takes place before the flow encounters the magnetosphere, and thus the appearance of a shock is expected. Both turbulence and rapid cooling shift the critical point upstream the flow.

In the case of rapid cooling, $t_{ff}/t_C \gg 1$, $T \to T_x$, so that $u_r/u_{ff} \approx 1/2$ (cf. (7.139) for an adiabatic flow), but the critical point lies above the Alfvén surface, and in the flow above the magnetosphere a free-fall zone appears. The ratio $f(u) = |u_r|/u_{ff}$ reaches maximum at $t_{ff}/t_C \approx 0.46$ for the typical temperature ratio $T_{cr}/T_x = 10$, and depending on the value of the factor $A = 0.8 \div 1.23$ (anisotropic or isotropic turbulence) it turns out to be equal to $f(u) = 0.5$–0.6.

7.8 Physical Conditions in the Shell

To form a hot shell around a NS magnetosphere, the matter downstream the bow shock at the gravitational capture radius should not cool down too rapidly and fall freely towards the magnetosphere. In other words, the cooling time of the gas heated up behind the shock t_{cool} should exceed the plasma free-fall time.

Behind the front of a strong shock, the gas heats up to the temperature

$$T_{ps} = \frac{3}{16}\mu_m \frac{v_w^2}{\mathcal{R}} \approx 1.36 \times 10^5 [\mathrm{K}] \left(\frac{v_w}{100\,\mathrm{km/s}}\right)^2 . \tag{7.150}$$

The radiation cooling time of a plasma is

$$t_{cool} = \frac{3kT}{2\mu_m n_e \Lambda} \tag{7.151}$$

where ρ is the plasma density, $n_e = Y_e\rho/m_p$ is the electron number density (in a fully ionized solar plasma, the molecular weight is $\mu_m = 0.6$ and the lepton number is $Y_e \approx 0.8$); Λ is the cooling function of collisional equilibrium plasma that can be approximated as (Raymond et al. 1976; Cowie et al. 1981)

$$\Lambda(T) = \begin{cases} 0, T < 10^4\,\mathrm{K} \\ 1.0 \times 10^{-24} T^{0.55},\ 10^4\,\mathrm{K} < T < 10^5\,\mathrm{K} \\ 6.2 \times 10^{-19} T^{-0.6},\ 10^5\,\mathrm{K} < T < 4 \times 10^7\,\mathrm{K} \\ 2.5 \times 10^{-27} T^{0.5},\ T > 4 \times 10^7\,\mathrm{K}. \end{cases} \tag{7.152}$$

Compton cooling becomes effective starting from the radius R_x at which the gas temperature T, determined from the hydrostatic formula (7.9), exceeds the Compton temperature of radiation T_x. The Compton cooling time (see (7.21)) is:

$$t_C \approx 1060[\mathrm{s}]\dot{M}_{16}^{-1}\left(\frac{R}{10^{10}\,\mathrm{cm}}\right)^2 . \tag{7.153}$$

Compton heating starts above some radius R_x at which $T_x = T$. For the temperature distribution in the shell according to formula (7.9), we find $R_x \approx 2 \times 10^{10}$ cm. Note that both Compton and photoionisation heating is controlled by the photoionization parameter ξ (Tarter et al. 1969; Hatchett et al. 1976)

$$\xi = \frac{L_x}{n_e R^2} . \tag{7.154}$$

In most parts of the accreting flow the density follows the law $n \sim R^{-3/2}$, therefore $\xi \sim R^{-1/2}$, and with account for the continuity equation, ξ does not depend on the X-ray luminosity. The characteristic value of ξ is:

$$\xi \approx 5 \times 10^5 f(u) R_{10}^{-1/2} . \tag{7.155}$$

Should the Compton processes be effective everywhere, such a high value of the photoionization parameter ξ would suggest that the plasma heats up to Compton temperatures of the order of several keV up to very large distances $\sim 10^{12}$ cm. However, at large distances the Compton time exceeds the characteristic gas accretion time:

$$\frac{t_C}{t_{accr}} = \frac{t_C f(u) u_{ff}}{R} \approx 20 f(u) \dot{M}_{16}^{-1} R_{10}^{1/2} , \tag{7.156}$$

which means that Compton heating is ineffective in the falling matter. Therefore, far from the magnetosphere the gas temperature is determined by the photoionization only and cannot exceed $T_{max} \approx 5 \times 10^5$ K (Tarter et al. 1969), which is much smaller than $T_x \sim 3$ keV.

The effective gravitational capture radius corresponding to the sound velocity of the gas in the photoionization heating region is

$$R_B^* = \frac{2GM}{c_s^2} = \frac{2GM}{\gamma \mathcal{R} T_{max}/\mu_m} \approx 3.5 \times 10^{12}[\mathrm{cm}]\left(\frac{T_{max}}{5 \times 10^5\,\mathrm{K}}\right)^{-1} . \tag{7.157}$$

Everywhere up to the shock, photoionization keeps the temperature at about $\simeq T_{max}$. The sound velocity corresponding to T_{max}, is about 80 km s^{-1}. If the stellar wind velocity exceeds 80 km s^{-1}, a bow shock arises near the Bondi radius with the post-shock temperature given by formula (7.150). If the stellar wind velocity is below this value, the shock disappears and accretion occurs from the region determined by the effective radius R_B^*.

The photoionization heating time at the effective Bondi radius 3×10^{12} cm is

$$t_{pi} \approx \frac{(3/2)kT_{max}/\mu_m}{(h\nu_{eff} - \zeta_{eff})n_\gamma \sigma_{eff} c} \approx 2 \times 10^4 [\mathrm{s}] \dot{M}_{16}^{-1} . \tag{7.158}$$

(here $h\nu_{eff} \sim 10\,\mathrm{keV}$ is the characteristic photon energy, ζ is the effective ionization potential, $\sigma_{eff} \sim 10^{-24}\,\mathrm{cm}^2$ is the typical photoionization cross-section and $n_\gamma = L_x/(4\pi R^2 h\nu_{eff} c)$ is the photon number density). The ratio of the photoionization time to the accretion time at the effective Bondi radius is

$$\frac{t_{pi}}{t_{accr}} \approx 0.07 f(u) \dot{M}_{16}^{-1} . \tag{7.159}$$

At stellar wind velocities $v_w > 80\,\mathrm{km\,s^{-1}}$ the shock arises near the classical Bondi radius R_B lying inside the effective Bondi radius R_B^*, determined by formula (7.157).

The radiation cooling time of plasma heated downstream the shock R_B is expressed through the velocity of the stellar wind being captured v_w as:

$$t_{cool} \approx 4.7 \times 10^4 [\mathrm{s}] \dot{M}_{16}^{-1} v_7^{0.2} . \tag{7.160}$$

The photoionization heating time behind the shock front can also be expressed through the wind velocity:

$$t_{pi} \approx 3.5 \times 10^4 [\mathrm{s}] \dot{M}_{16}^{-1} v_7^{-4} . \tag{7.161}$$

A comparison of these two characteristic times shows that at low wind velocities, radiative plasma cooling is important, and free-fall supersonic (Bondi) accretion onto the neutron star with conservation of the specific angular momentum of the accreting matter sets in.

Thus, at low wind velocities the plasma behind the bow shock front cools down and falls freely. When approaching the gravitational centre, the photoionization heating becomes important, and the plasma temperature attains the level $T_{max} \approx 5 \times 10^5$ K. Should this occur at the radius R_{pi} corresponding to $T_{max} < GM/(\mathcal{R} R_{\mathrm{pi}})$, the plasma with constant temperature T_{max} continues falling freely towards the magnetosphere, above which a shock is formed. However, if $T_{max} > GM/(\mathcal{R} R_{\mathrm{pi}})$, the subsonic settling accretion regime is possible even at low wind velocities.

For wind velocities $v_w \gtrsim 100\,\mathrm{km\,s^{-1}}$, the gas temperature behind the shock exceeds T_{max}, photoionization heating is unimportant, and the settling accretion regime may be established in the shell if the radiation cooling time is longer than the accretion time. By comparing these two time-scales, we obtain the critical mass accretion rate, depending on the stellar wind velocity, below which the settling regime is possible:

$$\dot{M}_{16}^{\ddagger} \lesssim 0.12 v_7^{3.2} . \tag{7.162}$$

Here we should stress the difference between the critical mass accretion rate $\dot{M}^{\ddagger}$ and the value $\dot{M}^{\dagger}$, obtained above. For $\dot{M} > \dot{M}^{\ddagger}$ the plasma rapidly cools down in the wind gravitational capture zone and falls freely towards the magnetosphere (unless photoionization heats it up above the virial temperature), whereas for $\dot{M} > \dot{M}^{\dagger} \simeq 4 \times 10^{16}\,\mathrm{g\,s^{-1}}$ determined by Eq. (7.38), a free-fall supersonic zone appears immediately above the magnetosphere.

7.9 X-ray Emission and Quasi-Periodic Pulsations from the Hot Shell

The spectra of X-ray pulsars are dominated by the emission formed in accretion columns near the surface of magnetized neutron stars. A hot optically thin shell above the magnetosphere generates proper thermal emission. However, if all gravitational energy of the accreting matter were released in such a shell, its X-ray luminosity would be reduced by a factor R_{NS}/R_A compared to the emission from the accretion column, i.e. it would be less than 1% of the total X-ray luminosity. In addition, the shell should scatter the X-ray emission from the accretion column, but for the scattering to be effective, the Comptonization parameter y should be of the order of one. The Thomson optical depth in the shell is obviously very small. Indeed, from the continuity equation, formula (7.16) for the Alfvén radius and Eq. (7.29) for the dimensionless factor $f(u)$ at the subsonic settling accretion, we find:

$$\tau_T = \int_{R_A}^{R_B} n_e(R)\sigma_T dR \approx 3.2 \times 10^{-3} \dot{M}_{16}^{8/11} \mu_{30}^{-2/11} .$$

Therefore, at temperatures near the magnetospheric boundary (see (7.9)) the Comptonization parameter y is small:

$$y = \frac{4kT}{m_e c^2} \tau_T \approx 2.4 \times 10^{-3} .$$

This means that the X-ray spectrum generated in the zone of main accretion energy release near the neutron star surface is not affected by scattering off electrons in the hot shell.

Large-scale convective motions in the shell occur on a specific time scale of the order of the free-fall time, which may give rise to features in the time variability power density spectra (for example, quasi-periodic oscillations, QPOs). QPOs have been detected in spectra of some X-ray pulsars (see e.g. Marykutty et al. 2010 and references therein). The expected QPO frequency should fall into the mHz range. Such QPOs were indeed detected in some cases (Sidoli et al. 2016a).

A stronger effect may be related to the appearance of a dynamical instability in the shell due to, for example, Compton cooling enhancement leading to a runaway increase of the accretion rate through the shell. This instability would give rise to a sharp increase in X-ray luminosity, as observed in SFXTs (see our discussion in Sect. 7.10 for more detail).

7.10 Bright Flares in Supergiant Fast X-ray Transients

In this section, we consider another possible application of the theory of subsonic settling accretion to bright outbursts observed in supergiant fast X-ray transients (Shakura et al. 2014a).

Supergiant Fast X-ray Transients (SFXTs) are a subclass of HMXBs associated with early-type supergiant companions (Pellizza et al. 2006; Chaty et al. 2008; Rahoui et al. 2008), and characterized by sporadic, short and bright X–ray flares reaching peak luminosities of 10^{36}–10^{37} erg s^{-1}. Most of them were discovered by INTEGRAL (Molkov et al. 2003; Sunyaev et al. 2003; Grebenev et al. 2003; Sguera et al. 2005; Negueruela et al. 2006). They show high dynamic ranges (between 100 and 10,000, depending on the specific source; e.g. Romano et al. (2011, 2014)) and their X-ray spectra in outburst are very similar to accreting pulsars in HMXBs. In fact, half of them have measured neutron star spin periods similar to those observed from persistent HMXBs (see Sidoli 2012 for a review).

The physical mechanism driving their transient behavior, related to the accretion by the compact object of matter from the supergiant wind, has been discussed by several authors and is still a matter of debate, as some of them require particular properties of the compact objects hosted in these systems (Grebenev and Sunyaev 2007; Bozzo et al. 2008), and others assume peculiar clumpy properties of the supergiant winds and/or special orbital characteristics (in't Zand 2005; Walter and Zurita Heras 2007; Sidoli et al. 2007; Negueruela et al. 2008; Ducci et al. 2009; Oskinova et al. 2012).

The typical energy released in a SFXT bright flare is about 10^{38}–10^{40} ergs (Shakura et al. 2014a), varying by one order of magnitude between different sources. That is, the mass falling onto the NS in a typical bright flare varies from 10^{18} g to around 10^{20} g.

The typical X-ray luminosity outside outbursts in SFXTs is about $L_{x,low} \simeq 10^{34}$ erg s^{-1} (Sidoli et al. 2008), and below in this section we shall normalize the luminosity to this value, L_{34}. At these low X-ray luminosities, the plasma entry rate into the magnetosphere is controlled by radiative plasma cooling. Further, it is convenient to normalize the typical stellar wind velocity from hot OB-supergiants v_w to 1000 km s^{-1} (for orbital periods of about a few days or larger the NS orbital velocities can be neglected compared to the stellar wind velocity from the OB-star), so that the Bondi gravitational capture radius is $R_B = 2GM/v_w^2 = 4\times10^{10}[\text{cm}]v_8^{-2}$ for a fiducial NS mass of $M_x = 1.5M_\odot$.

7.10.1 Magnetospheric Shell Instability

Let us assume that a quasi-static shell hangs over the magnetosphere around the NS, with the magnetospheric accretion rate being controlled by radiative plasma cooling. We denote the actual steady-state accretion rate as $\dot{M}_a$ so that the observed X-ray steady-state luminosity is $L_x = 0.1\dot{M}_a c^2$. Then from the theory of subsonic quasi-spherical accretion (Shakura et al. 2012) we know that the factor $f(u)$ (the ratio of the actual velocity of plasma entering the magnetosphere, due to the Rayleigh-Taylor instability, to the free-fall velocity at the magnetosphere, $u_{ff}(R_A) = \sqrt{2GM/R_A}$) reads (Shakura et al. 2013a, 2014b)

$$f(u)_{rad} \simeq 0.036\zeta^{7/11}L_{34}^{2/9}\mu_{30}^{2/27}\,. \tag{7.163}$$

(see also (7.37) above).

The shell is quasi-static (and likely convective). It is straightforward to calculate the mass of the shell using the density distribution $\rho(R) \propto R^{-3/2}$ (Shakura et al. 2012). Using the mass continuity equation to eliminate the density above the magnetosphere, we readily find

$$\Delta M \approx \frac{2}{3}\frac{\dot{M}_a}{f(u)}t_{ff}(R_B)\,. \tag{7.164}$$

Note that this mass can be expressed through measurable quantities $L_{x,low}$, μ_{30} and the (not directly observed) stellar wind velocity at the Bondi radius $v_w(R_B)$. Using (7.163) for the radiative plasma cooling, we obtain

$$\Delta M_{rad} \approx 8\times10^{17}[g]\zeta^{-7/11}L_{34}^{7/9}v_8^{-3}\mu_{30}^{-2/27}\,. \tag{7.165}$$

This simple estimate (7.165) shows that for a typical wind velocity near the NS of about 500 km s^{-1} the *typical* mass of the hot magnetospheric shell is around 10^{19} g, corresponding to 10^{39} ergs released in a flare if all the matter from the shell is accreted onto the NS. Variations in stellar wind velocity between different sources by a factor of ~2 would produce the one-order-of-magnitude spread in ΔM observed in bright SFXT flares.

As noted in Shakura et al. (2013a), if there is an unstable flow of matter through the magnetosphere, a large quantity of X-ray photons produced near the NS surface should rapidly cool down the plasma near the magnetosphere, further increasing the plasma fall velocity $u_R(R_A)$ and subsequently the NS accretion luminosity L_x. Therefore, in a bright flare the entire shell may fall onto the NS from the outer radius of the shell on the free-fall time scale $t_{ff}(R_B) \sim 1000\,\mathrm{s}$. Clearly, the shell will be replenished by new wind capture, so the flares will repeat as long as a rapid mass-entry rate into the magnetosphere is sustained.

7.10.2 Magnetized Stellar Wind as the Flare Trigger

We suggest that the shell instability described above can be triggered by a large-scale magnetic field sporadically carried by the stellar wind of the optical OB companion. Observations suggest that about $\sim$10% of hot OB-stars have magnetic fields up to a few kG (see Braithwaite (2013) for a review and discussion). It is also well known from Solar wind studies (see e.g. reviews Zelenyi and Milovanov (2004), Bruno and Carbone (2013) and references therein) that the Solar wind patches carrying tangent magnetic fields has a lower velocity (about $350\,\mathrm{km\,s^{-1}}$) than the wind with radial magnetic fields (up to $\sim 700\,\mathrm{km\,s^{-1}}$). Fluctuations of the density and velocity of stellar winds from massive stars are known from spectroscopic observations (Puls et al. 2008), and velocity fluctuations up to $0.1\,v_\infty \sim 200$–$300\,\mathrm{km\,s^{-1}}$ are typical.

The effect of the magnetic field carried by the stellar wind is twofold: first, it may trigger rapid mass entry to the magnetosphere via magnetic reconnection (a phenomenon well known in the Earth dayside magnetosphere, (Dungey 1961)), and secondly, the magnetized parts of the wind (magnetized clumps with a tangent magnetic field) have a lower velocity than the non magnetized ones (or the ones carrying the radial field). As discussed in Shakura et al. (2014a) and below, magnetic reconnection may increase the plasma fall velocity in the shell from inefficient, radiative-cooling controlled settling accretion with $f(u)_{rad} \sim 0.03$–0.1, up to the maximum possible free-fall velocity with $f(u) = 1$. In other words, during a bright flare subsonic settling accretion turns into supersonic Bondi accretion. The second factor (slower wind velocity in magnetized clumps with tangent magnetic field) strongly increases the Bondi radius $R_B \propto v_w^{-2}$ and the corresponding Bondi mass accretion rate $\dot{M}_B \propto v_w^{-3}$.

Indeed, we may write down the mass accretion rate onto the NS in the unflaring (low-luminosity) state as $\dot{M}_{a,low} = f(u)\dot{M}_B$ with $f(u)$ given by expression (7.163) and $\dot{M}_B \simeq \pi R_B^2 \rho_w v_w$. Eliminating the wind density ρ_w using the mass continuity equation, written for the spherically symmetric stellar wind from the optical star with power $\dot{M}_o$, and assuming a circular binary orbit, we arrive at $\dot{M}_B \simeq \frac{1}{4}\dot{M}_o\left(\frac{R_B}{a}\right)^2$. Using the well-known relation for the radiative wind mass-loss rate from massive hot stars $\dot{M}_o \simeq \epsilon \frac{L}{cv_\infty}$ where L is the optical star luminosity, v_∞ is the stellar wind velocity at infinity, typically 2000–$3000\,\mathrm{km\,s^{-1}}$ for OB stars and $\epsilon \simeq 0.4-1$ is the efficiency factor (Lamers et al. 1976); in the numerical estimates below we shall assume $\epsilon = 0.5$. It is also possible to reduce the luminosity L of a massive star to its mass M using the phenomenological relation $(L/L_\odot) \approx 19(M/M_\odot)^{2.76}$ (see e.g. Vitrichenko et al. 2007). Combining the above equations and using Kepler's third law to express the orbital separation a through the binary period P_b, we find for the X-ray luminosity of SFXTs in the non-flaring state

$$L_{x,low} \simeq 5\times 10^{35}[\mathrm{erg\,s^{-1}}] f(u) \left(\frac{M}{10M_\odot}\right)^{2.76-2/3} \left(\frac{v_\infty}{1000\,\mathrm{km\,s^{-1}}}\right)^{-1} \left(\frac{v_w}{500\,\mathrm{km\,s^{-1}}}\right)^{-4} \left(\frac{P_b}{10\mathrm{d}}\right)^{-4/3}, \quad (7.166)$$

which for $f(u) \sim 0.03$–0.1 corresponds to the typical low-state luminosities of SFXTs of $\sim 10^{34}\,\mathrm{erg\,s^{-1}}$.

It is straightforward to see that a transition from the low state (subsonic accretion with slow magnetospheric entry rate $f(u) \sim 0.03$–0.1) to supersonic free-fall Bondi accretion with $f(u) = 1$, due to a velocity decrease by a factor of two in the magnetized stellar wind would, for example, lead to a flaring luminosity of $L_{x,flare} \sim (10 \div 30) \times 2^5 L_{x,low}$. This shows that the dynamical range of SFXT bright flares ($\sim$300–1000) can be naturally reproduced by the proposed mechanism.

7.10.3 Conditions for Magnetic Reconnection Near the Magnetosphere

For magnetic field reconnection to occur, the time the magnetized plasma spends near the magnetopause should be at least comparable to the reconnection time, $t_r \sim R_A/v_r$, where v_r is the magnetic reconnection rate, which is difficult to assess from first principles (Zweibel and Yamada 2009). In real astrophysical plasmas the large-scale magnetic reconnection rate can be as high as $v_r \sim 0.03$–$0.07 v_A$ (Zweibel and Yamada 2009), and phenomenologically we can parametrize it as $v_r = \epsilon_r v_A$ with $\epsilon_r \sim 0.01$–0.1. The longest time-scale the plasma penetrating into the magnetosphere spends near the magnetopause is the instability time, $t_{inst} \sim t_{ff}(R_A) f(u)_{rad}$ (Shakura et al. 2012), so reconnection may occur if $t_r/t_{inst} \sim (u_{ff}/v_A)(f(u)_{rad}/\epsilon_r) \lesssim 1$. As close to R_A (from its definition) $v_A \sim u_{ff}$, we arrive at $f(u)_{rad} \lesssim \epsilon_r$ as a necessary condition for reconnection. According to (7.163), this condition is satisfied only at sufficiently low X-ray luminosities, pertinent to 'quiet' SFXT states. This explains why in HMXBs with convective shells at higher luminosity (but still lower than $4 \times 10^{36}\,\mathrm{erg\,s^{-1}}$, at which settling accretion is possible), reconnection from magnetized plasma accretion will not lead to shell instability, but only to a temporal establishment of the 'strong coupling regime' of angular momentum transfer through the shell, as discussed in Shakura et al. (2012) and below in Sect. 7.11. Episodic strong spin-ups, as observed in GX 301-2, may be manifestations of such 'failed' reconnection-induced shell instability.

Therefore, it seems likely that the key difference between steady HMXBs like Vela X-1, GX 301-2 (showing only moderate flaring activity) and SFXTs is that in the first case the effects of possibly magnetized stellar winds from optical OB-companions are insignificant (basically due to the rather high mean accretion rate), while in SFXTs with lower 'steady' X-ray luminosity, large-scale magnetic fields, sporadically carried by clumps in the wind, can trigger SFXT flaring activity via magnetic reconnection near the magnetospheric boundary. The observed power-law SFXT flare distributions, discussed in Paizis and Sidoli (2014), with respect to the log-normal distributions for classical HMXBs (Fürst et al. 2010), may be related to the properties of magnetized stellar wind and physics of its interaction with the NS magnetosphere (Shakura et al. 2014a; Sidoli et al. 2016b).

7.11 Angular Momentum Transfer to the Neutron Star Magnetosphere and Spin-Up/Spin-Down of X-ray Pulsars

Consider a quasi-static shell around a neutron star magnetosphere in which the subsonic settling regime is established. We stress that in this regime, the accretion rate onto the neutron star is determined by the gas density at the shell base, which is directly related to the gas density downstream the bow shock in the stellar wind in the gravitational capture region, and by the ability of the plasma to enter the magnetosphere through the Alfvén surface.

The rotation law in the shell depends on the treatment of the turbulent viscosity (see Sect. 7.5 if Prandtl's law for isotropic turbulence is used), and also on any possible turbulence anisotropy due to convection (Sect. 7.6). In the latter case, the anisotropy gives rise to more powerful turbulence in the radial direction compared to the tangential motions. Thus, as we have shown in Sects. 7.4 and 7.5, a series of solutions appears that describe the radial dependence of the angular velocity of matter in a convective shell. Below we will use pure power-law rotation:

$$\omega(R) \sim R^{-n} . \tag{7.167}$$

The quasi-Keplerian case corresponds to $n = 3/2$, and the isomomentum distribution to $n = 2$, which in some sense describe limiting cases of the possible solutions.

When approaching the bow shock, $R \to R_B$, the angular velocity of gas approaches the orbital angular velocity of the binary system (for simplicity, we consider circular orbits): $\omega \to \omega_B$. Near the bow shock the problem is no longer spherically symmetric, the characteristics of the flow may be very complicated (parts of the matter may, for example, bend the hot shell), and the solution should be sought for numerically. As there are no such solutions at present, we will assume that the assumption of power-law rotation in the shell is valid up to the bow shock located at the Bondi radius R_B:

$$R_B \simeq 2GM/(V_w^2 + v_{orb}^2) ,$$

where V_w is the stellar wind velocity at the location of the neutron star orbit, and v_{orb} is the neutron star orbital velocity.

This means that the angular velocity of the rotation of the shell at the magnetospheric boundary ω_m is related to the orbital angular velocity ω_B as

$$\omega_m = \tilde{\omega}\omega_B \left(\frac{R_B}{R_A}\right)^n . \quad (7.168)$$

(Here the numerical coefficient $\tilde{\omega} > 1$ takes into account the deviation of the rotation law in the shell near the magnetosphere from a pure power-law dependence, see above in Sects. 7.4 and 7.5.)

Let the neutron star magnetosphere rotate with the angular velocity $\omega^* = 2\pi/P^*$, where P^* is the neutron star spin period. The matter at the base of the shell rotates with the angular velocity ω_m, which, in general, differs from ω^*. If $\omega^* > \omega_m$, the plasma-magnetosphere interaction provides angular momentum transfer from the magnetosphere to the shell, and in the opposite case $\omega^* < \omega_m$—from the shell to the magnetosphere.

In the general case, the coupling of the matter with the magnetosphere can be moderate or strong. In the strong coupling regime, the toroidal magnetic field component B_t is proportional to the poloidal component B_p, and we can write $B_t \sim -B_p(\omega_m - \omega^*)t$, so that $|B_t|$ can grow up to $\sim |B_p|$. This regime may be realized around a rapidly rotating magnetosphere, when the NS angular velocity ω^* is comparable to or even exceeds the Keplerian angular frequency $\omega_K(R_A)$. In the latter case, the so-called propeller regime sets in. In the moderate coupling regime, plasma may enter the magnetosphere via instabilities at a rate faster than required for the magnetic field toroidal component to grow to the value of the poloidal component, therefore $B_t < B_p$.

7.11.1 The Case of Strong Coupling

Let us first consider strong coupling. In this case, powerful large-scale motions of gas in the shell may lead to turbulent diffusion of the magnetic field and its dissipation. This process is characterized by the turbulent diffusion coefficient of the magnetic field η_t. Then the toroidal magnetic field (see, for example, Lovelace et al. (1995) and references therein) is

$$B_t = \frac{R^2}{\eta_t}(\omega_m - \omega^*)B_p . \quad (7.169)$$

The turbulent magnetic diffusion coefficient is related to the kinematic viscosity coefficient: $\eta_t \simeq \nu_t$. The latter can be written in the form

$$\nu_t = \langle u_t l_t \rangle . \quad (7.170)$$

According to the phenomenological Prandtl's law connecting the mean characteristics of a turbulent flow (the velocity u_t, the characteristic spatial scale l_t and shear $\omega_m - \omega^*$) we have:

$$u_t \simeq l_t |\omega_m - \omega^*| \,. \tag{7.171}$$

In our case, the turbulence scale should be determined by the maximum scale of the energy pumping into turbulent motions from the rotating non-spherical magnetosphere surface. This scale is determined by the velocity difference between the rigidly rotating magnetosphere and the accreting matter which still does not interact with the magnetosphere, i.e. $l_t \simeq R_A$. This scale determines the turn velocity of the largest turbulent eddies, and on smaller scales a turbulence cascade develops. Plugging this scale into Eqs. (7.169)–(7.171) we find that in the strong coupling regime $B_t \simeq B_p$.

The torque arising from the plasma-magnetosphere interaction acts on the neutron star and changes its angular momentum according to the equation

$$I\dot{\omega}^* = \int \frac{B_t B_p}{4\pi} \varpi \, dS = \pm \tilde{K}(\theta) K_2 \frac{\mu^2}{R_A^3} \tag{7.172}$$

where I is the moment of inertia of the neutron star, ϖ is the distance to the rotational axis, and $\tilde{K}(\theta)$ is a numerical coefficient depending on the angle between the rotational axis and the magnetic dipole axis. The coefficient K_2 appears in (7.172) for the same reason as in Eq. (7.13). The positive sign (spin-up) corresponds to angular momentum transfer to the neutron star ($\omega_m > \omega^*$). The negative sign (spin-down) corresponds to angular momentum removal from the neutron star to the surrounding shell ($\omega_m < \omega^*$).

At the Alfvén radius, the matter enters the magnetosphere and acquires the angular velocity of the neutron star rotation. Then it falls freely onto the neutron star and supplies it with the angular momentum it acquired at the Alfvén radius R_A from the magnetospheric interaction. As a result, the neutron star spins up at a rate

$$I\dot{\omega}^* = +z\dot{M} R_A^2 \omega^* \tag{7.173}$$

where z is a numerical coefficient taking into account the specific angular momentum of the infalling matter. If the matter falls from the magnetospheric equator, $z = 1$; if the matter falls strictly along the neutron star spin axis, $z = 0$. If all the matter would enter evenly across a spherical magnetosphere, then we would have $z = 2/3$.

Finally, we find that the total torque applied to the neutron star in the strong coupling regime changes the neutron star spin at a rate

$$I\dot{\omega}^* = \pm \tilde{K}(\theta) K_2 \frac{\mu^2}{R_A^3} + z\dot{M} R_A^2 \omega^* \,. \tag{7.174}$$

Using (7.16), we may exclude $\dot{M}$ from this equation to obtain in the spin-up regime ($\omega_m > \omega^*$)

$$I\dot{\omega}^* = \frac{\tilde{K}(\theta)K_2\mu^2}{R_A^3}\left[1 + z\frac{4\gamma f(u)}{\sqrt{2}(\gamma-1)(1+\gamma m_t^2)\psi(\gamma, m_t)\tilde{K}(\theta)}\left(\frac{R_A}{R_c}\right)^{3/2}\right] \tag{7.175}$$

where $R_c^3 = GM/(\omega^*)^2$ is the corotation radius. In the spin-down regime ($\omega_m < \omega^*$) we find

$$I\dot{\omega}^* = -\frac{\tilde{K}(\theta)K_2\mu^2}{R_A^3}\left[1 - z\frac{4\gamma f(u)}{\sqrt{2}(\gamma-1)(1+\gamma m_t^2)\psi(\gamma, m_t)\tilde{K}(\theta)}\left(\frac{R_A}{R_c}\right)^{3/2}\right]. \tag{7.176}$$

Note that in both cases R_A should be smaller than R_c; if not, the propeller regime would set in, and accretion would stop. In the propeller regime $R_A > R_c$, the matter does not fall onto the neutron star surface, and there is no powerful generation of X-ray emission. In this case, the shell downstream the bow shock can cool down rapidly (see below) likely giving rise to the standard Illarionov-Sunyaev propeller regime (Illarionov and Sunyaev 1975), which is accompanied by the outflow of matter from the magnetosphere.

In both regimes (spin-up and spin-down), the neutron star angular velocity ω^* approaches the angular velocity of matter at the magnetospheric boundary, $\omega^* \to \omega_m(R_A)$. The difference between ω^* and ω_m is small, and therefore the second term in the square brackets in Eqs. (7.175) and (7.176) is much smaller than one. Also note that by approaching the propeller regime ($R_A \to R_c$) the accretion rate decreases, $f(u) \to 0$, the second term in the square brackets vanishes, and the evolution of the neutron star spin is determined solely by the braking torque $-\tilde{K}(\theta)\mu^2/R_A^3$. (In the propeller regime $\omega_m < \omega_K(R_A)$, $\omega_m < \omega^*$, $\omega^* > \omega_K(R_A)$). Therefore, the neutron star spins down until reaching the Keplerian frequency at the Alfvén radius. In this regime, the specific angular momentum of matter moving towards or outwards from the magnetosphere is, of course, conserved.

Near the equilibrium ($\omega^* \sim \omega_m$), relatively small fluctuations of the mass accretion rate $\dot{M}$ in the shell give rise to a very strong fluctuations in the pulsar frequency $\dot{\omega}^*$ since the toroidal component of the magnetic field can change the sign from $+B_p$ to $-B_p$. If the strong coupling regime can indeed happen in nature, this could be its distinctive feature. It is known (see, for example, Bildsten et al. (1997) and on-line Fermi/GBM data[4] that real X-ray pulsars sometimes display rapid transitions from spin-up to spin-down without a corresponding change in X-ray luminosity. It is not excluded that strong coupling may switch-on due to the magnetic field frozen in the plasma that has not yet entered the magnetosphere.

[4]http://gammaray.nsstc.nasa.gov/gbm/science/pulsars/lightcurves/.

Some thoughts regarding accretion of a magnetized plasma onto a rotating neutron star magnetosphere can be found in Ikhsanov and Beskrovnaya (2012).

7.11.2 The Case of Moderate Coupling

The strong coupling regime considered above can be realized in the limiting case where the toroidal magnetic field B_t reaches a maximum possible value $\sim B_p$ due to magnetic turbulent diffusion. Usually, the plasma coupling with the magnetosphere is mediated by various instabilities, whose characteristic growth time is insufficient for the toroidal time to increase significantly. As discussed above in Sect. 7.2.3, the shell at the magnetosphere is very hot, so without cooling the plasma turns out to be marginally stable with respect to the Rayleigh-Taylor instability (see, e.g., the model calculations in Arons and Lea (1976)).

The torque due to magnetic forces applied to the neutron star reads:

$$I\dot{\omega}^* = \int \frac{B_t B_p}{4\pi} \varpi dS \tag{7.177}$$

where B_t is the toroidal magnetic field component which arises if there is a difference between the angular velocity of matter ω_m and the magnetosphere angular velocity ω^*. On the other hand, there is a mechanical torque acting on the magnetosphere from the base of the shell caused by the turbulent stresses $W_{R\phi}$:

$$\int W_{R\phi} \varpi dS \,, \tag{7.178}$$

where the viscous turbulent stresses can be written as

$$W_{R\phi} = \rho \nu_t R \frac{\partial \omega}{\partial R} \,. \tag{7.179}$$

To specify the turbulent viscosity coefficient

$$\nu_t = \langle u_c l_t \rangle \,, \tag{7.180}$$

we assume that the characteristic scale of turbulence close to the magnetosphere is

$$l_t = \zeta_d R_A \,, \tag{7.181}$$

where we have introduced the dimensionless factor $\zeta_d \lesssim 1$, characterizing the size of the zone in which there is an effective exchange of angular momentum between the magnetosphere and the shell base. The characteristic velocity of the turbulent pulsations u_c is determined by the mechanism of turbulence in the plasma above the magnetosphere. In the case of strong convective motions in the shell, caused by heating of its base, $u_c \sim c_s$, where c_s is the sound speed.

Equating the torques (7.177) and (7.178) and allowing for (7.179) and (7.181), we get

$$\rho u_c \zeta_d R_A^2 \frac{\partial \omega}{\partial R} = \frac{B_t B_p}{4\pi} \tag{7.182}$$

We now eliminate the density from this expression using the pressure balance at the magnetospheric boundary and the expression for the temperature (7.9), and make the substitution

$$\frac{\partial \omega}{\partial R} = \frac{\omega_m - \omega^*}{\zeta_d R_A}. \tag{7.183}$$

Then we find the relation between the toroidal and poloidal components of the magnetic field in the magnetosphere:

$$\frac{B_t}{B_p} = K_2 \frac{\gamma}{\sqrt{2}(\gamma - 1)} \left(\frac{u_c}{u_{ff}}\right) \left(\frac{\omega_m - \omega^*}{\omega_K(R_A)}\right). \tag{7.184}$$

(Note that there is no dependence on the width of the layer characterized by the parameter ζ_d). Substituting (7.184) into (7.177), the spin-down rate of the neutron star may be written as:

$$I\dot{\omega}^* = K_1 K_2 \left(\frac{u_c}{u_{ff}}\right) \frac{\mu^2}{R_A^3} \frac{\omega_m - \omega^*}{\omega_K(R_A)}. \tag{7.185}$$

where $K_1 \sim 1$ is a constant arising from integration of the torques over the surface of the magnetosphere.

Using the definition of the Alfvén radius R_A (7.16) and the expression for the Keplerian frequency ω_K, we can write (7.185) in the form

$$I\dot{\omega}^* = Z\dot{M}R_A^2(\omega_m - \omega^*). \tag{7.186}$$

Here the dimensionless coefficient Z is

$$Z = K_1 \left(\frac{u_c}{u_{ff}}\right) \frac{1}{f(u)}. \tag{7.187}$$

Taking into account that the matter falling onto the neutron star brings the angular momentum $z\dot{M}R_A^2\omega^*$, we ultimately get

$$I\dot{\omega}^* = Z\dot{M}R_A^2(\omega_m - \omega^*) + z\dot{M}R_A^2\omega^*. \tag{7.188}$$

Here $0 < z < 1$ is a numerical coefficient which is $\sim 2/3$ if the matter enters across the magnetospheric surface with equal probability at different magnetospheric latitudes. Substituting $\omega_m(R_A) = \omega_B(R_B/R_A)^2$ for an iso-angular-momentum shell, we can rewrite the above equation in the form

$$I\dot{\omega}^* = Z\dot{M}\omega_B R_B^2 - Z(1 - z/Z)\dot{M}R_A^2\omega^* . \quad (7.189)$$

Substituting for the coupling coefficient Z, in the case of Compton cooling we can rewrite (7.188) in a form explicitly showing the spin-up (K_{su}) and spin-down (K_{sd}) torques:

$$\dot{\omega}^* = A\dot{M}^{\frac{7}{11}} - B\dot{M}^{3/11} = K_{su} - K_{sd} . \quad (7.190)$$

Here the spin-up/spin-down coefficients A and B do not explicitly depend on $\dot{M}$.

For a characteristic value of the accretion rate $\dot{M}_{16} \equiv \dot{M}/10^{16}$ g/s, the spin-up and spin-down torques read (in CGS units):

$$K_{su} \approx 5.29 \times 10^{-13}[\frac{\text{rad}}{\text{s}}^2]K_1 \left(\frac{u_c}{u_{ff}}\right) \zeta^{-\frac{7}{11}} \mu_{30}^{\frac{1}{11}} \left(\frac{v_8}{\sqrt{\delta}}\right)^{-4} \left(\frac{P_b}{10\text{d}}\right)^{-1} \dot{M}_{16}^{7/11} I_{45}^{-1} \quad (7.191)$$

$$K_{sd} \approx 5.36 \times 10^{-12}[\frac{\text{rad}}{\text{s}}^2](1 - \frac{z}{Z})K_1 \left(\frac{u_c}{u_{ff}}\right) \zeta^{-3/11} \mu_{30}^{13/11} \left(\frac{P^*}{100\text{s}}\right)^{-1} \dot{M}_{16}^{3/11} I_{45}^{-1} . \quad (7.192)$$

Here $I_{45} = I/10^{45}$ g cm^2 is the NS moment of inertia, and the dimensionless factor $\delta \lesssim 1$ takes into account the actual location of the gravitational capture radius.

Another approach to the problem of interaction of a quasi-spherically accreting magnetized plasma with rotating NS magnetospheres is presented in Ikhsanov et al. (2014).

7.12 Equilibrium Pulsars

For equilibrium pulsars we set $\dot{\omega}^* = 0$ and from (7.188) we get

$$Z_{eq}(\omega_m - \omega^*) + z\omega^* = 0 . \quad (7.193)$$

Close to equilibrium we may vary (7.188) with respect to $\dot{M}$. Variations in $\delta\dot{M}$ may in general be caused by changes in the density $\delta\rho$ as well as in the velocity of the stellar wind δv (and thus the Bondi radius). For density variations only we find (see Eq. (67) in Shakura et al. (2013b) for more detail)

$$Z_{eq,\rho} = \frac{I \frac{\partial \dot{\omega}^*}{\partial \dot{M}}|_{eq}}{\frac{4}{11}\omega^* R_A^2} \approx 2.52 \left(\frac{\frac{\partial \dot{\omega}^*}{\partial y}|_{y=1}}{10^{-12}}\right)\left(\frac{P^*}{100s}\right) \zeta^{-4/11} \dot{M}_{16}^{-7/11} \mu_{30}^{-12/11}. \tag{7.194}$$

On the other hand, by equating this value to the definition of the coupling coefficient Z (see (7.187) above), we can find the dimensionless combination of the theory parameters:

$$\Pi_0 \equiv \frac{K_1 \left(\frac{u_c}{u_{ff}}\right)}{\zeta^{3/11}} \approx 0.55 \left(\frac{\frac{\partial \dot{\omega}^*}{\partial y}|_{y=1}}{10^{-12}}\right)\left(\frac{P^*}{100s}\right) \dot{M}_{16}^{-3/11} \mu_{30}^{-13/11}. \tag{7.195}$$

The equilibrium period of an X-ray pulsar with known NS magnetic field can be found from (7.189) (or, which is the same, by equating the spin-up and spin-down torques from (7.191) and (7.192)):

$$P_{eq} \approx 1000[\mathrm{s}](1 - z/Z_{eq})\zeta^{4/11} \mu_{30,eq}^{12/11} \left(\frac{P_b}{10\mathrm{d}}\right) \dot{M}_{16}^{-4/11} \left(\frac{v_8}{\sqrt{\delta}}\right)^4. \tag{7.196}$$

In equilibrium, from this formula we may determine another dimensionless combination of the theory parameters:

$$\Pi_1 \equiv \frac{\left(1 - \frac{z}{Z_{eq}}\right)\zeta^{4/11}}{\delta^2} \approx 0.1 \left(\frac{P^*}{100s}\right)\left(\frac{P_b}{10\mathrm{d}}\right)^{-1} \dot{M}_{16}^{4/11} \mu_{30}^{-12/11} v_8^{-4}. \tag{7.197}$$

Because of the strong dependence of the equilibrium period on the (usually, poorly measurable) wind velocity, for pulsars with independently known magnetic fields μ, it is more convenient to estimate the wind velocity, assuming $P^* = P^*_{eq}$:

$$v_8 \approx 0.56 \left[\frac{\left(1 - \frac{z}{Z_{eq}}\right)\zeta^{4/11}}{\delta^2}\right]^{-1/4} \dot{M}_{16}^{1/11} \mu_{30,eq}^{-3/11} \left(\frac{P_*/100\mathrm{s}}{P_b/10\mathrm{d}}\right)^{1/4}, \tag{7.198}$$

which is only weakly dependent on $\dot{M}$ and the theory parameter Π_1.

In the possible case of mass accretion rate variations due to wind velocity changes only, the coupling coefficient $Z_{eq,v}$ reads (see Eq. (68) in Shakura et al. (2013b)):

$$Z_{eq,v} \approx 0.76 \left(\frac{\frac{\partial \dot{\omega}^*}{\partial y}|_{y=1}}{10^{-12}}\right)\left(\frac{P^*}{100s}\right) \zeta^{-4/11} \dot{M}_{16}^{-7/11} \mu_{30}^{-12/11} + \frac{7}{10}z. \tag{7.199}$$

Table 7.1 Parameters for the equilibrium X-ray pulsars

	Equilibrium pulsars	
Pulsar	GX 301-2	Vela X-1
Measured parameters		
P^*(s)	680	283
P_B(d)	41.5	8.96
v_w(km/s)	300?	700
μ_{30}	2.7	1.2
$\dot{M}_{16}$	3	3
$\frac{\partial \dot{\omega}}{\partial y}\|_{y=1}$(rad/s^2)	1.5×10^{-12}	1.2×10^{-12}
Derived parameters		
$f(u)\zeta^{-7/11}$	0.32	0.30
$Z_{eq}\zeta^{4/11}$	4.32	3.49
Π_0	1.28	1.11
$v_8 \Pi_1^{1/4}$(km/s)	530	800

Clearly, in this case the coupling is weaker. Below we will consider only the wind density variations. In principle, if $z > 0$ and $(\omega_m - \omega^*) > 0$, (7.193) implies that there can be no equilibrium at all – the pulsar can only spin-up. However, two well-measured equilibrium pulsars (see below) show that an equilibrium does exist, suggesting that in these objects $(\omega_m - \omega^*) < 0$.

To illustrate the theory outlined above, we show the measured and obtained model parameters of two well-known persistent X-ray pulsars, Vela X-1 and GX 301-2 (see Table 7.1).

It is clear from Table 7.1 that for Vela X-1 the observed and derived parameters are in good agreement, with the value of the dimensionless theory parameter $\Pi_0 \sim 1$, as expected from very general hydrodynamic similarity principles (Sedov 1959). It is remarkable that the parameter $\Pi_0 \sim 1$ in GX 301-2 as well, suggesting a common physics of hydrodynamic interactions in these objects. However, the observed wind velocity in GX 301-2 is inferred from observations to be around 300 km/s, which is almost twice as low as that derived using our theory. To obtain such a low velocity from (7.198), the dimensionless parameter Π_1 should be around 10, which is unrealistically high (in fact, this parameter should not be higher than (1). From this we conclude that in GX 301-2 the observed wind velocity is likely estimated far from the region interacting with the NS.

7.13 Non-equilibrium Pulsars

It is convenient to introduce the dimensionless parameter

$$y \equiv \frac{\dot{M}}{\dot{M}_{eq}} \tag{7.200}$$

where $\dot{M}_{eq}$ represents the accretion rate at which $\dot{\omega}^* = 0$:

$$\dot{M}_{eq} = \left(\frac{B}{A}\right)^{11/4} . \tag{7.201}$$

Equation (7.190) can be rewritten in the form

$$I\dot{\omega}^* = A\dot{M}_{eq}^{\frac{7}{11}} y^{\frac{7}{11}} \left(1 - y^{-\frac{4}{11}}\right) . \tag{7.202}$$

A plot of the function $\dot{\omega}^*(y)$ is shown schematically in Fig. 7.6. The function $\dot{\omega}^*(\dot{M})$ reaches minimum at $\dot{M} = \dot{M}_{cr}$:

$$\dot{M}_{cr} = \dot{M}_{eq} \left(\frac{3}{7}\right)^{\frac{11}{4}} , \tag{7.203}$$

In other words, $\dot{\omega}^*$ attains a minimum for the dimensionless parameter

$$y_{cr} = \left(\frac{3}{7}\right)^{\frac{11}{4}} < 1. \tag{7.204}$$

The minimum $\dot{\omega}^*$ for $y = y_{cr}$ (i.e. the maximum possible spin-down rate of the pulsar) is

$$I\dot{\omega}^*_{min} = -\frac{4}{3} A\dot{M}_{eq}^{\frac{7}{11}} y^{\frac{7}{11}} . \tag{7.205}$$

Numerically, the maximum spin-down rate at y_{cr} is

$$\dot{\omega}^*_{sd,min} \approx -1.12 \times 10^{-12}[\text{rad/s}^2](1 - z/Z)^{7/4} K_1 \left(\frac{u_c}{u_{ff}}\right) \mu_{30}^2 \left(\frac{v_8}{\sqrt{\delta}}\right)^3 \left(\frac{P^*}{100\,\text{s}}\right)^{-7/4} \left(\frac{P_b}{10\text{d}}\right)^{3/4} . \tag{7.206}$$

Then, from the condition $|\dot{\omega}^*_{sd}| \le |\dot{\omega}^*_{sd,min}|$ follows a lower limit on the neutron star magnetic field:

$$\mu_{30} > \mu'_{30,min} \approx 0.94 \left|\frac{\dot{\omega}^*_{sd}}{10^{-12}\text{rad/s}^2}\right|^{1/2} (1 - z/Z)^{-7/8} \left[K_1 \left(\frac{u_c}{u_{ff}}\right)\right]^{-1/2} \left(\frac{v_8}{\sqrt{\delta}}\right)^{-3/2} \left(\frac{P^*}{100\,\text{s}}\right)^{7/8} \left(\frac{P_b}{10\text{d}}\right)^{-3/8} . \tag{7.207}$$

Table 7.2 Parameters of non-equilibrium X-ray pulsars

	GX 1+4	SXP1062	4U 2206+54
Measured parameters			
P^*(s)	140	1062	5560
P_B(d)	1161	∼300[a]	19(?)
v_w(km/s)	200	∼300[b]	350
μ_{30}	?	?	?
$\dot{M}_{16}$	1	0.6	0.2
$\dot{\omega}^*_{sd}$	-2.34×10^{-11}	-1.63×10^{-11}	-1.1×10^{-13}
Derived parameters			
$\mu''_{30,min}$	≈2.4	≈10	≈0.6

[a] Estimate of the source's position in the Corbet diagram
[b] Estimate of typical wind velocity for Be X-ray binaries

At very low accretion rates $y \ll 1$ the spin-up torque K_{su} can be neglected, and the spin-down rate of the pulsar is

$$\dot{\omega}^*_{sd} \approx -0.54 \times 10^{-12}[\mathrm{rad/s^2}](1 - \tfrac{z}{Z})K_1\left(\tfrac{u_c}{u_{ff}}\right)\zeta^{-3/11} \mu_{30}^{13/11}\dot{M}_{16}^{3/11}\left(\tfrac{P^*}{100\,\mathrm{s}}\right)^{-1} . \tag{7.208}$$

From this we obtain a lower limit on the neutron star magnetic field that does not depend on the stellar wind velocity or the binary orbital period:

$$\mu_{30} > \mu''_{30,min} \approx 1.68 \left|\frac{\dot{\omega}^*_{sd}}{10^{-12}\mathrm{rad/s^2}}\right|^{11/13} (1 - \tfrac{z}{Z})^{-11/13}\left[K_1\left(\tfrac{u_c}{u_{ff}}\right)\right]^{-11/13}\zeta^{3/13} \dot{M}_{16}^{-3/13}\left(\tfrac{P^*}{100\,\mathrm{s}}\right)^{11/13} . \tag{7.209}$$

As an example, consider the steady spin-down behavior in several slowly rotating moderate-luminosity X-ray pulsars (GX 1+4, SXP 1062, 4U 2206+54) within the framework of the quasi-spherical settling accretion theory. The results are summarized in Table 7.2.

7.14 On the Possibility of the Propeller Regime

The very slow rotation of the neutron stars in X-ray pulsars GX 1+4, GX 301-2 and Vela X-1 which we consider here as examples, $\omega^*(R_A) < \omega_K(R_A)$, means that in these objects the propeller regime, in which the gas is ejected from the rotating magnetosphere with a parabolic velocity and the neutron star spins down, can hardly occur.

We start with estimating an important ratio of the viscous stresses ($\sim B_t B_p$) to the gas pressure ($\sim B_p^2$) at the magnetospheric boundary. This ratio is proportional to the ratio of the magnetic field components, B_t/B_p (see Eq. (7.184)), and is always less than one (in the moderate coupling regime). This implies that only large-scale convective motions with turbulent radial scaling of eddies can be present in the shell. When $\omega^* > \omega_K(R_A)$, a centrifugal barrier appears and accretion ceases. In this case, the maximum possible braking torque applied to the neutron star will be of the order of $\sim -K_2\mu^2/R_A^3$ because of the strong plasma-magnetosphere coupling. In this regime, the toroidal magnetic field component, B_t, is comparable to the poloidal, B_p. It can not be excluded that the hot shell with isomomentum angular rotation will be conserved also in this case, and that the angular momentum removal from the rotating magnetosphere will be mediated by this shell. If the characteristic cooling time of the shell plasma is shorter than the gas free-fall time, the shell disappears and, probably, a thin storage disc as considered in Syunyaev and Shakura (1977) will be formed. No accretion occurs through such a disc, it only mediates the angular momentum removal from the rotating magnetosphere.

7.15 Do Slow X-ray Pulsars Have Prograde or Retrograde Accretion Discs?

The analysis of real slow X-ray pulsars presented above suggests that they have convective magnetospheric shells with isomomentum angular rotation. Therefore, we will consider only the case with $\omega \sim R^{-2}$. Equation (7.189) implies that for $\dot{\omega}^* = 0$ the equilibrium spin frequency of the neutron star is

$$\omega_{eq}^* = \omega_B \frac{1}{1 - z/Z} \left(\frac{R_B}{R_A}\right)^2 . \tag{7.210}$$

We stress that such an equilibrium in our model is possible only if a settling accretion shell is present above the magnetosphere. At high accretion rates $\dot{M} > \dot{M}_* \simeq 4 \times 10^{16}\,\mathrm{g\,s^{-1}}$ accretion proceeds in the free-fall regime, and no hot shell is formed above the magnetosphere.

The equilibrium period of an X-ray pulsar in the quasi-spherical settling accretion regime is determined by formula (7.196):

$$P_{eq} \simeq 1000[\mathrm{s}]\mu_{30}^{12/11}(P_b/10\mathrm{d})\dot{M}_{16}^{-4/11} v_8^4 . \tag{7.211}$$

For comparison, in the case of standard disc accretion the equilibrium period is:

$$P_{eq,d} \approx 10[\mathrm{s}]\mu_{30}^{6/7}\dot{M}_{16}^{-3/7}\,. \tag{7.212}$$

Therefore, the long spin periods observed in some X-ray pulsars may be explained in the presence of an accretion disc only by assuming a very strong (magnetar-like) magnetic field of the neutron star. Another explanation based on retrograde accretion discs (i.e. those with angular momentum opposite to the orbital) around magnetospheres of such X-ray pulsars is also discussed in the literature (see, for example, Nelson et al. (1997) and references therein). A conversion of the torques due to the temporal formation of a retrograde accretion disc from the stellar wind can, in principle, lead to very long spin periods even in X-ray pulsars with standard magnetic fields. Such discs may be formed due to inhomogeneities in the captured stellar wind (Ruffert 1997, 1999). The observed torque reversal in some X-ray pulsars could be explained, in principle, by this mechanism. In the case of GX 1+4 with long-term stable spin-down it is highly unlikely to observe a stable retrograde disc on timescales much longer than the binary orbital period (see González-Galán et al. (2012) for more detail). For GX 301-2 and Vela X-1, the direct proportionality of the torques to the X-ray luminosity (see Figs. 7.4 and 7.5) also does not support the presence of a retrograde accretion disc (Fig. 7.6).

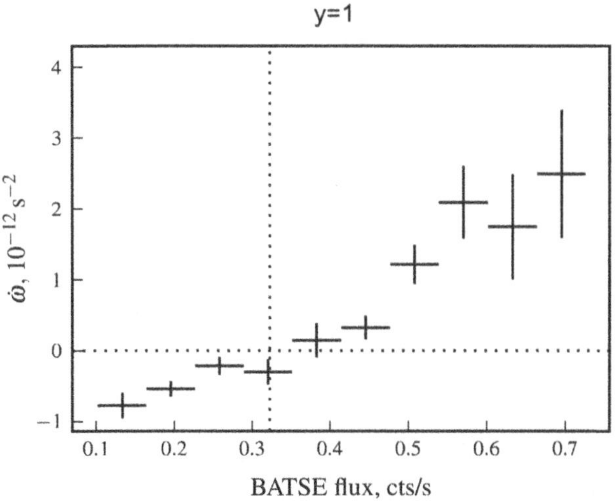

Fig. 7.4 Torque-luminosity correlation in GX 301-2, $\dot{\omega}^*$ as a function of BATSE data (20–40 keV pulsed flux) near the equilibrium frequency (Doroshenko et al. 2010). The assumed X-ray flux at equilibrium (in terms of the dimensionless parameter y) is shown by the vertical dotted line

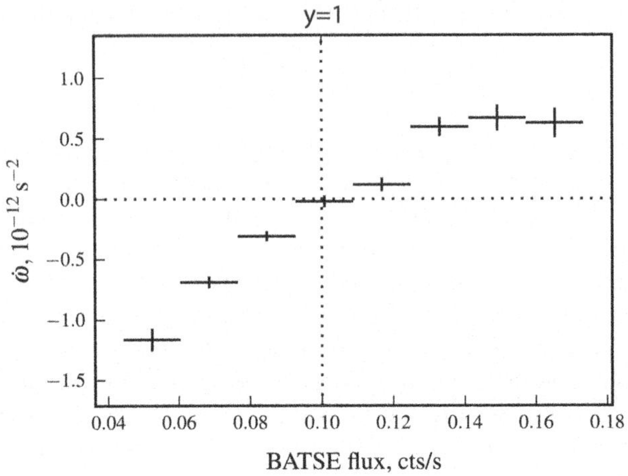

Fig. 7.5 The same as in Fig. 7.4 for Vela X-1 (V. Doroshenko, PhD Thesis, 2010, IAAT)

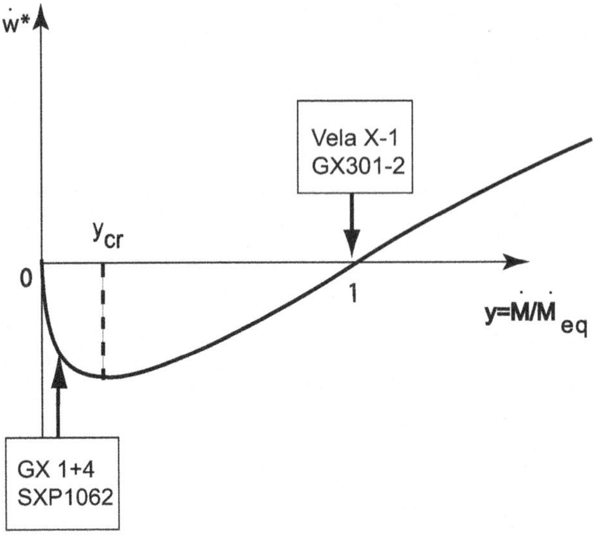

Fig. 7.6 Schematics of the dependence of $\dot{\omega}^*$ on the dimensionless accretion rate y. The figure shows the position in the diagram for equilibrium pulsars with $y \sim 1$ and for non-equilibrium pulsars at steady spin-down with $y < y_{cr}$

7.16 Conclusions

In Shakura et al. (2012) we constructed a theoretical model of quasi-spherical subsonic accretion onto slowly rotating magnetized neutron stars. In this model, the accreting matter is gravitationally captured from the stellar wind of the optical component and settles subsonically onto a rotating magnetosphere through a hot extended quasi-static gas shell. In the shell, large-scale convective motions occur, mediating the angular momentum transfer, and depending on the difference of the velocity of the gas and the magnetosphere at the magnetospheric boundary, spin-up or spin-down of the X-ray pulsar can be observed.

A detailed analysis and comparison with observations of the two slowly rotating X-ray pulsars GX 301-2 and Vela X-1, which both demonstrate correlative changes of the spin-up/spin-down torques with X-ray luminosity near the equilibrium period of the neutron star, likely suggests strongly anisotropic convection in their magnetospheric shells leading to rotation with a constant specific angular momentum, $\omega \sim R^{-2}$. A statistical analysis of long-period X-ray pulsars in Be-binary systems in the Small Magellanic Cloud (Chashkina and Popov 2012) also favors the rotation law $\omega \sim R^{-2}$. The accretion rate through the magnetospheric shell is determined by the ability of the plasma to enter the magnetosphere. The settling accretion regime, which enables angular momentum removal from the neutron star magnetosphere may be realized at low accretion rates $\dot{M} < \dot{M}^{\dagger} \simeq 4 \times 10^{16}$ g/c (X-ray luminosities $L < L^{\dagger} \simeq 4 \times 10^{36}\,\mathrm{erg\,s^{-1}}$). At higher accretion rates (and, correspondingly, at higher X-ray luminosities) rapid Compton cooling of the plasma above the magnetospheric boundary causes a free-fall gap to emerge above the magnetosphere, and the accretion becomes highly non-stationary.

Spin-up/spin-down observations of long-period X-ray pulsars (i.e. measurements of the torque $\dot{\omega}^*$, or $\partial\dot{\omega}^*/\partial\dot{M}$ near the torque reversal point) allow the basic dimensionless parameters of the model to be inferred as well as the neutron star magnetic field to be independently estimated. Such an analysis was carried out for the equilibrium X-ray pulsars GX 301-2 and Vela X-1 and suggest magnetic fields in agreement with estimations from cyclotron line measurements in these sources.

Measurements of the equilibrium pulsar period P^*, the orbital binary period P_b and the neutron star magnetic field estimate μ make it possible to estimate the stellar wind velocity v from the optical companion without using complicated spectroscopic measurements. For non-equilibrium pulsars, there is a maximum possible value of the spin-down at the accretion stage depending on P^*, P_b, μ and v. For such pulsars (e.g., GX 1+4, SXP 1062, 4U 2206+54) the observed spin-down rate and X-ray luminosity can be used to obtain a lower limit on the neutron star magnetic field, which in all cases is found to be close to the standard neutron star magnetic field $\sim 10^{12}$–10^{13} G and is in agreement with cyclotron line measurements.

In the model of subsonic quasi-spherical accretion, the observed long-term stable periods of spin-up or spin-down of the neutron stars in some X-ray pulsars can be quantitatively explained by a change of the mean accretion rate onto the neutron star (with a corresponding change in X-ray luminosity). Apparently, such variations are related to the properties of the stellar wind from the optical companion in these X-ray binaries.

The model predicts a specific behavior of the frequency variations $\delta\dot{\omega}^*$ on top of a steady spin-up or spin-down, as a function of the accretion rate variations $\delta\dot{M}$. There is a critical mass accretion rate, $\dot{M}_{cr}$, below which an anti-correlation of the frequency fluctuations $\delta\dot{\omega}^*$ with $\delta\dot{M}$ should be observed. This is the case in GX 1+4 at the long-term steady spin-down stage presently observed. Above this accretion rate, the frequency fluctuations $\delta\dot{\omega}^*$ relative to the mean value should correlate with the mass accretion rate fluctuations $\delta\dot{M}$. This is the case in the equilibrium X-ray pulsars Vela X-1 and GX 301-2 around the equilibrium period and in GX 1+4 at the steady spin-up stage. The model also gives a quantitative explanation of the relative amplitude and sign of the observed frequency fluctuations in GX 1+4.

References

Arons J, Lea SM (1976) Accretion onto magnetized neutron stars - structure and interchange instability of a model magnetosphere. Astrophys J 207:914–936. https://doi.org/10.1086/154562

Beskin VS (2010) MHD flows in compact astrophysical objects. Springer, Berlin. https://doi.org/10.1007/978-3-642-01290-7

Bildsten L, Chakrabarty D, Chiu J, Finger MH, Koh DT, Nelson RW, Prince TA, Rubin BC, Scott DM, Stollberg M, Vaughan BA, Wilson CA, Wilson RB (1997) Observations of accreting pulsars. Astrophys J Suppl 113:367–408. https://doi.org/10.1086/313060. ArXiv:astro-ph/9707125

Bisnovatyi-Kogan GS (1991) Rotational equilibrium of long-periodic X-ray pulsars. Astron Astrophys 245:528–530

Bondi H (1952) On spherically symmetrical accretion. Mon Not R Astron Soc 112:195. https://doi.org/10.1093/mnras/112.2.195

Bozzo E, Falanga M, Stella L (2008) Are there magnetars in high-mass X-ray binaries? The case of supergiant fast X-ray transients. Astrophys J 683:1031–1044. https://doi.org/10.1086/589990. ArXiv:0805.1849

Bradshaw P (1969) The analogy between streamline curvature and buoyancy in turbulent shear flow. J Fluid Mech 36:177–191. https://doi.org/10.1017/S0022112069001583

Braithwaite J (2013) The nature and origin of magnetic fields in early-type stars. ArXiv:1312.4755

Bruno R, Carbone V (2013) The solar wind as a turbulence laboratory. Living Rev Sol Phys 10(2). https://doi.org/10.12942/lrsp-2013-2

Burnard DJ, Arons J, Lea SM (1983) Accretion onto magnetized neutron stars - X-ray pulsars with intermediate rotation rates. Astrophys J 266:175–187. https://doi.org/10.1086/160768

Chashkina A, Popov SB (2012) Magnetic field estimates for accreting neutron stars in massive binary systems and models of magnetic field decay. New Astron 17:594–602. https://doi.org/10.1016/j.newast.2012.01.004. ArXiv:1112.1123

Chaty S, Rahoui F, Foellmi C, Tomsick JA, Rodriguez J, Walter R (2008) Multi-wavelength observations of Galactic hard X-ray sources discovered by INTEGRAL. I. The nature of the companion star. Astron Astrophys 484:783–800. https://doi.org/10.1051/0004-6361:20078768. ArXiv:0802.1774

Cowie LL, McKee CF, Ostriker JP (1981) Supernova remnant revolution in an inhomogeneous medium. I - numerical models. Astrophys J 247:908–924. https://doi.org/10.1086/159100

Davidson K, Ostriker JP (1973) Neutron-star accretion in a stellar wind: model for a pulsed X-ray source. Astrophys J 179:585–598. https://doi.org/10.1086/151897

Davies RE, Pringle JE (1981) Spindown of neutron stars in close binary systems. II. Mon Not R Astron Soc 196:209–224. https://doi.org/10.1093/mnras/196.2.209

Doroshenko V, Santangelo A, Suleimanov V, Kreykenbohm I, Staubert R, Ferrigno C, Klochkov D (2010) Is there a highly magnetized neutron star in GX 301-2? Astron Astrophys 515:A10

Doroshenko V, Santangelo A, Suleimanov V (2011) Witnessing the magnetospheric boundary at work in Vela X-1. Astron Astrophys 529:A52. https://doi.org/10.1051/0004-6361/201116482. ArXiv:1102.5254

Ducci L, Sidoli L, Mereghetti S, Paizis A, Romano P (2009) The structure of blue supergiant winds and the accretion in supergiant high-mass X-ray binaries. Mon Not R Astron Soc 398:2152–2165. https://doi.org/10.1111/j.1365-2966.2009.15265.x. ArXiv:0906.3185

Dungey JW (1961) Interplanetary magnetic field and the auroral zones. Phys Rev Lett 6:47–48. https://doi.org/10.1103/PhysRevLett.6.47

Elsner RF, Lamb FK (1977) Accretion by magnetic neutron stars. I - magnetospheric structure and stability. Astrophys J 215:897–913. https://doi.org/10.1086/155427

Fryxell BA, Taam RE (1988) Numerical simulation of nonaxisymmetric adiabatic accretion flow. Astrophys J 335:862–880. https://doi.org/10.1086/166973

Fürst F, Kreykenbohm I, Pottschmidt K, Wilms J, Hanke M, Rothschild RE, Kretschmar P, Schulz NS, Huenemoerder DP, Klochkov D, Staubert R (2010) X-ray variation statistics and wind clumping in vela X-1. Astron Astrophys 519:A37. https://doi.org/10.1051/0004-6361/200913981. ArXiv:1005.5243

Ghosh P, Lamb FK (1979) Accretion by rotating magnetic neutron stars. III - accretion torques and period changes in pulsating X-ray sources. Astrophys J 234:296–316. https://doi.org/10.1086/157498

González-Galán A, Kuulkers E, Kretschmar P, Larsson S, Postnov K, Kochetkova A, Finger MH (2012) Spin period evolution of GX 1+4. Astron Astrophys 537:A66. https://doi.org/10.1051/0004-6361/201117893. ArXiv:1111.6791

Grebenev SA, Sunyaev RA (2007) The first observation of AX J1749.1-2733 in a bright X-ray state-Another fast transient revealed by INTEGRAL. Astron Lett 33:149–158. https://doi.org/10.1134/S1063773707030024

Grebenev SA, Lutovinov AA, Sunyaev RA (2003) New outburst of IGR J17544-2619. The Astronomer's Telegram 192:1

Hatchett S, Buff J, McCray R (1976) Transfer of X-rays through a spherically symmetric gas cloud. Astrophys J 206:847–860. https://doi.org/10.1086/154448

Ikhsanov NR, Beskrovnaya NG (2012) Signs of magnetic accretion in X-ray pulsars. Astron Rep 56:589–594. https://doi.org/10.1134/S1063772912070037. ArXiv:1205.2846

Ikhsanov NR, Likh YS, Beskrovnaya NG (2014) Spin evolution of long-period X-ray pulsars. Astron Rep 58:376–385. https://doi.org/10.1134/S1063772914050035. ArXiv:1402.1029

Illarionov AF, Kompaneets DA (1990) A spin-down mechanism for accreting neutron stars. Mon Not R Astron Soc 247:219

Illarionov AF, Sunyaev RA (1975) Why the number of galactic X-ray stars is so small? Astron Astrophys 39:185

in't Zand JJM (2005) Chandra observation of the fast X-ray transient IGR J17544-2619: evidence for a neutron star? Astron Astrophys 441:L1–L4. https://doi.org/10.1051/0004-6361:200500162. ArXiv:astro-ph/0508240

Kluźniak W, Rappaport S (2007) Magnetically torqued thin accretion disks. Astrophys J 671:1990–2005. https://doi.org/10.1086/522954. ArXiv:0709.2361

Koh DT, Bildsten L, Chakrabarty D, Nelson RW, Prince TA, Vaughan BA, Finger MH, Wilson RB, Rubin BC (1997) Rapid spin-up episodes in the wind-fed accreting pulsar GX 301-2. Astrophys J 479:933–947. https://doi.org/10.1086/303929

Kompaneets A (1957) The establishment of thermal equilibrium between quanta and electrons. J Exp Theor Phys 4:730

Lamers HJGLM, van den Heuvel EPJ, Petterson JA (1976) Stellar winds and accretion in massive X-ray binaries. Astron Astrophys 49:327–335

Landau LD, Lifshitz EM (1987) Fluid mechanics, 2nd edn. Pergamon, Oxford. https://doi.org/10.1016/B978-0-08-033933-7.50001-5, https://www.sciencedirect.com/science/article/pii/B9780080339337500015

Lovelace RVE, Romanova MM, Bisnovatyi-Kogan GS (1995) Spin-up/spin-down of magnetized stars with accretion discs and outflows. Mon Not R Astron Soc 275:244–254. https://doi.org/10.1093/mnras/275.2.244. ArXiv:astro-ph/9412030

Marykutty J, Biswajit P, Jincy D, Kavila I (2010) Discovery of a 0.02 hz qpo feature in the transient X-ray pulsar ks 1947+300. Mon Not R Astron Soc 407(1):285–290. https://doi.org/10.1111/j.1365-2966.2010.16880.x

Molkov S, Mowlavi N, Goldwurm A, Strong A, Lund N, Paul J, Oosterbroek T (2003) Igr J16479-4514. The Astronomer's Telegram 176:1

Monin AS, I'Aglom AM (1971) Statistical fluid mechanics; mechanics of turbulence. MIT Press, Cambridge

Negueruela I, Smith DM, Reig P, Chaty S, Torrejón JM (2006) Supergiant fast X-ray transients: a new class of high mass X-ray binaries unveiled by INTEGRAL. In: Wilson A (ed) Proceedings of the "The X-ray Universe 2005". ESA SP-604, vol 1, p 165

Negueruela I, Torrejón JM, Reig P, Ribó M, Smith DM (2008) Supergiant fast X-ray transients and other wind accretors. In: Bandyopadhyay RM, Wachter S, Gelino D, Gelino CR (eds) AIP conference proceedings, vol 1010, pp 252–256. https://doi.org/10.1063/1.2945052

Nelson RW, Bildsten L, Chakrabarty D, Finger MH, Koh DT, Prince TA, Rubin BC, Scott DM, Vaughan BA, Wilson RB (1997) On the dramatic spin-up/spin-down torque reversals in accreting pulsars. Astrophys J Lett 488:L117–L120. https://doi.org/10.1086/310936. ArXiv:astro-ph/9708193

Oskinova LM, Feldmeier A, Kretschmar P (2012) Clumped stellar winds in supergiant high-mass X-ray binaries: X-ray variability and photoionization. Mon Not R Astron Soc 421:2820–2831. https://doi.org/10.1111/j.1365-2966.2012.20507.x. ArXiv:1201.1915

Paizis A, Sidoli L (2014) Cumulative luminosity distributions of supergiant fast X-ray transients in hard X-rays. Mon Not R Astron Soc 439:3439–3452. https://doi.org/10.1093/mnras/stu191. ArXiv:1401.6861

Parker EN (1963) Interplanetary dynamical processes. Interscience Publishers, New York

Pellizza LJ, Chaty S, Negueruela I (2006) IGR J17544-2619: a new supergiant fast X-ray transient revealed by optical/infrared observations. Astron Astrophys 455:653–658. https://doi.org/10.1051/0004-6361:20054436. ArXiv:arXiv:astro-ph/0605559

Postnov K, Oskinova L, Torrejón JM (2017) A propelling neutron star in the enigmatic Be-star γ Cassiopeia. Mon Not R Astron Soc 465:L119–L123. https://doi.org/10.1093/mnrasl/slw223. ArXiv:1610.07799

Pringle JE, Rees MJ (1972) Accretion disc models for compact X-ray sources. Astron Astrophys 21:1

Puls J, Vink JS, Najarro F (2008) Mass loss from hot massive stars. Astron Astrophys Rev 16:209–325. https://doi.org/10.1007/s00159-008-0015-8. ArXiv:0811.0487

Rahoui F, Chaty S, Lagage PO, Pantin E (2008) Multi-wavelength observations of Galactic hard X-ray sources discovered by INTEGRAL. II. The environment of the companion star. Astron Astrophys 484:801–813. https://doi.org/10.1051/0004-6361:20078774. ArXiv:0802.1770

Raymond JC, Cox DP, Smith BW (1976) Radiative cooling of a low-density plasma. Astrophys J 204:290–292. https://doi.org/10.1086/154170

Romano P, La Parola V, Vercellone S, Cusumano G, Sidoli L, Krimm HA, Pagani C, Esposito P, Hoversten EA, Kennea JA, Page KL, Burrows DN, Gehrels N (2011) Two years of monitoring supergiant fast X-ray transients with Swift. Mon Not R Astron Soc 410:1825–1836. https://doi.org/10.1111/j.1365-2966.2010.17564.x. ArXiv:1009.1146

Romano P, Krimm HA, Palmer DM, Ducci L, Esposito P, Vercellone S, Evans PA, Guidorzi C, Mangano V, Kennea JA, Barthelmy SD, Burrows DN, Gehrels N (2014) The 100-month Swift catalogue of supergiant fast X-ray transients. I. BAT on-board and transient monitor flares. Astron Astrophys 562:A2. https://doi.org/10.1051/0004-6361/201322516. ArXiv:1312.4955

Ruffert M (1997) Non-axisymmetric wind-accretion simulations. I. Velocity gradients of 3% and 20% over one accretion radius. Astron Astrophys 317:793–814. ArXiv:astro-ph/9605072

Ruffert M (1999) Non-axisymmetric wind-accretion simulations. II. Density gradients. Astron Astrophys 346:861–877. ArXiv:astro-ph/9903304

Sedov LI (1959) Similarity and dimensional methods in mechanics. Academic, New York

Sguera V, Barlow EJ, Bird AJ, Clark DJ, Dean AJ, Hill AB, Moran L, Shaw SE, Willis DR, Bazzano A, Ubertini P, Malizia A (2005) INTEGRAL observations of recurrent fast X-ray transient sources. Astron Astrophys 444:221–231. https://doi.org/10.1051/0004-6361:20053103. ArXiv:astro-ph/0509018

Shakura NI, Sunyaev RA (1973) Black holes in binary systems. Observational appearance. Astron Astrophys 24:337–355

Shakura NI, Sunyaev RA (1988) The theory of an accretion disk/neutron star boundary layer. Adv Space Res 8:135–140. https://doi.org/10.1016/0273-1177(88)90396-1

Shakura NI, Sunyaev RA, Zilitinkevich SS (1978) On the turbulent energy transport in accretion discs. Astron Astrophys 62:179–187

Shakura N, Postnov K, Kochetkova A, Hjalmarsdotter L (2012) Theory of quasi-spherical accretion in X-ray pulsars. Mon Not R Astron Soc 420:216–236. https://doi.org/10.1111/j.1365-2966.2011.20026.x. ArXiv:1110.3701

Shakura N, Postnov K, Hjalmarsdotter L (2013a) On the nature of 'off' states in slowly rotating low-luminosity X-ray pulsars. Mon Not R Astron Soc 428:670–677. https://doi.org/10.1093/mnras/sts062. ArXiv:1209.4962

Shakura NI, Postnov KA, Kochetkova AY, Hjalmarsdotter L (2013b) Quasispherical subsonic accretion in X-ray pulsars. Phys Usp 56:321–346. https://doi.org/10.3367/UFNe.0183.201304a.0337. ArXiv:1302.0500

Shakura N, Postnov K, Sidoli L, Paizis A (2014a) Bright flares in supergiant fast X-ray transients. Mon Not R Astron Soc 442:2325–2330. https://doi.org/10.1093/mnras/stu1027. ArXiv:1405.5707

Shakura NI, Postnov KA, Kochetkova AY, Hjalmarsdotter L (2014b) Theory of wind accretion. In: European Physical Journal Web of conferences, vol 64, p 2001. https://doi.org/10.1051/epjconf/20136402001. ArXiv:1307.3029

Sidoli L (2012) Supergiant fast X-ray transients: a review. In: Proceedings 9th INTEGRAL workshop. Published online at http://pos.sissa.it/cgi-bin/reader/conf.cgi?confid=176, id.11. ArXiv:1301.7574

Sidoli L, Romano P, Mereghetti S, Paizis A, Vercellone S, Mangano V, Götz D (2007) An alternative hypothesis for the outburst mechanism in supergiant fast X-ray transients: the case of IGR J11215-5952. Astron Astrophys 476:1307–1315. https://doi.org/10.1051/0004-6361:20078137. ArXiv:0710.1175

Sidoli L, Romano P, Mangano V, Pellizzoni A, Kennea JA, Cusumano G, Vercellone S, Paizis A, Burrows DN, Gehrels N (2008) Monitoring supergiant fast X-Ray transients with swift. I. Behavior outside outbursts. Astrophys J 687:1230–1235. https://doi.org/10.1086/590077. ArXiv:0805.1808

Sidoli L, Esposito P, Motta SE, Israel GL, Rodríguez Castillo GA (2016a) XMM-Newton discovery of mHz quasi-periodic oscillations in the high-mass X-ray binary IGR J19140+0951. Mon Not R Astron Soc 460:3637–3646. https://doi.org/10.1093/mnras/stw1246. ArXiv:1605.06356

Sidoli L, Paizis A, Postnov K (2016b) INTEGRAL study of temporal properties of bright flares in Supergiant Fast X-ray Transients. Mon Not R Astron Soc 457:3693–3701. https://doi.org/10.1093/mnras/stw237. ArXiv:1601.07000

Sunyaev RA, Grebenev SA, Lutovinov AA, Rodriguez J, Mereghetti S, Gotz D, Courvoisier T (2003) New source IGR J17544-2619 discovered with INTEGRAL. The Astronomer's Telegram 190:1

Syunyaev RA, Shakura NI (1977) Disk reservoirs in binary systems and prospects for observing them. Sov Astron Lett 3:138–141

Tarter CB, Tucker WH, Salpeter EE (1969) The interaction of X-ray sources with optically thin environments. Astrophys J 156:943. https://doi.org/10.1086/150026

Thorne KS, Blandford RD (2017) Modern classical physics optics, fluids, plasmas, elasticity, relativity, and statistical physics. Princeton University Press, Princeton

Vitrichenko EA, Nadyozhin DK, Razinkova TL (2007) Mass-luminosity relation for massive stars. Astron Lett 33:251–258. https://doi.org/10.1134/S1063773707040044

Walter R, Zurita Heras J (2007) Probing clumpy stellar winds with a neutron star. Astron Astrophys 476:335–340. https://doi.org/10.1051/0004-6361:20078353. ArXiv:0710.2542
Wasiutynski J (1946) Studies in hydrodynamics and structure of stars and planets. Astrophys Norvegica 4:1–497
Weymann R (1965) Diffusion approximation for a photon gas interacting with a plasma via the compton effect. Phys Fluids 8:2112–2114. https://doi.org/10.1063/1.1761165
Zeldovich YB (1981) On the friction of fluids between rotating cylinders. Proc R Soc Lond Ser A 374:299–312. https://doi.org/10.1098/rspa.1981.0024
Zelenyi LM, Milovanov AV (2004) REVIEWS OF TOPICAL PROBLEMS: fractal topology and strange kinetics: from percolation theory to problems in cosmic electrodynamics. Phys Usp 47:1. https://doi.org/10.1070/PU2004v047n08ABEH001705
Zweibel EG, Yamada M (2009) Magnetic reconnection in astrophysical and laboratory plasmas. Annu Rev Astron Astrophys 47:291–332. https://doi.org/10.1146/annurev-astro-082708-101726

Chapter 8
On the Properties of Velikhov-Chandrasekhar MRI in Ideal and Non-ideal Plasmas

Nikolay Shakura and Konstantin Postnov

Abstract In this chapter, conditions of the Velikhov-Chandrasekhar magneto-rotational instability (MRI) in ideal and non-ideal plasmas are examined. A linear WKB analysis of hydromagnetic axially symmetric flows shows that in the Rayleigh-unstable hydrodynamic case where the angular momentum decreases with radius, the MRI branch becomes stable, and the magnetic field suppresses the Rayleigh instability at small wavelengths. We investigate the limiting transition from hydromagnetic flows to hydrodynamic flows. The Rayleigh mode smoothly transits to the hydrodynamic case, while the Velikhov-Chandrasekhar MRI mode completely disappears without the magnetic field. The effects of viscosity and magnetic diffusivity in the plasma on the MRI conditions in thin accretion discs are studied. We find the limits on the mean free-path of ions allowing MRI to operate in such discs.

8.1 Introduction

In the end of the 1950s—beginning of the 1960s, E. Velikhov and S. Chandrasekhar studied the stability of sheared hydromagnetic flows (Velikhov 1959; Chandrasekhar 1960). In these papers, the magneto-rotational instability (MRI) in axisymmetric flows with magnetic fields was discovered. MRI arises when a relatively small seed poloidal magnetic field is present in the fluid. This instability was applied to astrophysical accretion discs in the influential paper by Balbus and Hawley (1991), and has since then been considered the major reason for the turbulence arising in accretion discs (see Balbus and Hawley 1998 for a review). Non-linear

N. Shakura
Sternberg Astronomical Institute, Lomonosov Moscow State University, Moscow, Russia

Kazan Federal University, Kazan, Russia

K. Postnov (✉)
Sternberg Astronomical Institute, Lomonosov Moscow State University, Moscow, Russia

National Research University Higher School of Economics, Moscow, Russia

N. Shakura (ed.), *Accretion Flows in Astrophysics*, Astrophysics and Space Science Library 454, https://doi.org/10.1007/978-3-319-93009-1_8

numerical simulations (e.g. Hawley et al. 1995; Sorathia et al. 2012; Hawley et al. 2013) confirmed that MRI can sustain turbulence and dynamos in accretion discs. However, semi-analytical and numerical simulations (see, for example, Masada and Sano 2008; Stone 2011; Hawley et al. 2013; Suzuki and Inutsuka 2014; Nauman and Blackman 2015) suggest that the total (Reynolds + Maxwell) stresses due to MRI are insufficient to cause effective angular momentum transfer in accretion discs, in terms of the phenomenological alpha-parameter α_{SS} (Shakura and Sunyaev 1973), giving rather low values $\alpha_{SS} \sim 0.01$–0.03. Note that from the observational point of view, the alpha-parameter can be reliably evaluated, e.g. from analysis of non-stationary accretion discs in X-ray novae (Suleimanov et al. 2008), dwarf-nova and AM CVn stars (Kotko and Lasota 2012), and turns out to be an order of magnitude higher than typically found in the numerical MRI simulations.

In this chapter we use the local linear analysis of MRI in the WKB-approximation by Balbus and Hawley (1991) to examine properties of MRI for different laws of differential rotation in weakly magnetized flows, $\Omega^2(r) \propto r^{-n}$, i.e. when the solution to the linearized MHD equations in the Boussinesq approximation is searched for in the form $\sim e^{i(\omega t - k_r r - k_z z)}$, where k_r, k_z are wave vectors in the radial and normal direction to the disc plane, respectively, in cylindrical coordinates.

In this approximation, the dispersion relation represents a biquadratic algebraic equation. A linear local analysis of unstable modes in this case was performed earlier (see, e.g., Balbus 2012). Here we emphasize the different behaviour of stable and unstable modes of this equation for different rotation laws of the fluid. We show that in the Rayleigh-unstable hydrodynamic case, where the angular momentum decreases with radius, the Velikhov-Chandrasekhar MRI does not arise, and the magnetic field suppresses the Rayleigh instability at small wavelengths.

Then we turn to the analysis of a non-ideal plasma characterized by a non-zero kinematic viscosity ν and magnetic diffusivity η. This problem has been addressed previously by different authors (see, e.g. Balbus and Hawley (1998), Sano and Miyama (1999), Ji et al. (2001), Balbus (2004), Islam and Balbus (2005), Pessah and Chan (2008), among others), aimed at studying various aspects of the MRI physics and applications. To keep the paper self-contained, we re-derive the basic dispersion relation in the general case and investigate its behaviour for different values of the magnetic Prandtl number $\mathrm{P_m} = \nu/\eta$ and the kinematic viscosity ν. Specifically, we consider the limitations implied by the viscosity in accretion discs with finite thickness, and find phenomenologically interesting constraints on the disc parameters where MRI can operate. Below we delineate the derivation of the dispersion equation for non-ideal plasma in the Boussinesq approximation for both adiabatic and non-adiabatic perturbations for different magnetic Prandtl numbers, $P_m = \nu/\eta$, and different values of the kinematic viscosity ν. Then we consider limitations on the viscosity in thin accretion discs in which MRI can operate.

8.2 Derivation of the Dispersion Equation for a Non-ideal Plasma

Here we generalize the derivation of the MRI dispersion equation (8.36) given in Kato et al. (1998) to the case of a non-ideal plasma with arbitrary kinetic coefficients ν and η (see also Ji et al. 2001).

The system of non-ideal MHD equations reads:

1) Mass conservation equation

$$\frac{\partial \rho}{\partial t} + \nabla \cdot (\rho \boldsymbol{u}) = 0\,, \tag{8.1}$$

2) Navier-Stokes equation including gravity force and Lorentz force

$$\frac{\partial \boldsymbol{u}}{\partial t} + (\boldsymbol{u}\nabla) \cdot \boldsymbol{u} = -\frac{1}{\rho}\nabla p - \nabla \phi_g + \frac{1}{4\pi\rho}(\nabla \times \boldsymbol{B}) \times \boldsymbol{B} + \nu \Delta \boldsymbol{u} \tag{8.2}$$

(here ϕ_g is the Newtonian gravitational potential),

3) Induction equation

$$\frac{\partial \boldsymbol{B}}{\partial t} = \nabla \times (\boldsymbol{u} \times \boldsymbol{B}) + \eta \Delta \boldsymbol{B}\,, \tag{8.3}$$

4) Energy equation

$$\frac{\rho \mathscr{R} T}{\mu}\left[\frac{\partial s}{\partial t} + (\boldsymbol{u}\nabla) \cdot s\right] = Q_{visc} - \nabla \cdot \boldsymbol{F} + \frac{\eta}{4\pi}[\nabla \times \boldsymbol{B}]^2\,. \tag{8.4}$$

where s is the specific entropy (per particle), $\mathscr{R}$ is the universal gas constant, μ is the molecular weight, T is the temperature, and terms on the right stand for viscous, energy flux $\boldsymbol{F}$ and Joule dissipation, respectively.

5) These equations should be completed with the equation of state for a perfect gas, which is convenient to write in the form:

$$p = K e^{s/c_V} \rho^\gamma\,, \tag{8.5}$$

where K is a constant, c_V is the specific volume heat capacity and $\gamma = c_p/c_V$ is the adiabatic index (5/3 for a monoatomic gas).

We will consider small axially symmetric perturbations in the WKB approximation with space-time dependence $e^{i(\omega t - k_r r - k_z z)}$, where r, z, ϕ are cylindrical coordinates. The unperturbed magnetic field is assumed to be purely poloidal: $\boldsymbol{B_0} = (0, 0, B_0)$. The velocity and magnetic field perturbations are $\boldsymbol{u} = (u_r, u_\phi, u_z)$ and $\boldsymbol{b} = (b_r, b_\phi, b_z)$, respectively. The density, pressure and entropy perturbations are ρ_1, p_1, and s_1 over the unperturbed values ρ_0, p_0, and s_0, respectively. To filter out magnetoacoustic oscillations arising from the restoring pressure force, we will use the Boussinesq approximation, i.e. consider incompressible gas motion $\nabla \cdot \boldsymbol{u} = 0$. In the energy equation we neglect Eulerian pressure variations, $p_1(t, r, \phi, z) = 0$, but Lagrangian pressure variations $\delta p(t, r(t_0), \phi(t_0), z(t_0))$ are non-zero. (Recall that for infinitesimally small shifts the perturbed gas parcel acquires the pressure equal to that of the ambient medium; see e.g. Spiegel and Veronis (1960) and Kundu et al. (2012) for discussion of the Boussinesq approximation).

In the linear approximation, the system of differential non-ideal MHD equations is reduced to the following system of algebraic equations.

a) The Boussinesq approximation for gas velocity $\boldsymbol{u}$ is $\nabla \cdot \boldsymbol{u} = 0$:

$$k_r u_r + k_z u_z = 0\,. \tag{8.6}$$

b) The radial, azimuthal and vertical components of the Euler momentum equation are, respectively:

$$i\omega u_r - 2\Omega u_\phi = ik_r \frac{p_1}{\rho_0} - \frac{\rho_1}{\rho_0^2}\frac{\partial p_0}{\partial r} + i\frac{c_A^2}{B_0}(k_r b_z - k_z b_r) - \nu k^2 u_r\,, \tag{8.7}$$

$$i\omega u_\phi + \frac{\kappa^2}{2\Omega}u_r = -i\frac{c_A^2}{B_0}k_z b_\phi - \nu k^2 u_\phi\,, \tag{8.8}$$

$$i\omega u_z = ik_z\frac{p_1}{\rho_0} - \frac{\rho_1}{\rho_0^2}\frac{\partial p_0}{\partial z} - \nu k^2 u_z \tag{8.9}$$

Here $k^2 = k_r^2 + k_z^2$ so that in the linear order $\nu \Delta \boldsymbol{u} \to -\nu k^2\{u_r, u_\phi, u_z\}$,[1] and we have introduced the unperturbed Alfvén velocity $c_A^2 = B_0^2/(4\pi\rho_0)$.

To specify the density perturbations ρ_1/ρ_0, we need to address the energy equation. First, we consider adiabatic perturbations, i.e. require that

$$\frac{\partial s}{\partial t} + (\boldsymbol{u} \cdot \nabla)s = 0\,. \tag{8.10}$$

For small density perturbations from Eq. (8.5) we obtain for entropy perturbations

$$\frac{s_1}{c_V} + \gamma\frac{\rho_1}{\rho_0} = 0\,, \tag{8.11}$$

and after substituting this into Eq. (8.10) we get

$$i\omega\gamma\frac{\rho_1}{\rho_0} + u_z\frac{\partial \ln p\rho^{-\gamma}}{\partial z} + u_r\frac{\partial \ln p\rho^{-\gamma}}{\partial r} = 0 \tag{8.12}$$

[1]Here we neglect terms $\sim (k_r/r)$ compared to terms $\sim k^2$, see also discussion in Acheson (1978).

(cf. Eq. (122) in Balbus and Hawley 1998). Hence in the absence of entropy gradients we obtain

$$\frac{1}{\rho_0}\frac{\partial \rho_1}{\partial t} = 0\,. \tag{8.13}$$

Consider now the more general case of *non-adiabatic* linear perturbations. To do this, we need to specify the right-hand side of the energy equation (8.4). Let us start with the last term. Writing for the magnetic field $\boldsymbol{B} = \boldsymbol{B_0}+\boldsymbol{b}$ and taking into account that for the unperturbed field $\nabla \times \boldsymbol{B_0} = 0$, we see that the Joule dissipation term is quadratic in magnetic field perturbations $\boldsymbol{b}$, so we exclude it from consideration. The heat flux divergence is

$$\nabla \cdot \boldsymbol{F} = \nabla(-\kappa_T \nabla T) = -\kappa_T \Delta T\,, \tag{8.14}$$

where κ_T is the temperature conductivity coefficient. From the equation of state for an ideal gas written in the form $p = \rho \mathscr{R} T/\mu$, we find for small perturbations with zero Eulerian pressure variations $p_1/p_0 = 0$

$$\frac{\rho_1}{\rho_0} = -\frac{T_1}{T_0}\,, \tag{8.15}$$

i.e. in the axially symmetric waves considered here the density variations are in counter-phase with the temperature variations.

The viscous dissipative function Q_{visc} can be written as $Q_{visc} = \rho\nu\Phi$, where the function Φ in polar coordinates is

$$\begin{aligned}\Phi = 2&\left[\left(\tfrac{\partial u_r}{\partial r}\right)^2 + \left(\tfrac{1}{r}\left(\tfrac{\partial u_\phi}{\partial \phi}\right) + \tfrac{u_r}{r}\right)^2 + \left(\tfrac{\partial u_z}{\partial z}\right)^2\right] \\ &+ \left[r\tfrac{\partial}{\partial r}\left(\tfrac{u_\phi}{r}\right) + \tfrac{1}{r}\tfrac{\partial u_r}{\partial \phi}\right]^2 + \left[\tfrac{1}{r}\tfrac{\partial u_z}{\partial \phi}\right]^2 \\ &+ \left[\tfrac{\partial u_r}{\partial z} + \tfrac{\partial u_z}{\partial r}\right]^2 - \tfrac{2}{3}(\nabla \cdot \boldsymbol{u})^2\,.\end{aligned} \tag{8.16}$$

All terms but one in this function are quadratic in small velocity perturbations; this term has the form:

$$\nu\rho\left(\frac{\partial u_\phi}{\partial r} - \frac{u_\phi}{r}\right)^2\,. \tag{8.17}$$

Writing for the azimuthal velocity $u_\phi = u_{\phi,0} + u_{\phi,1}$ (here for the purposes of this paragraph and only here we specially mark the unperturbed velocity with index 0, not to be confused with our notations u_ϕ for perturbed velocity in Eqs. (8.7)–(8.8) above and below). Thus we obtain for the viscous dissipation

$$Q_{visc} = \nu\rho r\frac{d\Omega}{dr}\left[r\frac{d\Omega}{dr} - 2ik_r u_{\phi,1} - 2\frac{u_{\phi,1}}{r}\right] + \text{quadratic terms}. \tag{8.18}$$

Here $\Omega = u_{\phi,0}/r$ is the angular (Keplerian) velocity of the unperturbed flow. The first term in parentheses describes the viscous energy release in the unperturbed Keplerian flow. For this unperturbed flow we have

$$\frac{\partial s_0}{\partial t} = \nu\mu\frac{[r(d\Omega/dr)]^2}{\mathscr{R}T} = \frac{9}{4}\nu\mu\frac{\Omega^2}{\mathscr{R}T}. \tag{8.19}$$

Thus, the entropy of the unperturbed flow changes along the radius. However, on a scale of the order of or smaller than the disc thickness, the entropy gradient can be neglected. The second term in Eq. (8.18) vanishes if $k_r = 0$, i.e. we consider two-dimensional perturbations with only $k_z \neq 0$. As a result, the energy equation with zero entropy gradients in the Boussinesq limit becomes

$$\frac{\rho_0\mathscr{R}T_0}{\mu}s_1 = -2ik_r\nu\rho_0 r\frac{d\Omega}{dr}u_{\phi,1} - \kappa_T k^2 T_0\frac{T_1}{T_0}. \tag{8.20}$$

Like in the linearized equation $\nabla \cdot \boldsymbol{u} = 0$, here we have neglected the term $u_{\phi,1}/r$. By substituting Eqs. (8.11) and (8.15) into Eq. (8.20), we find the relation between the density variations and u_ϕ in the Boussinesq limit with zero entropy gradients:

$$\frac{\rho_1}{\rho_0}\left(i\omega c_p + \frac{\kappa_T k^2}{\rho_0\mathscr{R}/\mu}\right) = \frac{2ik_r\nu r(d\Omega/dr)}{\mathscr{R}T_0/\mu}u_\phi \tag{8.21}$$

Here $c_p = \gamma c_V = \gamma/(\gamma-1)$ is the specific heat capacity (per particle) at constant pressure.

To describe the effects of thermal conductivity, it is convenient to introduce the usual dimensionless Prandtl number:

$$\Pr \equiv \frac{\nu\rho_0 C_p}{\kappa_T}. \tag{8.22}$$

(Here $C_p = c_p\mathscr{R}/\mu$). Substituting Eq. (8.22) into Eq. (8.21) yields:

$$\frac{\rho_1}{\rho_0} = \frac{\gamma/(\gamma-1)}{(i\omega + \nu k^2/\Pr)}\frac{2ik_r\nu r(d\Omega/dr)}{\mathscr{R}T_0/\mu}u_\phi \tag{8.23}$$

It is straightforward to include the density perturbations in the non-adiabatic case (8.21) in the analysis. This significantly complicates the final dispersion equation (see Eq. (8.32) below). We stress again that the two-dimensional case with $k_r = 0$ produces the dispersion relation for small local perturbations which is *exact* even in the case of non-adiabatic perturbations.

c) The three components of the induction equation with account for $\eta \Delta \boldsymbol{B} \to -\eta k^2 \{b_r, b_\phi, b_z\}$ read:

$$i\omega b_r = -i B_0 k_z u_r - \eta k^2 b_r \,, \tag{8.24}$$

$$i\omega b_\phi = -i B_0 k_z u_\phi + r\frac{d\Omega}{dr} b_r - \eta k^2 b_\phi \,, \tag{8.25}$$

$$i\omega b_z = i B_0 k_r u_r - \eta k^2 b_z \,. \tag{8.26}$$

Following Kato et al. (1998), we express all perturbed quantities through u_z:

$$u_r = -\frac{k_z}{k_r} u_z \,, \tag{8.27}$$

$$u_\phi = \frac{k_z}{k_r} \frac{\frac{\kappa^2}{2\Omega}(i\omega + \eta k^2)^2 + c_A^2 k_z^2 r \frac{d\Omega}{dr}}{\left[(i\omega + \nu k^2)(i\omega + \eta k^2) + c_A^2 k_z^2\right](i\omega + \eta k^2)} u_z \,, \tag{8.28}$$

$$\frac{b_r}{B_0}(i\omega + \eta k^2) = i\frac{k_z^2}{k_r} u_z \,, \tag{8.29}$$

$$\frac{b_\phi}{B_0}(i\omega + \eta k^2) = -i k_z u_\phi + \frac{i r \frac{d\Omega}{dr}}{(i\omega + \eta k^2)} \frac{k_z^2}{k_r} u_z \,, \tag{8.30}$$

$$\frac{b_z}{B_0}(i\omega + \eta k^2) = -i k_z u_z \,, \tag{8.31}$$

The system of linear equations (8.6) and (8.27)–(8.31) contains the equation $\nabla \cdot \boldsymbol{b} = 0$. Indeed, by multiplying Eqs. (8.29) and (8.31) by k_r and k_z, respectively, and summing up the obtained equations, we get $k_r b_r + k_z b_z = 0$. Substituting Eqs. (8.27)–(8.31) into Eq. (8.7) and rearranging the terms, we arrive at the dispersion relation (8.48).

The dispersion relation in the general case of non-adiabatic perturbations with $k_r \neq 0$, i.e. with non-vanishing density perturbations ρ_1 (see Eq. (8.21)) is:

$$\omega_{**}^4 + \left(\frac{k_z}{k}\right)^2 \left[\left(i\omega + \eta k^2\right)^2 \kappa^2 + c_A^2 k_z^2 (\kappa^2 - 4\Omega^2)\right]$$
$$\left[1 - \frac{\gamma - 1}{\gamma} \frac{i k_r}{(i\omega + \nu k^2/\mathrm{Pr})} \left(A - \frac{k_r}{k_z} B\right)\right] = 0 \,, \tag{8.32}$$

where ω_{**} is determined as

$$\omega_{**}^2 = -(i\omega + \nu k^2)(i\omega + \eta k^2) - c_A^2 k_z^2 \,. \tag{8.33}$$

The coefficients A and B are:

$$A = \nu \left(\frac{d\ln\Omega}{d\ln r}\right)\left(\frac{1}{p_0}\frac{dp_0}{dr}\right); \quad B = \nu \left(\frac{d\ln\Omega}{d\ln r}\right)\left(\frac{1}{p_0}\frac{dp_0}{dz}\right). \tag{8.34}$$

Although the terms with A and B arising from the viscous dissipation function are proportional to $(k_r/r)(\nu/\omega)$ and $(k_r^2/k_z r)(\nu/\omega)$, they are retained in our analysis since for large viscosities they may become comparable to or even higher than one. The expression in the square brackets in Eq. (8.32) above can be rewritten in the equivalent form:

$$\left[1 + \frac{\gamma-1}{\gamma}\frac{i\nu}{(i\omega+\nu k^2/\mathrm{Pr})}\left(\frac{k_r}{k_z}\right)\frac{d\ln\Omega/d\ln r}{\mathcal{R}T_0/\mu}(k_z g_{r,eff} - k_r g_z)\right], \tag{8.35}$$

where $g_{r,eff} = -1/\rho_0(dp_0/dr)$ and $g_z = -1/\rho_0(dp_0/dz)$ are the effective radial and vertical gravity accelerations in the unperturbed flow, respectively. Clearly, for $k_r = 0$ we return to Eq. (8.48) with $k = k_z$. Note that for $k_r \neq 0$ Eq. (8.32) is a fifth-order algebraic equation. For perturbations with $k_r = 0$ this equation becomes a fourth-order algebraic equation, which already has exponentially growing MRI modes. For completeness, it would be desirable to investigate this five-order equation. However, in the absence of a magnetic field Eq. (8.32) turns into a third-order algebraic equation. As we show in Shakura and Postnov (2015), one of the Rayleigh modes in this case can become exponentially unstable at long wavelengths even in the Rayleigh-stable case of Keplerian rotation.

8.3 Linear Analysis for an Ideal Fluid

The dispersion relation for local small axially symmetric disturbances in the simplest case of an ideal fluid without entropy gradients reads (see also Balbus and Hawley 1991; Kato et al. 1998):

$$\omega_{*^4} - \left(\frac{k_z}{k}\right)^2 \kappa^2 \omega_{*^2} - 4\Omega^2\left(\frac{k_z}{k}\right)^2 k_z^2 c_A^2 = 0\,. \tag{8.36}$$

Here

$$\omega_{*^2} = \omega^2 - c_A^2 k_z^2\,, \tag{8.37}$$

$k^2 = k_r^2 + k_z^2$,

$$\kappa^2 = 4\Omega^2 + r\frac{d\Omega^2}{dr} \equiv \frac{1}{r^3}\frac{d\Omega^2 r^4}{dr} \tag{8.38}$$

is the epicyclic frequency, and

$$c_A^2 = B_0^2/(4\pi\rho_0) \tag{8.39}$$

is the unperturbed Alfvén velocity squared. The initial magnetic field B_0 is assumed to be purely poloidal (directed along the z-coordinate) and homogeneous.

The solution to the biquadratic equation (8.36) has the form:

$$\omega^2 = \left(\frac{k_z}{k}\right)^2 \left[c_A^2 k^2 + \frac{\kappa^2}{2} \pm \sqrt{\frac{\kappa^4}{4} + 4\Omega^2 c_A^2 k^2}\right]. \tag{8.40}$$

We will examine solutions to this equation by assuming $k_z^2/k^2 \equiv k_z^2/(k_r^2 + k_z^2) = const$, i.e. the direction of the wave vector in the $r - z$ plane is conserved, which is not restricting our analysis. Depending on the sign of the root ω^2, one of three modes can exist: the stable oscillating mode for $\omega^2 > 0$, the indifferent equilibrium (neutral) mode for $\omega^2 = 0$, or the exponentially growing mode for $\omega^2 < 0$.

According to the classical Rayleigh criterion (Lord Rayleigh 1916), if the epicyclic frequency $\kappa^2 > 0$ (in this case the angular momentum in the flow increases with radius), the equilibrium is stable. If $\kappa^2 < 0$ (the angular momentum decreases with radius), the equilibrium is unstable. If $\kappa^2 = 0$ (the angular momentum does not change with radius), the equilibrium is indifferent.

8.3.1 The Ideal MHD Case

Let us start with discussing the behaviour of different modes of the dispersion relation (8.36) in the ideal MHD case. It is instructive to investigate the asymptotics of these modes with decreasing (but non-zero) seed magnetic field (see Sect. 8.3.2 for more detail about the limiting transition for a vanishing magnetic field).

If a magnetic field is present, there are five different types of solutions to Eq. (8.40) depending on how the angular velocity (angular momentum) changes with radius.

Case 1 $\kappa^2 > 4\Omega^2$, $n < 0$. In this case there are two stable modes (see Fig. 8.1), which at large k^2 (in the short-wavelength limit) approach the asymptotic behaviour $\omega^2 = (k_z/k)^2 c_A^2 k^2$. With decreasing (but non-zero) seed magnetic field amplitude B_0 (and corresponding unperturbed Alfvén velocity c_A), one mode tends to the classical Rayleigh branch $\omega_R^2 = (k_z/k)^2\kappa^2$ (the horizontal dashed line in Fig. 8.1), and the second mode tends to the neutral branch $\omega_{VC}^2 \to 0$.

Case 2 $0 < \kappa^2 < 4\Omega^2$, $0 < n < 4$. In this case the Rayleigh mode ω_R^2 behaves almost in the same way as in case 1 (upper curves in Fig. 8.2). For the mode ω_{VC}^2 (lower thick curves in Fig. 8.2) the instability arises in the interval: $0 < k^2 c_A^2 <$

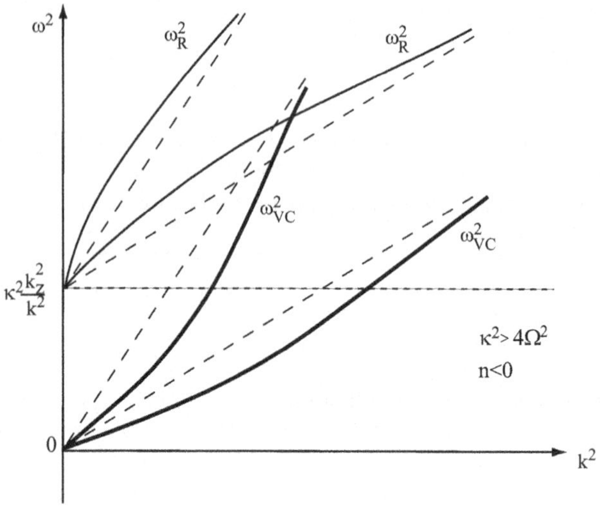

Fig. 8.1 Schematic behaviour of two branches of the dispersion equation (8.36) (the 'Reynolds mode' ω_R^2, thin curves, and the 'MRI mode' ω_{VC}^2, thick curves) for two values of the Alfvén velocity c_A^2 (two values of the seed magnetic field B_0). The dashed straight lines show the asymptotic behaviour of the solutions at large k^2: $\omega^2 = (k_z/k)^2 c_A^2 k^2$. The smaller the seed magnetic field, the flatter the slope of the asymptotes. Case 1, where the angular velocity and angular momentum increasing with radius ($\kappa^2 > 4\Omega^2$; $n < 0$)

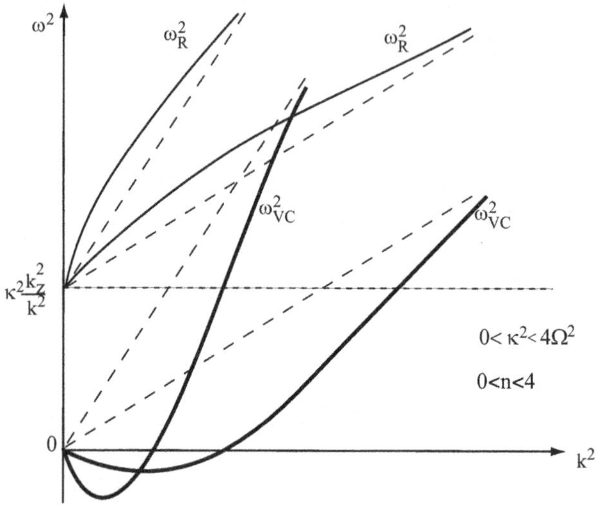

Fig. 8.2 The same as in Fig. 8.1 for the case of decreasing angular velocity with radius but increasing angular momentum ($0 < \kappa^2 < 4\Omega^2$; $0 < n < 4$) (case 2)

$n\Omega^2$. It is in this case that the MRI instability occurs in a Keplerian accretion disc with $n = 3$ and $\kappa = \Omega$. With decreasing B_0 the critical wave number separating the stable and unstable behaviour

$$k_{cr}^2(\omega^2 = 0) = n\frac{\Omega^2}{c_A^2} \tag{8.41}$$

tends to infinity. The maximum instability growth rate characterized by the minimum of the mode ω_{VC}^2 occurs at

$$k_{max}^2 = \frac{n(8-n)}{16}\frac{\Omega^2}{c_A^2}. \tag{8.42}$$

By substituting Eq. (8.42) into Eq. (8.40), we find for the MRI mode

$$\omega_{VC,max}^2 = -\frac{n^2}{16}\left(\frac{k_z}{k}\right)^2\Omega^2 = -\frac{n}{8-n}\left(\frac{k_z}{k}\right)^2 c_A^2 k_{max}^2. \tag{8.43}$$

With decreasing (but non-zero) B_0 and c_A^2, $\omega_{VC}^2(k_{max}^2) \to -0$ as $k_{max}^2 \to \infty$.

Case 3 $\kappa^2 = 0$, $n = 4$. In this case (see Fig. 8.3) both the Rayleigh mode ω_R^2 and the MRI mode ω_{VC}^2 leave zero with infinite derivatives (positive and negative for the Rayleigh and MRI modes, respectively). In the presence of a finite seed magnetic field, the ω_{VC}^2 mode displays the MRI. As B_0 becomes small (but non-zero), both modes asymptotically approach the neutral mode $\omega^2 \to 0$.

Case 4 $\kappa^2 < 0$, $4 < n < 8$. In this case (see Fig. 8.4) in the absence of a magnetic field the instability according to the Rayleigh criterion emerges (the bottom dashed

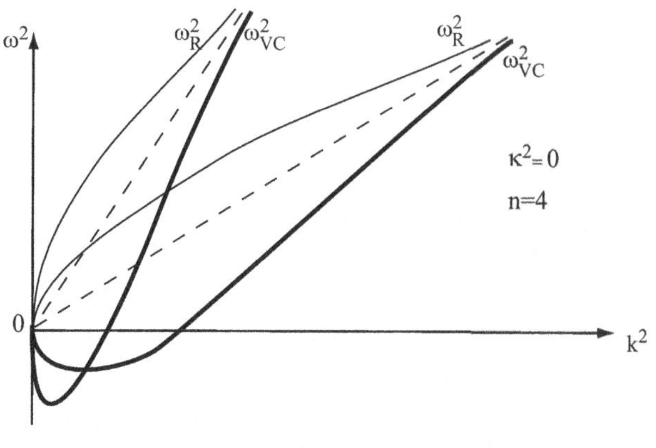

Fig. 8.3 The same as in Fig. 8.1 for the case of constant angular momentum ($\kappa^2 = 0$; $n = 4$) (case 3). Both the Rayleigh and MRI branches have infinite derivatives $d\omega^2/dk^2$ at $k^2 = 0$

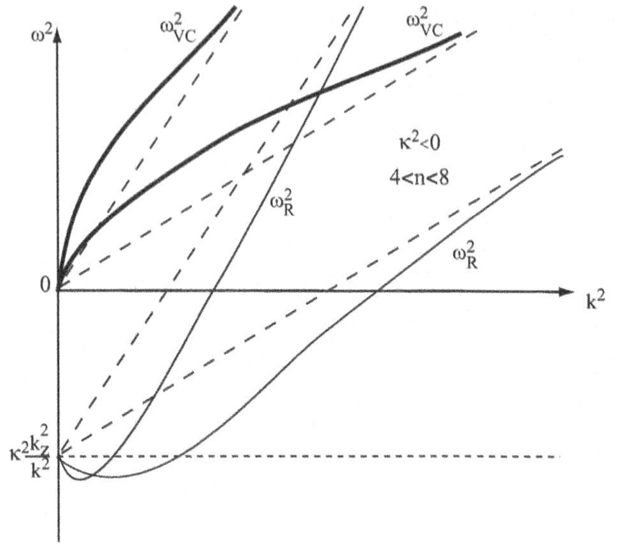

Fig. 8.4 The same as in Fig. 8.1 for the case of decreasing angular momentum ($\kappa^2 < 0$; $4 < n < 8$) (Case 4). Instability according to the Rayleigh criterion occurs. The Rayleigh branch has a negative derivative at $k^2 = 0$

horizontal line in Fig. 8.4) with $\omega_R^2 = \kappa^2(k_z/k)^2$. If a magnetic field is present, the Rayleigh instability is stabilized by the magnetic field at $k^2 > k_{cr}^2$ (bottom thin curves in Fig. 8.4). Note that k_{cr}^2 and k_{max}^2 here are the same as in Case 2. While similar to the MRI mode, this is now *the Rayleigh mode* ω_R^2 that is unstable and reaches a maximum growth rate $\omega_{R,max}^2$ determined by Eq. (8.43). In contrast, *the Velikhov-Chandrasekhar mode* ω_{VC}^2 (upper thick curves in Fig. 8.4) remains stable *at all wavenumbers*, and with decreasing (but non-zero) magnetic field $\omega_{VC}^2 \to +0$. We stress again that the difference between the Rayleigh and MRI modes is due to their different asymptotic behaviour as $B_0 \to +0$: the Rayleigh mode is unstable and behaves as $\omega_R \to -\kappa^2 k_z^2/k^2$, unlike the stable Velikhov-Chandrasekhar mode.

Case 5 $\kappa^2 < 0$, $n > 8$. The only difference of this case from Case 4 is that the Rayleigh mode ω_R^2 leaves zero with a positive derivative (bottom thin curves in Fig. 8.5).

8.3.2 On the Behaviour of MRI for a Vanishing Magnetic Field

The transition to the purely hydrodynamic case without magnetic field should be treated separately. Let us consider asymptotic solutions (8.40) for a vanishing magnetic field. In the leading order in c_A the two branches of the dispersion relation are:

$$\omega_R^2 \simeq \left(\frac{k_z}{k}\right)^2 \left[\kappa^2 + c_A^2 k^2 \left(1 + \frac{4\Omega^2}{\kappa^2}\right)\right], \qquad (8.44)$$

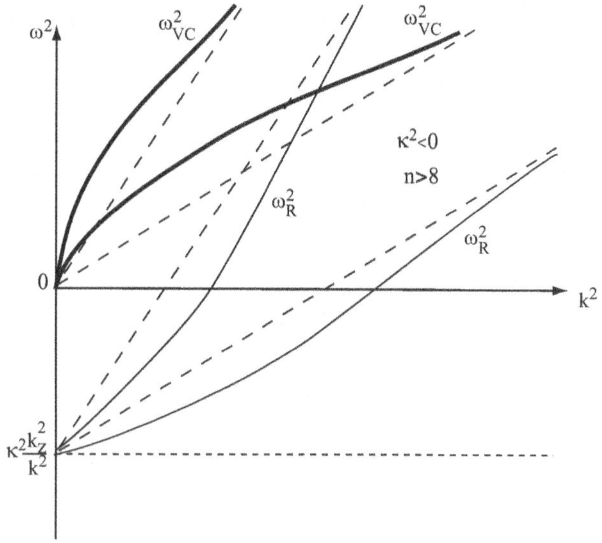

Fig. 8.5 The same as in Fig. 8.4 for the case ($\kappa^2 < 0$; $n > 8$) (case 5 in the text); the Rayleigh branch has a positive derivative at $k^2 = 0$

which we have referred to as the Rayleigh mode since in the absence of a magnetic field it tends to the classical Rayleigh mode $\omega_R^2 = (k_z/k)^2\kappa^2$, and

$$\omega_{VC}^2 \simeq {k_z}^2 c_A^2 \left(1 - \frac{4\Omega^2}{\kappa^2}\right), \tag{8.45}$$

which we have refered to as the Velikhov-Chandrasekhar mode and which is manifestly unstable for Keplerian motion ($\kappa^2 = \Omega^2$).

Note that unlike for the Rayleigh mode, setting the magnetic field to zero in Eq. (8.45) leads to a paradoxical result: $\omega_{VC}^2 = 0$. This 'neutral mode' is fictitious, it does not exist in the purely hydrodynamic case. To see this, let us write down the linearized system of perfect fluid equations in the Boussinesq approximation (see (8.6)–(8.9) and Eq. (8.13) in Sect. 8.2):

$$\begin{aligned} k_r u_r + k_z u_z &= 0 \\ i\omega u_r - 2\Omega u_\phi &= i k_r \tfrac{p_1}{\rho_0} \\ i\omega u_\phi + \tfrac{\kappa^2}{2\Omega} u_r &= 0 \\ i\omega u_z &= i k_z \tfrac{p_1}{\rho_0} \end{aligned} \tag{8.46}$$

It is easy to find the dispersion relation in this case:

$$\omega^2 = \left(\frac{k_z}{k}\right)^2 \kappa^2, \tag{8.47}$$

which is the classical Rayleigh branch. No neutral mode $\omega^2 = 0$ arises. The neutral mode $\omega = 0$ does exist in the purely hydrodynamic case but only for a specific choice of radial perturbations with $u_r = u_z = k_z = 0$ and $-2\Omega u_\phi = ik_r(p_1/\rho_0)$ (see (8.46)). The odd mode $\omega^2 = 0$ arising in the limiting transition to a vanishing magnetic field formally appears from Eq. (8.36) since the fourth order of this dispersion relation is entirely due to the expression in square brackets $\sim \omega^2$ in the denominator of Eq. (8.28), which for the case $B = 0$ is cancelled out by the expressions in brackets $\sim \omega^2$ in the numerator.

Similarly, no smooth transition to the hydrodynamic case occurs if viscosity is included (see below). The absence of a smooth transition to the ideal hydrodynamic case when $B \to 0$ was first noted by Velikhov (1959). At the same time, the transition to the classical Rayleigh mode with vanishing magnetic field proceeds smoothly.

8.4 Linear Analysis for a Fluid with Viscosity and Magnetic Diffusivity

Let us now consider the more general case of a non-ideal viscous fluid with finite electric conductivity characterized by the kinematic viscosity coefficient ν and resistivity (magnetic diffusivity) η. Naturally, in problems with viscosity and magnetic diffusivity there is no initial steady state. The angular momentum is redistributed by viscosity on the time scale $\tau_\nu \sim R^2/\nu$, and the magnetic field changes on the magnetic diffusion time scale $\tau_\eta \sim R^2/\eta$, where R is the characteristic size of the system. Everywhere below we will assume these timescales to be extremely long compared to the Keplerian rotation time and the characteristic instability growth time, if conditions are suitable for the latter to arise. The dispersion relation in this case can be derived following the local linear analysis of MRI performed, e.g. in the monograph by Kato et al. (1998), taking into account viscosity and conductivity in the WKB-approximation (see Sect. 8.2, with zero density perturbations Eq. (8.13)):

$$\omega_{**}^4 + \left(\frac{k_z}{k}\right)^2 \left[\left(i\omega + \eta k^2\right)^2 \kappa^2 + c_A^2 k_z^2 (\kappa^2 - 4\Omega^2)\right] = 0\,, \tag{8.48}$$

where ω_{**}^2 is determined by Eq. (8.33).

The dispersion relation (8.48) is identical to the one derived for a rotating liquid metal annulus in the incompressible limit (Ji et al. 2001).[2] This equation was also derived and mathematically analysed in Pessah and Chan (2008). However, that paper focused on the application of the MRI mode to the calculations of the

[2]Note that those authors searched for a stable differential rotation law between cylinders with given viscosity and electric conductivity while we are investigating conditions for MRI in a viscous, electrically conducting flow in a gravitational field with given differential rotation law.

Reynolds and Maxwell stresses in the differentially rotating flow. In what follows we shall discuss the constraints on MRI modes in astrophysical accretion discs, where the free-path length of particles (and hence the viscosity) is limited by the disc thickness.

The magnetic Prandtl number is introduced as $\mathrm{P_m} = \nu/\eta$. Using the standard expressions for ν and η for fully ionized hydrogen plasma from Spitzer (1962), we readily find

$$\mathrm{P_m} \approx 3.4 \times 10^{-28} \frac{T^4}{\rho \ln \Lambda_{eH} \Lambda_{pH}}, \tag{8.49}$$

where T is the temperature, ρ is the density and Λ_{eH} and Λ_{pH} are electron and proton Coulomb logarithms, respectively.

As was shown by Balbus and Henri (2008), the magnetic Prandtl number can be of the order of one in the inner parts of accretion discs around neutron stars and black holes.

8.4.1 The Case of the Magnetic Prandtl Number $\mathrm{P_m}$=1

Here we will discuss the exact analytic solution to Eq. (8.48) for the important particular case $\mathrm{P_m} = 1$ (which can be derived, for example, from the general analytic solution found in Pessah and Chan 2008) and obtain restrictions on the maximum mean free-path length of ions in accretion discs at which MRI disappears due to non-ideality effects.

The exact solution to Eq. (8.48) for $\mathrm{P_m} = 1$ is

$$\omega = i\nu k^2 \pm \sqrt{\left(\frac{k_z}{k}\right)^2 \left[c_A^2 k^2 + \frac{\kappa^2}{2} \pm \sqrt{\frac{\kappa^4}{4} + 4\Omega^2 c_A^2 k^2}\right]}. \tag{8.50}$$

Here the plus sign before the second square root corresponds to the Rayleigh branch, and the minus sign corresponds to the Velikhov-Chandrasekhar (MRI) branch. We shall examine below the MRI branch only.

We note that the first square root in this equation contains the solutions (8.40) to Eq. (8.36):

$$\omega = i\left(\nu k^2 - \sqrt{-\omega_{\nu=0}^2}\right). \tag{8.51}$$

(Recall that for regions with MRI $\omega^2 < 0$). Also note that like in the ideal MHD case considered above in Sect. 8.3.2, here there is no smooth transition from the MRI mode to the hydrodynamic case with viscosity with a vanishing magnetic field. As can be straightforwardly derived from Eqs. (8.6) to (8.9) in Sect. 8.2, the dispersion relation for the hydrodynamic case with viscosity reads:

$$(i\omega + \nu k^2)^2 + \left(\frac{k_z}{k}\right)^2 \kappa^2 = 0\,. \tag{8.52}$$

While the Rayleigh mode (with a positive sign before the second square root in Eq. (8.50)) tends to the mode given by Eq. (8.52) when magnetic field is vanishing, the MRI mode (with a negative sign before the second square root in Eq. (8.50)) completely disappears (there is no mode $i\omega + \nu k^2 = 0$ without magnetic field, unless $k_z = 0$).

Below we will consider the case $k_z = k$, i.e. with $k_r = 0$. For further analysis it is convenient to rewrite the dispersion relation (8.48) in the dimensionless form. We introduce the dimensionless variables:

$$\tilde{\omega} \equiv \omega/\Omega; \quad \tilde{k} \equiv \frac{c_A k}{\Omega}; \quad \tilde{\kappa}^2 \equiv \kappa^2/\Omega^2; \quad \tilde{\nu} \equiv \frac{\nu\Omega}{c_A^2}\,. \tag{8.53}$$

For Keplerian discs the dimensionless epicyclic frequency is $\tilde{\kappa}^2 = 1$. In dimensionless variables, the solution to Eq. (8.48) takes the form:

$$\tilde{\omega} = i\left(\tilde{\nu}\tilde{k}^2 \pm \sqrt{-\tilde{k}^2 - \frac{1}{2} \mp \sqrt{\frac{1}{4} + 4\tilde{k}^2}}\right)\,. \tag{8.54}$$

Of the four solutions to Eq. (8.54) we choose the one for the MRI mode:

$$\tilde{\omega} = i\left(\tilde{\nu}\tilde{k}^2 - \sqrt{-\tilde{k}^2 - \frac{1}{2} + \sqrt{\frac{1}{4} + 4\tilde{k}^2}}\right)\,. \tag{8.55}$$

Now we find the neutral point $\tilde{\omega} = 0$. Squaring Eq. (8.55) twice, we obtain the equation for the critical wavenumber $\tilde{k}_{cr}$ separating unstable ($\tilde{k} < \tilde{k}_{cr}$) and stable ($\tilde{k} > \tilde{k}_{cr}$) perturbations:

$$\tilde{\nu}^4\tilde{k}^6 + 2\tilde{\nu}^2\tilde{k}^4 + (1 + \tilde{\nu}^2)\tilde{k}^2 - 3 = 0\,. \tag{8.56}$$

Without viscosity we recover the old result: $\tilde{k}_{cr}^2 = 3$ (see Eq. (8.41)). It is easy to check that for the dimensionless viscosity $\tilde{\nu} = 4/5$ the neutral point is $\tilde{k}_{cr} = \sqrt{15/16}$, i.e. here the neutral point coincides with the maximum wavenumber k_{max} at which maximum MRI growth occurs in the inviscid case (see Eq. (8.42) above). For a large dimensionless viscosity $\tilde{\nu} \gg 1$, the asymptotic solution to Eq. (8.55) reads

$$\tilde{k}_{cr} \simeq \frac{\sqrt{3}}{\tilde{\nu}}\,. \tag{8.57}$$

Therefore, at arbitrarily high viscosity there exists an interval of wavenumbers $0 < \tilde{k} < \tilde{k}_{cr}$ where MRI is still at work, but the MRI increment here is very small.

Actually, in realistic models of accretion discs with finite thickness H we should take into account that there is a lower limit for k corresponding to the obvious restriction on the maximum perturbation wavelength $\lambda < 2H$:

$$k = \frac{2\pi}{\lambda} > \frac{\pi}{H} \equiv k_{min} \,. \tag{8.58}$$

Therefore, in dimensionless variables we obtain the MRI condition in the form:

$$\tilde{k}_{min} \leq \tilde{k} \leq \tilde{k}_{cr} \,. \tag{8.59}$$

It is also convenient to change from the disc thickness H to the characteristic thermal velocity in the disc c_s, since in accretion discs the hydrostatic equilibrium along the vertical coordinate yields

$$c_s = \Pi \Omega H \tag{8.60}$$

where Π is a numerical coefficient. For example, in the standard geometrically thin Shakura-Sunyaev α-disc $\Pi = 1/\sqrt{4\Pi_1} \simeq 1/\sqrt{20}$ (see Ketsaris and Shakura 1998). Thus, in an inviscid fluid $\tilde{k}_{cr} = \sqrt{3}$, $\tilde{k}_{min} = \pi\Pi(c_A/c_s)$, and the MRI condition Eq. (8.59) takes the form

$$\pi\Pi\left(\frac{c_A}{c_s}\right) \leq \sqrt{3} \,. \tag{8.61}$$

Essentially, this is the well-known condition that for MRI to operate the seed magnetic field should not exceed some critical value.

In a non-ideal plasma the MRI condition Eq. (8.61) becomes

$$\pi\Pi\left(\frac{c_A}{c_s}\right) \leq \tilde{k}_{cr} \,. \tag{8.62}$$

Note that $\tilde{k}_{cr}$ decreases with $\tilde{\nu}$. For example, if $\tilde{\nu}$ is high, Eq. (8.57) implies very small values of $\tilde{k}_{cr}$ and, correspondingly, very low c_A giving rise to MRI with uninterestingly small increments. The schematic behaviour of the MRI mode at non-zero viscosity is shown in Fig. 8.6. For an arbitrary finite viscosity $\tilde{\nu}$ the neutral point $\tilde{\omega}(\tilde{k}_{cr})$ separates exponentially growing small perturbations $\propto exp(i\omega t)$ (the lower part of Fig. 8.6 where Im$\tilde{\omega} > 0$) from exponentially decaying ones (the upper part of Fig. 8.6). At zero viscosity, however, the function $\tilde{\omega}(\tilde{k})$ (the curve labeled by $\tilde{\nu} = 0$) ends at the point $\tilde{k}_{cr} = \sqrt{3}$, since in this case at $\tilde{k} \geq k_{cr}$, $\tilde{\omega}$ becomes purely real and small perturbations start to oscillate.

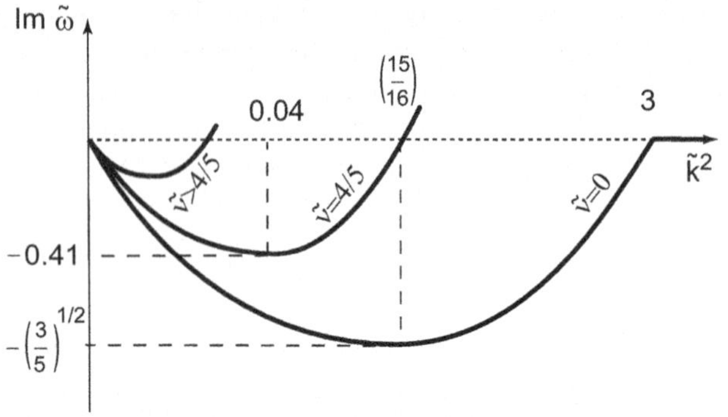

Fig. 8.6 Schematics of the influence of viscosity on the MRI condition $0 < \tilde{k} < \tilde{k}_{cr}$. Shown are curves of the imaginary part of $\tilde{\omega}$ as a function of the dimensionless wave number $\tilde{k}^2$. With increasing viscosity, the MRI interval shifts to the left and shrinks (see also Fig. 1 in Pessah and Chan 2008)

In the case of high viscosity it is convenient to express the ratio c_A/c_s through the dimensionless viscosity $\tilde{\nu}$. Using the conventional definition of the viscosity coefficient $\nu = c_s l$, where l is the effective mean-free path of ions with account for the Coulomb logarithm, and our convention for the thermal velocity in the disc (8.60) introduced above, we find:

$$\tilde{\nu} \equiv \nu \frac{\Omega}{c_A^2} = \frac{1}{\Pi} \left(\frac{c_s}{c_A} \right)^2 \left(\frac{l}{H} \right) . \tag{8.63}$$

Finally, we obtain the MRI condition in the convenient form:

$$\frac{l}{H} \leq \frac{1}{\pi^2 \Pi} \tilde{\nu} \tilde{k}_{cr}^2 . \tag{8.64}$$

In the particular case P_m=1 we may explicitly find $\tilde{\nu}\tilde{k}_{cr}^2$ from Eq. (8.55):

$$\tilde{\nu}\tilde{k}_{cr}^2 = \sqrt{-\tilde{k}_{cr}^2 - \frac{1}{2} + \sqrt{\frac{1}{4} + 4\tilde{k}_{cr}^2}} , \tag{8.65}$$

so that condition (8.64) takes the form:

$$\frac{l}{H} \leq \frac{1}{\pi^2 \Pi} \sqrt{-\tilde{k}_{cr}^2 - \frac{1}{2} + \sqrt{\frac{1}{4} + 4\tilde{k}_{cr}^2}} . \tag{8.66}$$

(This formula should be used when $\nu \neq 0$, i.e. when $\tilde{k}_{cr}^2 < 3$). Consider first the case of low viscosities where $\tilde{k}_{cr}^2 \approx 3$. By introducing the small parameter $\epsilon = 3 - \tilde{k}_{cr}^2 \ll 1$ and expanding the left-hand side of Eq. (8.66) in ϵ, we obtain

$$\frac{l}{H} \leq \frac{1}{\pi^2 \Pi} \sqrt{\frac{41}{49}\epsilon} \,. \tag{8.67}$$

Now consider the special case where $\tilde{k}_{cr}$ coincides with the wavenumber of maximum MRI increment in the ideal fluid: $\tilde{k}_{cr} = \tilde{k}_{max} = \sqrt{\frac{15}{16}}$ (see Eq. (8.42)). This is realized at $\tilde{\nu} = 4/5$. Here we find the limit

$$\left(\frac{l}{H}\right) \leq \frac{1}{\pi^2 \Pi} 0.75 \approx 0.34. \tag{8.68}$$

Finally, in the high-viscosity limit for P_m=1 $\tilde{\nu} \gg 1$, substituting the asymptotic (8.57) into Eq. (8.64) making use of the expression for dimensionless viscosity (8.63) we obtain

$$\left(\frac{l}{H}\right) \leq \frac{\sqrt{3}}{\pi} \left(\frac{c_A}{c_s}\right), \quad P_m = 1, \tilde{\nu} \gg 1 \,. \tag{8.69}$$

Note that this constraint is insensitive to the disc vertical structure parameter Π. This condition can be checked for particular microphysics plasma properties in different thin Keplerian discs.

8.4.2 *The Case of an Arbitrary Magnetic Prandtl Number*

The generalization of the above analysis to an arbitrary Prandtl number is straightforward. First, for given P_m and $\tilde{\nu}$ we solve the dimensionless Eq. (8.48) to find $\tilde{k}_{cr}(\tilde{\nu}, P_m)$. at the neutral point where $\tilde{\omega}(\tilde{k}_{cr}) = 0$.

To do this, it is convenient, for the sake of brevity, to introduce the new dimensionless variables

$$y \equiv \tilde{k}^2, \quad X = i\tilde{\omega} + \tilde{\nu} y \tag{8.70}$$

and rewrite dimensionless dispersion relation (8.48) in the equivalent form:

$$X^4 + 2\frac{1-P_m}{P_m}\tilde{\nu} y X^3 + \left[\left(\frac{1-P_m}{P_m}\right)^2 \tilde{\nu}^2 y^2 + 2y + 1\right] X^2 +$$

$$\left[\frac{1-P_m}{P_m}\tilde{\nu} y(y+1)\right] X + \left(\frac{1-P_m}{P_m}\right)^2 \tilde{\nu}^2 y^2 + y^2 - 3y = 0 \,. \tag{8.71}$$

(Here we assumed Keplerian discs with $\tilde{\kappa} = 1$ and used the condition $k_z/k = 1$). Noticing that at the neutral point determined by the condition $\tilde{\omega}(y_{cr}) = 0$ we have $X = \tilde{\nu} y_{cr}$, we arrive at the equation for y_{cr}:

$$y_{cr}\left[\tilde{\nu}^4 y_{cr}^3 + \tilde{\nu}^2 y_{cr}(2y_{cr}\mathrm{P_m} + 1) + \mathrm{P_m}^2(y_{cr} - 3)\right] = 0\,. \tag{8.72}$$

At $\mathrm{P_m} = 1$ this equation, of course, coincides with Eq. (8.56). The non-trivial real solution to the cubic equation in the square brackets of Eq. (8.72) reads:

$$y_{cr} \equiv \tilde{k}_{cr}^2 = \mathscr{A} - \frac{2\mathrm{P_m}}{3\tilde{\nu}^2} - \frac{1}{\mathscr{A}}\left(\frac{1}{3\tilde{\nu}^2} - \frac{\mathrm{P_m}^2}{9\tilde{\nu}^4}\right), \tag{8.73}$$

where

$$\mathscr{A} = \left[\left(\frac{1}{27\tilde{\nu}^6} + \frac{2\mathrm{P_m}^2}{27\tilde{\nu}^8} + \frac{\mathrm{P_m}^3}{\tilde{\nu}^8} + \frac{9\mathrm{P_m}^4}{4\tilde{\nu}^8} + \frac{\mathrm{P_m}^4}{27\tilde{\nu}^{10}} + \frac{\mathrm{P_m}^5}{9\tilde{\nu}^{10}}\right)^{1/2} + \right.$$
$$\left. \frac{\mathrm{P_m}}{3\tilde{\nu}^4} + \frac{3\mathrm{P_m}^2}{2\tilde{\nu}^4} + \frac{\mathrm{P_m}^3}{27\tilde{\nu}^6}\right]^{1/3}. \tag{8.74}$$

At high dimensionless viscosities there is an asymptotic to the solution (8.73) for $\mathrm{P_m}/\tilde{\nu}^2 \ll 1$:

$$y_{cr} = \tilde{k}_{cr}^2 \approx \frac{3\mathrm{P_m}^2/\tilde{\nu}^2}{1 + \mathrm{P_m}^2/\tilde{\nu}^2} = \frac{3\mathrm{P_m}^2}{\tilde{\nu}^2} + \mathscr{O}\left(\frac{\mathrm{P_m}^2}{\tilde{\nu}^2}\right)^2. \tag{8.75}$$

Note that this asymptotic may also be found in Pessah and Chan (2008) (their Eq. (97)) and for small $\mathrm{P_m}$ can be derived for Keplerian rotation and $k = k_z$ from Eq. (3) in Ji et al. (2001).

Thus, the general MRI condition for arbitrary non-ideal plasma (8.62) takes the form:

$$\left(\frac{c_A}{c_s}\right) \le \frac{1}{\pi\Pi}\tilde{k}_{cr}(\tilde{\nu}, \mathrm{P_m})\,. \tag{8.76}$$

The result of a calculation of $\tilde{k}_{cr}$ for a range of magnetic Prandtl numbers $\mathrm{P_m}$ and dimensionless viscosities $\tilde{\nu}$ can be found in Pessah and Chan (2008) (see e.g. their Figs. 6 and 7) and is illustrated in Fig. 8.7.

In the limiting case of high dimensionless viscosities $\mathrm{P_m}/\tilde{\nu}^2 \ll 1$, which may be realized in the outer parts of thin Keplerian accretion discs (see Eq. (8.49) above), using asymptotic (8.75) and definition (8.63), we find the restriction on the mean-free path of ions in the disc

$$\left(\frac{l}{H}\right) \le \frac{\sqrt{3}\mathrm{P_m}}{\pi}\left(\frac{c_A}{c_s}\right), \quad \mathrm{P_m}/\tilde{\nu}^2 \ll 1\,. \tag{8.77}$$

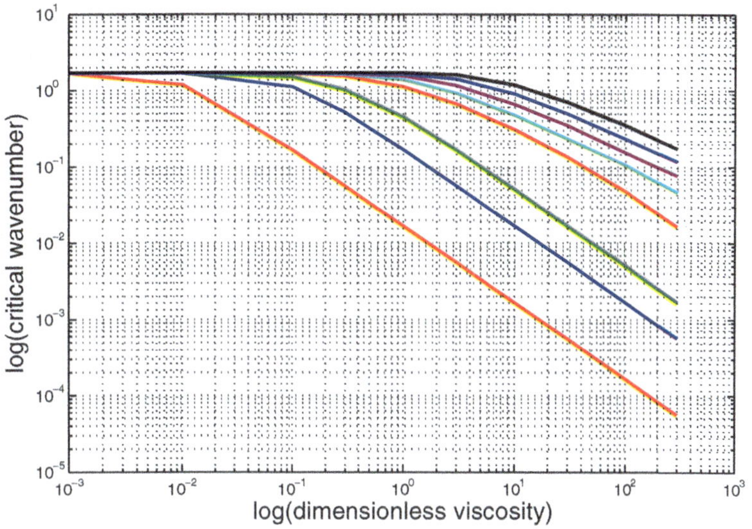

Fig. 8.7 The dimensionless critical wavenumber $\tilde{k}_{cr}$ as a function of the dimensionless viscosity coefficient $\tilde{\nu}$ for different magnetic Prandtl numbers $\mathrm{P_m}$. Lines from bottom to top correspond to $\mathrm{P_m}$=0.01, 0.1, 0.3, 3, 10, 30, 100, 300

which is the generalization of Eq. (8.69) to an arbitrary magnetic Prandtl number. Using the expression for the dimensionless viscosity (8.63), the condition for the power-law asymptotic $\mathrm{P_m}/\tilde{\nu}^2 \ll 1$ can be recast to the inequality

$$\mathrm{P_m}/\tilde{\nu}^2 \ll 1 \Leftrightarrow \left(\frac{l}{H}\right)^2 \gg \Pi \mathrm{P_m} \left(\frac{c_A}{c_s}\right)^4 . \tag{8.78}$$

Therefore, the MRI condition can be written in terms of the interval for l/H in a Keplerian disc as

$$\sqrt{\Pi \mathrm{P_m}} \left(\frac{c_A}{c_s}\right)^2 \ll \left(\frac{l}{H}\right) \leq \frac{\sqrt{3}\mathrm{P_m}}{\pi} \left(\frac{c_A}{c_s}\right) . \tag{8.79}$$

8.5 Conclusions

In this chapter we have extended the original analysis of MRI in ideal MHD plasmas carried out by Balbus (2012). First, we emphasize that hydromagnetic flows in which the angular momentum increases or decreases with radius are different from the point of view of MRI development. In the classical Rayleigh-unstable case where the angular momentum decreases with radius, the Velikhov-

Chandrasekhar MRI mode is stable, while the Rayleigh mode is unstable (see Figs. 4, 5); the magnetic field stabilizes the Rayleigh mode in the short-wavelength limit. When the angular momentum in the flow increases with radius, MRI arises at long wavelengths (small wave numbers k, see Fig. 8.2). However, the local WKB approximation should be applied with caution at long wavelengths. At long wavelengths, the *Ansatz* for the solution should rather be used in the global form $f(r)e^{i(\omega t-k_r r-k_z z)}$. Note that in the original papers by Velikhov and Chandrasekhar, they analyzed the linear stability of magnetized flows between cylinders exactly in that approximation (see also Sano and Miyama (1999) for a global analysis of perturbations in an inviscid magnetized proto-planetary disc with non-zero magnetic diffusivity).

Further, in the phenomenologically interesting case of thin Keplerian accretion discs, viscosity may restrict MRI growth. This situation may arise in the inner parts of an accretion disc. Indeed, at high temperatures the mean free path of ions $l \sim T^2$ can become comparable to the characteristic disc thickness H at $H < r$ (thin discs). This means that the flow should be treated kinetically (see, for example, recent 2.5D hybrid calculations (Shirakawa and Hoshino 2014) or the discussion of MRI in rarefied astrophysical plasmas with Braginskii viscosity in Islam and Balbus 2005). The weak seed magnetic field does not grow under these conditions, i.e. the high ion viscosity can suppress MRI. Clearly, this interesting regime requires further study.

At large magnetic Prandtl numbers $\mathrm{P}_m \gg 1$, which may be present in the innermost parts of accretion discs around neutron stars and black holes, the kinematic viscosity ν is much larger than the magnetic diffusivity η. In this case the plasma may become collisionless, and the hydrodynamic description fails. Our analysis shows that, in principle, a collisionless regime (where the ion mean-free path is comparable to or larger than the disc thickness, $l \sim H$) may be established in Keplerian discs even for magnetic Prandtl numbers $\mathrm{P}_m \simeq 1$ (see Eq. (8.79)).

We have also obtained the dispersion relation for local small perturbations in the Boussinesq limit for non-adiabatic perturbations (see Eq. (8.32)). This is a fifth-order algebraic equation, in contrast to the fourth-order dispersion relation for adiabatic perturbations or non-adiabatic perturbations with $k_r = 0$ in a non-ideal plasma (8.48). Also note that when the density perturbations are expressed through the entropy gradients (see Eq. (2.2h) in Balbus and Hawley 1991), the frequency appears in the denominator but the final dispersion relation (2.5) in Balbus and Hawley (1991) remains a fourth-order equation in ω, even when taking the entropy gradients into account. Apparently, the difference is due to the fact that in the case of non-adiabatic perturbations the density variations are proportional to the azimuthal velocity perturbations u_ϕ (see our Eq. (8.21)) and not to u_z and u_r as in the case considered by Balbus and Hawley (1991). The analysis of the effect of non-adiabatic perturbations deserves a separate study and will be addressed in a future work.

Perturbations with $k_r = 0$ represent waves propagating along the z-coordinates, and when their wavelength is comparable to the disc thickness, the WKB approximation becomes problematic. Perturbations with $k_z = 0$ propagate along the r-coordinate, which is much larger than the disc thickness for thin accretion discs.

However, for such perturbations with $k = k_r$ and $k_z = 0$, the second term in Eqs. (8.48) and (8.32) vanishes, and therefore from Eq. (8.33) we find two perturbation modes

$$\omega_1 = i\nu k^2, \quad \omega_2 = i\eta k^2, \tag{8.80}$$

i.e. decaying standing waves for any seed magnetic field. This may suggest that in poloidal magnetic fields purely radial perturbations with $k = k_r$ do not grow. The situation is different when an azimuthal magnetic field is present. This case should be considered separately and has been investigated for a range of astrophysical applications in other works (see, e.g., Acheson 1978; Sano and Miyama 1999; Ruediger et al. 2014; Kirillov et al. 2014).

We conclude that in thin Keplerian accretion discs the addition of viscosity may strongly restrict the MRI conditions once the mean free path of ions becomes comparable to the disc thickness. This limitation should be taken into account in the direct numerical simulations of MRI in astrophysical accretion discs.

References

Acheson DJ (1978) On the instability of toroidal magnetic fields and differential rotation in stars. R Soc Lond Philos Trans Ser A 289:459–500. https://doi.org/10.1098/rsta.1978.0066

Balbus SA (2004) Viscous shear instability in weakly magnetized, dilute plasmas. Astrophys J 616:857–864. https://doi.org/10.1086/424989, astro-ph/0403678

Balbus SA (2012) On the behaviour of the magnetorotational instability when the Rayleigh criterion is violated. Mon Not R Astron Soc 423:L50–L54. https://doi.org/10.1111/j.1745-3933.2012.01255.x

Balbus SA, Hawley JF (1991) A powerful local shear instability in weakly magnetized disks. I - Linear analysis. II - Nonlinear evolution. Astrophys J 376:214–233. https://doi.org/10.1086/170270

Balbus SA, Hawley JF (1998) Instability, turbulence, and enhanced transport in accretion disks. Rev Mod Phys 70:1–53. https://doi.org/10.1103/RevModPhys.70.1

Balbus SA, Henri P (2008) On the magnetic Prandtl number behavior of accretion disks. Astrophys J 674:408–414. https://doi.org/10.1086/524838, 0706.0828

Chandrasekhar S (1960) The stability of non-dissipative Couette flow in hydromagnetics. Proc Natl Acad Sci 46:253–257. https://doi.org/10.1073/pnas.46.2.253

Hawley JF, Gammie CF, Balbus SA (1995) Local three-dimensional magnetohydrodynamic simulations of accretion disks. Astrophys J 440:742. https://doi.org/10.1086/175311

Hawley JF, Richers SA, Guan X, Krolik JH (2013) Testing convergence for global accretion disks. Astrophys J 772:102. https://doi.org/10.1088/0004-637X/772/2/102, 1306.0243

Islam T, Balbus S (2005) Dynamics of the magnetoviscous instability. Astrophys J 633:328–333. https://doi.org/10.1086/447762, astro-ph/0504666

Ji H, Goodman J, Kageyama A (2001) Magnetorotational instability in a rotating liquid metal annulus. Mon Not R Astron Soc 325:L1–L5. https://doi.org/10.1046/j.1365-8711.2001.04647.x, astro-ph/0103226

Kato S, Fukue J, Mineshige S (eds) (1998) Black-hole accretion disks. Kyoto University Press, Kyoto

Ketsaris NA, Shakura NI (1998) On the calculation of the vertical structure of accretion discs. Astron Astrophys Trans 15:193. https://doi.org/10.1080/10556799808201769

Kirillov ON, Stefani F, Fukumoto Y (2014) Local instabilities in magnetized rotational flows: a short-wavelength approach. J Fluid Mech 760:591–633. https://doi.org/10.1017/jfm.2014.614, 1401.8276
Kotko I, Lasota JP (2012) The viscosity parameter α and the properties of accretion disc outbursts in close binaries. Astron Astrophys 545:A115. https://doi.org/10.1051/0004-6361/201219618, 1209.0017
Kundu PK, Cohen IM, Dowling DR (2012) Fluid mechanics, 5th edn. Academic Press, Boston. http://dx.doi.org/10.1016/B978-0-12-382100-3.10032-0. http://www.sciencedirect.com/science/article/pii/B9780123821003100320
Lord Rayleigh (1916) Proc R Soc A 93:143
Masada Y, Sano T (2008) Axisymmetric magnetorotational instability in viscous accretion disks. Astrophys J 689:1234–1243. https://doi.org/10.1086/592601, 0808.2338
Nauman F, Blackman EG (2015) Sensitivity of the magnetorotational instability to the shear parameter in stratified simulations. Mon Not R Astron Soc 446:2102–2109. https://doi.org/10.1093/mnras/stu2226, 1409.2442
Pessah ME, Chan Ck (2008) Viscous, resistive magnetorotational modes. Astrophys J 684:498–514. https://doi.org/10.1086/589915, 0801.4570
Ruediger G, Schultz M, Stefani F, Mond M (2014) Diffusive MHD instabilities: beyond the Chandrasekhar theorem. ArXiv e-prints 14070240 1407.0240
Sano T, Miyama SM (1999) Magnetorotational instability in protoplanetary disks. I. On the global stability of weakly ionized disks with ohmic dissipation. Astrophys J 515:776–786. https://doi.org/10.1086/307063
Shakura N, Postnov K (2015) A viscous instability in axially symmetric laminar shear flows. Mon Not R Astron Soc 448:3707–3717. https://doi.org/10.1093/mnras/stv262, 1502.01888
Shakura NI, Sunyaev RA (1973) Black holes in binary systems. Observational appearance. Astron Astrophys 24:337–355
Shirakawa K, Hoshino M (2014) Asymmetric evolution of magnetic reconnection in collisionless accretion disk. Phys Plasmas 21(5):052903. https://doi.org/10.1063/1.4875739
Sorathia KA, Reynolds CS, Stone JM, Beckwith K (2012) Global simulations of accretion disks. I. Convergence and comparisons with local models. Astrophys J 749:189. https://doi.org/10.1088/0004-637X/749/2/189, 1106.4019
Spiegel EA, Veronis G (1960) On the Boussinesq approximation for a compressible fluid. Astrophys J 131:442. https://doi.org/10.1086/146849
Spitzer L (1962) Physics of fully ionized gases. Interscience, New York
Stone JM (2011) Recent results from simulations of the magnetorotational instability. In: Bonanno A, de Gouveia Dal Pino E, Kosovichev AG (eds) IAU symposium, vol 274, pp 422–428. https://doi.org/10.1017/S174392131100740X
Suleimanov VF, Lipunova GV, Shakura NI (2008) Modeling of non-stationary accretion disks in X-ray novae A 0620-00 and GRS 1124-68 during outburst. Astron Astrophys 491:267–277. https://doi.org/10.1051/0004-6361:200810155, 0805.1001
Suzuki TK, Inutsuka Si (2014) Magnetohydrodynamic simulations of global accretion disks with vertical magnetic fields. Astrophys J 784:121. https://doi.org/10.1088/0004-637X/784/2/121, 1309.6916
Velikhov EP (1959) Sov Phys. J Exp Theor Phys 36:1398

Index

N. Shakura (ed.), *Accretion Flows in Astrophysics*, Astrophysics and Space Science Library 454, https://doi.org/10.1007/978-3-319-93009-1

The manufacturer's authorised representative in the EU is Springer Nature Customer Service Centre GmbH, Europaplatz 3, 69115 Heidelberg, Germany. If you have any concerns regarding our products, please contact ProductSafety@springernature.com

Printed and bound by CPI Group (UK) Ltd, Croydon, CR0 4YY
30/09/2026
02225228-0018